Automotive Ergonomics

Automotive Ergonomics

Edited by

Brian Peacock

General Motors Corporation

and

Waldemar Karwowski

University of Louisville

Taylor & Francis
London · Washington, DC

UK	Taylor & Francis Ltd, 4 John St, London WC1N 2ET
USA	Taylor & Francis Inc., 1900 Frost Road, Suite 101, Bristol PA 19007

Copyright © Taylor & Francis Ltd 1993

All rights reserved. No part of this publication may be reproduced, stored in a retrieval system, or transmitted, in any form or by any means, electronic, electrostatic, magnetic tape, mechanical, photocopying, recording or otherwise, without the prior permission of the copyright owner.

British Library Cataloguing in Publication Data

A catalogue record for this book is available from the British Library

ISBN 0 7484 0005 2

Library of Congress Cataloging in Publication Data are available

Cover design by Amanda Barragry

Typeset by Santype International Limited, Netherhampton Road, Salisbury, Wilts SP2 8PS

Printed in Great Britain by Burgess Science Press, Basingstoke on paper which has a specified pH value on final paper manufacture of not less than 7·5 and is therefore 'acid free'.

Contents

Preface vi

Contributors ix

1 Modeling the human-equipment interface
  K. H. E. Kroemer 1

2 Occupant packaging
  R. W. Roe 11

3 Computer-aided ergonomics design of automobiles
  J. M. Porter, K. Case, M. T. Freer and M. C. Bonney 43

4 Visual aspects in vehicle design
  C. M. Haslegrave 79

5 Automotive seat design for sitting comfort
  H. M. Reynolds 99

6 Ergonomic guidelines for design of a passenger car trunk
  W. Karwowski, J. W. Yates and N. Pongpatana 117

7 Physical aspects of car design: Occupant protection
  M. R. Lehto and J. P. Foley 141

8 Vision and perception
  P. L. Olson 161

9 Human factors considerations in the design of vehicle headlamps and signal lamps
  M. Sivak and M. Flannagan 185

10 Indirect vision systems
   M. Flannagan and M. Sivak 205

11 The effects of age on driving skill cognitive–motor capabilities
   G. E. Stelmach and A. Nahom 219

12 Design and evaluation of symbols for automobile controls and displays
   P. Green 237

13	Role of expectancy and supplementary cues for control operation *W. W. Wierwille and J. McFarlane*	269
14	Visual and manual demands of in-car controls and displays *W. W. Wierwille*	299
15	Informational aspects of car design: Navigation *J. F. Antin*	321
16	Age, display design and driving performance *D. Imbeau, W. W. Wierwille and Y. Beauchamp*	339
17	Driver mental workload *R. E. Schlegel*	359
18	Models that simulate driver performance with hand controls *A. C. Graesser and W. Marks*	383
19	Informational aspects of vehicle design: A systems approach to developing facilitators *L. F. Laux and D. L. Mayer*	401
20	Unintended acceleration: Human performance considerations *R. A. Schmidt*	431
21	The older driver and passenger *D. B. D. Smith, N. Meshkati and M. M. Robertson*	453
Postscript:	Future challenges for automotive ergonomics *B. Peacock*	473
Index		479

Preface

The purpose of this book is to present discussions of contemporary topics in automobile design. Contributions were solicited from academic researchers who have made major contributions to the theory, techniques and applications of automotive ergonomics. The views expressed by the authors do not necessarily reflect the opinions of the automobile industry, whose ergonomics specialists have also published significantly in these areas. Many of these contributions may be found in the technical and professional automotive literature. The value of the book is that it maintains a focus on ergonomics and human performance issues, without the constraints imposed by marketing, engineering, cost and aesthetic design considerations. The reality of human factors practice in automobile design is a blend of many viewpoints, with expressed or predicted customer wants and behaviors sometimes taking precedence over functional needs. Also the customer is a very variable entity, in many dimensions, thus the aspiration of human factors must be to please most of the people, most of the time.

This volume contains a balance of physical and informational topics. Sometimes these issues can be dealt with in isolation, as in basic applications of occupant packaging. However, in most situations physical and informational issues are inextricably intertwined as in seat comfort, control design, unwanted acceleration and serviceability. The book also addresses theories of human interactions with the automobile, empirical research, models of human performance, design methods and descriptions of particular populations and situations. Some chapters focus on human characteristics, whereas others concentrate on the design of the vehicle. Thus the reader is presented with the full spectrum of ergonomics activity—from the abstract, where the car simply represents an example of a technology with which people interact, to the pragmatic, where the focus is on particular vehicle parameters and situations.

An initial chapter by Kroemer describes some of the general issues of anthropometric data, modeling and application. Roe, Porter *et al.* and Haslegrave present contrasting techniques and applications of anthropometry to the critical issues of occupant packaging and vision. The complex, perennial problem of seat design and seat comfort is addressed by Reynolds, and Karwowski *et al.* deal with the biomechanical parameters of trunk design. The sensitive topic of occupant protection is addressed by Lehto and Foley. The assessment by Olsen of human vision characteristics leads naturally into the discussions by Sivak and Flannagan of lighting and mirrors. The bridge between the physical and cognitive domains is pursued by Stelmach and Nahom, who focus on the motor skills of older drivers.

Wierwille and McFarlane provide an analysis of direction of motion stereotypes as applied to the design of a selection of secondary controls. Schlegel, and Graesser and Marks present a mental workload and modeling approach to the interaction of drivers with secondary controls. Wierwille, Antin and Imbeau *et al.* discuss the human factors issues related to the design and use of both traditional and contemporary display systems. Green discusses the theory, design and evaluation of vehicle symbols followed by an analysis by Laux and Mayer of the important topic of instructions, warnings and labels. Schmidt provides some insight into the controversial subject of unwanted acceleration and Smith *et al.* close the book with a survey of the particular challenges of older drivers.

This collection of papers, therefore, covers a very wide range of human factors topics in automotive design. The reader will gain insights into both human factors methodology and particular aspects of vehicle design. Thus the book may be used as a text, as a design guide or simply as a means of understanding the scope and complexity of ergonomics in vehicle design.

Contributors

Jonathan F. Antin
Assistant Professor
AT & T Room 1549
3330 West Friendly Avenue
Greensboro
North Carolina, 27410
USA

Yves Beauchamp
Engineering Department
Universite du Quebec a Trois Rivieres
Trois Rivieres (Quebec)
Canada

M. C. Bonney
Department of Production Engineering
and Production Management
University of Nottingham
UK

K. Case
Department of Manufacturing
Engineering
Loughborough University
UK

Michael Flannagan PhD
The University of Michigan
Transportation Research Institute
2901 Baxter Road
Ann Arbor
Michigan, 48109-2150
USA

James P. Foley
Purdue University
School of Industrial Engineering
West Lafayette
Indiana, 47907
USA

M. T. Freer
SAMMIE CAD Ltd
Loughborough
USA

Arthur C. Graesser
Department of Psychology and the
Institute for Intelligent Systems
Memphis State University
Memphis
Tennessee, 38152
USA

Paul Green
Transportation Research Institute
University of Michigan
2901 Baxter Road
Ann Arbor
Michigan, 48109-2150
USA

Christine M. Haslegrave
Institute for Occupational Ergonomics
University of Nottingham
Nottingham
UK

Daniel Imbeau
Engineering Department
Universite du Quebec a Trois Rivieres
Trois Rivieres (Quebec)
Canada

Waldemar Karwowski
Center for Industrial Ergonomics
Department of Industrial Engineering
University of Louisville
Louisville
Kentucky, 40292
USA

K. H. E. Kroemer
Professor and Director
Industrial Ergonomics
Research Laboratory
Human Factors Engineering Center
Industrial and Systems Engineering
Department
Virginia Polytechnic Institute
and State University
Blacksburg
Virginia, 24061-0118
USA

Lila F. Laux
Rice University
Houston
Texas
USA

Mark R. Lehto
Purdue University
School of Industrial Engineering
West Lafayette
Indiana, 47907
USA

William Marks
Department of Psychology and the
Institute for Intelligent Systems
Memphis State University
Memphis
Tennessee, 38152
USA

David L. Mayer
Rice University
Houston
Texas
USA

Najmedin Meshkati
Human Factors
Institute of Safety and
Systems Management
University of Southern California
USA

Ariella Nahom
Exercise and Sport Research Institute
Arizona State University
Tempe
Arizona
USA

Paul L. Olson
3545 Sulgrave
Ann Arbor
Michigan, 48105
USA

Brian Peacock
Manufacturing Ergonomics Center
Advanced Engineering Staff
General Motors Technical Center
30300 Mound Road
Warren
Michigan, 48090-9040
USA

N. Pongpatana
Center for Industrial Ergonomics
University of Louisville
Louisville
Kentucky, 40292
USA

J. M. Porter
Department of Human Sciences
Loughborough University
Loughborough
UK

Michelle M. Robertson
Human Factors
Institute of Safety and
Systems Management
University of Southern California
USA

Herbert M. Reynolds
Department of Biomechanics
Michigan State University
USA

R. W. Roe
Systems Engineering
General Motors Corporation
1151 Crooks Road
Troy
Michigan, 48084
USA

Robert E. Schlegel
School of Industrial Engineering
The University of Oklahoma
Norman
Oklahoma, 73019
USA

Contributors

Richard A. Schmidt
Department of Psychology
University of California,
Los Angeles
Los Angeles
California, 90024-1563
USA

Michael Sivak
The University of Michigan
Transportation Research Institute
2901 Baxter Road
Ann Arbor
Michigan, 48109-2150
USA

David B. D. Smith
Human Factors
Institute of Safety and
Systems Management
University of Southern California
USA

George E. Stelmach
Exercise and Sport Research Institute
Arizona State University
Tempe
Arizona
USA

Walter W. Wierwille
Vehicle Analysis and
Simulation Laboratory
Virginia Polytechnic Institute and
State University
Blacksburg
Virginia
USA

J. W. Yates
Exercise Physiology Laboratory
University of Louisville
Louisville
Kentucky, 40292
USA

1

Modeling the human-equipment interface

K. H. E. Kroemer

Abstract

The designer, engineer and tester of automobiles utilizes computerized mathematically-formulated models of the human (as driver or occupant) to design a proper shell and interface so that the human is safe, comfortable and competent in driving the vehicle. Furthermore, such models of the human are important for design and evaluation of work-places at which automobiles are assembled or repaired. Current models of the human have come a long way from simplistic assumptions, such as represented by static two-dimensional templates. Yet realistic behavior of the human body, either in passive reaction to external forces or in active action to perform activities, is still incompletely understood and modeled. This chapter provides an overview of the state of the art, and enumerates future challenges in modeling the human interfacing with vehicles or other work-places.

Introduction

Models of the human interfacing with a vehicle or other equipment can have very practical applications, for example as 'drafting templates', or they may be of rather theoretical interest, such as in the evaluation of 'fit' of a product, or in the simulation of a crash. Ergonomists/human factors engineers often use models, which are paradigms (or patterns) which describe or imitate, in a systematic and well-organized manner, the appearance and the behavior of the human, often in some stressful situation. While there are many usages of the word model, it is used here according to the following definition:

> A model is a mathematical/physical system, obeying specific rules and conditions, whose behavior is used to understand a real (physical, biological, human-technical, etc.) system to which it is analogous in certain respects.

Several aspects of this definition deserve attention. First, the model is 'obeying specific rules and conditions'. This means that the model is itself restricted, for example a simple design template that is only two-dimensional or of only one single-percentile size, or being able to be executed only on certain computer systems.

Second, the model is 'analogous' to the real system 'in certain respects'. This means that the model is restricted in its validity (or fidelity), with its boundaries often so tight that they barely overlap the actual conditions. Thus, the model's internal limitations and its limits of applicability need to be carefully kept in mind.

Types of models

One distinguishes between 'normative' and 'descriptive' models. A normative model assumes that there is necessarily some sort of a 'normal' appearance or behavior, which is perfect, ideal, in a standardized way, in a non-varying manner; in other words, that there is one singular appearance or behavior which the model represents. Thus, a normative model is often deterministically constructed, presuming that the effects of variables within the system, or acting upon the system from the outside, can be clearly predicted and hence modeled. A normative model is used to predict 'normal' behavior to which actually observed behavior is compared.

On the other hand, a descriptive model is one that reflects variations in behavior due to variations (often stochastic) of internal or external variables. Such descriptive models are often used for 'simulation', exercising a model to imitate actual conditions and behaviors.

While in the past most models were physical (such as templates or manikins), they are now often mathematical and computerized. A mathematical model has the advantage of being precise, formal and often general; the variables can be manipulated easily, and parameters in the equations assumed freely. Disadvantages of mathematical models may be in their rigid mathematical structure, often including equations the nature of which needs to be presumed and cannot be changed without changing the model (see below). Thus, some mathematical models 'fit' reality poorly, often being either too general or too specific. Furthermore, if the boundary conditions are not explicitly and carefully stated, computerized mathematical models easily can be extrapolated inappropriately.

Often, models are classified by the academic disciplines in which they are primarily used. Thus, a number of anatomical models exist, usually relying on specific anthropometric and biomechanical formulae. A large subset consists of physiological models, primarily as they reflect metabolic or circulatory events within the body, usually related to the environment or to conditions of work. Anatomical/biomechanical/physiological models comprise the majority of 'engineering body models' used, for example, in the design of spacecraft, aircraft or automobiles.

Inadequate models

Some models seem to have been developed by persons who know very well how to manipulate the computer system, but too little about the human and how the human functions with and within the system. Their models of the human are likely to be inaccurate, unrealistic and overly simplified—but they may 'work well' in terms of the model mathematics. Predictions based on such a human-machine system are likely to be 'off', particularly in extreme situations where the human (or the machine) behaves in unusual manners for which the model has no appropriate algorithms.

A related fallacy is the assumption, born from the desire to keep the model simple, that the behavior will be 'linear', meaning if one variable (say, workload) increases, the associated activities of a dependent variable (say, speed of human activities) will increase linearly as well. Many human behavior traits, however, do not respond linearly to changes in the load. If, therefore, a system is based on linear algorithms, then the system behavior will fail to be truly descriptive (or predictive) the more extreme (non-linear) the conditions are.

One set of such models depicts motions of the human based on observations of how such movements were performed under certain conditions. Recreation of these motions is called 'animation' which may simplify or exaggerate motions as in cartoons. Smooth as many of these animations appear to be, they are 'true' only in the most superficial way, and even then usually only for that situation in which the behavior was previously observed.

Misuse of models

A typical example of the usefulness, and the possible misuse, of models was demonstrated recently. In a 1986 dissertation, a biomechanical model of the human body was described which meant to explain the stresses on the spinal column while performing lifting tasks. The model was a major step forward from previous models because it included more details and attempted to explain dynamic activities and their effects on the body while previous models were static in nature. However, a thorough review of the dissertation showed that, altogether, many assumptions were made in the development of the model, including the following:

— dimensions of the 50th percentile male,
— movement at constant velocity (after initial acceleration and before final deceleration),
— body segments treated as cylinders,
— curvature of the lumbar section considered constant under all conditions,
— joint locations taken from erect standing posture,
— constant lever arm of posterior back muscles about the spinal column assumed, in particular a constant lever arm of the abdominal muscles, at 10 per cent of stature,
— all involved muscle forces sum to minimal total effort (no co-activation).

Obviously, these assumptions are overly simplistic, in fact unrealistic and limit the model validity immensely. Yet, the temptation is strong to overlook or disregard some of the basic model assumptions, and the limitations which they impose on the validity of the model, in order to expand the application of the model to wider boundaries. Thus, in 1988, the model was described (not by the original author) in a shortened text with the titillating title 'A Knowledge-Based Model of Human . . . Capability'. In this publication, the application of the model to a variety of actual working conditions was proposed, some of which appear to be outside the inherent boundaries of the original model.

A typical example of this overexertion is the application of models based on static (isometric) muscle strength data to highly dynamic situations.

Another problem lies in the feeding of incorrect data to the model. Not able to evaluate the correctness of the input data, the model spits out results anyhow—'garbage in, garbage out'. A related problem is hidden under the euphemism 'fitting input data to the model'. This may simply mean that the data format needs to fit input format requirements—in this case, there is no problem. However, if fitting data really means a modification of their content, their meaning, their 'behavior', then such fitting is really data falsification. To avoid these problems the model user needs to be knowledgeable, able to judge the appropriateness of the model for the situation, to judge the input data for their validity and must refrain from applying model outputs to conditions outside the model constraints.

Incomplete, unrealistic, false model structures lead to incomplete, unrealistic and false conclusions drawn from the model behavior. Validation is one way to check whether the model (re)presents reality. Validation of the model means, in the simplest sense, feeding 'true' data into the model and comparing the model output (prediction) with the behavior of the 'true' system. If the model does not describe reality correctly, then internal algorithms and/or the basic structure of the model are insufficient. Not to assess the validity of a model is like buying an airplane or car without testing it.

Reviewing the past and guiding the future of modeling the 'human at work'

The feasibility of developing an integrated ergonomic model of the 'human at work' was the topic of a two-day workshop in June 1985, convened by the Committee on Human Factors of the National Research Council (Kroemer *et al.*, 1988; Kroemer, 1989a, 1989b). In 1988, the NATO Research Group 9 organized a workshop on applications of human performance models to system design (McMillan *et al.*, 1989). In 1989, a review of computer models of the human body was conducted for the German Department of Defense (Aune and Juergens, 1989).

The specific objectives of the 1985 workshop were to:

—assess the usefulness of current anthropometric, biomechanical and interface models;
—identify critical points of compatibility and disparity among existing models;
—review of the feasibility of using these models for the development of an integrated ergonomic model;
—recommend research approaches for the development of such an integrated model.

The prospectus of the workshop limited its scope to three major classes of models:

1. anthropometric models, i.e, representations of static body geometry;
2. biomechanical models, i.e., representing physical activities of the body in motion, for which anthropometric data are the primary inputs; and
3. interface models, i.e., specific combinations of anthropometric and biomechanical models with regard to their interfacing with the technological system (the 'machine'), representing human-technology interactions.

The major findings of the workshop were:

Anthropometric models

In the past, human body models have been mostly physical in forms of templates, manikins and dummys, but future development is likely to concentrate on computer analogs of the human body. Such models need exact anthropometric information in order to be accurate representation of body size, shape and proportions. In the USA, anthropometric information is most often drawn from the anthropometric data bank at the US Air Force's Armstrong Aerospace Medical Research Laboratory (CSERIAC, 1990). This repository contains many survey results of military samples, but the information on civilian populations is weak because no comprehensive anthropometry study of the civilian population has ever been undertaken in any western country.

The existing anthropometric information does not contain three-dimensional body data, but only univariate descriptors. Furthermore, many of these univariate dimensions lack a common reference system to which the individual measurements are related. This fact causes much conceptual and practical difficulty in the development of computer models of the human body size. Hence, various techniques for three-dimensional anthropometric data acquisition have been proposed, including stereophoto techniques, and the laser as a distance measuring device. Mathematical-statistical techniques need to be developed that collect, organize, and summarize as well as display the huge number of collected data. Surface definition has been much improved by 'facet algorithms' which allow a complete topographic description of the body surface.

Of course, the current use of landmarks and reference points on the body, now often palpated below the skin, needs to be modified for the use of photographic or laser measuring techniques. This poses the question of whether traditionally measured dimensions can be compared with body dimensions gathered by newer technologies.

Biomechanical models

Most models simulate the body as a series of rigid lengths, in two or three dimensions, reacting to external impulses, forces and torques. Many of the early models were built to describe body displacements as a result of externally applied vibrations and impacts, to study body segment positions at work or in motion, such as gait, or to predict static forces and torques that can be applied to outside objects. A separate set of models describes the stresses in human bones resulting from external loads. These are often combined with models of body articulations (knee, hip and intervertebral joints), a difficult task particularly because of the involvement of many muscles and ligaments, and the consideration of elastic or plastic properties of human tissues.

Very little is achieved yet with respect to the true internal activation of muscles and the resulting loading of joints, bones and connective tissue. For example, the simultaneous use of agonistic and antagonistic muscles (co-activation) is neither well understood nor modeled. Hence, the loads on joints are calculated simply from the resultant force, and therefore may render the internal loading incorrectly, i.e., as too small. Consideration of muscle dynamics is only beginning (Marras, 1989; Kroemer et al., 1990).

Nevertheless, a large variety of models exists which differ in inputs, outputs, model structure, optimization, etc. (Aune and Juergens, 1989). An extensive table, prepared by Marras and King and contained in the Proceedings of the 1985 Workshop, lists the model types, their input and output variables, and particularly their underlying assumptions. This list shows not only the successes made in modeling, but also indicates the often severe restrictions in model coverage, usually making the applicability and validity of models limited to a few given cases and conditions.

Incorporating cognitive characteristics is still untouched by biomechanical modelers. People are information processors who can modify the interaction with their musculo-skeletal system; for example, in life-threatening danger, one can short-circuit internal protective mechanisms and is capable of exhibiting usually 'impossible' actions. Such cognitive control processes are virtually nonexistent in biomechanical models at present.

It appears that progress in biomechanical modeling is currently more hindered by our limited basic understanding of the human body and mind rather than by computational abilities.

Human-machine interface models

Anthropometric and biomechanical models combine to the next higher level in the hierarchical structure, i.e, to interface models. Interface models describe the interactions between the modeled person with the equipment in a human-machine system.

While the origin of such models is difficult to determine, the first published models in today's sense of the term appeared in the late 1960s and in the 1970s. Further developments are usually known by their acronyms, such as ATB, COMBIMAN, CREW CHIEF (CSERIAC, 1990), BOEMAN, CAPE, CAR, PLAID-TEMPUS and SAMMIE. These models were discussed in the 1985 and 1988 workshops. They represent the state of the art as follows:

— the models are specific to given designs, purposes and characteristics;
— their usefulness is basically limited by their anthropometric and biomechanical components. Predictive models of the effects of the dynamics of either their workstations, their tasks, or of the modeled human, are not yet available;
— effects of stress, motivation, fatigue or injuries are not adequately quantified, hence not modeled. Furthermore, the effects of environmental conditions on human performance usually are not included;
— validity of the models is largely unknown.

Future model developments

For future development, three major guiding tenets are apparent:

1 There is a need for an integrated model of the human, of its performance characteristics and limitations, and of its interactions with technological systems while doing 'real' tasks;
2 The development of such an integrated model of the human body is feasible now, and is becoming easier;
3 An integrated ergonomic model will guide future research and improve engineering applications.

Two approaches to the development of an integrated model were discussed in the 1985 workshop. The first relies on the development of one 'supermodel' which integrates the best qualities of all other models. Current interface models such as COMBIMAN, CREW CHIEF, PLAID-TEMPUS, and SAMMIE follow largely the supermodel approach. But these models are not compatible with each other, due to different data formats, different model complexity, different model theories and techniques, and the use of different computer systems.

An alternate to the 'integrated' model is the continued use of specific limited models, or 'modules'. Yet, if one attempts to link together modules, for instance those describing body size and the dimensions of its surroundings, or of body behavior and movement of the structures encapsulating it, one finds that such building-block process of joining compatible modules requires 'translators' between them. This linking would be much facilitated by a standard structure, i.e, by some supermodel.

Whether the future approach is that of a supermodel or of the modular type, general research needs must be fulfilled. Some Research Recommendations expressed in the COHF/NRC Workshop are:

RR1: Establish the objectives, procedures and outline for the development of an integrated ergonomic model, including a supermodel strategy;
RR2: Review and integrate existing anthropometric and biomechanical databases;
RR3: Develop submodels and modular groups;
RR4: Develop generic interfaces between human models and workstation models;
RR5: Develop methods and criteria for the validation of the ergonomic model.

Conclusions

It is of concern to note that today (1990) the 1985 recommendations are still as true as they were half a decade earlier. Yet, given the advances made in defining and measuring human biodynamics and in the use of the computer as a tool, and considering the now widespread use of computer-aided design, modeling the 'human working with equipment' such as the driver in an automobile is important and useful. Designers use models of the human now, and will use them increasingly in the future. Such models must reflect solid knowledge, however, based on results of research that, e.g. in dynamics, still needs to be done. For example, it is known among computer experts ('animators') that human motion does not follow optimization with respect to a single criterion, such as energy (Lee et al., 1990). Thus fairly complex 'objective functions' and algorithms must be developed to make models realistic; an 'animated' model will not do, it must be anthropomorphous.

Design of vehicles, and of workstations and work procedures, has come a long way from the simplistic two-dimensional templates (which still serve some purposes for simple tasks). Certainly, a wealth of information has been collected, and has been expressed on paper, for example in the various SAE standards and recommendations. Yet many rely on certain underlying assumptions which simplify reality, or in fact tend to proliferate existing conditions. For example, a drawing manikin is rather inflexible with respect to postures, clothing or simply changes in body dimensions. Or, a measured reach envelope or field of vision depend on certain presumed sitting postures and reflect current locations of structures and controls which are not necessarily true in the future. Another example is the presumption of static postures used at work, such as when assembling automobiles, which are inadequate or even misleading if rapid changes in postures and body motions occur.

Over-simplification and over-reliance on existing conditions can be overcome by basing body models on knowledge about changing body contours, varying biomechanical properties, on predicted behavior including the results of cognitive and emotional processes. In the physiological and biomechanical domains, much progress has been made towards such predictive (instead of normative) modeling, and the presumption of a static condition has given way to the recognition that people are dynamic, not static.

For the automotive industry, both in the design of new vehicles and in the design of workstation and work procedures, such 'biodynamic' (in contrast to biostatic) modeling yields many gains and advantages. For example, better restraining systems have been designed both on the basis of experimental results and of biodynamic modeling. Useful arrangement of controls and of control panels is better understood now than just a few years ago. Assembly work has been made much safer and less injurious, to a large extent owing to the development of biomechanical models of the body, discussed elsewhere in this volume. Yet some design tasks are still a matter of experience and trial and error, instead of being solved by computer-based modeling, including the old problems of ingress and egress, or of maintenance and repair jobs. Regarding maintenance, much effort is spent by the armed forces on modeling of related activities and the appropriate design of vehicles and aircraft to facilitate such work. The results of these modeling efforts will doubtlessly be directly applicable to the automotive industry.

Yet progress in such advanced modeling requires some systematic knowledge that does not yet exist. For example, little is known about cognitive processes, decision making, or about 'instinctive actions' in the case of an emergency. Thus, much 'basic research' needs to be done still until proper modeling is possible.

Certainly, the better, more complete, and more realistic the model of the human interfacing with the equipment is, the better the 'fit' of the equipment to the human. Given the complexities of the human body (and mind), one would expect that paper-based information and physical models (such as templates and dummies) will be replaced soon by computer-based analogs which incorporate complex and variable information about the human. This will facilitate the design of equipment (vehicle, workstation or hand tool) for ease of use, and allow for safe operating and working procedures.

References

Aune, I. A. and Juergens, H. W., 1989, *Computer Models of the Human Body* (in German) (Ergonomics Studies, Report no. 27) Koblenz, Germany: Federal Office for Defense Technology and Acquisition.

CSERIAC, 1990, Wright-Patterson Air Force Base, OH: Armstrong Aerospace Medical Research Laboratory-HE.

Kroemer, K. H. E., 1989a, A Survey of Ergonomic Models of Anthropometry, Human Biomechanics, and Operator-Equipment Interfaces, in *Proceedings of the Human*

Factors Society 33rd Annual Meeting, pp. 571–5, Santa Monica, CA: Human Factors Society.

Kroemer, K. H. E., 1989b, A Survey of Ergonomic Models of Anthropometry, Human Biomechanics, and Operator-Equipment Interfaces, in McMillan, G. R., Beevis, D., Salas, E., Strub, M. H., Sutton, R. and Van Breda, L. (Eds) *Applications of Human Performance Models to System Design*, pp. 331–9, New York, NY: Plenum.

Kroemer, K. H. E., Marras, W. S., McGlothlin, J. D., McIntyre, D. R. and Nordin, M., 1990, Assessing Human Dynamic Muscle Strength, *International Journal of Industrial Ergonomics*, **6**, 199–210.

Kroemer, K. H. E., Snook, S. H., Meadows, S. K., and Deutsch, S. (Eds), 1988, *Ergonomic Models of Anthropometry, Human Biomechanics, and Operator-Equipment Interfaces*, Washington, DC: National Academy Press.

Lee, P., Wei, S., Zhao, J., and Badler, N. I., 1990, Strength Guided Motion, *Computer Graphics*, **24**, 253–62.

Marras, W. S., 1989, Toward an Understanding of Internal Trunk Responses to Dynamic Trunk Activity, in Kroemer, K. H. E., McGlothlin, J. D. and Bobick, T. G. (Eds) *Manual Material Handling*, pp. 23–33, Akron, OH: American Industrial Hygiene Association.

McMillian, G. R., Beevis, D., Salas, E., Strub, M. H., Sutton, R. and Van Breda, L. (Eds), 1989, *Applications of Human Performance Models to System Design*, New York, NY: Plenum.

2

Occupant packaging

R. W. Roe

Introduction

One of the first steps in designing a new motor vehicle is to create the 'occupant envelope'. This procedure involves establishing the required interior space, and arranging interior and structural components (seats, controls, displays, air ducts, steering column, roof rails, etc.) in a manner that is consistent with driver and passenger safety, comfort, convenience and accommodation. At the beginning of a new design program the number of occupants to be packaged and the vehicle type (sporty, luxury, utility, etc.) are defined by a marketing or product planning group. Spatial requirements for the occupants are then built around these definitions.

In many vehicles, *passenger* spatial compromises are purposely made. For example, the rear seat in a four passenger 'sporty' car or compact economy sedan may not comfortably accommodate large segments of the population for leg room and head room, nor is it expected to. The frequency and importance of rear seat use are not sufficiently marketable to compromise vehicle mass, fuel economy, cost and particularly for the sporty car, styling appeal. The *driver's* compartment, however, has more demanding concerns. Drivers in a wide range of sizes, proportions and skills must be able to position themselves comfortably for vision and control operation regardless of vehicle type, size, styling and cost. The application of appropriate occupant packaging methodologies is necessary to create vehicles that will be successful and meet these user needs.

The automotive occupant packaging process relies on a human factors database created through years of research and practical application. The research, conducted both in the laboratory and on the road, has defined locations for the driver's eye and head locations, hand reach, preferred seat positions, and other workspace related measures. Many of the research studies involved hundreds of human subjects and have resulted in standardized drafting templates and computer aided design procedures that are used to develop seating packages for combined populations of male and female vehicle users. The side view seating arrangement shown in Figure 2.1 is a representation of the occupants' envelope in a typical vehicle package.

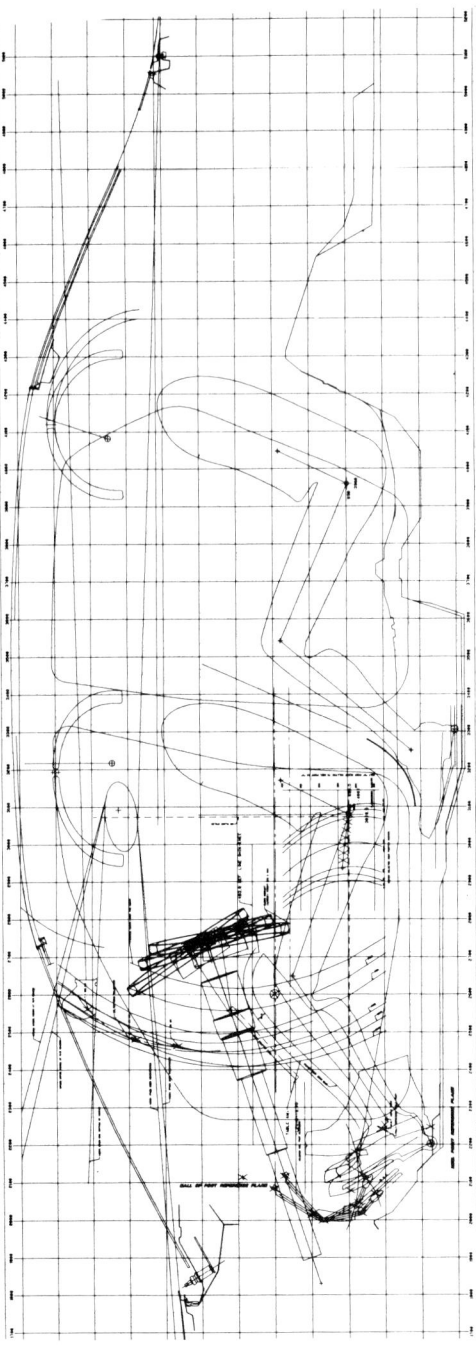

Figure 2.1. Typical seating arrangement in side view

Workspace anthropometry

An old and well-established branch of anthropology called anthropometry provides the methodology for defining the sizes of vehicle drivers and passengers. Distinction needs to be made with both the definition and application of the *conventional static anthropometric measurements* used by anthropologists and the *functional anthropometric measurements* used by engineers and designers (Kyropoulos and Roe, 1970). This distinction requires that anthropometric measurements be grouped into two categories as follows:

1 *Conventional static measurements*. These measurements are the 'classic' measurements taken by anthropologists on the human body with the subjects in rigid, standardized positions. They are typically lengths, widths, heights and circumferences. These measurements include the following:
   Standing height,
   Seated height,
   Seated eye height,
   Upper leg length,
   Knee height,
   Seat length,
   Upper and lower arm length,
   Reach (total arm length),
   Shoulder width,
   Hip or seat width,
   Weight.
   These measurements are referenced to the non-deflecting horizontal and vertical surfaces supporting the subject. Conventional static measurements are shown in Figure 2.2.
2 *Functional task oriented measurements*. These measurements are taken with the human body at work, in motion, or in workspace attitudes and are typically expressed as three-dimensional x, y, z coordinates for workspace or body landmarks (reference points) and statistically defined accommodation envelopes.

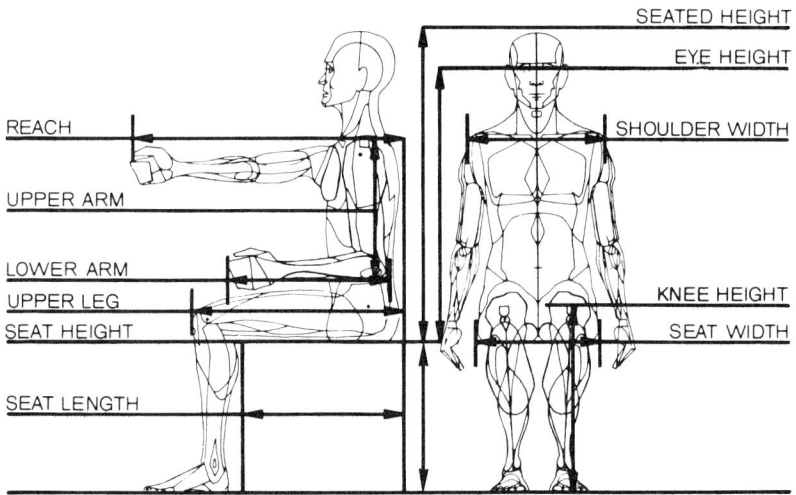

Figure 2.2. Conventional static anthropometric measurements

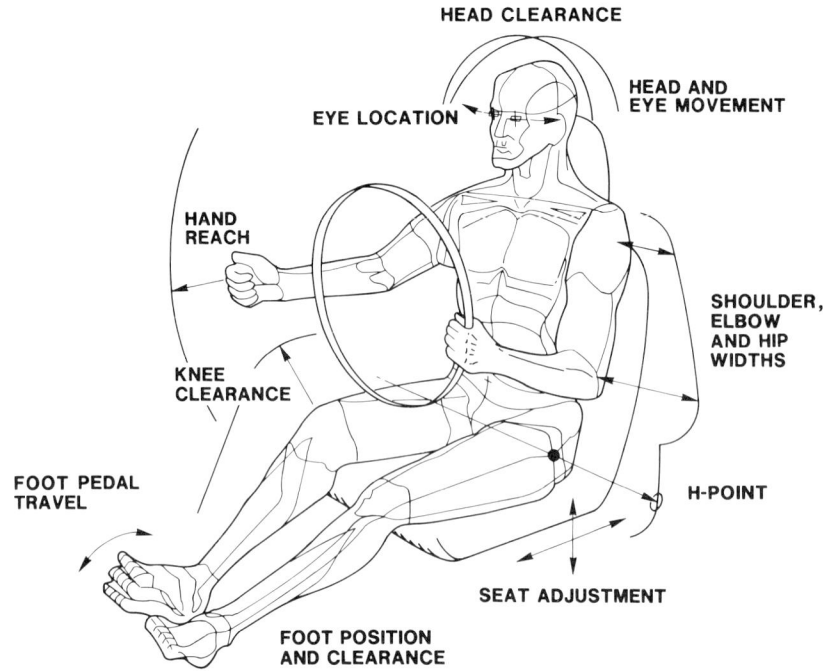

Figure 2.3. Functional task oriented anthropometric measurements

Although they are more difficult to obtain, these measurements are needed to describe driver and passenger spatial accommodations, entering, operating and exiting vehicles. Functional task oriented measurements are shown in Figure 2.3.

Since the seat and seat configurations in cars and trucks are quite different from the rigid positions used to obtain the conventional static dimensions, few of these static anthropometric measurements can be applied directly to automotive design. For example, total arm length, defined as reach, does not correlate well with reach capability in an actual vehicle workspace. Factors such as adjusted seat position, seat deflection, shoulder articulation, and lean allowed by slack in a shoulder harness (if one is worn), affect a driver's reach capability. Because of their repeatability, the conventional anthropometric measurements are those universally taken for statistical studies on human size, proportion and growth. These data are useful for the selection of test subjects used in the workspace studies to allow accurate representation of user populations with workable sample sizes.

Functional task oriented anthropometric measurements are referenced to human body landmarks called the H-point (hip pivot), BOF (ball of foot) and AHP (accelerator heel point) as shown in Figure 2.4 (see definitions in Appendix I). These body landmarks relate the occupant to components in the vehicle interior such as the seat, foot controls and floor. Functional measurements are described as accommodation envelopes that include locations in three dimensions for:

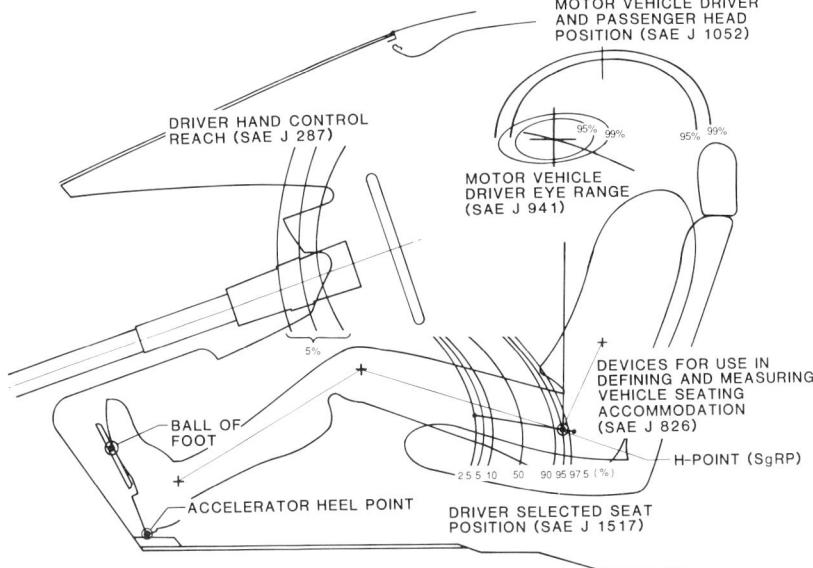

Figure 2.4. Functional task oriented SAE accommodation design tools

1 Left and right eye (eyellipse),[1]
2 Head (top, front, back and side contours),
3 Hand reach (fingertip, finger grasp and hand grasp envelopes),
4 Seat position (horizontal and vertical H-point travel).

These functional measurements are described in detail by various Society of Automotive Engineers' Practices (SAE, 1990) and are shown in Figure 2.4.[2] They will be described in more detail as the chapter progresses.

Spatial clearances and operational requirements for head, hands and feet and body widths at the shoulder, elbow and hip can in some cases be derived from the conventional static anthropometric data. These measures must be referenced to the body landmarks and three-dimensional coordinates discussed above. For example, the foot is related to the ball of foot and accelerator heel point, whereas hip, elbow and shoulder width are related to the H-point location. A number of accommodation related measurements based on these considerations are described in SAE J1100 and will be presented later in the chapter.

Statistics of anthropometric measurements

Anthropometric measurements, when taken from a representative population sample, follow the 'normal' bell-shaped or Gaussian[3] distribution. This distribution curve can be described in terms of its average value and its dispersion about this average (Arkin and Colton, 1970; see Figure 2.5).

An extremely useful statistic for designers and engineers is the percentile. If 100 people are ordered from least to greatest for any given measurement, the

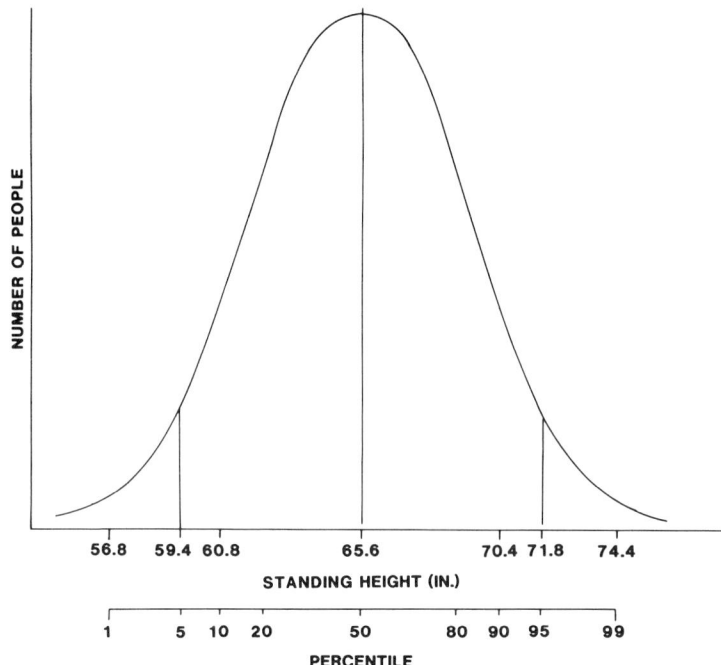

Figure 2.5. Normal distribution curve

first percentile is that exceeded by 99 per cent of the group; the 70th percentile is that exceeded by 30 per cent of the group, and so on. Figure 2.6 portrays a representative sample of 100 men and women ordered by stature to describe percentile visually. The 50th percentile, or median, is one kind of average. Another statistical average, the mean, is essentially the same as the median for large, normally distributed samples. Percentiles provide a basis for estimating the proportion of a group accommodated or inconvenienced by a specific design as well as a basis for selecting design limits to define desired accommodation ranges (Hertzberg, 1955).

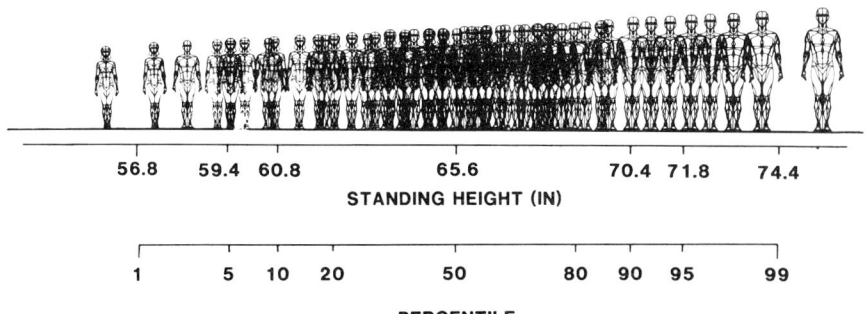

Figure 2.6. Pictorial of a representative sample of 100 people, rank ordered by stature

Table 2.1. Tabulation of the HES data by percentile

Anthropometric measurements of 6672 Americans—Male (M)—Female (F)—18–79 years (1960–1962)

%tile	Standing Ht.		Sitting Ht. Normal		Sitting Ht. Erect 1		Seated Eye Height 2		Upper Leg Length		Knee Height		Seat Length		Popliteal Ht. 3		Seat Width		Elbow Room Height		Elbow Room 4		Weight	
	M	F	M	F	M	F	M	F	M	F	M	F	M	F	M	F	M	F	M	F	M	F	M	F
1	60.4	57.2	31.9	29.5	27.5	26.4	20.3	19.5	18.3	17.1	16.5	16.1	14.9	13.1	11.5	11.7	11.5	11.7	6.3	6.1	13.0	11.4	112	93
2	62.4	57.6	31.0	29.0	32.4	30.1	27.9	26.8	20.8	20.0	18.8	17.3	17.0	16.4	15.1	13.4	12.0	12.0	7.0	6.5	13.3	11.9	118	98
3	63.0	58.2	31.2	29.2	32.9	30.4	28.3	27.1	21.1	20.1	19.1	17.5	17.1	16.6	15.3	13.6	12.1	12.2	7.1	6.9	13.4	12.1	122	101
4	63.3	58.6	31.4	29.4	33.1	30.6	28.5	27.2	21.2	20.2	19.2	17.7	17.2	16.9	15.4	13.8	12.1	12.2	7.3	7.1	13.6	12.2	124	102
5	63.6	59.0	31.6	29.6	33.2	30.9	28.7	27.4	21.3	20.4	19.3	17.9	17.3	17.0	15.5	14.0	12.2	12.3	7.4	7.1	13.7	12.3	126	104
10	64.5	59.8	32.2	30.2	33.8	31.4	29.3	27.8	21.8	20.9	20.0	18.2	17.9	17.3	16.0	14.2	12.5	12.7	8.0	7.6	14.3	12.9	134	111
20	66.0	61.1	32.9	31.0	34.4	32.2	30.0	28.4	22.3	21.3	20.4	18.6	18.4	17.9	16.4	14.7	13.1	13.3	8.5	8.2	15.0	13.5	144	118
30	66.8	61.8	33.3	31.5	34.9	32.6	30.5	28.7	22.7	21.7	20.7	19.1	18.8	18.2	16.7	15.1	13.4	13.6	8.9	8.5	15.5	14.1	152	125
40	67.6	62.4	33.7	31.9	35.3	33.1	30.9	29.1	23.0	22.1	21.1	19.3	19.2	18.9	17.0	15.4	13.7	14.0	9.2	8.9	16.0	14.6	159	131
50	68.3	62.9	34.1	32.3	35.7	33.4	31.3	29.3	23.3	22.4	21.4	19.6	19.5	18.9	17.3	15.7	14.0	14.3	9.5	9.2	16.5	15.1	166	137
60	68.8	63.7	34.5	32.7	36.0	33.8	31.7	29.6	23.6	22.6	21.7	19.8	19.8	19.2	17.6	16.0	14.3	14.7	9.8	9.5	17.0	15.6	173	144
70	69.7	64.4	34.8	33.1	36.5	34.2	32.0	29.8	23.9	22.9	22.0	20.1	20.1	19.5	17.8	16.3	14.6	15.1	10.2	9.7	17.5	16.3	181	152
80	70.6	65.1	35.3	33.6	36.9	34.6	32.5	30.2	24.4	23.4	22.4	20.5	20.5	19.9	18.2	16.6	14.9	15.6	10.6	10.1	18.1	17.1	190	164
90	71.8	66.4	35.9	34.1	37.6	35.2	33.0	30.7	24.8	24.0	22.9	21.0	21.0	20.6	18.8	17.0	15.5	16.4	11.0	10.7	19.0	18.3	205	182
95	72.8	67.1	36.6	34.7	38.0	35.7	33.5	31.0	25.2	24.6	23.4	21.5	21.6	21.0	19.3	17.5	15.9	17.1	11.6	11.0	19.9	19.3	217	199
96	72.9	67.5	36.7	34.8	38.2	35.8	33.7	31.2	25.4	24.8	23.6	21.6	21.8	21.2	19.5	17.6	16.0	17.4	11.7	11.1	20.1	19.6	220	205
97	73.3	67.9	36.9	35.0	38.4	35.9	33.8	31.3	25.6	24.9	23.7	21.8	21.9	21.4	19.7	17.7	16.3	17.7	11.9	11.4	20.5	20.0	225	210
98	73.8	68.4	37.1	35.3	38.7	36.2	34.0	31.5	25.9	25.1	23.9	21.9	22.1	21.7	19.8	17.9	16.6	18.1	12.0	11.7	20.8	20.5	230	219
99	74.6	68.8	37.6	35.7	38.9	36.6	34.3	31.8	26.3	25.7	24.1	22.4	22.7	22.0	20.0	18.0	17.0	18.8	12.5	11.9	21.4	21.2	241	236

Notes:
1 Normal Sitting Height—Subject normally relaxed with head in Frankfort plane.
2 Seated Eye Height—Not recorded in this survey. Data was determined by using the difference between standing height and seated heights of previous data.
3 Popliteal Height—Measured on seated subject vertical from seat to adjusted footboard. May be considered as seat height.
4 Elbow Room—Measured on seated subject from elbow to elbow, arms held tightly against sides with hands on lap.

This data is tabulated from original statistical worksheets used in the preparation of Weight, Height, and Selected Body Dimensions of Adults, US Dept of Health, Education and Welfare, Public Health Service publication No. 1000-Series 11, No. 8.

Sources of anthropometric data

Military populations (Air Force and Army) in the United States have been well represented in anthropometric surveys since 1944. Such select groups as Air Force fighter pilots, bombardiers, navigators, gunners, Army drivers, inductees, separatees, WAF basic trainees, Air Force pilots (WASPs) and Army and Air Force nurses have been measured. Unfortunately, these data do not represent the consumer population because of the following:

1 Age range of subjects measured,
2 Physical condition (in training),
3 Nature of sample selection (upper and lower height and weight limits for acceptance in service).

A study called the Health Examination Survey (HES) was conducted from 1960 to 1962 by the United States Public Health Service (Stoudt et al., 1965). The sample of 6672 people was carefully selected to represent racially, geographically and socio-economically the non-military and non-institutionalized American adult population between 18 and 79 years old (see Table 2.1). This survey considered 14 body measurements, including seated heights, leg lengths and torso breadths. Body size data from this study has been used until recently to define the driving population. Since it is known that people have been getting larger (Stoudt, 1978) it may be necessary to use updated data for future applications.

A more recent study on weight and height was conducted in the United States from 1971 through 1974 on men and women 18 to 79 years old (Abraham et al., 1979). This study was called HANES, for Health and Nutritional Examination Survey.

While the HANES study was primarily concerned with health and nutrition, the standing height and weight provided a useful update and comparison to the earlier HES data (see Table 2.2). At first glance, this comparison suggested that the tendency for increasing body size (secular trend) was continuing. In

Table 2.2. HES and HANES standing height data

	Standing Height (inches)					
	Male			Female		
Age	1960–62 HES mean height	1971–74 HANES mean height	HANES minus HES	1960–62 HES mean height	1971–74 HANES mean height	HANES minus HES
18–74 years	68.2	69.0	+0.8	63.0	63.6	+0.6
18–24 years	68.7	69.7	+1.0	63.8	64.3	+0.5
25–34 years	69.1	69.6	+0.5	63.7	64.1	+0.4
35–44 years	68.5	69.1	+0.6	63.5	64.1	+0.6
45–54 years	68.2	68.9	+0.7	62.9	63.6	+0.7
55–64 years	67.4	68.3	+0.9	62.4	62.8	+0.4
65–74 years	66.9	67.3	+0.4	61.4	62.3	+0.9

1978 Dr Howard Stoudt at Michigan State University estimated that standing height was increasing at the rate of 10 mm per decade and was expected to continue but at a decreasing rate (Stoudt, 1978).

Both the HES and HANES data were examined in detail by Dr Larry Schneider at the University of Michigan Transportation Research Institute (Schneider *et al.*, 1983). This examination matched the actual age of the groups (the 20-year-old HES age group data corresponds to the 30-year-old HANES age group data and so on) and considered effects of aging (biological decreases in stature with increasing age). The analysis shows discrepancies and incompatibilities between the two studies. The conclusion was that the secular increase over recent generations has been rather modest, even though it is inconsistent with intuitive observations of military data and personal observations of children growing taller than their parents.

It appears that we are near the end of increasing body size in the United States for the collective user population. The reasons are summarized as follows:

1 Improvements in nutrition and health care have stabilized, reducing the growth rate increase.
2 Because of improved health care, elderly persons who become smaller in size with age will comprise a greater proportion of the adult product user population.

In conclusion, the original 10 mm per decade growth rate has been revised to 10 mm or less per every two decades (Schneider *et al.*, 1983). This revision along with shifts in age distribution suggest that the 1970 HANES database is acceptable to define a 1990 population.

Figure 2.7. General Motors anthropometer

Anthropometers

Anthropometers are device used to obtain physical body size measurements of people. Hardware to obtain the conventional static measurements usually consist of clumsy-to-handle calipers and scales used on flat hard seats with vertical backs. In the early 1960s, General Motors Design Staff developed a semi-automated anthropometer to facilitate measuring people (Kyropoulos and Roe, 1970). An improved version, shown in Figure 2.7, was built in 1970 and is still in use today. Measurements are taken in a manner that replicates the methods used by anthropologists for 14 body-size measurements taken in the HES survey. To date, the primary use of this device has been to achieve the following:

1 Measure designers and engineers to provide data on their own body measurements and percentiles so they will have a clearer understanding of user population variations;
2 Select test groups to be used in dynamic automotive workspace studies. This allows test group stratification on selected body measurements such that a representation of the user population can be achieved with a workable sample size. The HES data updated to the HANES height and weight is generally used.

A consideration is to identify user populations for specific vehicle types. Some SAE practices consider different proportions of male and female users for truck design applications. A desirable database could be obtained by taking anthropometric measures at marketing clinics, since the participants have already been identified as users of specific vehicle types.

Selection of design limits

The term 'design limit' is used to express what percentage of a population is to be accommodated. To exemplify this concept, assume a driver seated in an automobile. The driver's task is to operate the pedals and steering wheel comfortably and be able to see. The seat adjusts fore and aft to account for the variations in driver size. The question is, where should the seat track be placed and how much should it adjust?

Seat position data gathered from expected users in a vehicle of concern may be used to describe seat travel usages (seat location and travel is defined by the H-point and its movement per SAE J826 and J1100). Typically this type of data fits a normal Gaussian distribution. When seat position is plotted versus percentile, a fairly straight line portion comprises approximately the central 90 per cent of the sample (see Figure 2.8). A horizontal seat adjustment of 168 mm, positioned properly, will accommodate this range. At each end there is a shoulder of sharply increased slope representing the remaining 5 per cent of the users who would be inconvenienced if additional travel were not available. The steeper slopes at the extremes reprsent areas that accommodate a progressively smaller percentage of the users. An additional 34 mm adjustment (20 per cent adjustment increase) at either end of the seat adjustment accommodates only another 4 per cent of the users. It can be seen that the ends of the accommodation range for the middle 90 per cent (5 per cent to 95 per cent) are at points of diminishing returns. Selection of the design limit percentile to be

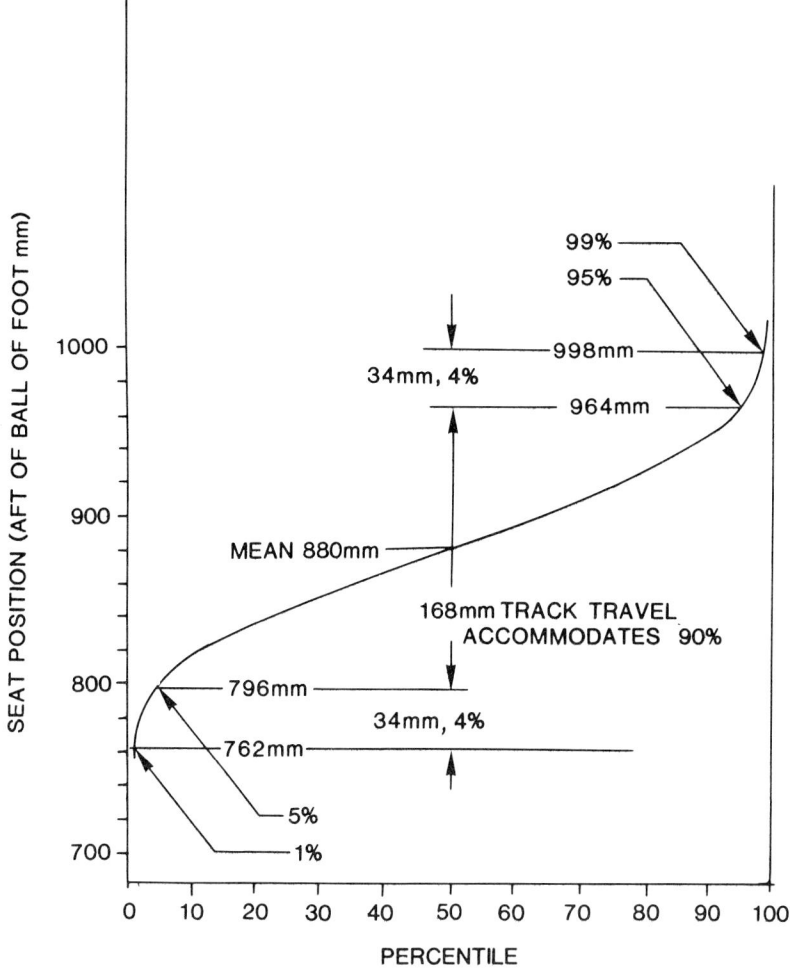

Figure 2.8. Section of design limits from automotive workspace data on driver selected seat position

used, however, may involve a number of considerations that involve safety, comfort, cost and reliability.

Manikin percentile models versus task oriented percentile models

In the occupant packaging process, it may seem that applying the conventional anthropometric measurements is all that is needed to describe accommodation requirements. For example, fifth percentile reach could be defined by placing an analogue of a fifth percentile female (a manikin) on a vehicle drawing and measuring reach. This is a problematic approach for the following reasons:

1 It is difficult to define a 'fifth percentile female'. While there tends to be a correlation between body measurements, there is no assurance that a manikin defined on the basis of stature would have fifth percentile arms and legs (Robinette and McConville, 1981).
2 Since the user population is both male and female, one must consider both sexes in proportion to use (passenger car design considers a 50/50 male to female user ratio). The use of two different populations, male and female, confounds the design limit selection for a normal distribution.
3 It is difficult to position a manikin on a drawing. Both seat deflection and posture affect occupant location. It is also difficult to define body articulation. People tend to be more flexible and adaptable than manikins.
4 The interaction of anthropometric variables with the workspace cannot always be predicted. Reach for a small fifth percentile person may not be less than a larger percentile person. A conventional adjustable seat may offset driver position such that a large driver may have a more restrictive reach than a small driver (Kyropoulos and Roe, 1970).

The concerns cited above also apply to other accommodation issues (eye, head and seat position) in automotive workspaces with adjustable seats.

As previously mentioned, some simplistic models can be considered using conventional anthropometric measurements. These measurements include body breadths and head, hand and foot sizes that are generally unaffected by the interactions described above.

Rather than consider percentile manikins or selection of a given percentile person, a task oriented statistical approach has been used by the Society of Automotive Engineers to develop function anthropometric models (computer aided design tools and drafting templates). These functional task oriented models describe the probabilistic locations of body landmarks (eyes, head, hands, torso and feet of populations of drivers in typical vehicle workspaces). These probabilistic locations express design limits in percentile levels of accommodation. Table 2.3 provides an interesting comparison between task oriented percentile models and percentile manikin models.

Functional anthropometry in the vehicle workspace

Though most of the conventional anthropometric data have limits for direct use in automotive design, the data can be used to structure test groups and simplify the collection of functional task oriented data. A relatively small sample of 50 to 100 subjects can be selected on the basis of stature, weight and/or other key conventional anthropometric measurements to represent a much larger sample of the user population.

A typical example was the Society of Automotive Engineers Controls Reach Study (Hammond and Roe, 1972a). A representative sample of male and female drivers were tested in specially constructed hand control reach checking fixtures that allowed for restrained and unrestrained reach (with and without shoulder belts). Hand reach to controls was measured along the longitudinal x-axis at a number of stations left and right of the steering wheel, above the floor and in front of the driver. Occupant-to-vehicle-interfaces at the accelerator pedal (heel point) and seat (H-point) were used in each of the test fixtures to reference reach performance to the workspace (see Figure 2.9). The study resulted in fifth percentile reach envelopes defining 95 per cent accommoda-

Table 2.3. Driver workspace design and evaluation models

Task oriented percentile models	Manikin oriented models
Uses anthropometric data to define a user population stratified across stature, weight and age.	Uses anthropometric data to define 95 per cent male and fifth percentile female dimensions.
Conducts tests for specific tasks (reach, eye location) in vehicles to develop statistical models defining spatial requirements.	Uses manikins and/or selected large (95 per cent) males and small (5 per cent) females to define spatial requirements.
Assumes that sample populations define user requirements.	Assumes that specified large males and small females comprehend all user requirements.
Assumes user needs are expressed by a central tendency (usually a normal Gaussian distribution) with exclusions at both ends, for example 95 per cent accommodation with 2.5 per cent excluded at each end.	Assumes that 2-D (or 3-D) manikins can predict or model human requirements Assumes that a given percentile person is definable from the sum of the parts.
Many levels of accommodation are described, but only for the task studied.	Many tasks are measured and evaluated, but only for a defined small and large user.
Results in well defined *statistical* model that defines accommodation levels for *specific tasks*.	Results in questionably defined *geometric* manikin models that pedict accommodation for only *two extreme percentile people*.

tion for various male/female population ratios in a variety of seating arrangements. The typical shape of the envelope is shown in Figure 2.10. The data were used to develop SAE J287 Driver Hand Controls Reach (SAE, 1990) and provide a means of describing hand reach envelopes in a variety of vehicle types ranging from sport cars to light trucks. Other SAE task oriented accommodation design tools developed in a similar manner were illustrated in Figure 2.4.

Accommodation practices

The SAE Driver's Eye Range (SAE J941), Selected Seat Position (SAE J1517), Driver and Passenger Head Position (SAE J1052) and Driver Hand Control

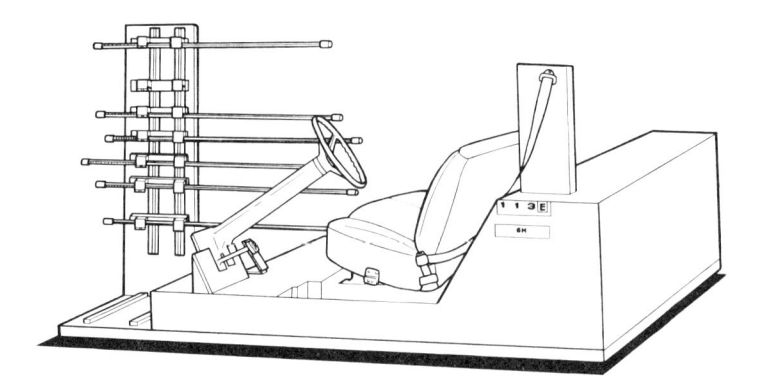

Figure 2.9. SAE control reach measuring fixture

Reach (SAE J287) are functional task oriented accommodation design tools. When used in conjunction with human body landmarks such as the H-point (SAE J826) and Vehicle Dimensioning Procedures (SAE J1100), vehicle seating packages can be developed that describe desired levels of accommodation (90, 95 and 99 per cent).

The accurate application of these tools in driver workspace design requires use in accordance with detailed procedures described in the practices. Misapplication of the occupant packaging tools, or use of an improperly located 'reference point' can compromise the vehicle package.

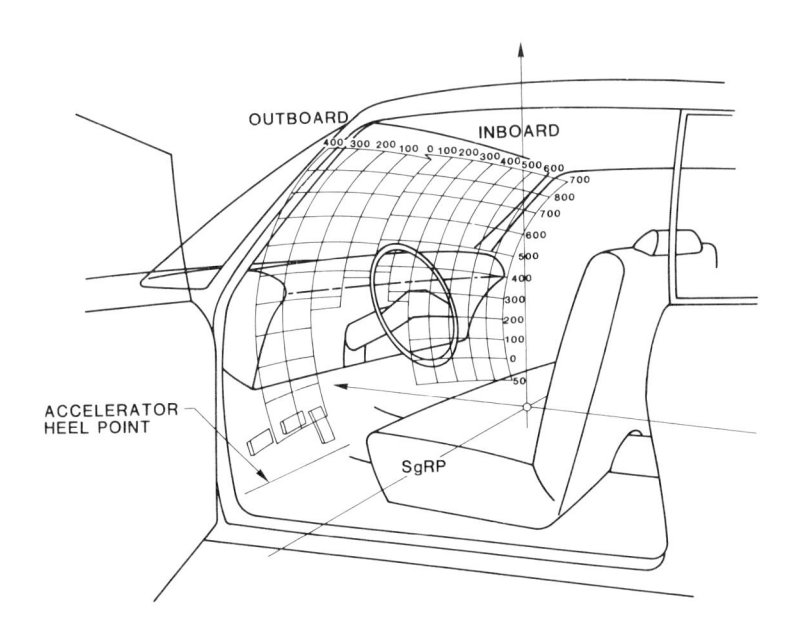

Figure 2.10. Control reach envelope in the vehicle workspace

Occupant packaging

The tools used in occupant packaging can be grouped in three areas:

1. Task oriented accommodation design tools. These are statistical models of driver and passenger body landmarks (eye, top of head, front of head, back of head, H-point and hand) while seated in the vehicle workspace. The tools, shown in Figure 2.4, take the form of two-dimensional templates, curves or envelopes and are referenced to a specified H-point, the SgRP or SgRP related reference points, the Ball of Foot or Heel Point (see definitions in Appendix I). These tools are statistical predictions of driver and passenger space requirements, not exact engineering measures. They are also based on 'present day' driver accommodation systems where the seats adjust and the pedals are 'fixed'.
2. H-point machine and H-points (SAE J826). The machine, shown in Figure 2.11, is a 'yardstick' used to define the location of a reference point on a seat. This reference point, the H-point, takes into account the deflection of the seat contour when body weight is applied. The H-point machine, when used with SAE J1517 Driver Selected Seat Position, links accommodation tools to the driver workspace with the definition of a specific H-point location, the SgRP. The SgRP defined at 95 per cent accommodation provides accurate statistical location of the eyellipse and head position contours (Roe, 1989).

   H-points for fixed passenger seats are a simpler matter since seat travel is not involved. Chair height and specific clearance requirements from SAE J1100 establish these H-point locations.
3. Vehicle dimensioning procedures (SAE J1100). These are standardized measurements that provide a database for analysis and comparison of vehicle interior space.[4] These measures are defined as lengths (L), widths (W), heights (H) and volumes (V) and provide additional information for positioning occupants and components in the vehicle workspace. Some of these measures are shown in Figure 2.12.

The tools are used by automobile manufacturers to develop designs that accommodate as large a percentage of the driving population as practicable.

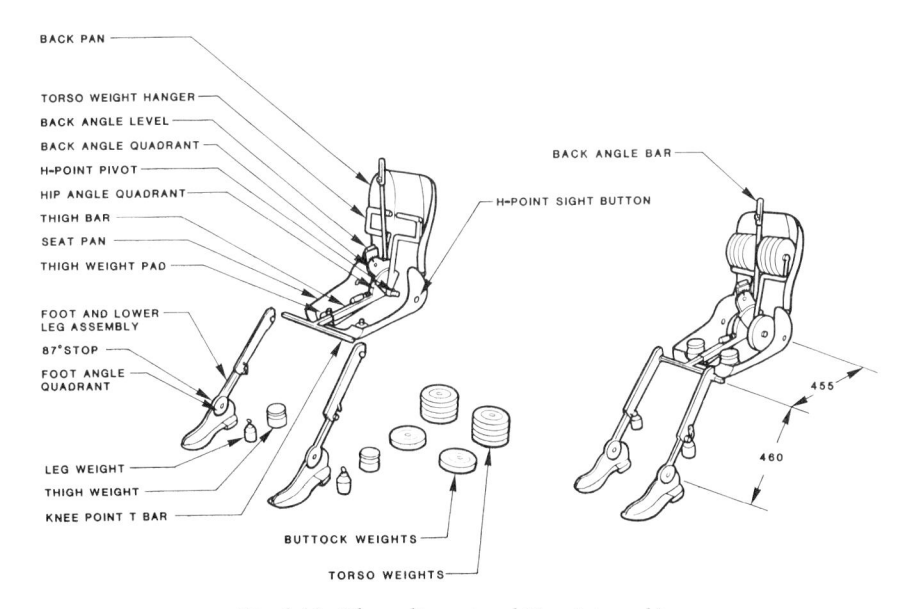

Fig. 2.11. Three-dimensional H-point machine

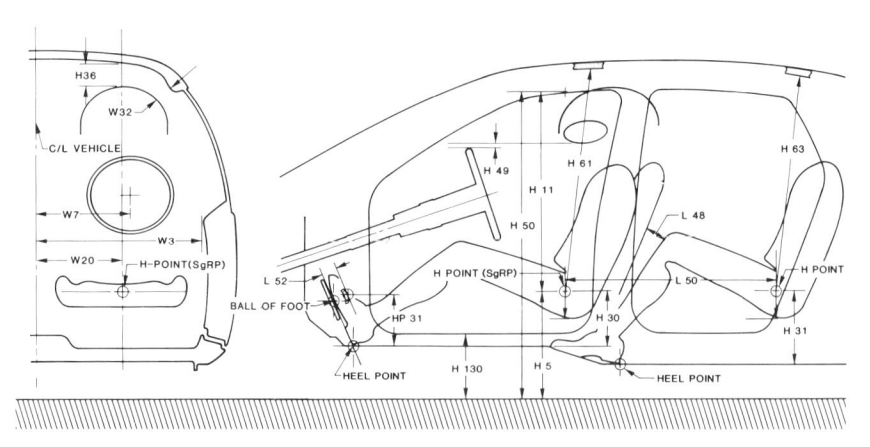

Figure 2.12. Dimensions in SAE J1100 that define interior space and access

As new data become available, the tools and practices are updated. The *SAE Handbook* (SAE, 1990) is published annually to account for these revisions. Many of these tools and other related SAE practices have been adopted by the International Standards Organization (ISO) for world-wide use.

Development of a seating package

User needs and wants will define vehicle design requirements for interior size and configuration. A sports car may seat only two people side by side at low chair heights[5] to maintain small frontal area for good aerodynamics, low center of gravity for good handling and a low profile for desired styling characteristics. At the opposite extreme is a family van requiring space for seven to eight people at higher chair heights for efficient use of space, a flat load floor for cargo utility and large door openings for ease of access. Chair height, number of passengers and projected use are key factors in establishing a seating package.

The packaging engineer develops the architecture for a vehicle literally from the ground up (see Figure 2.13). Ground clearance and underbody structure determine the interior floor height (driver and passenger heel point height). The accelerator pedal hardware with considerations for operational clearances is located relative to the toeboard structure. The driver's ball of foot and accelerator foot plane angle are established at the accelerator pedal as functions of the desired chair height. The driver selected seat position curves are set in place to define requirements for seat track travel accommodation. The accommodation range selected (between the foremost and rearmost track travel extremes) may vary depending on safety, comfort and convenience factors. The SgRP (design H-point) is established, typically at 95 per cent accommodation for dimensional comparison consistency and proper statistical positioning of the eyellipse and head position contours (Roe, 1989). Sight lines are drawn from the eyellipse to necessary visual targets. Roof clearances are measured to the head position contours. The reach envelopes are placed relative to the heel point as a function of the package configuration. Controls rearward of the

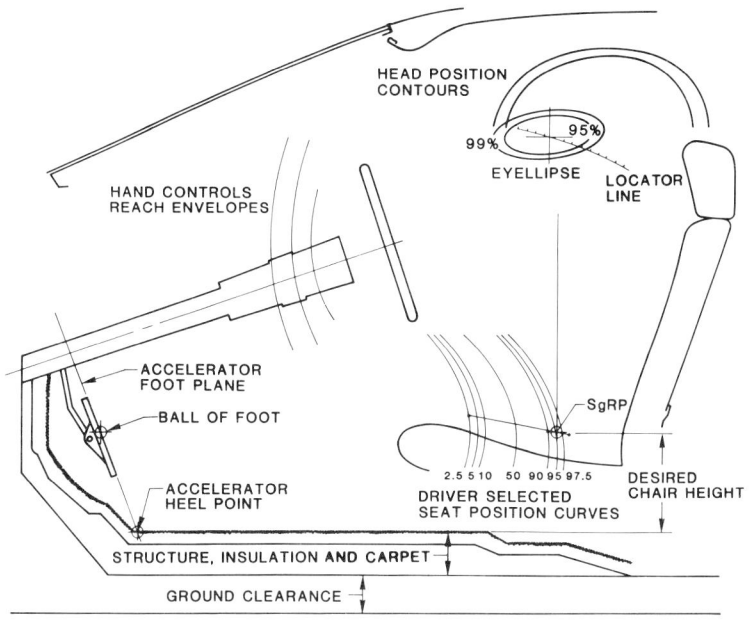

Figure 2.13. Seating package development

reach envelopes will be assured of greater than 95 per cent reach accommodation.

It is important to note that occupant locations and accommodation in the seating package are established prior to the use of two-dimensional or three-dimensional manikins (H-point machines). These machines do not define accommodation but do facilitate location of reference points that reflect where people sit in seats such that seats can be positioned properly and task oriented accommodation tools can be set in place to define and measure accommodation.

Since some occupant packaging requirements are highly subjective, market segment related or confounded by many variables, it is difficult to define these minimum or recommended dimensions using the accommodation tools. In this case vehicle spatial comparisons are made using measurement procedures defined in SAE J1100. These measures may include:

Opening sizes for entry and egress,
Lift over heights for luggage access,
Hip and shoulder room,
Distances between H-points,
Chair heights,
Steering wheel location,
Heights to ground,
Cargo dimensions,
Luggage volume.

Automotive design groups maintain dimensional data files on vehicles, past, present and future to provide spatial comparisons. The Motor Vehicle Manufacturers Association publishes dimensional data using SAE J1100 definitions

Table 2.4. MVMA vehicle data sheet

MVMA Specifications
METRIC (U.S. Customary)
Vehicle Dimensions

Vehicle Line _____
Model Year __1990__ Issued __Jan, 1990__ Revised(●) _____

Body Type	SAE Ref. No.	2-door sedan

⌀ Front Compartment

SgRP front, "X" coordinate	L31	1494 (58.8)
Effective head room	H61	938 (36.9)
Max. eff. leg room (accelerator)	L34	1094 (43.1)
SgRP to heel point	H30	177 (7.0)
SgRP to heel point	L53	917 (36.1)
Back angle	L40	25°
Hip angle	L42	98°12'
Knee angle	L44	134°04'
Foot angle	L46	87°
Design H-point front travel	L17	189 (7.4)
Normal driving & riding seat track trvl.	L23	189 (7.4)
Shoulder room	W3	1348 (53.1)
Hip room	W5	1308 (51.5)
Upper body opening to ground	H50	1207 (47.5)
Steering wheel maximum diameter*	W9	375 (14.8)
Steering wheel angle	H18	20°05'
Accel. heel pt. to steer. whl. cntr	L11	506 (19.9)
Accel. heel pt. to steer. whl. cntr	H17	548 (21.6)
Undepressed floor covering thickness	H67	25 (1.0)

⌀ Rear Compartment

SgRP Point couple distance	L50	616 (24.3)
Effective head room	H63	867 (34.1)
Min. effective leg room	L51	688 (27.1)
SgRP (second to heel)	H31	265 (10.4)
Knee clearance	L48	-112 (-4.4)
Shoulder room	W4	1299 (51.1)
Hip room	W6	1146 (45.1)
Upper body opening to ground	H51	N.A.
Back angle	L41	28°30'
Hip angle	L43	82°45'
Knee angle	L45	65°51'
Foot angle	L47	106°48'
Depressed floor covering thickness	H73	20 (0.8)

Luggage Compartment

Usable luggage capacity [L (cu. ft.)]	V1	———
Liftover height	H195	809 (31.9)

Interior Volumes (EPA Classification)

Vehicle class	Subcompact
Interior volume index (cu. ft.)	87
Trunk / cargo index (cu. ft.)	11

for vehicles manufactured in the United States each year. A sample of the MVMA data sheet is shown in Table 2.4. Typical interior spatial dimensions, some listed in Table 2.4, are pictured in Figure 2.12.

In many instances single dimensions are insufficient to describe complex interactions, therefore it is necessary to make two-dimensional or three-dimensional spatial comparisons. Usually front and/or side view drawings are prepared, showing sections through the occupant and/or vehicle (referenced to the SgRP, ball of foot, heel point or vehicle ground lines). Comparisons are then made by overlaying the proposed design with the present or competitive design to study spatial differences. A typical side view door opening profile comparison for evaluating vehicle entrance is shown in Figure 2.14.

As the total package develops on paper, styling themes are conceived and a seating buck is constructed to provide a three-dimensional representation of the vehicle interior. Although the seating buck is constructed primarily of wood, the seats are fully trimmed and sprung. All primary controls such as the steering wheel, foot pedals and transmission selector are in place. All doors open and close with access to the luggage compartment. The seating buck is not drivable and is therefore limited to static evaluation.

The three-dimensional H-point machine is used as a design/build validation tool. Each seated position in the buck is checked to ensure that the two-dimensional layout dimensions (H-point locations defining occupant location) are maintained. An iterative process follows; as changes are made in the two-dimensional package drawings and the seating buck, H-point checks are made to confirm build accuracy.

This checking procedure continues through the development and manufacturing process to ensure that the seats are built properly and that vehicles coming off the production line are providing the accommodation levels initially intended.

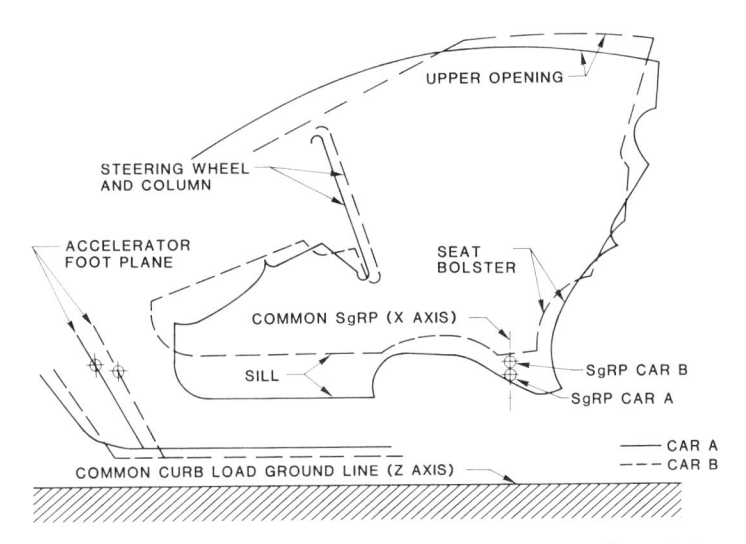

Figure 2.14. Side view door opening and profile comparison for vehicle entrance evaluation

Many manufacturers have developed additional task oriented tools for further definition of driver and passenger spatial requirements. General Motors for example, has developed driver shin knee profiles (Roe, 1975), preferred steering wheel locations and occupant width envelopes. In many instances such tools are used and validated by their developer. Then supporting research and procedures are presented as an SAE paper ultimately to become a Recommended SAE Practice utilized by the total industry.

Future research and new packaging tools

Basic occupant packaging tools (the H-point machine, eyellipses, selected seat position curves, head position contours and hand controls reach envelopes) for passenger cars were developed with 1960s vehicles and 1960s anthropometry. Both the vehicle and the driver have changed. It should be noted, however, that heavy duty truck practices were researched and revised in the 1980s and that many ongoing revisions have been made to passenger car practices.

The appropriate anthropometry to be used for defining accommodation in the driver workspace was discussed in a previous section, 'Sources of anthropometric data'. One may expect that the statistical increase in weight and physical size of the user population could shift driver seat positions rearward and eye locations upward.

Increased use of power brakes and power steering have reduced the amount of effort required of the driver to operate the vehicle, allowing a more relaxed driving posture. This, along with added accommodation features such as tilt wheels, bi-directional seat adjustments, seat back recliners and easily reached pod and/or column mounted primary controls, could affect driving positions. Seats are now more countoured with variations in lumbar support which could 'fix' the occupant laterally but affect vertical and longitudinal head and eye location.

Recent, but as yet unpublished studies, conducted by General Motors at the University of Michigan Transportation Research Institute (UMTRI) using the accepted 1970s anthropometric data and contemporary vehicles, have shown that the 'coulds' suggested above are generally true. The preliminary findings also suggest that there is a need to define and measure seat and packaging variables that affect driver location in the vehicle better. The SAE committee activities responsible for the accommodation practices[6] are aware of these findings and are monitoring research at UMTRI funded by the Motor Vehicles Manufacturing Association and MVMA member companies. This research was started in 1990 and will continue over a three-year period. Areas of concern are as follows:

1 Assembly and analysis of all available recent nonproprietary occupant positioning data already collected by MVMA member companies;
2 Collection of occupant positioning data from additional vehicles as needed to cover the full range of vehicle types manufactured;
3 Study of the H-point machine and its ability to measure accurately seating and packaging variables that affect driver positioning and comfort.

While SAE Practices are continually updated as research becomes available from the individual automobile companies, the results of this effort will likely produce changes in the accommodation tools and their application methodology.

Appendix I

A description of SAE occupant packaging practices

H-point machine (SAE J826; see Figures 2.11 and 2.15)

The three-dimensional H-point machine or manikin has 50th percentile male weight and body contour but is used with 95 per cent male legs. Use of this weighted and articulated manikin establishes a reliable design reference point, the H-point, that takes into account the deflection of the seat contour when body weight is applied. This results in a reference location *intrinsic to the seat* but defined at a specified seat location in the workspace.

The two-dimensional H-point machine is a side view centerline representation of the three dimensional H-point machine.

Use of these tools (manikins) assumes that seat comfort can be divided into two components:

1. Attitude comfort (measure of comfort due to occupant position in the car);
2. Seat comfort (the degree of support that a seat provides the occupant).

The manikin is related to *attitude* comfort only. The foot, knee, hip and back angles provide an indication of relative position and interior space. Small hip and knee angles reflect restricted space. The back angle provides an indication of seat back inclination, and the foot angle an indication of toeboard or pedal angle relative to seat position.

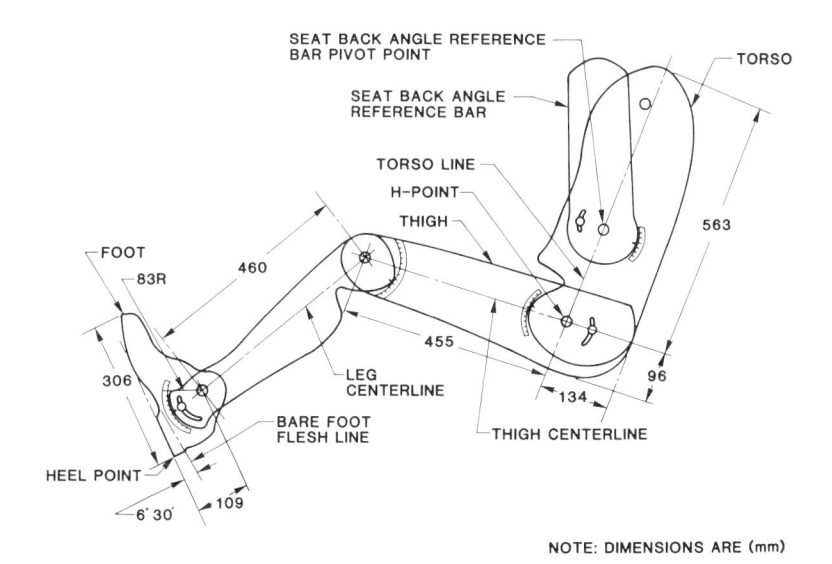

Figure 2.15. Two-dimensional H-point machine

The three-dimensional H-point machine is used as a developmental tool to ensure that a proposed design places the H-point at its intended location in a vehicle or buck. This closes the loop in the design process and ensures that a seating buck, development vehicle or production vehicle, as built, meets design intent.

The two-dimensional H-point machine is used as drafting tool to facilitate the development of seating arrangement drawings.

H-point (SAE J826)

A point which simulates the pivot center of the human torso and thigh, and provides a landmark reflecting where people sit in a seat. *The H-point is intrinsic to the seat.*

Seating reference point SgRP (SAE J1100)

A specific H-point near or at the rear of the seat track travel path that provides a landmark for positioning both the two and three-dimensional H-point machine and task oriented accommodation design tools. The SgRP is usually located at a specific per cent accommodation (95 per cent) based on the selected seat position curves. *The SgRP is intrinsic to the workspace.*

The definition of SgRP from SAE J1100 (SAE, 1990) is:

the manufacturer's design reference point, a unique design H-point which:

(a) Establishes the rearmost normal design driving or riding position of each designated seating position, which includes consideration of all modes of adjustment, horizontal, vertical and tilt, in a vehicle;
(b) Has X, Y, Z coordinates established relative to the designed vehicle structure;
(c) Simulates the position of the pivot center of the human torso and thigh;
(d) Is the reference point employed to position the two-dimensional drafting template with the 95th percentile leg described in SAE J826?

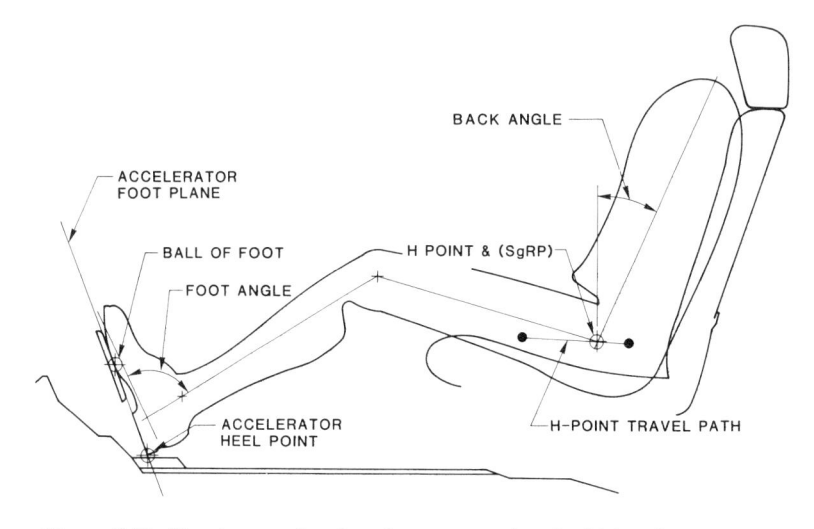

Figure 2.16. H-point travel path and occupant related vehicle reference points

H-point travel path (SAE J1100; see Figure 2.16)

The path the H-point travels through the full extent of seat adjuster travel. The SgRP must be within this path.

Driver selected seat position (SAE J1517; see Figure 2.17)

Statistical predictions of driver adjustable seat positions (H-point travel) in the vehicle workspace aft of the ball of foot and above the accelerator heel point (Philippart *et al.*, 1984; Philippart and Kuechenmeister, 1985). Curves are described at 2.5, 5, 10, 50, 95 and 97.5 per cent accommodation levels as functions of chair heights (H30) ranging from 100 to 420 mm.

The curves are based on a 50/50 male/female population mix in typical motor vehicle seating packages. The driver selected seat position curves are commonly used to define the seating reference point (SgRP) during the development of the driver seating package. They are also used to establish seat track travel (driver leg reach accommodation). By definition, a seat track travel that extends the H-point rearward from the 2.5 per cent curve to the 97.5 per cent curve would predict accommodation for the middle 95 per cent of the US driving population. An H-point on the 95 per cent curve does not represent where a 95th percentile male would sit, but a location forward of which 95 per cent of the population would sit. Two different sets of curves have been developed, one for passenger cars, vans and light trucks (class A vehicles) and a second for heavy duty trucks (class B vehicles). Curves for class B vehicles are described for three truck driver populations with male/female ratios of 50/50, 75/25 and 90/10 to 95/5.

Eyellipse (SAE J941; see Figure 2.18)

Eye location contours for 99 per cent and 95 per cent sight line accommodation developed by the statistical analysis of photogrammetric data of driver eye locations in passenger cars representing a population mix of United States licensed drivers with a male to female ratio of one to one (Meldrum, 1965).

The eyellipses are perimeters of envelopes formed in side and plan view by an infinite number of planes dividing the eye positions so that P per cent of the eyes are one side of the plane and 100–P per cent are on the other side (see Figure 2.19). The resulting envelopes are ellipses by statistical definition since

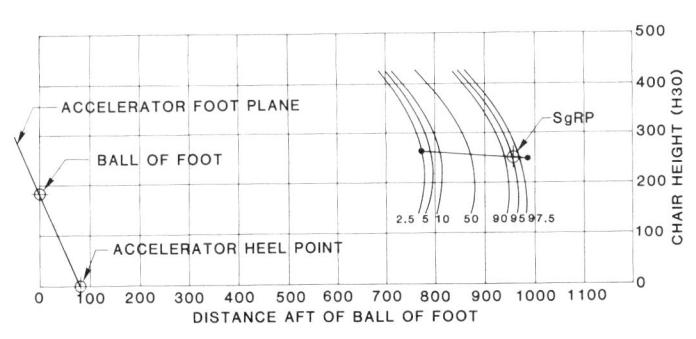

Figure 2.17. Driver selected seat position curves for passenger cars, vans and light trucks

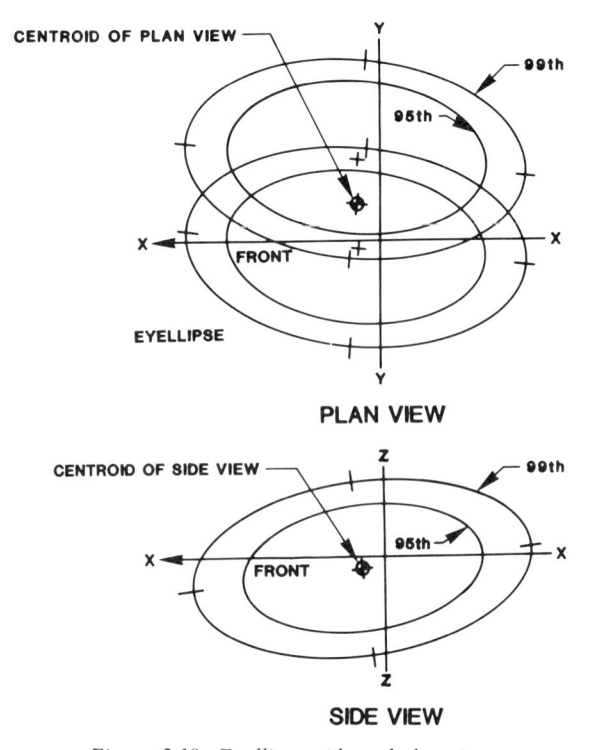

Figure 2.18. Eyellipse, side and plan view

the data were normally distributed. Each of the planes used to make up an ellipse cuts off a different (but overlapping) set of eye points. The result is that, in a two-dimensional view, the 99 per cent eyellipse contour encloses 93 per cent of the total eye locations and the 95 per cent eyellipse contour encloses only 74 per cent of the total. This curiosity presents no problem in application

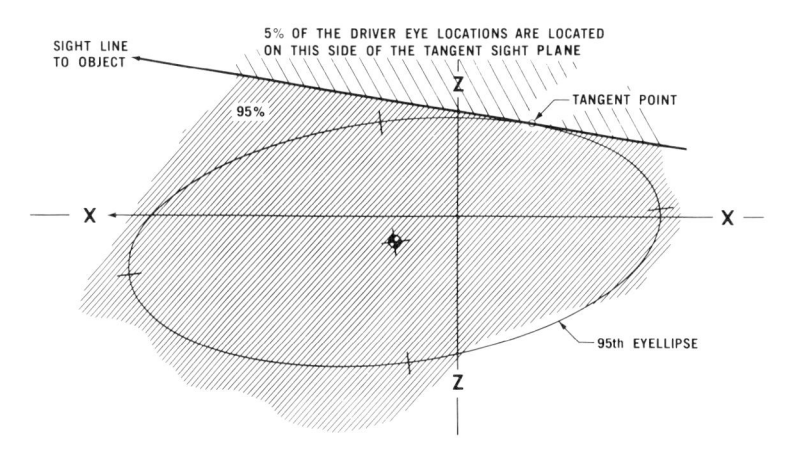

Figure 2.19. Eyellipse tangent cutoff sight planes

since each individual plane (sight line) represents accommodation for the selected 99 or 95 per cent of the users. The 99 and 95 per cent ellipses formed by the planes or sight lines are often called 'tangent cutoff contours'.

The original study involved drivers with straight-ahead viewing tasks without head turn (Meldrum, 1965). A subsequent study was conducted on head and eye movement for targets located at various lateral angles in the forward field of view (Devlin and Roe, 1968). This study indicated that targets within 30 degrees, to the left or right, resulted in little head movement. The eyellipse, without modification, is applicable to direct viewing of targets within plus or minus 30 degrees as described above and limited only to determining target presence beyond 30 degrees. Direct viewing beyond 30 degrees involves head and eye movement that changes sight line geometry and effective interpupillary distance. A method of modifying the eyellipses to account for lateral targets beyond 30 degrees and up to 90 degrees is described in SAE J1050 (SAE, 1990).

Other studies have provided data on the effect of seat back angle on eye location in various seating packages (Roe, 1975; Hammond and Roe, 1972b). The resulting locator line positions the eyellipse as a function seat back angle (L40).

The original eye location study in passenger cars indicated that drivers tended to lean outboard. Plan view location of the eyellipse is based on seating package geometry that shifts the eyellipse outboard of the centerline of the seat as a function of shoulder room (W3) and centerline of steering wheel (W7).

The SAE study, 'Describing The Truck Driver's Workspace' (Philippart and Kuechenmeister, 1985) provided eye locations in three heavy duty truck cab configurations. Data from this study supported the size and shape of the passenger car eye contours regardless of the proportion of males and females in the user population. However, the side view slope (angle) and location had to be changed for application to heavy truck seating characteristics (class B vehicles). Three locator line equations are provided to position the truck eyellipses for a range of back angles (L40) between 11 and 18 degrees; one for each of three population ratios of males to females (50/50, 75/25, and 90/10 to 95/5).

Proper application of the eyellipse on a vehicle seating layout involves drawing sight lines from the eyellipse outward to points of interest in accordance with a desired vision task (Meldrum, 1965). This procedure allows the vehicle to be configured to ensure that 99 per cent of the drivers will be accommodated (or 95 per cent depending upon which eyellipse is employed).

New passenger car driver positioning studies are in progress that suggest revisions to eyellipse lateral positioning and reduction in the number of eyellipses as a function of seat travel lengths. These revisions should be published in 1993.

Driver and passenger head position contours (SAE J1052; see Figure 2.20)

Two-dimensional shapes that describe seated vehicle occupant head postions at 95 and 99 per cent accommodation levels in side and rear view. The driver head position contours with seat travel, as with the eyellipses, apply to drivers in horizontally adjustable seats. The head position contours without seat travel apply to occupants in fixed seats. Two different location procedures are

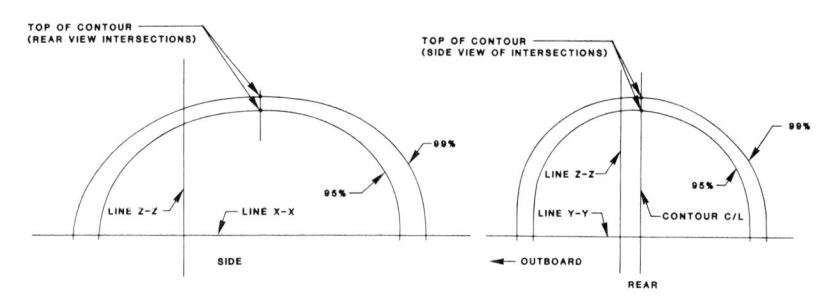

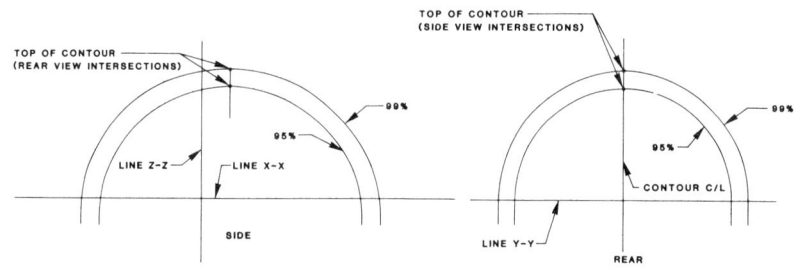

Figure 2.20. Driver and passenger head position contours

used, one for passenger cars, vans and light trucks (class A vehicles) and a second for heavy trucks (class B vehicles).

The head position contours were developed by plotting an average head profile (including hair) with a reference eye location about the perimeters of the side and rear view eye ellipses. Both fixed seat travel and rear view eye ellipses had to be developed from original eye location data to achieve head profiles for the desired views and seating conditions (Roe, 1975). These contours can be placed at all seated positions on a vehicle seating layout to determine clearances and interferences to 99 per cent and 95 per cent head space.

Locator lines (SAE J941, J1052; see Figure 2.21)

(1) The eyellipse and head contour locator line—adjustable seat. A locus of points (in one degree increments) that is used to locate the (x-x and z-z) datum lines of the eyellipse and the driver head position contour in horizontally adjustable seats with seat back angles from 5 to 40 degrees.

(2) The head contour locator line—fixed seat. A locus of points used to locate the passenger head position contours in fixed seats with back angles from 5 to 45 degrees.

Hand controls reach envelopes (SAE J287; see Table 2.5)

Hand reach envelopes developed using data acquired from test subjects performing reach tasks in test fixtures simulating a range of actual vehicle configurations (Hammond and Roe, 1972a).

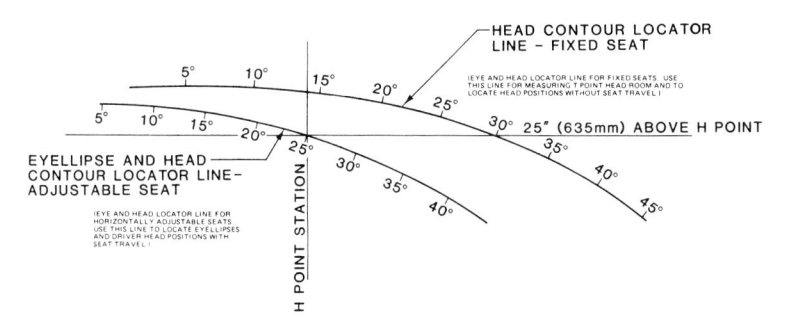

Figure 2.21. Eyellipse and head position locator lines

The test subjects included equal numbers of men and women selected to represent the (United States) driving population on the basis of stature and age (HES data) and were tested both with and without upper torso restraint (a diagonal non-extending shoulder strap attached separately to the lap belt). The envelopes constructed using the non-extending shoulder and lap belt are meant to define a restrained reach while the envelopes constructed using the lap belt are meant to describe unrestrained reach. The hand reach envelopes are three-dimensional surfaces described in table form and can be referenced to a particular vehicle seating configuration according to procedures described in the practice. Only the 5 per cent boundaries are presented in the tables (locations 95 per cent of the drivers can reach beyond). While the tables define finger grasp reach, adjustments are given for and full hand grasp and finger extended reach. Tables are provided for three different proportions of male/female user populations in passenger cars, multi-purpose passenger vehicles, and light and medium trucks. This practice is not applicable to heavy trucks.

Body base grids (SAE J182; see Figure 2.22)

The basic orthogonal reference system used on motor vehicles to measure points of interest (eye location, H-point, centerline of vehicle, etc.).

Accommodation tool reference point (SAE J1517)

A procedure describing a reference point that is used for locating the eyellipse and head position contours in heavy trucks (class B vehicles). While a referencing procedure is described for passenger cars, it has not as yet been utilized. The accommodation tool reference point represents the mean H-point location for adjustable driver seats as a function of chair height (H30).

Vehicle dimensions (SAE J1100)

The definition of a set of uniform exterior and interior dimensions for passenger cars, multipurpose passenger vehicles and trucks. All dimensions are defined in relation to the three-dimensional reference system described in SAE

Table 2.5. Hand controls reach envelope

HAND REACH ENVELOPE VEHICLE RANGE: (+0.25 < G < +0.74) POPULATION MIX: 50/50 MALES-TO-FEMALES
(mm)

HORIZONTAL REACH FORWARD OF THE HR REFERENCE PLANE AT STATIONS LOCATED LATERALLY FROM THE CENTERLINE OF OPERATOR (C L/O) AND AT ELEVATIONS ABOVE THE REARMOST H POINT. THE ENVELOPE DESCRIBES A 95% LEVEL OF PERFORMANCE OF A DRIVER POPULATION COMPOSED OF 50% MALE AND 50% FEMALE DRIVERS WEARING A TYPE 2a RESTRAINT.

ELEVATION ABOVE H PT (mm)	STATIONS OUTBOARD OF C L/O (mm)										STATIONS INBOARD OF C L/O (mm)										
	400.	300.	250.	200.	100.	50.	0.		0.	50.	100.	200.	250.	300.	400.	500.	600.
800.	401.	453.	471.	486.	507.	514.	520.		521.	531.	535.	525.	513.	498.	460.	412.	
700.	474.	518.	534.	545.	561.	567.	572.		575.	590.	595.	586.	576.	563.	531.	489.	
600.	529.	566.	579.	589.	599.	601.	600.		612.	628.	635.	632.	624.	614.	585.	544.	476.
500.	564.	596.	608.	617.	623.	618.	607.		632.	648.	658.	662.	658.	650.	622.	578.	511.
450.	575.	604.	616.	626.	629.	620.	602.		635.	651.	662.	671.	669.	663.	634.	587.	521.
400.	582.	608.	621.	630.	631.						662.	677.	678.	672.	642.	592.	528.
350.	584.	608.	622.	632.	629.						657.	679.	682.	678.	646.	591.	530.
300.	582.	604.	618.	629.	624.						648.	678.	683.	680.	646.	585.	528.
250.	576.	595.	611.	624.	616.						635.	672.	681.	679.	642.	575.	522.
200.	565.	583.	600.	614.	604.						617.	664.	676.	675.	633.	559.	512.
100.	531.	546.	567.	585.								635.	655.	657.	605.	515.	480.
0.	481.	494.	520.	542.								593.	621.	625.	561.	451.	
-100.	415.	426.	458.	486.								536.	573.	581.	501.	369.	

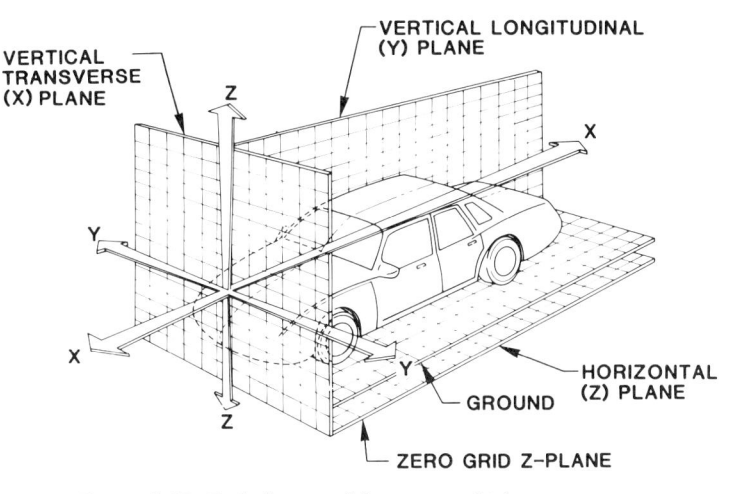

Figure 2.22. Body base grid for motor vehicle measurement

J182. The dimensions are grouped according to length (L), Width (W), Height (H), and areas of interest as follows:

Prefix letter:

W — Width dimensions
H — Height dimensions
PD — Passenger distribution dimensions
L — Length dimensions
S — Surface area dimensions
SD — Seat facing direct dimensions
V — Volume dimensions

Number:

1–99	Interior dimensions
100–199	Exterior dimensions
200–299	Cargo or luggage dimensions
300–399	Interior dimensions—unique to trucks and MPVs
400–499	Exterior dimensions—unique to trucks and MPVs
500–599	Cargo dimensions—unique to trucks and MPVs

The dimensions code W3, for example, is for front shoulder room, measured internally between the trimmed door surface on the 'x' plane through the SgRP at the height between the belt line and 254 mm above the SgRP (See Figure 2.12).

Chair height (H30; see Figures 2.12, 2.13)
The vertical dimension from the driver's SgRP (H-point) to the floor at the accelerator pedal (Accelerator heel point).

Accelerator heel point (See Figures 2.4, 2.12, 2.13, 2.16)
The lowest point at the intersection of the manikin heel and the depressed floor covering with the shoe on the undepressed accelerator pedal. The foot angle (L46) is at a minimum of 87 degrees with the manikin H-point at the

SgRP. For vehicles with SgRP to heel (H30) greater than 405 mm, the accelerator pedal may be depressed as specified by the manufacturer. If the depressed pedal is used, the foot must be flat on the accelerator pedal.

Ball of foot (see Figures 2.4, 2.12, 2.13, 2.16)
A point on a straight line tangent to the bottom of the manikin's shoe parallel to the y-plane 203 mm from the accelerator heel point.

Accelerator foot plane (see Figures 2.13, 2.16)
A plane passing through the accelerator heel point (AHP) and the ball of foot (BOF) that is normal to the y-plane (side view plane).

Back angle (L40; see Figure 2.16)
The angle between the torso line (of the two-dimensional drafting template) and the vertical line through the SgRP.

Foot angle (L46; see Figure 2.16)
The angle between the lower leg centerline (on the two-dimensional drafting template) and a straight line on the foot between the ball and heel of the bare foot. This angle is locked at 87 degrees to establish the accelerator foot plane.

Seating buck

A structure (usually of wood construction) that represents a vehicle interior package including: floor pan, body side, cowl, roof structure, steering wheel, pedals (functional), seats (front and rear with proposed adjustment), instrument panel, console, shifter and hand brake. The seating buck is a three-dimensional representation that allows hands-on study of interior space relationships. It is developed from the seating arrangement drawing.

Seating arrangement drawing (Package drawing; see Figure 2.1)

A side view and front view layout of occupant seating describing front and rear compartment space. It includes all of the task oriented accommodation design tools and many surfaces of the actual vehicle interior and exterior. The side view of the driver's seat will show the two-dimensional manikin penetrating the front seat contour, the manikin's hip point (H-point) at the SgRP, and H-point movement provided by the seat track travel.

Notes

1 'Eyellipse' is a contraction of the words 'eye' and 'ellipse'.
2 SAE Recommended Practices of concern in *The Handbook* are:
   Devices for Use in Defining and Measuring Vehicle Seating Accommodation, SAE J826;
   Motor Vehicle Dimension, SAE J1100;
   Motor Vehicle Driver's Eye Range, SAE J941;
   Motor Vehicle Driver and Passenger Head Position, SAE J1052;
   Accommodation Tool Reference Point, SAE J1516;
   Driver Selected Seat Position, SAE J1517;

Driver Hand Control Reach, SAE J287;
Motor Vehicle Fiducial Marks, SAE J182;
Describing and Measuring the Driver's Field of View, SAE J1050;
Truck Driver Stomach Position, SAE J1522.
3 After Carl Friedrich Gauss (1777–1855).
4 SAE J1100 contains a great number of vehicle dimension definitions, both interior and exterior. Many are unrelated to occupant packaging but important to overall vehicle design.
5 Chair height (H30) defines seat height above the floor (See Figure 2.12).
6 These are Human Accommodations and Design Devices Standards Committee and Driver Vision Standards Committee of the SAE Human Factors Engineering Committee.

References

Abraham, S., Johnson, C. L. and Najjar, M. F. 1979, *Weight and Height of Adults 18–74 Years of Age, United States 1971–1974*, Washington, DC: Division of Health Examination Statistics, US Department of Health, Education and Welfare.

Arkin, H. and Colton, R., 1970, *Statistical Methods*, New York: Barnes and Noble.

Devlin, W. A. and Roe, R. W. 1968, The eyellipse and considerations in the driver's forward field of view, SAE paper 680105, January, Warrendale, PA: Society of Automotive Engineers, Inc.

Hammond, D. and Roe, R., 1972a, SAE controls reach study, SAE paper 720199, January, Warrendale, PA: Society of Automotive Engineers, Inc.

Hammond, D. C. and Roe, R. W., 1972b, Driver head and eye positions, SAE paper 720200, January, Warrendale, PA: Society of Automotive Engineers, Inc.

Hertzberg, H. T. R., 1955, Some contributions of applied physical anthropology to human engineering, *Annals of the New York Academy of Science*, **63**, 616–29.

Kyropoulos, P. and Roe, R., 1970, The application of anthropometry to automotive design, Troy, Michigan: General Motors, Automotive Safety Seminar.

Meldrum, J. F., 1965, Automobile driver eye position, SAE paper 650464, May, Warrendale, PA: Society of Automotive Engineers, Inc.

Philippart, N. L. and Kuechenmeister, T. J., 1985, Describing the truck driver's workspace, SAE paper 852317, December, Warrendale, PA: Society of Automotive Engineers, Inc.

Philippart, N. L., Roe, R. W., Arnold, A. and Kuechenmeister, T. J. 1984, Driver selected seat position model, SAE paper 840508, February, Warrendale, PA: Society of Automotive Engineers, Inc.

Robinette, K. and McConville, J., 1981, An alternative to percentile models, SAE paper 810217, February, Warrendale, PA: Society of Automotive Engineers, Inc.

Roe, R., 1975, Describing the driver's work space: Eye, head, knee and seat positions, SAE paper 750356, February, Warrendale, PA: Society of Automotive Engineers, Inc.

Roe, R., 1989, Data to support revisions to SAE J941, SAE J1100, SAE J1516 and SAE J1517, SAE Vision Subcommittee, 15 June, Warrendale, PA: Society of Automotive Engineers, Inc.

SAE, 1990, *1990 SAE Handbook, On Highway Vehicles and Off Road Highway Machinery*, Vol. 4, Warrendale, PA: Society of Automotive Engineers, Inc.

Schneider, L. W., Robbins, D. H., Pflug, M. A. and Snyder, R. G., 1983, *Anthropometry of Motor Vehicle Occupants*, Vol. 1, December, Ann Arbor, MI: University of Michigan Transportation Research Institute.

Stoudt, H. W., 1978, Are people still getting bigger? SAE paper 780280, May, Warrendale, PA: Society of Automotive Engineers, Inc.

Stoudt, H. W., Damon, A. and McFarland, R., 1965, *Weight, Height, and Selected Body Dimensions, United States 1961–62*, U.S. Department of Health, Education and Welfare: Public Health Service Publication Series II, No. 8.

3

Computer-aided ergonomics design of automobiles

J. M. Porter, K. Case, M. T. Freer and M. C. Bonney

Computer-aided design and the design process

Computer-aided design (CAD) can refer to any kind of activity which utilizes a computer to assist in the creation, modification, presentation and analysis of a design (Majchrzak *et al.*, 1987). Groover and Zimmers (1984) present a useful, although greatly simplified, overview of the typical design process showing the contributions of both the designers and the CAD system during the main phases of design (see Figure 3.1).

Many manufacturing companies were first introduced to CAD because it is a means of rapidly producing and modifying engineering drawings for detail design. While this 'electronic drawing board' approach has been a relatively painless introduction to CAD, it is only the tip of the iceberg in terms of the functionality potentially available to the design team (i.e. engineers, stylists, ergonomists etc.). The aerospace industry has led the way since the early 1970s in using CAD as a design analysis tool for the representation of complex three-dimensional shapes and for the production of manufacturing information. The automotive industry has similarly undergone a transformation to the extent that one major British manufacturer has the stated intentions of performing all design work using three-dimensional systems and of requiring all its suppliers to have the facility to receive 'drawings' in electronic form. Similarly, the electronics industry is probably the leader in the use of its own products for the design, simulation of operation, manufacture, assembly and testing of new products.

In mechanical design great strides have been made in providing the link between design (CAD) and manufacture (CAM), often referred to as CAD/CAM. For example, it is often possible to extract geometric information from a CAD system and pass it to a CAM system for the addition of machining information before post-processing and transmission to the machine tool.

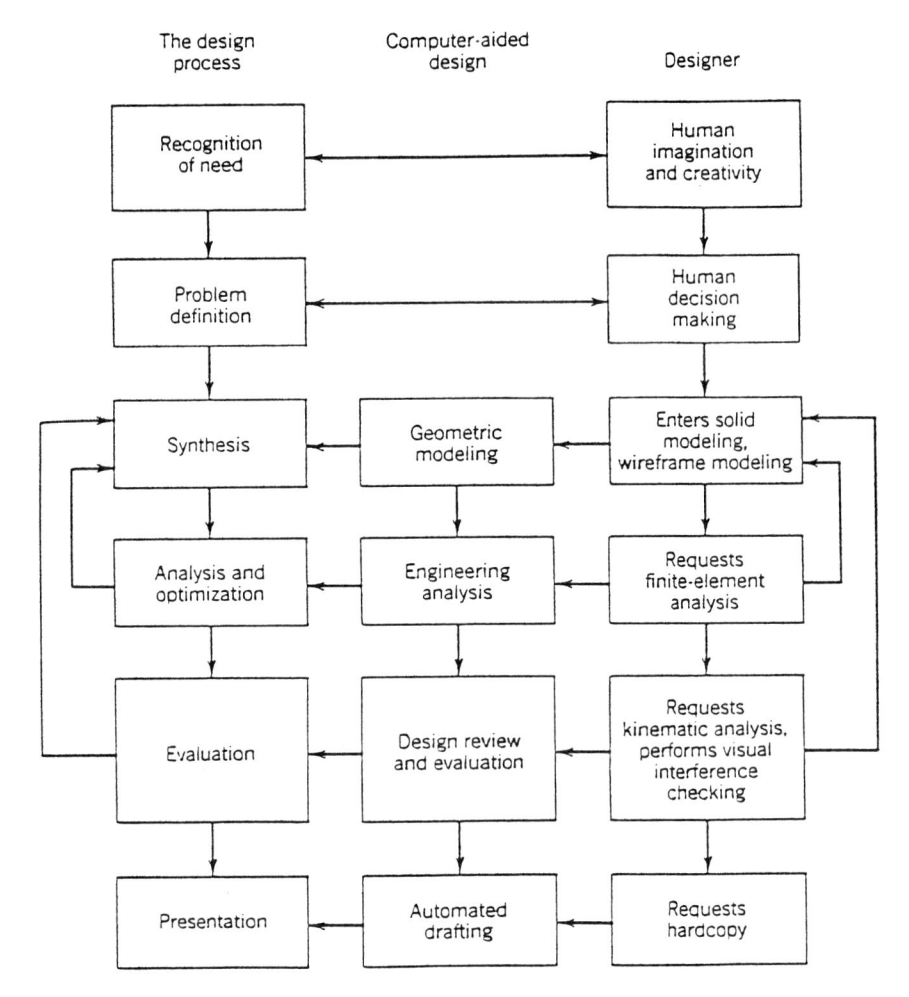

Figure 3.1. Contributions of the designer and CAD system in the design process (Source: Groover and Zimmers, 1984, by permission of Prentice Hall, Englewood Cliffs, NJ)

Ergonomics in the CAD age

The ergonomist may find schematics of the engineering design process (such as Figure 3.1) to be sadly lacking as they often take a reductionist approach (i.e. divide the problem into separate parts with separate solutions) rather than a systems approach which considers the human operator as an integral component of the system to be designed and lays emphasis on the suitability of all components for the functions allocated to them for the achievement of the overall purpose of the system.

Ergonomist's views of the design include those given by Singleton (1971), Meister (1982) and Bailey (1982). They are broadly in agreement and

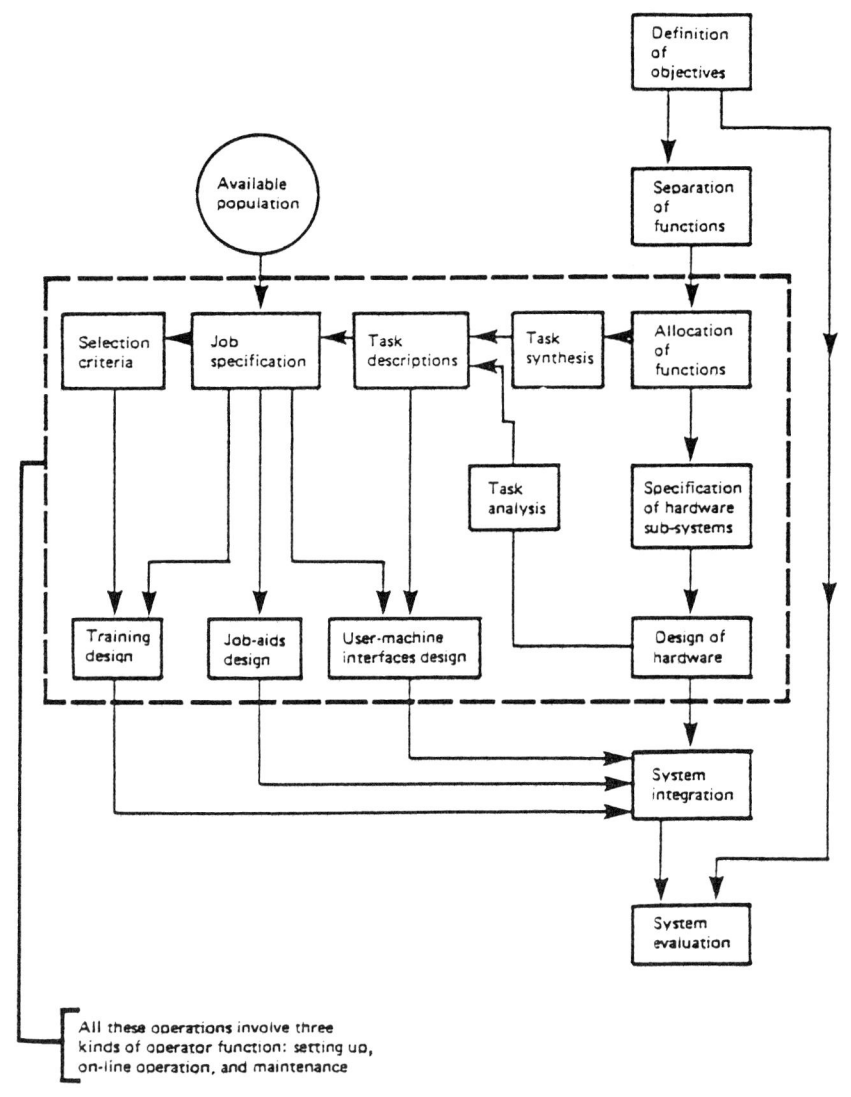

Figure 3.2. The role of human factors specialists in systems design
(Source: Singleton, 1971, by permission of the publishers, Butterworth-Heinemann Ltd)

Figure 3.2 shows the detailed schematic presented by Singleton (1971). The area where current ergonomics CAD systems have the most to offer ergonomists is the physical design of the man-machine interface. With the CAD/CAM systems available today it is quite possible for a design to progress from the concept stage right through to the manufacturing stage without any recourse to full size or even scaled models to perform necessary evaluations. This process helps to reduce the timescale of product design considerably although questions must be raised concerning the quality of design.

Traditionally, products were created by craftsmen who, as individuals, were responsible for the design and manufacture of their products, for example chairs, shoes and musical instruments. These craftsmen could modify their basic design to meet the needs for each new customer. For example, a chair could be made wider and lower to suit a short and heavy customer. If problems arose with a particular design, either in manufacture or use, then the craftsmen would receive swift feedback and be able to make any necessary modifications. In this way the design was often seen to evolve or become more 'finely tuned' to the needs of the end user or customer. With advent of CAD/CAM systems there is a risk that this evolutionary aspect of product design will suffer because the feedback from the end users will not be available until after the product's design and manufacture have been finalized. A successful design needs to find the optimum compromises within the variety of constraints that are imposed upon it from the financial, legislative, engineering, styling, manufacturing and end users' viewpoints. Unfortunately, most CAD/CAM systems provide little or no information concerning the needs of the end user and there is considerable danger that design decisions are only made to satisfy the other more tangible engineering and financial constraints. This can result in decisions which the customer finds unsatisfactory. For example, why do many car manufacturers force their customers to adopt a twisted posture when driving by placing the seat, steering wheel and pedals out of line with the straight ahead position? Why do these manufacturers actually design out a significant percentage of their potential customers by not allowing sufficient headroom or set and steering wheel adjustment? If problems such as these are evident now with cars that were designed before the advent of CAD/CAM then there must be worries for the future. If a major advantage of CAD/CAM is to reduce the timescale of product design then the costly and time-consuming process of constructing mock-ups will certainly decline. There is a danger that the ergonomics input will then take the form of working to approved guidelines and meeting legislation, both of which do not necessarily result in the optimum design.

It is essential that the ergonomics input to a product takes place throughout the design process but nowhere is it more important than at the concept and early development stages of design. Basic ergonomics criteria such as the adoption of comfortable and effective postures need to be satisfied very early on. If these criteria are not thoroughly assessed then there is usually only very limited scope for modifications later on as all the other design team members will have progressed too far to make major changes without considerable financial and time penalties.

The ideal solution to help ensure that the finished product meets the ergonomics requirements is to conduct the traditional mock-up stage using computer graphics. A prerequisite of this possibility is that we can successfully model the end user to a satisfactory level of accuracy.

In the concept stage the design team will need mainly dimensional information describing the future users' needs based upon the variation of body size and postures which will be adopted to meet the task requirements and to maximize comfort. The computer modeling of people (known as 'man-modeling') therefore needs to include anthropometric variation and to permit the postural evaluation of workstations based upon criteria related to fit, reach, vision and comfort. Such man-modeling CAD systems exist and, in some cases, have existed since the late 1960s well before the advent of CAD/CAM.

Brennan and Fallon (1990) consider that such CAD systems greatly enhance the abilities of the ergonomist and have the potential to place the ergonomist in a more directive, as opposed to supportive, role in the system design process.

Man-modeling CAD systems

A wide variety of such systems have been developed. Their functionality, flexibility and success have also varied considerably. Kinematic modeling enables the spatial evaluation of workplaces where either the human operator or parts of the physical environment are placed in different positions over time—the type of modeling which is the focus of this chapter. In order to represent these spatial relationships the man model needs to be suitably 'enfleshed', preferably using solid modeling as this allows evaluative features such as hidden lines or interference checking between two solids in three-dimensional space. Kinetic, or dynamic, modeling is usually associated with assessing the body's response to large external forces such as those experienced in car crash simulations. In this type of modeling more consideration may be given to body segment parameters such as mass, center of gravity and moments of inertia.

Descriptions of the earlier kinematic systems (i.e. BOEMAN, BUFORD, CAR, COMBIMAN, CYBERMAN and SAMMIE) are given in the review papers by Dooley (1982) and Rothwell (1985) and most of the currently available systems (i.e. COMBIMAN, CREW CHIEF, ErgoSHAPE, ErgoSPACE, MINTAC, SAMMIE, TADAPS and WERNER) are presented in Karwowski *et al.*, (1990). Other systems include FRANKY and ANYBODY. Brief details of each system are presented below.

ANYBODY (see Figure 3.3) is a three-dimensional ergonomics template or stencil of the human form which is used in conjunction with the CADKEY workplace modeling system which can run on an IBM AT. It is marketed by IST GmbH in Germany. The templates are available for men and women in a variety of sizes (females: fifth and 50th percentiles; males: 50th and 95th percentiles) and shapes (ectomorph, mesomorph and endomorph). The anthropometric data are taken from DIN 33402 part 2 which presents information for the German public although 50th percentile models are available for other databases such as those presented in Bodyspace (Pheasant, 1986). The developers suggest that these templates can be linearly scaled to represent other body dimensions. No published studies on the development or application of ANYBODY have been found.

BOEMAN (see Figure 3.4) was one of the earliest man models and was in use in 1969 by the Boeing Corporation, Washington. The system was not interactive as graphics terminals were not generally available at that time. It presumably fell into disuse as more interactive systems became available.

BUFORD (see Figure 3.5) offers a simple model of an astronaut, with or without a space suit, and was developed by Rockwell International, California. The model consists of an assemblage of individual body segments with no linkage system so the segments must be moved one by one to simulate working postures. The system is not generally available.

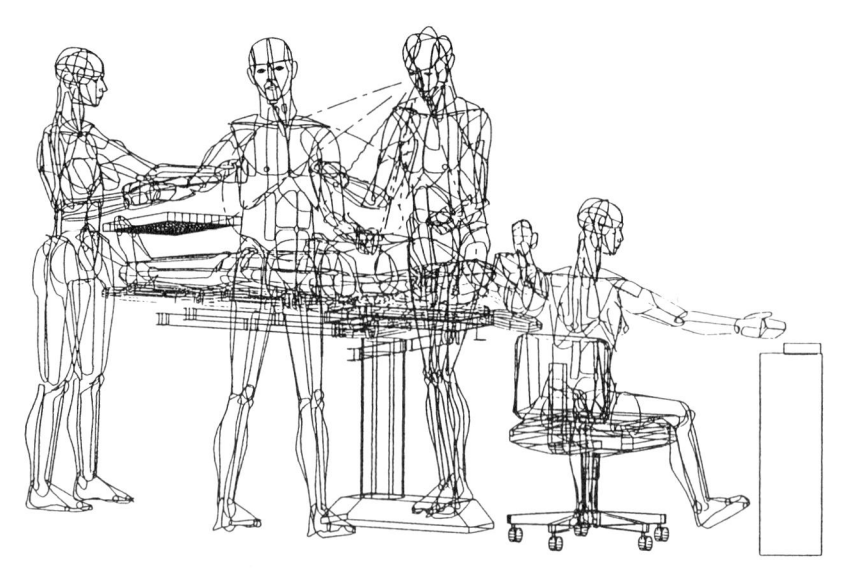

Figure 3.3. ANYBODY
(Source: IST GmbH, West Germany)

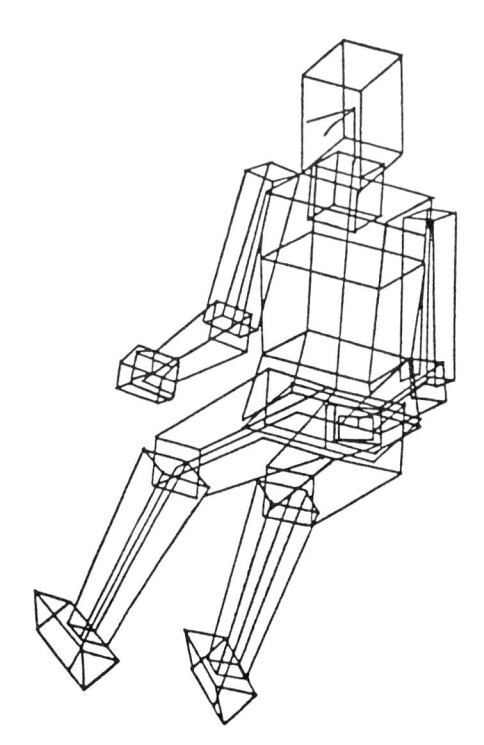

Figure 3.4. BOEMAN
(Source: Dooley, 1982, © IEEE)

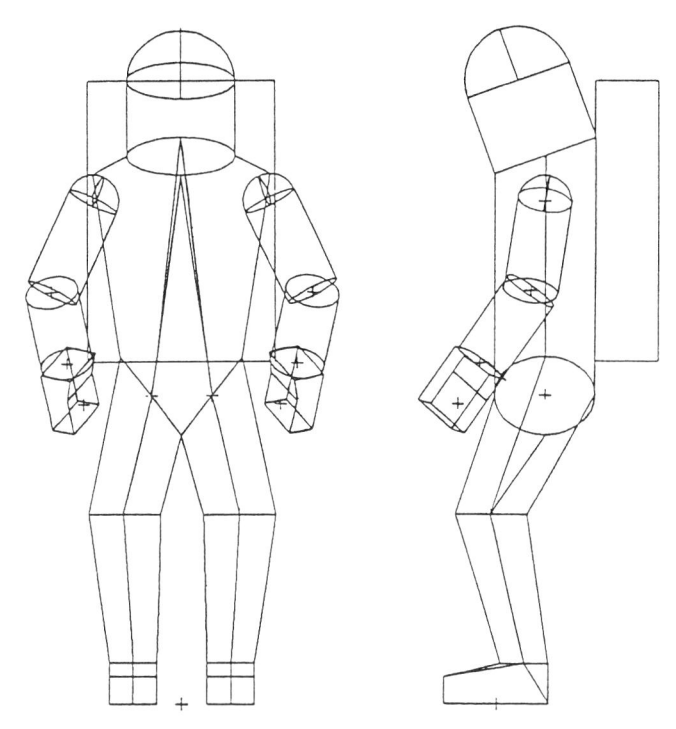

Figure 3.5. BUFORD: A model of an astronaut wearing a spacesuit
(Source: Dooley, 1982, © IEEE)

CAR stands for Crew Assessment of Reach and was developed by Boeing Aerospace Corporation for use by the Naval Air Development Center in the USA. The system has no graphical display and has been designed specifically for the in-house assessment of reach in aircraft crew stations.

COMBIMAN (see Figure 3.6) stands for COMputerized BIomechanical MAN Model and was developed by the Armstrong Aerospace Medical Research Laboratory. It is a three-dimensional model of an aircraft pilot which is used to evaluate the physical accommodation of pilots in aircraft crewstations. COMBIMAN has two methods for dimensioning the pilot model. The first is to specify one or two critical dimensions, the remainder being generated using appropriate regression equations. The second method involves the specification of dimensions for a variety of pilot models. This latter method is useful in the assessment of multivariate accommodation which will be discussed later. The current anthropometric databases consist of military pilots and women. The pilot model is constructed using an array of small interconnected triangles. This complexity can be automatically reduced to just the profile view and any essential features from any viewing angle. Postures can be set by specifying individual joint angles or by using task related commands. COMBIMAN can produce pilot visibility plots to meet military standards (MIL-STD-850), reach tests can be conducted for three types of controls, six types of clothing, three types of harnessing and seven

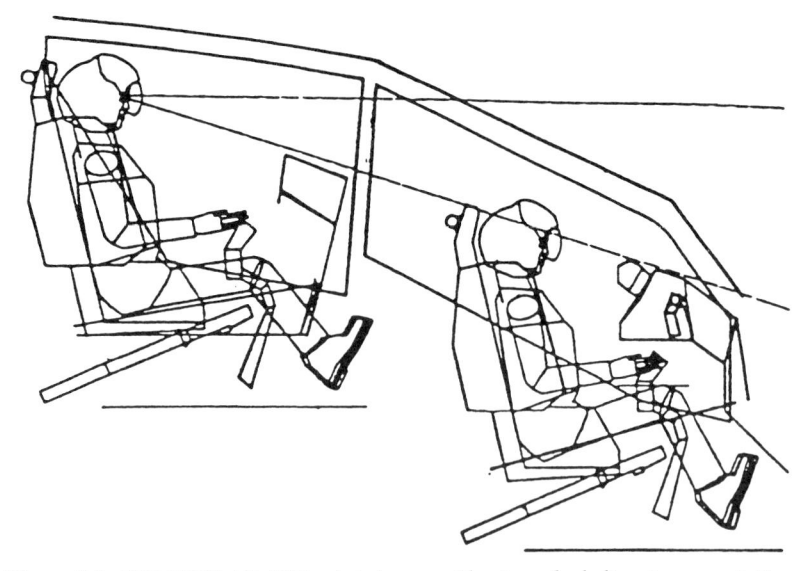

Figure 3.6. COMBIMAN. This plot shows a side view of a helicopter crewstation
(Source: McDaniel, 1990)

reaching planes, and strength predictions can be made for seated pilots for both hand and foot controls (based upon measurements of over 1000 subjects). COMBIMAN has been distributed to the major aerospace industries since 1978. Further details can be found in McDaniel (1990).

CREW CHIEF (see Figure 3.7) is a three-dimensional model of a maintenance technician and was developed by the Armstrong Aerospace Medical Research Laboratory and the Human Resources Laboratory. Much of CREW CHIEF's functionality is based upon that incorporated in COMBIMAN. CREW CHIEF can generate 10 sizes of model (five male and five female) with four types of clothing and 12 initial postures. It can be used to assess physical access for reaching into confined spaces as well as visual accessibility and strength analysis. CREW CHIEF became generally available for use in 1988. Further details can be found in McDaniel (1990).

CYBERMAN (see Figure 3.8) stands for CYBERnetic MAN Model and was developed by the Chrysler Corporation in 1974 for use in design studies of car interiors. As there are no constraints on the choice of joint angles the man model's usefulness for in-depth ergonomics evaluation is rather limited. This system is not generally available.

ErgoSHAPE (see Figure 3.9) offers a two-dimensional manikin which runs within the AutoCAD system. Being based on a microcomputer limits the scope of both the manikin and the workplace models. The models can be viewed from four viewpoints (left, right, top and front), the manikin can be constructed from up to nine segments, the basic postures are standing and sitting and the manikins are available in fifth, 50th and 95th percentiles, and a user-specified size, for both men and women. The anthropometric database is based on Finnish, North European and North American populations although the manikin can be linearly scaled as required. The system also

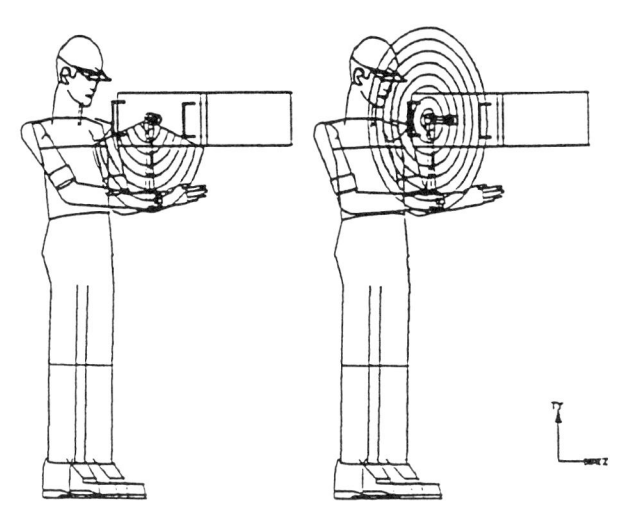

Computer-aided ergonomics design of automobiles 51

Figure 3.7. CREW CHIEF: Rotation of ratchet wrench is limited by handles on the box (*left plot*). The use of an extension rod between the ratchet and socket results in unobstructed rotation (*right plot*)

(Source: McDaniel, 1990)

Figure 3.8. CYBERMAN
(Source: Dooley, 1982, © IEEE)

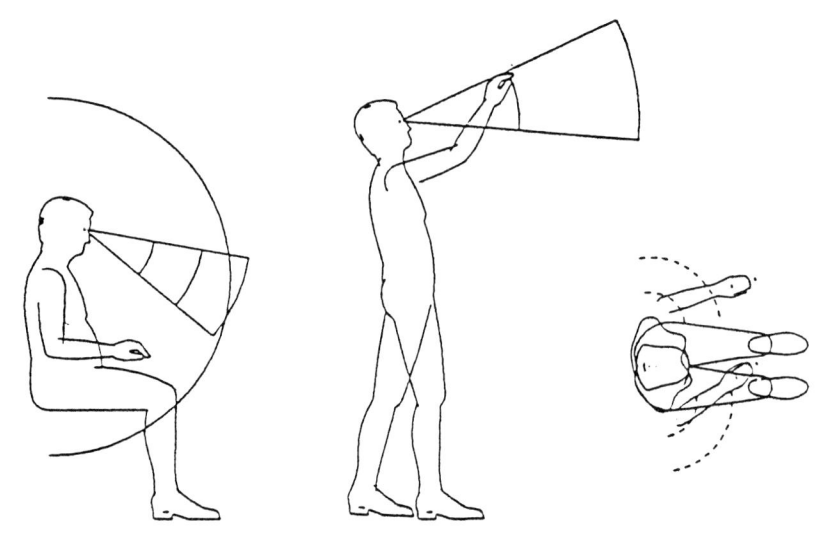

Figure 3.9. ErgoSHAPE: two-dimensional manikins in various postures
(Source: Launis and Lehtelä, 1990)

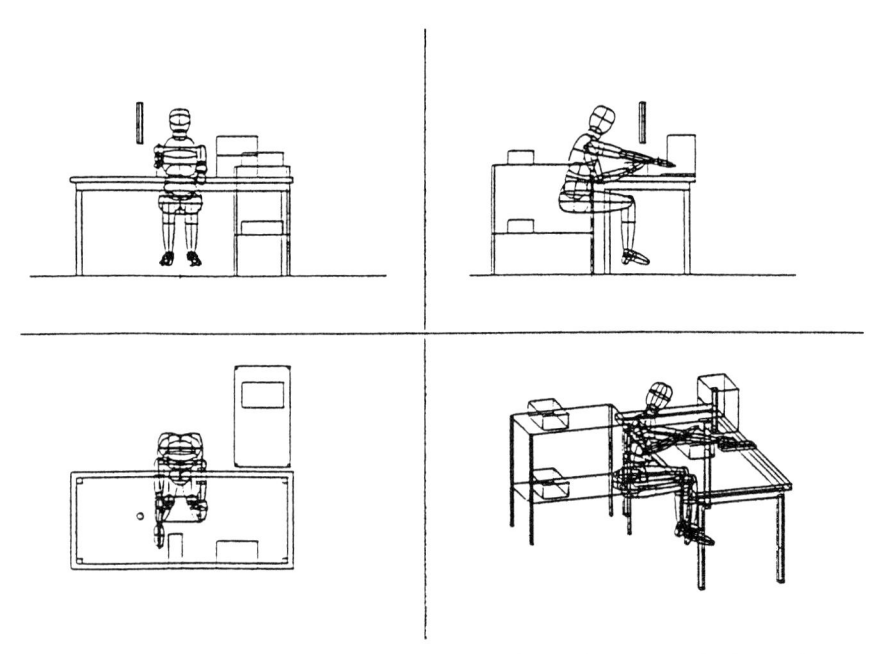

Figure 3.10. ErgoSPACE: four views (front, side, plan and perspective) of a workstation evaluation
(Source: Launis and Lehtelä, 1990)

permits the evaluation of postural stress resulting from vertical loads and provides recommendation charts giving guidance in various design areas. Further details can be found in Launis and Lehtelä (1990).

ErgoSPACE (see Figure 3.10) is a three-dimensional man model with its own workplace modeling facilities. The system was developed to comply with the restrictions of microcomputers and therefore the graphic presentations of the man model and the workplace are greatly simplified. The man model has 17 joints and, in order to attain a reasonable response time, a stick model (i.e. the man model's link structure) representation is used for moving the model. The model can subsequently be enfleshed using an ellipsoidal wire frame. The anthropometric database for ErgoSPACE is identical to that for ErgoSHAPE. Users of the ErgoSPACE system have found it to be of rather limited use in designing workplaces because of its over reduced unnatural and confusing visual appearance. For further details see Launis and Lehtelä (1990).

FRANKY (see Figure 3.11) was developed by GIT (Society of Engineering Technology) in Essen and it was hailed as being a German SAMMIE. The three-dimensional man model has four different sizes, based upon DIN 33416, although its anthropometric dimensions can be easily changed. A detailed hand model has been modeled for close-up views. The CAD package ROMULUS is used to construct the physical models. FRANKY can be used to assess fit, reach and vision in ways similar to SAMMIE. The closure of the GIT in 1987 is likely to have limited the development of FRANKY and the

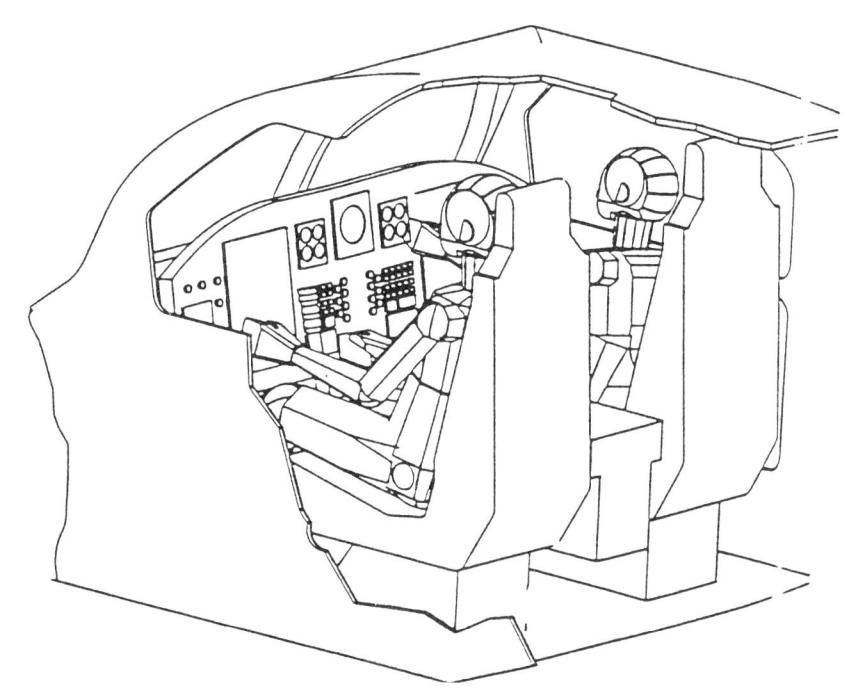

Figure 3.11. FRANKY: This plot shows a perspective view of a cockpit with hidden lines removed

(Source: Elias and Lux, 1986)

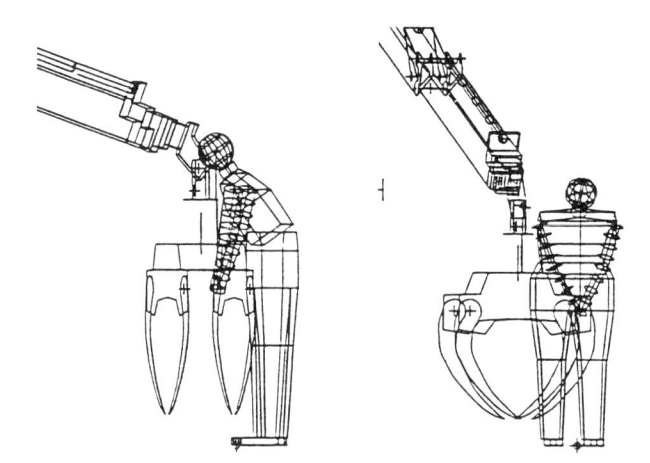

Figure 3.12. MINTAC. This plot shows the operator lubricating grease nipples of a tractor's grabbing mechanism
(Source: Kuusisto and Mattila, 1990)

software is not generally available. Further details can be found in Elias and Lux (1986).

MINTAC (see Figure 3.12) stands for Man Machine INTerACtion and was developed in 1984–5 by the Kuopio Regional Institute of Occupational Health and the University of Oulu. MINTAC was developed for the Computervision CAD/CAM system. The three-dimensional man model is based upon the anthropometric database published by Dreyfuss (1967) for the American civilian population, although the model is adjusted to simulate the wearing of winter clothes in order to evaluate difficult working postures encountered in Finnish agriculture and forestry. The simple man model contains six links: lower links (one rigid block which can be selected from a choice of 13 postures), back, upper arms and forearms. The man model was designed to be compatible with the OWAS working posture analysis system (Karhu et al., 1977) although it is considered that MINTAC is not appropriate for widespread use because of its simplified posture and suitability only for the analysis of heavy work. Further details of this system and some other Finnish systems are given in Kuusisto and Mattila (1990).

SAMMIE (see Figure 3.13) stands for System for Aiding Man-Machine Interaction Evaluation. This system has been developed at Nottingham University and subsequently at Loughborough University. SAMMIE has been used extensively as a consultancy tool by its developers since the mid 1970s and SAMMIE CAD Ltd are now marketing their latest software worldwide as well as offering their ergonomics design services. This system is described in considerable detail later in this chapter but, briefly, it is an extremely versatile three-dimensional system comprising a man model of completely variable anthropometry, with comfort and maximum angles for each of its 17 joints, together with a functional workplace modeler. The system runs on engineering workstations such as SUN, APOLLO and Silicon Graphics and contains a suite of sophisticated ergonomics and workplace modeling facilities. Recent

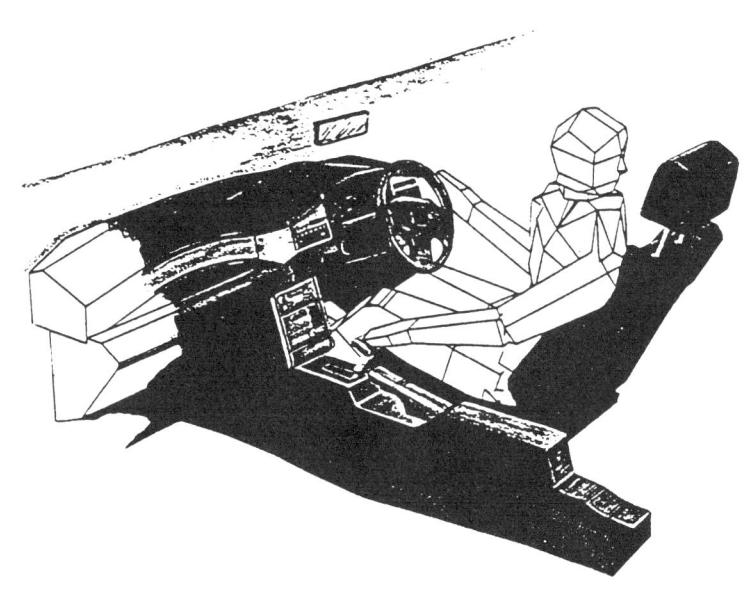

Figure 3.13. SAMMIE. This plot shows a postural analysis for the task of reversing a car

descriptions of the SAMMIE system include Porter *et al.*, (1990); Case *et al.*, (1990b) and Case *et al.*, (1990a).

TADAPS (see Figure 3.14) stands for Twente Anthropometric Design Assessment Program System which has been developed for use withVAX computers at the University of Twente in the Netherlands. This system is based upon ADAPS, developed by Delft University of Technology for the PDP-11 computer in the late 1970s (see Post and Smeets, 1981). The basic man model consists of 24 segments although this amount can be reduced or extended to suit the intended application. The system includes its own workplace modeler and the whole system appears to be similar to SAMMIE in concept, although it is not as fully developed. The anthropometric database comprises Dutch men, women and 4-year-old boys as well as American pilots and it is relatively easy to create models for other populations. All percentiles can be chosen although the man model is linearly scaled out of the 50th percentile proportions and it is not clear whether individual body segments can be set to different percentile values. TADAPS offers a prediction of the compression and shear force of the intervertebral disc L5-S1 for various postures and external loads. For further details see Westerink *et al.*, (1990).

WERNER (see Figure 3.15) has been developed at the Institute of Occupational Health at the University of Dortmund. It is implemented on an Astari ST personal computer. The three-dimensional man model consists of 19 segments, each of which is defined by simple solids most of which are ellipsoids. A convex hull is constructed over these solids to define a silhouette of the man model. The man model's anthropometry appears to be based only on the German National Standard DIN 43116. WERNER communicates with AutoCAD to provide its three-dimensional workspace modeling features. Further details can be found in Kloke (1990).

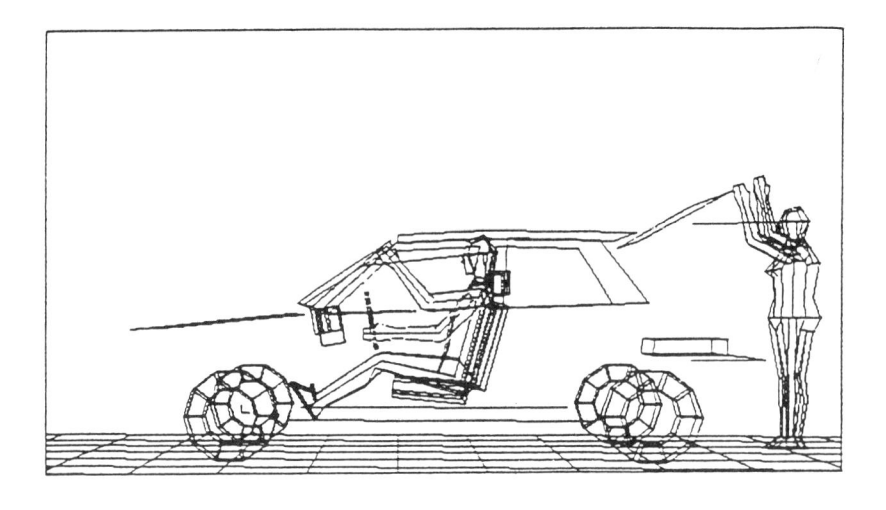

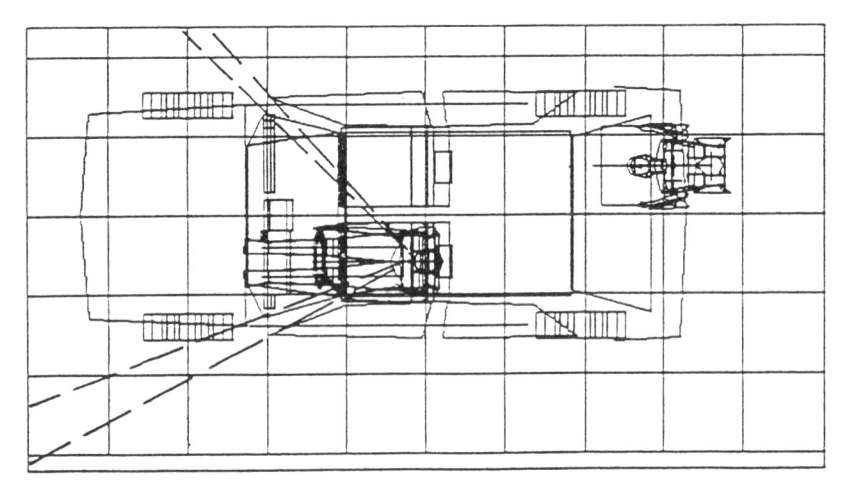

Figure 3.14. TADAPS: These two plots show an example of the analysis of reach (top) and vision (bottom)

(Source: Westerink *et al.*, 1990)

A comparison of these man-modeling CAD systems is provided in Table 3.1. The differences between the systems have been examined in terms of a number of features: the complexity of the man model; whether the joint angles are constrained to possible angles or whether impossible postures could be inadvertently set; the anthropometric databases available; the extent of control over the size of individual body segments (fixed, linear scaling possible or direct control with data); the ability of the workplace modeler to provide functional modeling (e.g. several components can be modeled and collectively called 'driver's door' which can be rotated as one unit around the hinge point), hidden line views (lines behind solids removed); mirror views or reflections; whether the system can assess the man model's reach, fit or vision (the

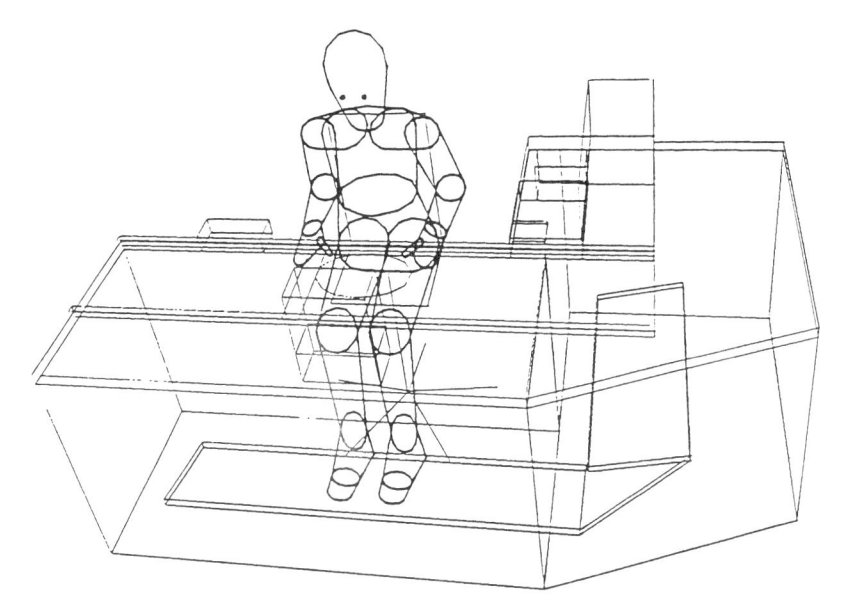

Figure 3.15. WERNER: An evaluation of a cash desk workstation
(Source: Kloke, 1990)

flexibility of these assessments varies considerably from simple reach to a point to automated volumetric reach, from visually inspected clearance to automated intersecting solid detection routines and from the display of eyepoint location to the display of perspective man's views).

Advantages of man-modeling CAD systems

Traditional methods used by ergonomists/human factors engineers for the specification of dimensional information for equipment and workstation design include the use of published or in-house recommendations and guidelines, anthropometry, two-dimensional plastic manikins (full size or scaled to be used in conjunction with engineering drawings), mock-ups, user questionnaires and user trials. Man-modeling CAD systems do offer significant advantages to the designer with respect to issues such as fit, reach, vision and task related posture; these issues are discussed below.

Time

While recommendations can provide information rapidly they are rarely sufficiently user and task specific to be used unreservedly. Detailed information of current designs from user questionnaires and user trials, or of new designs using mock-ups with selected subjects, can take weeks or, more usually

Table 3.1. Comparison of

System	2D/3D	Joint Constraints	Man-Model Anthropometry Database	Segment Percentiles	Workplace Modeler Within System?	2D/3D	Functional Modeling
ANYBODY	3D	No	German DIN 33402, Bodyspace	Overall linear scaling from 50th percentile	No (CADKEY)	3D	No
BOEMAN	3D	Yes	North American civilians and flying personnel	50th percentile only	Yes	3D	?
BUFORD	3D	No	North American	Linear scaling from 50th percentile	No (Computervision)	3D	?
COMBIMAN	3D	Yes	North American military pilots and women	Any percentile from data	Yes	3D	?
CREW CHIEF	3D	Yes	US Air Force maintenance technicians	Limited no. of body sizes available	No	3D	?
CYBERMAN	3D	No	2D SAE manikin, American civilians	Any percentile from data	No	3D	?
ErgoSHAPE	2D	No	Finnish, North American, European	Limited no. of set body sizes, linear scaling possible, user specified sizes	No (AutoCAD)	2D	No
ErgoSPACE	3D	No	Finnish, North American, European	?	Yes	3D	No
FRANKY	3D	Yes	German DIN 33416	Limited no. of set body sizes, user specified sizes	No (Romulus)	3D	?
MINTAC	3D	No	North American	5, 50 and 95th percentiles	No (Computervision)	3D	?
SAMMIE	3D	Yes, normal and absolute	Wide variety of databases	Any percentile from data	Yes	3D	Yes
TADAPS	3D	Yes	Dutch civilians, North American pilots	Linear scaling from 50th percentile	Yes	3D	No
WERNER	3D	No	German DIN 43116	?	No (AutoCAD)	3D	No

months, to acquire. Man-modeling CAD systems provide a means of simulating the user and task specification using computer graphics and anthropometry instead of users/subjects of an existing/new design. This results in a tremendous saving in time, in some cases a reduction from a few months to a few days.

Time is always the designer's enemy but this is especially true for the ergonomist who often has to respond very rapidly to prototype designs produced by engineers and stylists. If the ergonomist has to wait for the production of full size mock-ups before commencing the evaluation then this results in a long delay to the design process or, more likely, the design development continues unabated without the benefit of a timely ergonomics input.

Cost

'Time is money', so any reduction in development times should be advantageous from the financial point of view. In addition, the production of mock-ups is an expensive process, even for simple models constructed from wood,

Computer-aided ergonomics design of automobiles

man-modeling CAD systems

Workplace Modeler		Ergonomics Features			
Hidden Lines	Mirrors/ Reflections	Reach	Fit	Vision	Hardware Systems
No	No	Visually assessed	Visually assessed	Indirect by drawing sight lines	IBM AT
?	No	Reach baskets defined	Interference detection	Eyepoint depicted, visual interferences identified	?
?	No	Visually assessed	Visually assessed	No	Computervision
No	No	Success/failure indicated, failure distance specified	Visually assessed	Sight lines. Aitoff projections	IBM
No	No	Automatic reposturing to achieve reach and fit		Eyepoint depicted	Computervision
No	No	Visually assessed reach envelopes	Visually assessed	Eyepoint depicted	CDC
No	No	Visually assessed	Visually assessed	Sight lines	Microcomputers
No	No	Visually assessed	Visually assessed	No	?
Yes	No	Visually assessed or reach point analysis	Visually assessed	Perspective views from models eyepoints	?
?	No	Reach point analysis	Visually assessed	Eyepoint depicted	Computervision SUN
Yes	Yes	Visually assessed or reach point, reach areas and reach volumes analysis available	Visually assessed or interference detection	Perspective views from man model's eyepoints. Aitoff projections	SUN, APOLLO, Silicon graphics
Yes	No	Visually assessed or anchor point facility available to lock hand or foot onto control	Visually assessed	Sight lines	VAX
No	No	Visually assessed	Visually assessed	Eyepoint depicted	Astari ST

plastic, etc. The cost of a man-modeling system such as SAMMIE can be less than the cost of making just one full size mock-up. Fitting trials or similar studies also incur extra costs for space rental, staff costs, subject payment and so on. As CAD systems enable the ergonomics input to be provided much earlier in the design process, this reduces the likelihood of expensive or unfeasible modifications being necessary at later stages.

Accuracy

Ergonomics CAD systems are not intended to be used in isolation during the design process. They should be regarded as useful tools which can be used to predict the ergonomics problems that are likely to arise with a particular design or to explore several alternative designs to identify the superior one(s). Recommendations and results from marketing surveys and product evaluations can still be consulted. The greatest validity will usually be provided when real people are asked to perform real tasks with a real product for realistic durations in a real environment. Unfortunately, this information is often only available, if at all, after the production and sale of the product. The results of such an evaluation may highlight any postural problems, as a result of users' fit, reach and vision, as well as assessing comfort, convenience, behavior and performance measures. Ergonomics CAD systems can accurately predict problems of fit, reach and visions given that an appropriate anthropometric database and detailed task requirements are available for the prototype design. These predictions are mainly geometric in nature and result in the generation of the expected range or variety of working postures. The assessment of the consequences, in terms of comfort and/or performance, of these postures depends upon the CAD operator's knowledge base for its accuracy. Posture assessment is made considerably easier when research has identified optimum postures or a range of postures for specified tasks. For example, Rebiffé (1969) and Grandjean (1980) present comfort ranges for the various joint angles which should minimize the onset of pain or discomfort in the joints, ligaments and muscles while driving.

Man-modeling CAD systems have the potential to offer more accuracy than recommendations, guidelines and two-dimensional manikins because they can be made to simulate specific user groups and tasks and they can assess the postural consequences in three dimensions. While two-dimensional manikins are, even today, frequently used by many manufacturers it seems that only a few designers are fully aware of their limitations. One obvious problem is that most manikins only depict a side view of a person, resulting in only a two-dimensional evaluation of the design. This may explain why so many vehicle manufacturers design driving packages with the steering wheel and/or pedals offset from the seat or driver's centre-line, which is not only uncomfortable and a possible precursor of chronic health problems but it is also suspected of being a contributory cause of incidents of 'unintended acceleration' (Porter, 1989; see also Schmidt, this volume).

Two-dimensional manikins are often used without a knowledge of their origin in terms of user type (e.g. military/civilian data, age range, nationality, size, date of survey) and application (e.g. 'erect' or 'slumped' sitting height, percentile values for individual body segments). For example, we have found that several British vehicle manufacturers who market their vehicles in the United States were unaware that the sitting height (vertical distance from the compressed seat surface to the top of the head) of their 95th percentile adult male manikin was 50 mm shorter than the 95th percentile erect sitting height recorded by the National Health survey conducted in the United States some 30 years ago (Stoudt et al., 1965). According to the results of this North American survey, the manikin actually had a 60th percentile erect sitting height. A further problem arises because the 95th percentile manikin is often used to define the height of the roof lining and the sunroof is subsequently designed to actually encroach around 25 mm into this already limited headroom. This results in an even larger percentage of the potential market being inadvertently 'designed out'.

Another major problem with two-dimensional manikins is that they can be used in a very simplistic way. For example, designers may have 50th and 95th percentile adult male mankins and a fifth percentile adult female manikin. The nature of these manikins gives support to the notion that people come either tall and long-limbed, short and short-limbed or somewhere in between. It has been repeatedly demonstrated that this is not true and that the intercorrelation between body dimensions is rather poor (e.g. Haslegrave, 1980). For example, Haslegrave (1986) reported that seated shoulder height varies from 30.6 to 39.5 per cent of stature which means that among men of average stature, their shoulder height may differ by as much as 122 mm. Furthermore, this torso proportion was found to have virtually no correlation with either stature or weight.

Figure 3.16 shows the percentile values for a number of body dimensions recorded from a small sample of British automotive engineers by the Vehicle Ergonomics Group at Loughborough University. This figure clearly shows that people vary considerably in their body proportions and that very few people can be expected to be consistently around 95th, 50th or fifth percentile for more than a few dimensions. Statistically, it is not possible for an individual to be 95th or fifth percentile in all vertical body dimensions and still be 95th or fifth percentile in stature. McConville (1978) demonstrated with data provided by Clauser *et al.*, (1972) that if a person's stature is broken down into 14 vertical dimensions then the sum of all fifth percentile values is a diminutive

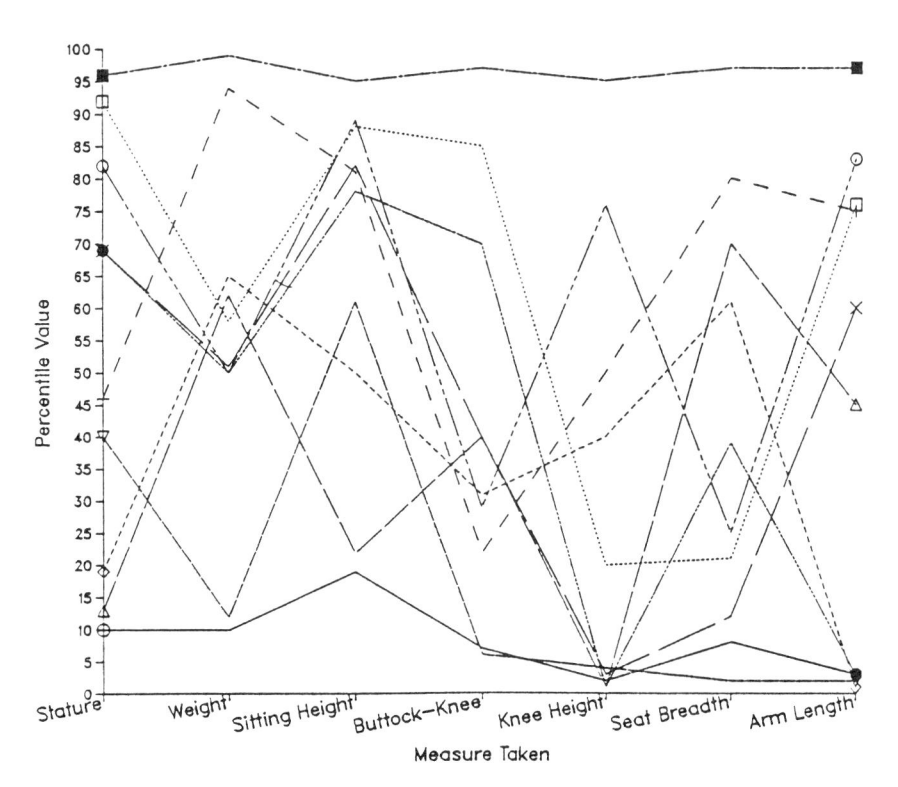

Figure 3.16. Percentile values for a variety of dimensions from a sample of British automotive engineers (n = 10)

1234 mm (48.6″) compared to the fifth percentile stature of 1524 mm (60.0″). Conversely, the sum of all 14 dimensions set at 95th percentile values is a staggering 2022 mm (79.6″) compared to the 95th percentile stature of 1722 mm (67.8″). The use of this approach would be very conservative indeed. Clearly this poses a disturbing problem for the manikin designer as no 'percentile man' can exist with the exception of the statistically acceptable '50th percentile man' as 50th percentile values only are additive. However, the 'average man' concept has been known for a long time to be a fallacy. For example, Hertzberg (1960) describes the results of the study by Daniels and Churchill (1952) which showed that nobody in a sample of 4000 flying personnel was loosely 'average' (i.e. man $\pm$ 15 per cent of the sample) in the 10 body dimensions examined.

The manikin designer can resort to other techniques to ensure that the manikins are statistically correct. One method is to measure a group of men or women who are fifth or 95th percentile in both stature and weight and to calculate the median values of all other dimensions among the group (Haslegrave, 1986). These median values are additive, allowing the manikin to remain statistically 'correct'. One problem with this approach is that, unless the total sample is very large, the number of people who fall into the two extreme categories is likely to be quite small. For example, only five and 11 people were selected in Haslegrave (1986) to be fifth and 95th percentile, respectively, for both stature and weight. Such small groups may provide a distorted picture if any one individual was atypical to any significant extent. Another approach advocated by Robinette and McConville (1982) is based upon the use of regression equations to derive component dimensions for a person of a given percentile stature. These regression predicted values are additive and this method has the advantage that predictions from first to 99th percentile stature or weight can be easily calculated and that the method does not require a huge sample of subjects.

Whichever method is chosen to define a variety of statistically 'correct' manikins, there is still the problem of estimating the percentage of people accommodated by a particular design. A common mistake made by many automotive manufacturers is to use the fifth percentile female stature and 95th percentile male stature manikins to assess a driving package, assuming that if both of these manikins can be accommodated then so can 95 per cent of the adult population. This assumption is incorrect as it implies that those people 'designed out' because either their sitting height, hip breadth or leg length, for example, are greater than 95th percentile male values are all the same people. Similarly, all those with sitting eye height or leg length smaller than fifth percentile female values are assumed to be the same individuals. As these dimensions are not strongly correlated then these assumptions are incorrect. A study of air crew selection standards and design criteria analysis reported by Roebuck *et al.*, (1975, p. 268) illustrates the problem perfectly as it was shown that nearly half of the air crew were 'designed out' when the fifth to 95th percentile range was used on a large number of body dimensions (in this case 15 dimensions). Even limiting the number of dimensions to just seven (sitting height, eye height-sitting, shoulder height-sitting, elbow rest height, knee height, forearm-hand length and buttock-leg length) 'designed out' over 30 per cent of the available air crew.

This brief overview of the inherent problems with the use of manikins indicates the benefits that can arise from using three-dimensional man-modeling CAD systems with variable anthropometry. For example, the CAD operator has complete control of the dimensions or percentile values of the man model (assuming the system offers this facility) and can interactively change them in a matter of seconds. A variety of anthropometric databases for different nationalities may be available, allowing the rapid creation of a large number of three-dimensional models. The operator can select to use models of known percentile values or use median or regression-predicted values for individual components. Additional man models can be constructed. For example, a model of 99th percentile male stature with short arms and long legs for such a stature can be used to determine the rearmost position required of an adjustable steering wheel. Unfortunately data at this level of detail are not commonly presented in surveys; one of the exceptions being the anthropometric survey of Royal Air Force Aircrew by Simpson and Hartley (1981). The facilities provided by the computer man model encourage the designer to think more clearly about the use of anthropometric data in design.

One further option is potentially available to a computer man-modeling system and this is the assessment of the percentage of the target population that will be accommodated by a particular workstation design. This could be accomplished by two methods. First, the computer system could model all the individuals recorded in a relevant anthropometric survey and automatically position them in the workstation according to a range of predefined postures. Evaluations of fit, reach and vision could also be conducted automatically resulting in the identification of those individuals who failed to complete any of these tests successfully. The second approach would be to use the Monte Carlo method where appropriate survey data for individuals was not available. This method is briefly described in Churchill (1978) and would require the means, standard deviations and correlation coefficients for the various body dimensions.

The importance of ensuring that people can be comfortably accommodated and able to operate controls and view displays easily cannot be overstressed. The earlier example of cockpit design showed the alarming percentage of aircrew who would be 'designed out' if the crew stations were designed according to fifth to 95th percentile range for important body dimensions. While the accommodation criteria (e.g. sight lines for head-up displays, knee clearance when ejecting, control operation in high 'g' environments) for pilots are considerably more demanding than for cars, car manufacturers often appear to overlook the fact that cars need to accommodate not just the drivers but whole families, including children and elderly relatives. Cars are often chosen to meet functional requirements such as this and many manufacturers may be unaware of the extent to which sales could be improved if fewer people were 'designed out'. A recent study by the Vehicle Ergonomics Group at Loughborough University (Porter *et al.*, 1990) demonstrated the advantages of providing highly adjustable driving packages. This study interviewed 1000 randomly selected drivers at 3 motorway service stations and concluded that those drivers with highly adjustable driving packages reported less discomfort, even though they travelled greater distances and for longer periods of time, than those with more basic packages.

Communication

The quality of a design is greatly influenced by the extent to which the compromises between the various constraints are successful. In order to achieve good compromises it is an important prerequisite that all the constraints have been identified at an early stage in the design. All too often the ergonomics issues have not been thoroughly researched early enough to enable optimum compromises to be made. Ergonomics is all about human variation and, as such, it is not always satisfied by existing legislation. Take, for example, EEC legislation concerning rearwards field of view in mirrors. This legislation requires that specified areas can be viewed from the legislative eye points which are two points vertically above the lowest, rearmost H-point (the H-point is the equivalent of a human's hip joint and it is measured using a standard SAE H-point manikin which has molded shapes to represent a person's back, buttocks and thighs which are weighted to simulate an average adult male. The seat is placed in the lowest and rearmost position). These legislative eyepoints take no consideration of the distribution of real driver's eyepoints as a function of seat adjustment and sitting eye height and also ignore the fact that the eyes move considerably when turning the head to look in the door mirrors. In addition, the legislative fields of view do not specify that areas adjacent to the vehicle should be visible in the mirrors (e.g. for buses, the area alongside the curb is of prime importance and must be visible to the driver to avoid potential accidents when leaving bus stops).

Communication is also of considerable importance to the ergonomist as it is to every member of the design team. The ergonomist can work most effectively when in collaboration with other members rather than acting as a critic assessing the efforts of others in producing the prototype. The collaboration is encouraged by man-modeling CAD systems as the ergonomics input is brought forward to an earlier stage and the system can often act as a focal point for the design team before detailed engineering development work commences. Several design ideas can be rapidly assessed from an ergonomic viewpoint giving added flexibility to the design process.

Description of the SAMMIE system

The SAMMIE system has benefited enormously from the detailed feedback provided by its extensive use as a consultancy and research tool, both by the system developers and others. The latest version of SAMMIE is highly interactive and the computing power now available in even the base model engineering workstations (e.g. SUN, APOLLO, Silicon Graphics) ensures a response time that could only have been dreamed of a few years ago.

It is perhaps worth pointing out that the version of SAMMIE currently distributed by SAMMIE CAD Ltd is markedly different from the system which was available through PRIME. Historically, both systems originated from Nottingham University. In 1978 the Research Council funding of SAMMIE ceased as it was deemed to have reached the stage where industry could support its further development. Consequently the software rights were passed to the British Technology Group for exploitation.

The software was modified by this group, the primary difference being the abolition of the menu interface. This system was marked by Compeda and subsequently by PRIME. During this period the Nottingham team continued to develop their original version supported by the income generated from consultancy work. This work moved to Loughborough University in the early 1980s and in 1986 the originators set up SAMMIE CAD to develop and market their own system on a variety of computer hardware. As PRIME no longer support their version of SAMMIE, the details of SAMMIE presented in this chapter will refer to the current version developed by SAMMIE CAD only.

Workplace modeling

A boundary representation form of solid modeling (Requicha, 1970) is used to enable the system to be highly interactive whilst maintaining a sufficiently accurate three-dimensional model. It allows solid evaluations such as hidden line removal and interference checking without producing unduly complex models. Models are typically constructed from the variety of primitive shapes available such as cuboids, prisms, cylinders, spheres, meshes, polyprisms and solids of revolution. These primitives can be specified interactively by stating their unique name and dimensions (e.g. width, depth, height, radius etc. as required). Irregular solids can also be modeled. Although truly curved surfaces are not available, this has never been a cause for concern from an ergonomic point of view as sufficient accuracy can be obtained from a multi-faceted model. A reflection facility is available so that mirror image copies can be made so, for example, only one side of a car's exterior needs to be defined manually.

The workplace modeler in SAMMIE is particularly strong in its ability to specify logical or functional relationships between components of a model. This is an important prerequisite for the evaluation stages where doors, seats, levers, pedals and so on all need to be operated or adjusted by the man models in order to check fit, reach and vision (see Figure 3.17).

Figure 3.17. Functional modeling is an important facility within the SAMMIE system as it allows the simulation of task requirements

Models, or individual subsets of components, can be shifted and rotated with respect to the local (e.g. headrest), owner (e.g. backrest) or global axis system and modified (e.g. alter dimensions or change the logical structure with other components) interactively which enables the rapid and flexible simulation of models at the concept stage in design when various configurations are being explored.

Man-modeling

The three-dimensional man-modeling facility is based upon and completely integrated with the workplace modeling system. The man model is constructed from 17 pin joints and 21 straight rigid links encased in solid modules of 'flesh'. The link lengths and joint constraints are all data-driven and can be easily controlled by the user. A large variety of body shapes are available using the somatotyping technique devised by Sheldon (1940) which allows general body shapes to be described on the three seven-point scales of endomorphy, mesomorphy and ecotomorphy (fatness, muscularity and thinness). Algorithms are used to determine the vertex locations within a fixed topological description of each body segment shape. The ability to vary the joint to joint dimensions, the flesh shape and the joint constraints allows the creation of as many different man models as desired depending upon the evaluation requirements.

Interacting with SAMMIE

The designer interacts via a menu driven interface using a mouse as the main input device. Each menu, of which there are nearly 30, typically contains between 10 and 20 separate commands grouped according to their functions.

The most frequently used menus (i.e. VIEW, DISPLAY, WORKPLACE, MASTER) can be accessed at any time while the remainder can be found as a subset of the 'MASTER' and 'MAN' menus. There is a 'RETURN' command which enables the immediate return to a previous menu. The menu structure is shallow being only a maximum of three layers deep. This menu structure works very efficiently as it allows the designer to integrate facilities quickly and it is an asset for both naive and skilled users as all the commands are displayed, acting as a job aid.

A brief description of the main menus is given below:

VIEW menu includes commands for locating the viewing point and the center of interest, selecting plane parallel projections or perspective views, changing the scale or acceptance angle of the displayed views and saving and retrieving the above parameters for any chosen views.

DISPLAY menu enables the various components of the models to be displayed selectively as and when required. This helps to keep the screen display as simple as possible with detail only when and where it is required. Colours can be specified for particular model components as a further aid to clarity.

WORKPLACE menu allows the positioning of the model components in three-dimensional space either singly or in combination depending upon the level chosen in the model's hierarchical data structure. For example, the mirror on the driver's door could be adjusted, or the door could be opened or the car as a whole could move forwards. Explicit instructions can be inputted or components can be 'dragged' on the screen using the mouse.

MASTER menu is the main menu through which access is gained to the remainder of the other menus.

INPUT menu enables the interactive construction of model components using a query and response format. Data input from a disk file can also be accepted.

OUTPUT menu allows the output of various forms of file, model and plot data.

GEOMETRY EDITOR menu enables the interactive modification of the geometry of model components so that, for example, a display area can be made larger or smaller in any of its dimensions.

STRUCTURE EDITOR menu provides interactive control of the functional/logical relationships between model components. For example, this allows the man model to become a subset of a selected equipment model, such as the car seat, enabling both items to be moved in three-dimensional space as one unit when necessary.

HIDDEN LINES menu allows the creation of hidden line views, by deleting those lines behind solids, and the presentation of saved hidden line views.

MODIFICATIONS menu allows constraints to be placed on the position or orientation of model components so that seats, doors, pedals etc., can only operate in a realistic manner. This menu also enables the creation of simple functional commands such as 'LEFT DOOR OPEN' which can be used instead of the typical commands used in the WORKPLACE menu (i.e. LEFT DOOR LOCAL Z POSITIVE ROTATE 90).

MIRRORS menu allows convex, concave or plane mirrors to be created from the chosen face of any solid. Reflected views are displayed on the face of the mirror is seen from the man model's viewpoint (see Figure 3.18). The mirror modeling facility

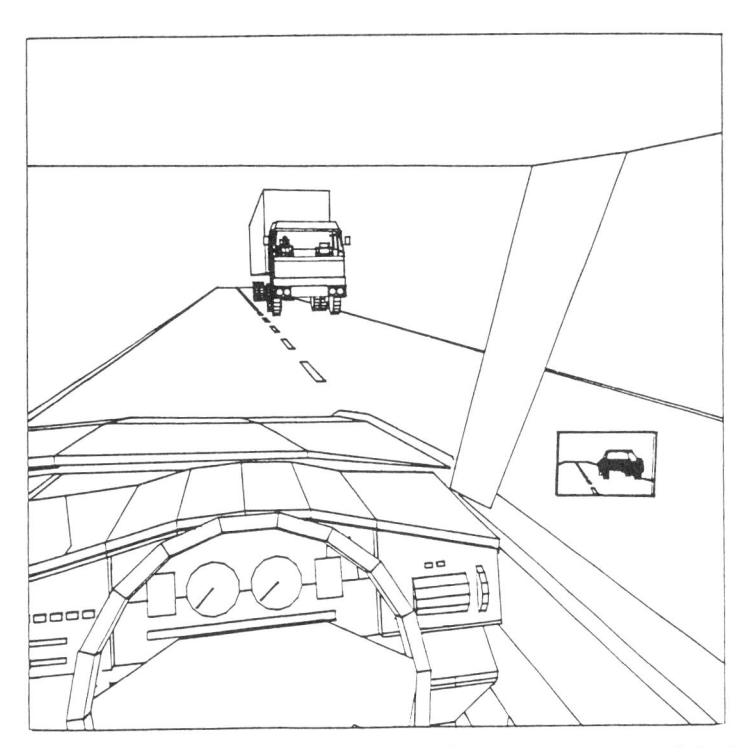

Figure 3.18. Driver's predicted view of external traffic environment, internal displays and door mirror reflection

has been used extensively in the design of mirrors for commercial vehicles (see Case et al., 1980, for details).

MEASUREMENTS menu is a geometric measuring facility which enables minimum distances between any two solids to be calculated.

INTERFERENCE menu can be used to determine whether any chosen model components are intersecting. Any such solids are highlighted by flashing.

CONFIGURATION menu allows the user to have control over the format of the screen display and the output of messages.

DATA STRUCTURE menu allows the user to interrogate and manipulate the data structure of the various models which have been created. Its primary purpose is for debugging.

MAN menu is the header menu for a suite of man-model related menus which are described below.

ANTHROPOMETRY menu enables the various link lengths to be independently or collectively altered using either a specified percentile value or actual dimensions (see Figure 3.19). This menu also allows the user to access the various stored databases (for different nationalities, age groups, sex etc.) in order to specify changes to the man model's anthropometry. These databases are under the user's control and they can be extended by providing the necessary means and standard deviations for the various link lengths.

SOMATOTYPE menu provides the user with control over the flesh shape with 76 options available.

JOINT ANGLES menu allows the incremental or absolute setting of the various joint angles using flexion/extension, abduction/adduction and medial/lateral rotation as appropriate. A listing of the current posture is available. Joint constraint data inform the user whether a given joint is within 'normal' limits or within 'maximum' limits. Any request which would place the joint outside the pe-set constraints is not modeled and the user is informed accordingly. The choice of 'normal' and 'maximum' values is at the user's discretion depending upon the task characteristics.

POSTURES menu provides several 'basic' postures, such as standing, sitting, crouching, crawling, etc., which can then be modified to suit the specified task requirements using the JOINT ANGLES menu or the REACH menu. Any posture can be saved and retrieved in this menu.

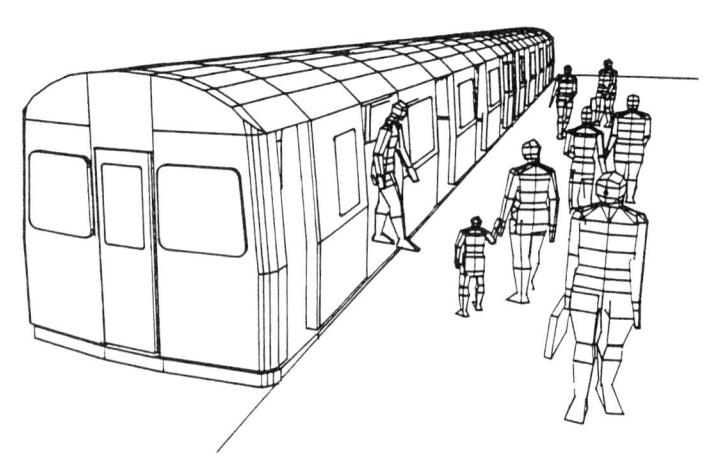

Figure 3.19. Variable anthropometry allows the construction of a wide range of body sizes and shapes

MAN'S VIEW menu provides perspective views from the man model's left, right or mean eye-point. The angle of view can be specified and the direction of view can be controlled by either asking the man model to look up/down or left/right with the eyes and/or head (both under joint constraints for normal and maximum angles) or by specifying the origin, vertex, edge or face of a model component that the user wishes to become the man model's center of interest (i.e. placed centrally on the screen only if the head and eye angles are within the joint constraints). Aitoff projections are an alternative method for assessing vision and this facility is available on the VIEW menu.

REACH menu provides a variety of automated methods to determine the reach capabilities of the man model for both feet and hands. In the reach point assessment the user can specify either the precise spatial location of the point to be reached, the object itself or an incremental movement from the current location of the hand or foot. A suitable algorithm selects the limb posture to be displayed by minimizing the extension of the joints away from their neutral positions and by preferring the greater extension of distal links to those that are more proximal. This algorithm does not necessarily predict the likely limb posture to be adopted by a human but it does confirm whether or not the reach attempt will be successful for a man model of given size and joint constraints. If the reach attempt fails, the relevant limb remains in its pre-test position and the user is informed of the failed attempt together with the distance by which it failed. An alternative approach is to produce reach contours which are overlaid on any surface of the chosen model component (in any orientation) as an aid to assessing suitable positions for controls. An extension of this approach is to produce reach volumes whereby reach is assessed over a number of imaginary surfaces parallel to either the frontal, sagital or transverse planes of the man model, or any other specified orientation, which is particularly useful in concept design. All reach evaluations only explore the range of joint angles in the limb itself. So, for example, hand reach evaluations involve the shoulder, elbow and wrist joints only. If appropriate to the task, the user can alter the man model's overall posture (e.g. bending forwards, twisting sideways) to extend the reach capabilities although the consequences of excessive bending and twisting upon musculoskeletal health, performance and fatigue must be considered carefully.

CHANGE OPERATOR menu allows the user to construct additional man models and to select the currently active model for evaluations of reach or vision.

Interfacing with other systems

SAMMIE has been written using standardized computer languages (Fortran 77 and C) and can be implemented on computer systems using Unix, the *de facto* standard operating system. Considerable development effort has been put into the implementation of computer graphics standards within SAMMIE. The various issues are discussed in detail in Case *et al.*, (1991). Data exchange between CAD systems of similar function is fairly well established in two-dimensional systems using the Initial Graphics Exchange Specification (IGES; see Smith *et al.*, 1988). However, standards are only just beginning to become effective for solid modeling systems such as SAMMIE. Typically, one might wish to use SAMMIE as a design originating concept tool, the results of which can then be passed on to other CAD/CAM systems for development or to plotting and visualization packages. It would also be useful if SAMMIE could receive geometric information from other three-dimensional CAD systems, evaluate it and return it to the originating system.

SAMMIE can create PHIGS archive file, Post Script and Computer Graphics Metafile (CGM, see Mumford and Liddell, 1988) output suitable for producing hard copy on pen and laser plotters or passing geometric information to various systems such as business graphics, color enhancement, cartography and desktop publishing. SAMMIE currently can generate output files complying with IGES 4.0 which can be used to pass orthographic projections to a drafting package for dimensioning and annotation. The forthcoming IGES 5.0 should allow transfer of all the geometric aspects of a SAMMIE model to other systems although this would not include the non-geometric aspects such as the functionality of the workplace and man models.

A general methodology for using SAMMIE in automobile design

SAMMIE has been used extensively in the design and evaluation of automobiles including cars, tractors, fork-lift trucks, buses, vans and heavy goods vehicles. Although each project had different aims, constraints and solutions, it is possible to describe the stages of a typical evaluation methodology.

Model the vehicle

The vehicle needs to be modeled to a level of accuracy that is sufficient for both the ergonomics evaluation and any intended presentations. Care must be taken to model the seat (or H-point) and its range of adjustment, controls and their points of rotations and/or travel, display areas, mirror surfaces and rotation points, 'A', 'B', and 'C' posts, header rail and internal headroom to a high level of accuracy. Although most other aspects of the vehicle do not need to be modeled unless they have a bearing upon the ergonomics evaluations, it is often wise to make the model as realistic as possible in order to give a stronger sense of reality when presenting the results of evaluations to the design team and management. Figures 3.13 and 3.20 show a very detailed car

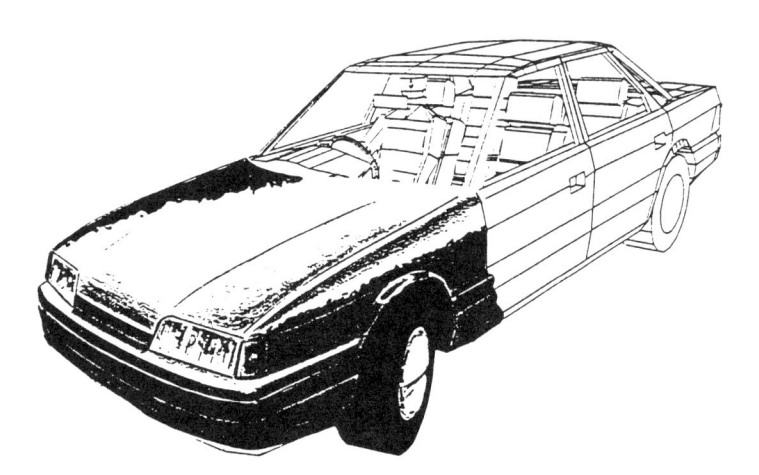

Figure 3.20. Detailed vehicle model intended for presentation

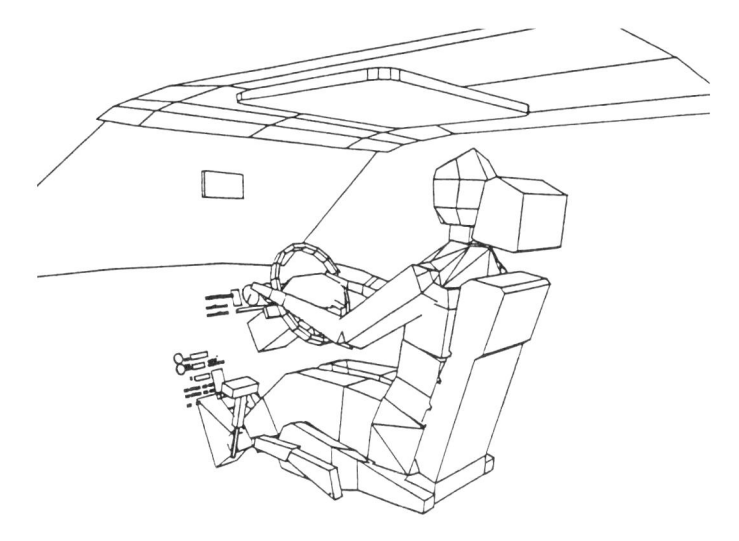

Figure 3.21. Simplistic, but accurate, modeling is sufficient to perform ergonomics evaluations

model which took five days to construct from engineering drawings whereas Figure 3.21 shows the minimum level of modeling often used for consultancy work which takes a fraction of the time (in this case less than one hour) but still gives quality results.

Model the current/potential user population

A prerequisite is the availability of a relevant anthropometric database in terms of the nationality, occupational group, age range and sex distribution of the target population. Automotive manufacturers are now marketing their products to a large number of countries worldwide and it makes sense to consider these various markets at the design stage for future vehicles. The design team will need to consider whether variations of the driving package will be made available for these different markets (i.e. the seat adjustment range remains the same but the seat is mounted closer to the pedals for the Japanese market) or whether just the one driving package will exist (presumably with increased levels of adjustment to cater for the wider range of body sizes to be accommodated). SAMMIE is very useful for aiding decisions such as this.

The designer needs to construct several three-dimensional man models of various sizes, shapes and proportions bearing in mind those issues discussed earlier concerning the accuracy of manikins. In order to avoid the cumulative 'designing out' of too many potential users, it may be wise to model from first to 99th percentile values. These extreme individuals are often difficult to find when selecting subjects for user trials. The ability to specify exactly the dimensions you want for a man model, and get it in seconds, is quite an advantage.

Specify the task requirements

The task requirements should be specified for the driving task (e.g. reach and operate the controls, view the displays and view defined areas of the external environment either directly or by mirrors while maintaining a comfortable posture and meeting legislation where appropriate). In addition, other tasks should be considered such as ingress, egress, loading the trunk and maintenance operations (both for the user and the service mechanic). As SAMMIE is a general purpose system it can examine ergonomics issues in the manufacture of the automobile equally well, particularly where employees need to work in confined spaces or with robots.

Evaluate task efficiency and postural comfort

Depending upon the progress of the design, SAMMIE can either be used to specify control and display locations for future development or it can be used to evaluate a proposed or existing package. In the first case, the various man models are positioned with their joint angles within the recommended ranges for the particular vehicle type, based on published literature or in-house knowledge. Given that the pedals are likely to be fixed, the man models are positioned according to the accelerator heel point. Other constraints such as leg room for rear passengers, headroom for styling and whether the package will include an adjustable steering wheel will comprise the postures which can be adopted by the man models. If the vehicle is to be a luxury saloon then it would seem logical that postural considerations should outweigh other constraints although this is rarely the case with current vehicles. The ranges of joint angles can be used to minimize or maximize the control adjustability required. For example, male drivers with long legs and short arms can be expected to reach the steering wheel with fairly straight arms whereas female drivers with short legs and fairly long arms would have their elbows at an acute angle. This adaptability allows a large proportion of the user population to drive the vehicle with a fixed steering wheel. However, it should not be forgotten that optimum comfort levels and efficiency will only be realized if a fully adjustable steering wheel is provided enabling all drivers to have their preferred arm posture.

When SAMMIE is used to evaluate a proposed or existing package then the man models' postures are governed by the task requirements and the analysis focuses on which tasks are impossible (exceeds absolute joint constraints) or difficult (exceeds comfort range).

As stated earlier, SAMMIE does not evaluate task efficiency or comfort directly. The system provides postural data which can form the basis of usefully accurate predictions of efficiency and comfort by a user with the appropriate skills and/or knowledge base. We have used SAMMIE as a consultancy tool since the mid 1970s and our predictions and design proposals have been well-received by clients and end users.

Liaise within the design team

The early communication of problems identified by SAMMIE within the design team is essential to help ensure that optimum compromises can be made. In our consultancy experience it is extremely productive for those team

Figure 3.22. Prediction of a helicopter pilot's view

members with responsibility in the problem areas to examine the issues together and to explore potential solutions with access to SAMMIE and other CAD systems as required. This promotes collaboration and the ergonomist can take a pro-active, rather than just a reactive, role.

Another advantage with SAMMIE, is that the optimum ergonomics specification can be modeled and viewed in perspective with color rendering by a

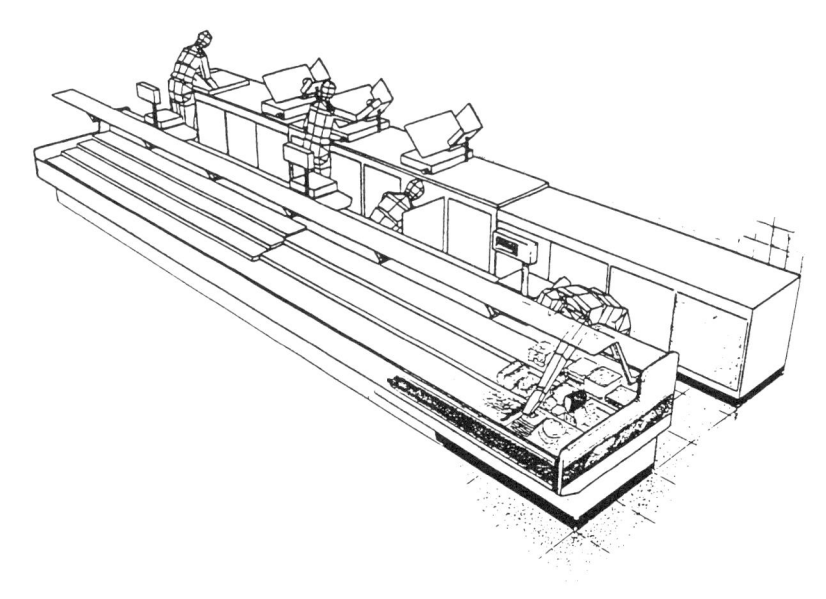

Figure 3.23. Evaluation of a delicatessen counter

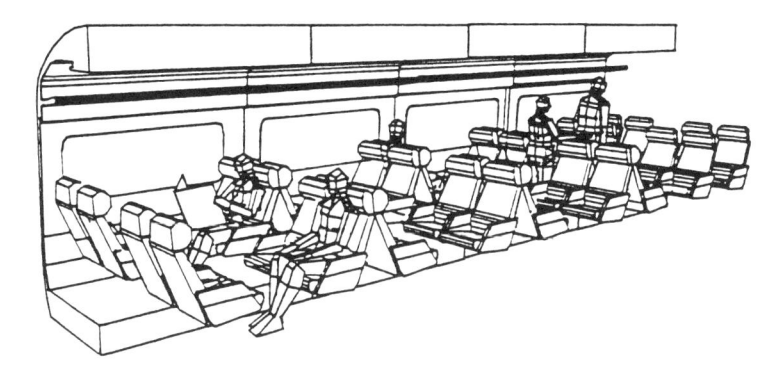

Figure 3.24. Interior layout of passenger vehicles

stylist (SAMMIE will soon have color surface shading available as a standard feature). This focuses attention on the tangible characteristics of the ergonomics design, such as the proposed control and display layout and required adjustment ranges, rather than presenting the ergonomics specification as list of requirements (e.g. the viewing distance to displays should be x mm with a look down angle of y degrees, the displays should not be obscured by the steering wheel etc.).

Full size prototype testing

The use of SAMMIE does not mean that user trials are no longer necessary as SAMMIE can only consider the geometric aspects of a design. It would be expected that full size prototypes (ride and drive vehicles) would be available as early as possible so that the basic driving package can be confirmed and that other issues can be explored such as the seat comfort, ventilation, control 'feel', graphics style and so on.

Other application areas for SAMMIE

As SAMMIE is a general purpose system it has been used in a large number of application areas in addition to automotive design, including aircraft, helicopters, ships, submarines, bank dealing rooms, computer workstations, control rooms, cashier workstations, delicatessen counters, etc. Details of SAMMIE applications have been presented in Bonney *et al.*, (1979), Case *et al.*, (1980, 1986, 1990a, 1990b), Case and Porter (1980), Levis *et al.*, (1980), Porter and Case (1980) and Porter *et al.*, (1980, 1990, 1991). Some recent application models are shown in Figures 3.22–3.24.

References

Bailey, R. W., 1982, *Human Performance Engineering: A Guide for System Designers*, Englewood Cliffs, NJ, Prentice-Hall.

Bonney, M. C., Blunsdon, C., Case, K. and Porter, J. M., 1979, Man-machine interaction in work systems, *International Journal of Production Research*, **17**, 6, 619–29.

Brennan, L. and Fallan, E. F., 1990, The contribution of CAD to the enhancement of the ergonomist's role in the design process, in Karwowski, W., Genaidy, A. M. and Asfour, S. S. (Eds) *Computer-Aided Ergonomics*, pp. 501–11, London: Taylor & Francis.

Case, K., Bonney, M. C. and Porter, J. M. 1991, Computer graphics standards for man modeling, *Computer Aided Design*, **23**, 4, 257–68.

Case, K., Bonney, M. C., Porter, J. M. and Freer, M. T., 1990a, Applications of the SAMMIE CAD System in Workplace Design, in Haslegrave, C. M., Wilson, J. R., Corlett, E. N. and Manenica, I. (Eds) *Work Design in Practice*, pp. 119–27, London: Taylor & Francis.

Case, K. and Porter, J. M., 1980, SAMMIE—A Computer-Aided Ergonomics Design System, *Engineering*, **220**, 21–25.

Case, K., Porter, J. M. and Bonney, M. C., 1986, SAMMIE: A Computer Aided Design Tool for Ergonomists, *Proceedings of the Human Factors Society Annual Conference*, Dayton, OH, pp. 694–8.

Case, K., Porter, J. M. and Bonney, M. C., 1990b, SAMMIE: A Man and Workplace Modelling System, in Karwowski, W., Genaidy, A. and Asfour, S. S. (Eds) *Computer-Aided Ergonomics*, pp. 31–56, London: Taylor & Francis.

Case, K., Porter, J. M., Bonney, M. C. and Levis, J. A., 1980, Design of Mirror Systems for Commercial Vehicles, *Applied Ergonomics*, **11**, 4, 199–206.

Clauser, C. E., Tucker, P. E., Reardon, J. A. and McConville, J. T., 1972, *Anthropometry of Air Force Women*, AMRL-TR-70-5, Aerospace Medical Research Laboratories, Wright-Patterson Air Force Base, OH.

Daniels, G. S. and Churchill, E., 1952, *The 'Average Man'?* T.N. WCRD 53-7, Wright Air Development Centre, Wright-Patterson Air Force Base, OH.

Dooley, M., 1982, Anthropomorphic modelling programmes—A survey, *IEEE Computer Graphics and Applications*, **2**, 17–25.

Dreyfuss, H., 1967, *The Measure of Man: Human Factors in Design*, 2nd edn, New York: Whitney Library of Design.

Elias, H. J. and Lux, C., 1986, Gestatung ergonomisch optimierter Arbeitsplatze und Produkte mit Franky und CAD (The design of ergonomically optimized workstations and products using Franky and CAD), *REFA Nachrichten*, **3**, 5–12.

Grandjean, E., 1980, Sitting Posture of Car Drivers from the Point of View of Ergonomics, in Oborne, J. and Levis, J. A. (Eds) *Human Factors in Transport Research*, Vol. 2, pp. 205–13, London: Academic Press.

Groover, M. P. and Zimmers, E. W. Jr., 1984, *CAD/CAM Computer-Aided Design and Manufacturing*, Englewood Cliffs, NJ, Prentice-Hall.

Haslegrave, C. M., 1980, Anthropometric Profile of the British Car Driver, *Ergonomics*, **23**, 436–67.

Haslegrave, C. M., 1986, Characterising the anthropometric extremes of the population, *Ergonomics*, **29**(2), 281–301.

Hertzberg, H. T. E., 1960, Dynamic anthropometry of working positions. Human Factors, 2, 147–55.

Karhu, O., Kansi, P. and Kuorinka, I., 1977, Correcting Working Postures in Industry: A Practical Method for Analysis, *Applied Ergonomics*, **8**, 199–201.

Karwowski, W., Genaidy, A. M. and Asfour, S. S. (Eds), 1990, *Computer-Aided Ergonomics*, London: Taylor & Francis.

Kloke, W. B., 1990, WERNER: A personal computer implementation of an extensive anthropometric workplace design tool, in Karwowski, W., Genaidy, A. M. and Asfour, S. S. (Eds) *Computer-Aided Ergonomics*, pp. 57–67, London: Taylor & Francis.

Kuusisto, A. and Mattila, M., 1990, Anthropometric and biomechanical man models in computer-aided ergonomic design structure and experiences of some programs, in Karwowski, W., Genaidy, A. M. and Asfour, S. S. (Eds) *Computer-Aided Ergonomics*, pp. 104–14, London: Taylor & Francis.

Launis, M. and Lehtelä, J., 1990, Man models in the ergonomic design of workplaces with the microcomputer, in Karwowski, W., Genaidy, A. M. and Asfour, S. S. (Eds) *Computer-Aided Ergonomics*, pp. 68–79, London: Taylor & Francis.

Levis, J. A., Smith, J., Porter, J. M. and Case, K., 1980, The impact of computer-aided design on pre-concept package design and evaluation, in Oborne, D. A. and Levis, J. A. (Eds) *Human Factors in Transport Research*, Vol. 1, pp. 356–64, London: Academic Press.

Majchrzak, A., Chang, T-C., Barfield, W., Eberts, R. and Salvendy, G., 1987, *Human Aspects of Computer-Aided Design*, London: Taylor & Francis.

McConville, J. T., 1978, Anthropometry in Sizing and Design, in *Anthropometric Source Book, Volume I: Anthropometry for Designers*, NASA Reference Publication 1024, Washington, DC: Scientific and Technical Information Office.

McDaniel, J. W., 1990, Models for ergonomic analysis and design: COMBIMAN and CREW CHIEF, in Karwowski, W., Genaidy, A. M. and Asfour, S. S. (Eds) *Computer-Aided Ergonomics*, pp. 138–56, London: Taylor & Francis.

Meister, D., 1982, The role of human factors in system development, *Applied Ergonomics*, **13**(2), 119–24.

Mumford, A. and Skall, M. (Eds), 1988, *CGM in the Real World*, Springer, Berlin.

Pheasant, S., 1986, *Bodyspace*, London: Taylor & Francis.

Porter, J. M., 1989, *Unintended acceleration and the design of foot pedals*, Unpublished report, Vehicle Ergonomics Group, Department of Human Sciences, Loughborough University.

Porter, J. M., Almeida, G., Freer, M. and Case, K., 1991, The design of supermarket workstations to reduce the incidence of musculo-skeletal discomfort, in Queinnec et al. (Eds) *Designing for Everyone and Everybody*, London: Taylor & Francis.

Porter, J. M. and Case, K., 1980, SAMMIE—Can cut out the prototypes in ergonomic design, *Control and Instrumentation*, **12**, 1, 28–29.

Porter, J. M., Case, K. and Bonney, M. C., 1980, A Computer-Generated Three Dimensional Visibility Chart, in Oborne, E. J. and Levis, J. A. (Eds) *Human Factors in Transport Research*, Vol. 1, pp. 365–73, London: Academic Press.

Porter, J. M., Case, K. and Bonney, M. C., 1990, Computer Workspace Modelling, in Wilson, J. A. and Corlett, E. N. (Eds) *Evaluation of Human Work: A Practical Ergonomics Methodology*, pp. 472–99, London: Taylor & Francis.

Porter, J. M., Porter, C. S. and Lee, V. J. A., 1990, *Car driver discomfort survey*, Unpublished report, Vehicle Ergonomics Group, Department of Human Sciences, Loughborough University.

Post, F. H. and Smeets, J. W., 1981, ADAPS: Computer-aided anthropometrical design, *Tijdschrift voor Ergonomic*, **6** (4), 11–18 (in Dutch).

Rebiffé, R., 1969, Le Siège du conducteur: Son adaptation aux exigence fonctionneles et anthropométriques, in Grandjean, E. (Ed.) *Sitting Posture*, London: Taylor & Francis.

Requicha, A. A. G., 1970, Representations for rigid solids: theory, methods and systems, *Computer Surveys*, **12**.

Robinette, K. M. and McConville, J. T., 1982, An alternative to percentile models, 810217, pp. 938–46, Warrendale, PA: Society of Automotive Engineers.

Roebuck, J. A., Kroemer, K. H. E. & Thomson, W. G. 1975, *Engineering Anthropometry Methods*, John Wiley & Sons.

Rothwell, P. L., 1985, Use of man-modelling CAD systems by the ergonomist, in Johnson, P. and Cook, S. (Eds) *People and Computers: Designing the Interface*, Cambridge University Press, pp. 199–208.

Sheldon, W. H., 1940, *The varieties of human physique*, New York: Harper & Bros.

Simpson, R. E. and Hartley, E. V., 1981, *Scatter diagrams based on the anthropometric survey of 2000 Royal Air Force aircrew (1970/71)*, Royal Aircraft Establishment Technical Report 81017, Farnborough, Hampshire.

Singleton, W. T., 1971, Systems Design, *Applied Ergonomics*, **2**, 3.

Smith, B., Rinaudot, G. R. and Wright, T., 1988, *Initial Graphics Exchange Specification (IGES) Version 4.0*, US Department of Commerce.

Stoudt, H. W., Damon, A., McFarland, R. and Roberts, J., 1965, *National Health Survey 1962: Weight, Height and Selected Body Dimensions of Adults, United States 1960–1962*, Public Health Service Publication no. 1000 Series 11, no. 8, Washington, DC: US Government Printing Office.

Westerink, J., Tragter, H., Van Der Star, A. and Rookmaaker, D. P., 1990, TADAPS: A three-dimensional CAD man model, in Karwowski, W., Genaidy, A. M. and Asfour, S. S. (Eds) *Computer-Aided Ergonomics*, pp. 90–103, London: Taylor & Francis.

4

Visual aspects in vehicle design

Christine M. Haslegrave

Introduction

Vision is obviously crucial in the driving task, with most of the information received by the driver coming through the visual sense. Connolly (1964) reviewed the visual capabilities which are needed for driving, showing the importance of visual acuity, motion perception sensitivity, fields of view of each eye, color sensitivity, eye movements and head movements, and also discussed the aspects of visual monitoring which are involved in a variety of driving maneuvers. He showed further how these aspects may be affected by task factors such as illumination, glare and vehicle vibration. However, despite a vast amount of research, there is still much to learn of what a driver really needs to see in order to drive efficiently and safely.

At night and in wintry conditions with rain, fog or snow, the driver's view of the road ahead may be minimal, yet he or she still manages to drive the vehicle, find the route and arrive at the desired destination. The minimum criteria for being able to drive are provided (not necessarily adequately) by indications of the direction and boundaries of the road and of the presence of other vehicles. However, driving under poor visibility conditions is stressful for the driver and results in a significantly increased risk of an accident. Moreover, it is only possible by reducing the normal driving speed, so that the degraded road information can be perceived and interpreted in time to make corrections to headway or to take any necessary action to avoid colliding with other vehicles.

The effects can often be seen when a car driver is followed by a truck in fog; the truck driver is impatient to pass, and is cursed by the car driver for being reckless. In fact, the truck driver may have a much clearer view of the road, over the top of low-lying banks of fog, and can thus drive faster with equal safety. It is therefore apparent that the position of the driver and design of the vehicle play a large part in determining what the driver is able to see, so that these are factors in defining the required field of view in addition to the driver's visual capabilities. In vehicle design both human and engineering factors have to be integrated, and the minimum requirements for the driver's view of the road environment have to be assessed in relation to parameters of the vehicle's body structure. In other words, the field of view requirements need to be translated into design specifications which are compatible with the criteria and terminology used in engineering design.

Driver's visual needs

The first step in the design must be to establish the minimum specification which gives an adequate field of view for the whole range of potential drivers of the particular vehicle. In the simplest terms, the clear view ahead has to be sufficient for the driver to stop, if necessary, under emergency braking conditions. For example, if a child stepped into the road, the driver would not be able to avoid a collision unless the child were first seen beyond this minimum stopping distance. Minimum stopping distance is dependent on the speed at which the vehicle is travelling, perhaps 12 m at 50 kph or 31 m at 80 kph for a car with normal braking capability of 0.8 g (7.8 ms^{-2}). The corresponding braking distances for a truck with a typical braking capability of 0.6 g (5.9 ms^{-2}) would be 16 m at 50 kph and 42 m at 80 kph. Thus a truck driver would need to see the child well before the car driver, since he/she needs a longer stretch of road in which to stop the heavier vehicle.

In reality, however, drivers need much longer views of the road in order to anticipate and prepare for avoiding action, so that the specification might be defined as a clear view to the horizon on a flat road. Views close to the vehicle are equally important when turning to left or right or when maneuvering the vehicle. A view of the head of a small child (around 0.9–1.0 m height) might well be taken as the minimum criterion for objects which must be visible when driving in residential areas. In this case again, the requirement is more stringent for truck design than for car design, since the view of the child will disappear more quickly below the level of the windscreen in the truck driver's field of view due to the higher cab of the truck.

The view to the rear of the vehicle, obtained mainly through the mirrors (indirect view), provides information on passing vehicles, and on the position of vehicles close to the rear when the driver is proposing to change lane or is completing a passing maneuver.

The need for direct views of other areas around the vehicle, or of different parts of the road environment, may depend on its operating conditions. For instance, bus drivers need a wide view of the front corners close to the bus where boarding or alighting passengers may be. Truck drivers may have greater need of views to the rear of the vehicle for reversing or maneuvering into delivery and loading bays, which present particular problems for articulated vehicles. As a result, both buses and trucks are often fitted with ancillary mirrors to give views of areas which are important for their special needs but not necessary for drivers of cars or other road vehicles.

These examples set a few of the parameters for considering the driver's field of vision, taking account of both technical and operational aspects of the vehicle. A complete specification of the view of road environment is obviously more complicated, in that the driver needs to see changes in the road alignment (delineated by topography and street furniture as well as by road markings), other vehicles (oncoming vehicles moving at similar or faster speeds in the opposite direction and so increasing the necessary distance of view), and warning or information signs placed beside the road. Road signs can also be mounted on high posts and overhead gantries, so that adequate upward vision is important in addition to forward, rearward and side vision.

The most thorough review of the driver's visual needs for both direct and mirror vision from road vehicles is contained in a report describing a series of studies by the Ford Motor Company (SAE, 1973b). These studies investigated many aspects such as the location of traffic signals and signs in the United States, pedestrian interaction with vehicles, driver behavior at road intersections and in response to vehicles approaching from the rear, and the detection of objects in the peripheral visual field. The data was compiled to produce a set of composite visual targets representing the zones which are most important for a driver to be able to see in the forward, side and rearward (mirror) fields of view, the limits essentially being defined as viewing angles from the position of the driver's eyes.

Legislation, as described later, has attempted to use these criteria to define the minimum requirements for fields of view provided by road vehicles (as well as for a few other off-road or special purpose vehicles). Vehicle designers may, however, wish to consider other aspects beyond the minimum requirements and to carry out more comprehensive evaluations of the field of view provided by their company's product and to compare its performance with that of competitors. The measurement techniques which have been developed for legislative testing are equally applicable for these purposes, but it would be necessary to define additional criteria for assessing the various aspects of the view provided.

Specifying the field of view

The driver's view is of course three-dimensional, and so needs to be defined in geometrical terms. It is simplest to define the space around the driver as two hemispheres: one below a plane through the driver's eyes and one above this plane. The field of view can then be assessed in terms of a view downwards towards the road, and a view upwards primarily concerned with signs and signals of various types, as shown in Figure 4.1. If the driver cannot see anything below a horizontal line of sight, he/she cannot see the ground at all, and a view of the far distance (horizon) is represented by a horizontal plane passing through the driver's eye. Objects on the road such as pedestrians or

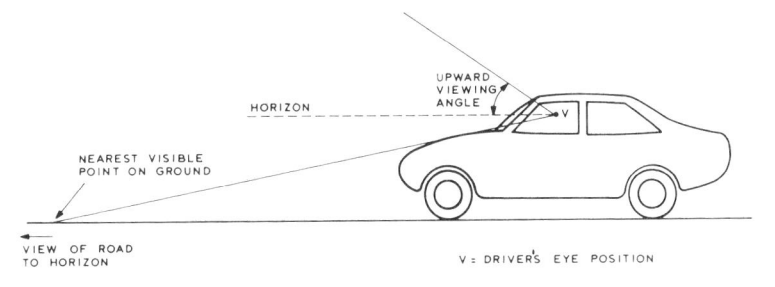

Figure 4.1. Specification of the driver's forward field of view

82   C. M. Haslegrave

other vehicles are in the lower hemisphere, while traffic warning and information signals or overhead signals are usually already in the upper hemisphere when they need to be seen or read.

Direct field of view

The visual field of the human eye is complex, limited by anatomical and optical factors, some of which are discussed later. However, it can be represented by sightlines drawn from the eye to all the points which can be seen, collectively defining the visible field of view. The limits to the field of view can therefore be defined by the sightlines at the boundaries, which are often represented as viewing angles or as planes originating from an eye point in space. Alternatively, in order to simplify evaluations, the field of view may be defined as the area visible on a specified plane, usually either as a ground plan or as a vertical cross-section through the field of view. Both types of definition are illustrated in Figure 4.1, and legislative requirements can be found expressed in either terms.

Mirror field of view

The driver's view through a mirror can be represented in the same way, but here the image is bounded by the frame of the mirror, which limits the reflected view which can be seen. The image boundaries are determined by the mirror dimensions, location of the mirror with respect to the driver's eye and optical characteristics of the mirror (specifically whether it is a plane or convex mirror). The views provided of following and passing vehicles are illustrated in Figure 4.2.

For practical purposes, the available field of view in a mirror can be varied by adjusting the position of the mirror and by adjusting the tilt of the mirror with respect to the direction of view. As the mirror is moved further from the driver, the area which can be seen becomes smaller and the size of any object in the reflected image (or angle subtended at the eye) also becomes smaller and less easily seen. Exterior mirrors therefore tend to be placed on the A-pillar close beside the driver rather than the front wing of the vehicle, but this location has the disadvantage of placing them further away from the direct forward line of sight and may force the driver to turn eyes or head to look at the mirror view.

When adjusting the mirror itself, different parts of the view to the rear are displayed as the mirror is tilted towards or away from the driver's sightline to the mirror. This allows each driver to adjust the mirror to give his/her best

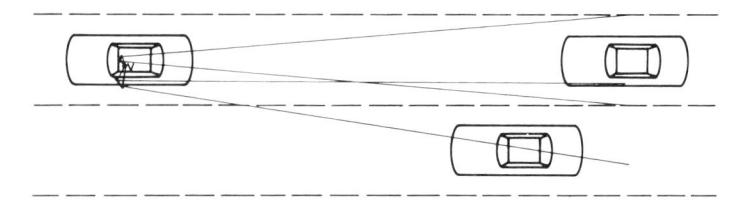

Figure 4.2. Rearward field of view in the interior and side mirrors

view of the road to the rear, but the adjustability permits extra degrees of freedom in defining or specifying the field of view provided and there is no unique orientation which can be specified for the purposes of evaluation. During mirror tests such an adjustment has to be made to give an 'optimum' view at the eyepoint specified for the test. It is worth noting that this often requires compromises to obtain the maximum depth and the maximum width simultaneously for the rearward view in the mirror, and thus involves a degree of judgment on the part of the tester which is not required for the more closely defined parameters of the direct field of view.

Driver eye locations

In order to determine the sightlines forming the boundaries to the field of view, it is necessary to know where the driver's eyes are in space. Every driver has a slightly different view of the road. A survey of car drivers travelling on British roads found that the heights of car drivers' eyes above the road surface ranged from 0.87 m–1.28 m with a mean of 1.14 m (Haslegrave, 1979). However, comparison with earlier data indicated that there had been a considerable lowering of the eye position in the previous 15 years due to changes in car styling, so that such dimensions cannot be regarded as fixed indefinitely.

The range of eye positions inside a vehicle for any driving population is also considerable. There is a range of perhaps 160 mm or more in seated eye height, although this range can be reduced slightly if height adjustment is available for the driver's seat, although this option is not yet commonly available in many vehicles. Seat fore/aft adjustment further complicates the problem of establishing where the driver's eyes are positioned. This position has a large element of preference and, for instance, many tall drivers choose to have their seats further forward than might be predicted from their leg lengths.

In order to address this problem, the SAE J941 Eyellipse was developed as a drafting tool to define the range of eye positions within the driving population (SAE, 1985; Devlin and Roe, 1968). It was based on a study of the eye positions of US drivers in a selection of saloon cars, which found that the distribution of eye positions in space closely approximated an ellipsoid. The eyellipse was derived from this distribution and consists of a plan view and side view of driver's right and left eye ranges.

It does not however represent the eye positions in the sense of defining percentile boundaries containing a given percentage of eye positions of the population. Instead, a pair of eyes is located by constructing a tangent to the eyellipse which represents a sightline to the object viewed. The tangent then represents a 'cut-off' line below which (or to one side of which) lies a given percentage of eye locations: thus 95 per cent of eye locations lie to one side of any tangent to the 95th percentile eyellipse. In other words, the tangent shows what percentage of the population is cut off by (or unable to view) that particular sightline.

The use of the eyellipse is illustrated in Figure 4.3. It is applicable for horizontal eye turn angles of up to 30 degrees, but its use was extended to take account of head turn, assuming a pivot point located 3.875 inches (98 mm) to the rear of the mid-point between the eyes by the procedure described in Devlin and Roe (1968) and SAE (1969). The eyellipse is located within the

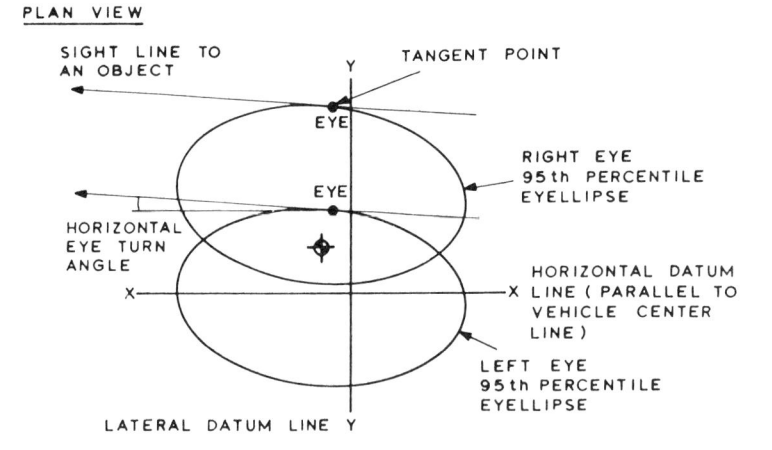

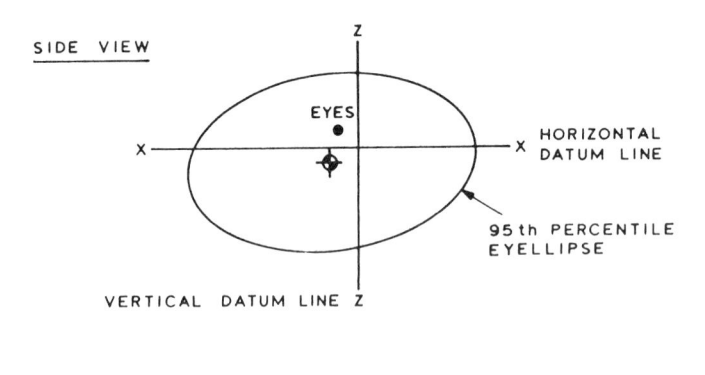

Figure 4.3. Using the eyellipse to define the driver's eye point

interior of a vehicle in relation to the H-point defined by the SAE J826 manikin (SAE, 1987). More detailed instructions for use of eyellipse can be found in SAE Recommended Practice J941 (SAE, 1985) and SAE Recommended Practice J1050 (SAE, 1977).

No similar eyellipses have yet been defined for other national populations, but the eyellipse is widely used by vehicle designers concerned with European and Australian markets as well as in the United States. It should perhaps be remembered that this data may not be adequate for assessing visual fields for drivers in Asian countries and in many other parts of the world where populations are not as tall or where seat adjustments are quite different.

A few studies have been made of drivers' eye positions while turning to look directly to the side or rear of the vehicle (reported in SAE SP-381; SAE, 1973b). These measurements were, however, taken for very small samples, and so there is only limited information on suitable eyepoints for assessing views out of the side window or to the rear for reversing a car.

Driver vision

Human view is binocular and acuity varies over the visual field, being most acute in the foveal region and poor in peripheral vision. The boundaries between these visual zones are not clearly delineated, and there is relatively little population data to establish the effect of individual variability.

The concept of a 'sightline' originating from a fixed eyepoint is of course a simplification of human vision. The driver in reality can turn both eyes and head to gain a wider field of view, and moreover can make use of peripheral vision to see objects or movement even without turning the eyes. However, although peripheral vision may be important to the driving task, it is seldom considered in evaluations of the fields of view provided by vehicles which are normally concerned with the central vision provided at each eye fixation.

Other aspects of human vision also need to be considered when making a proper assessment of available fields of view for the driving task. SAE J985 (SAE, 1967) reviews the most important human vision capabilities, covering the influence of visual acuity, motion perception and anatomical constraints (a fuller discussion may be found in Connolly, 1964, if desired). All of these capabilities are important in the highly complex driving task, but do not necessarily have to be accurately established to provide good fields of view in vehicle design.

In the horizontal plane, the binocular field of view extends some 120 degrees, as shown in Figure 4.4, but vision is sharp over a fairly small area directly ahead so that the eyes need to be turned to focus on objects outside this foveal area. According to SAE J985 (SAE, 1967), the eyes generally only turn by about 30 degrees before the head is turned, which can comfortably give a further 45-degree view to either side.

In the vertical plane, eye movement is comfortable within 15 degrees above or below the horizontal, although the eye can be rotated up to 45 degrees upward or 65 degrees downward if necessary. The head can easily be inclined 30 degrees upwards or downwards. Thus, by use of eye and head movement, the driver has an extensive direct field of view and, in practice, his/her view of the road environment is usually restricted by the vehicle structure rather than by anatomical constraints. However, the two interact and evaluations of the field of view need to take account of normal behavior in terms of eye and head movements.

It is obviously desirable that the driver should concentrate his/her attention on the road directly ahead, glancing away from the road for as short a period as possible. Thus the need for eye and head turn should generally be kept to a minimum (eye turn within ± 15 degrees if possible), which means that mirrors and instrument panel displays should be placed as close to the driver's forward line of sight as possible.

In order to determine the view provided by binocular vision, the view from each of the driver's eyes has to be considered separately. Binocular vision has very little effect on the view of distant objects, although it contributes to the sense of distance and of motion, but can have a considerable effect on obscur-

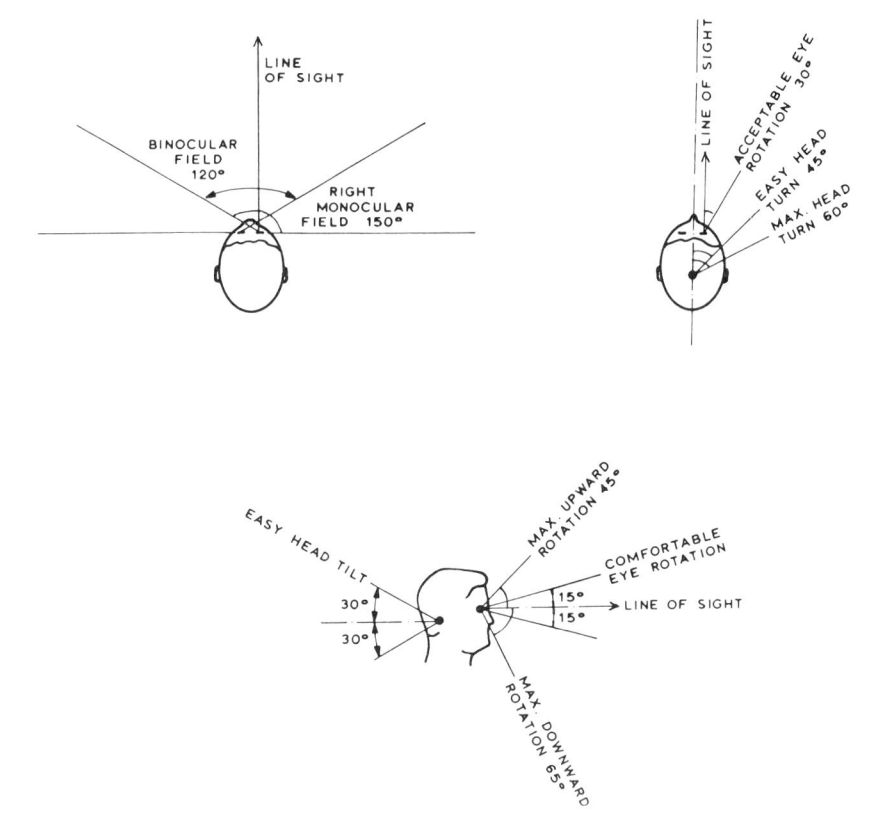

Figure 4.4. Limits of visual field, eye and head movement
(From SAE J985 (SAE, 1967))

ation caused by objects in the near field of view. In a vehicle, this mainly affects the design of the A-pillars and windscreen area, as seen in Figure 4.5. If the effective width of the pillar is less than the distance between the eyes, the blind spot behind the pillar will only obscure a portion of the road beyond, more distant objects being visible.

The forward field of view is usually considered adequate if all is visible to at least one eye (ambinocular view) even though some parts are obscured for one eye or the other (monocular views). Obscurations in the field of view, caused by obstructions such as an A-pillar, need to be assessed in terms of binocular view—the area which is not visible to either eye. The limits to the direct field of view are then determined by the sightlines from the left eye to the left side of the obstruction and from the right eye to the right side of the obstruction.

The limits to a mirror view are established by the ambinocular vision, where the boundaries are formed by sightlines of the left eye looking to the right side of the mirror and the right eye looking to the left side of the mirror. However, the more important parts of the road environment should be clearly visible to both the driver's eyes, which applies equally to direct and mirror views.

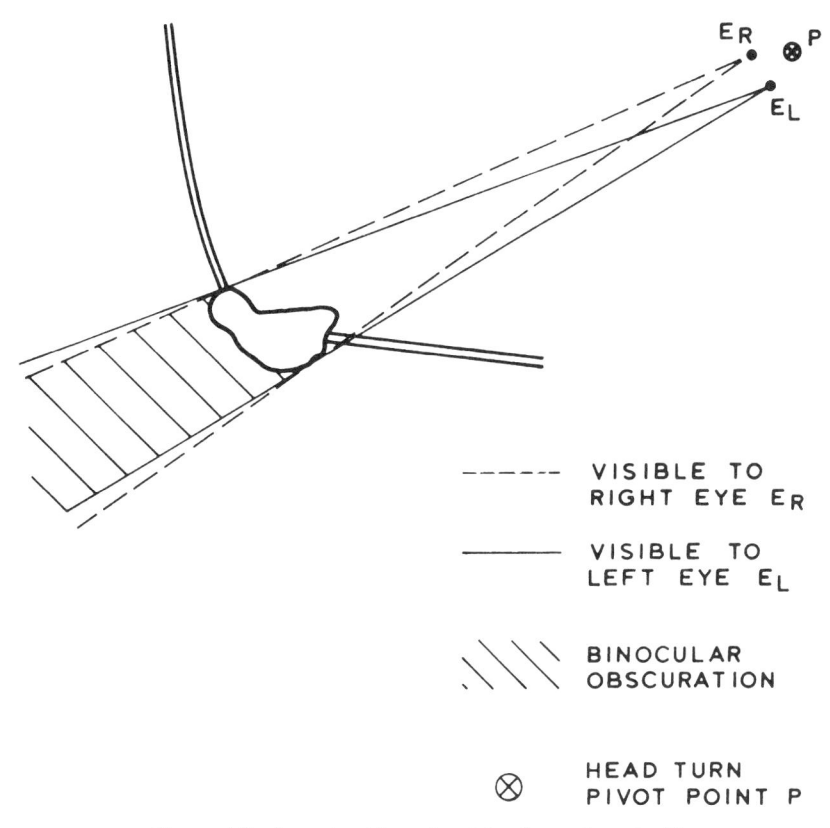

Figure 4.5. Assessing binocular vision beyond an A-pillar

Design considerations and legislation

The main field of view considerations for vehicle designers are:

1. to define the field of view provided for the driver, establishing any obscuration caused by the body structure and fixtures such as windscreen wipers or by the mirrors themselves;
2. to choose suitable interior and exterior mirrors, and to determine where they are best mounted to maximize the field of view while avoiding unnecessary obscuration of the direct view.

The driver's visual needs, as discussed earlier, are important in assessing the need for (or benefit of) possible improvements, but legislation is a primary determinant in vehicle design, and well-established methods of assessing fields of view have been developed for ensuring compliance.

It is therefore appropriate first to discuss the relevant regulations and ways in which vision criteria have been specified. The legislation does not attempt to define fields of view for individual drivers, but in most cases specifies areas which must be visible from a defined eye point (representing a typical driver and driving position). There has been relatively little discussion of the degree to which such a composite driver can represent the design limits for the driving population, perhaps because the regulations have been successful in providing adequate fields of view.

Several European and US regulations for cars are indicated in Table 4.1, while similar national regulations exist in other countries. Those for trucks and buses may differ, particularly in relation to fitting of exterior mirrors and mirror convexity. There are differences in requirements between the regulations, due to differences in specification of criteria or test techniques, and vehicle manufacturers may have to take account of several regulations if they are selling in worldwide markets.

Terminology and reference frame

Most regulations define the vision requirements in terms of a three axis coordinate reference frame which is aligned with the vehicle's longitudinal center plane (as defined by SAE J182a; SAE, 1973a). This frame is also aligned with the body reference lines used by vehicle designers; the x-axis is the vehicle's longitudinal axis, the y-axis and z-axis are the lateral and vertical axes through the reference origin.

The horizon or point of infinity on the road is established by sightlines in the horizontal plane through the specified eye point.

Eye location

Either two eyes or a single eye position may be specified for assessing the required fields of view. Where a single eye point is specified, this usually represents the mid-point between the two eyes, although other vision origin points may also be used.

The two eyes, often known as E-points, are usually located at an interpupillary width of 65 mm and rotate about a neck pivot point (or P-point)

Table 4.1. *European and US regulations specifying a car driver's fields of view*

European Community	Directive 71/127/EEC (as amended by Directives 79/795/EEC, 82/205/EEC and 88/321/EEC) Rear-view mirrors of motor vehicles
	Directive 77/649/EEC (as amended by Directives 81/643/EEC and 88/366/EEC) Field of vision of motor vehicle drivers
	Directive 78/317/EEC Windscreen wiper and washer systems
	Directive 78/318/EEC Defrosting and demisting systems
USA	Federal Motor Vehicle Safety Standard 111 Rear view mirrors
	Federal Motor Vehicle Safety Standard 103 Defrost/demist system
	Federal Motor Vehicle Safety Standard 104 Wash/wipe system

98 mm to the rear of the mid-point between the eyes to simulate head turn in the horizontal plane. EEC Directive 71/127 specifies that the eye points should be at a height of 635 mm vertically above the vehicle R-point or H-point (which is located by the procedures of ISO 6549; ISO, 1980). This establishes the seat in the rearmost lowest driving position, and so would represent a driver with 95th percentile leg length while the eye height above this approximates to the 50th percentile. US regulations specify the location of the eyes of a 95th percentile male, as defined by the SAE Eyellipse tool (SAE, 1977). In order to comply with FMVSS 111 mirror requirements, it is sensible to use the tangent to the eyellipse furthest from the mirror, which generates the smallest rearward field of view.

Direct field of view

Forward field of view is covered by EEC Directive 77/649 (as amended by EEC Directives 81/643 and 88/366), which considers several aspects: the limits on the downward angle of view determining the nearest visible points in front of the vehicle, the upward angle of view in the windscreen, and the binocular obscuration caused by the A-pillars. For assessing the pillar obscuration, account is taken of both eye and head turn.

Mirror view

Most interior mirrors currently in use give a view of approximately 20 degrees in width, although this may sometimes be limited by the dimensions of the rear screen rather than by the width of the mirror itself. Exterior mirrors generally provide about 10 degrees width of field, although this can be reduced in the area close to the rear of the vehicle which is often obscured by the contours of the bodywork or the exterior trim.

The regulations relating to rear view mirrors are EEC Directive 71/127 (as amended by a series of subsequent directives) in Europe and Federal Motor Vehicle Safety Standard FMVSS 111 in the United States. For each mirror they specify a minimum area on a ground plan of the road which must be visible to the driver (as shown in Figure 4.6). Thus the nearest visible point seen in the interior mirror should be not further than 60 m to the rear of the driver's eyes (Directive 71/127/EEC) or 200 ft (FMVSS 111) from the rear of the vehicle. The mirror must have sufficient width to give the driver a 20 m wide view (Directive 71/127/EEC) or 20 degrees horizontal angle of view (FMVSS 111) beyond these points. In the exterior mirror on the driver's side, the required field extends rearward from a vertical plane (stretching between ground level and the horizon) 10 m or 35 ft behind the driver's eyes (Directive 71/127/EEC, FMVSS 111) and is bounded by a plane through the widest point on the side of the vehicle and a second plane 2.5 m or 8 ft (Directive 71/127/EEC, FMVSS 111) away from this.

The EEC Directive also sets further requirements for mirrors which are fitted on the passenger's side (nearside), although this is not covered by US regulations. Other special mirrors may also be required by separate standards for trucks and buses in order to improve views of blind spots to the side of the vehicles, which may necessitate overhead views or the use of high convexity mirrors (called approach mirrors or close proximity mirrors).

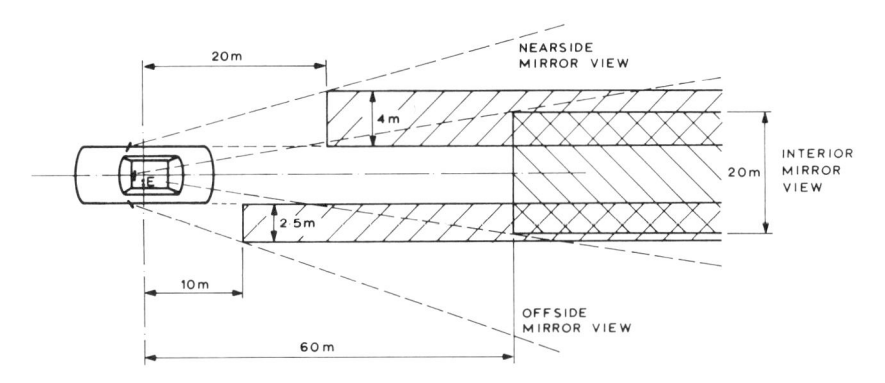

Figure 4.6. Minimum fields of view specified by ground plan of EEC Directive 71/127 (amended) for a left hand drive car

Measuring and assessing fields of view

Direct measurement
Three main techniques have been used for direct measurement of fields of view. The simplest method is to have an observer sitting in the vehicle and describing the view but, as a test technique, this has obvious difficulties of standardization for both repeatability and reproducibility.

More usefully, the view can be photographed directly by a camera placed in the position of the driver's eye. This technique has the advantage of providing a permanent record, but presents some problems as well. First, it is necessary to superimpose two photographs to assess the binocular field of view. A more serious difficulty is obtaining a realistic photographic representation of human vision, where the image of the three-dimensional world is collected on the retina of the eye. This can be approximated by using a pinhole camera, but ordinary camera lenses do not realistically represent the human view. With any photographic method, the three-dimensional field of view is obviously presented as a two-dimensional record. A reference grid therefore needs to be defined on the photograph for analysis purposes, in order to relate the image to the original three-dimensional space.

A third technique uses one or two lights to represent the driver's eyes (as illustrated in the tests in Figure 4.7). Any objects present in the field of view (such as the windscreen surround, B-pillars or windscreen wipers) obscure the light beam and cast shadows beyond, defining the boundaries to the driver's view. The shadows can then be used to generate a ground plan showing areas which are visible in either the forward or the direct rearward field of view, and can also be used to calculate angles of view or of obscuration in any vertical or horizontal plane. A reference grid is still needed in order to analyse the visual record. Other versons of this technique have used collimated optical beams,

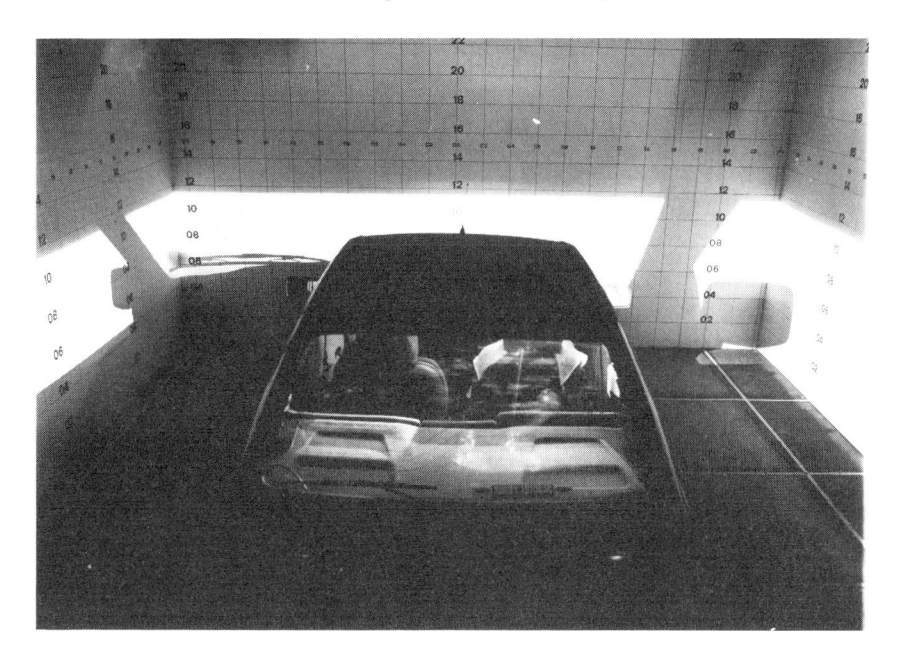

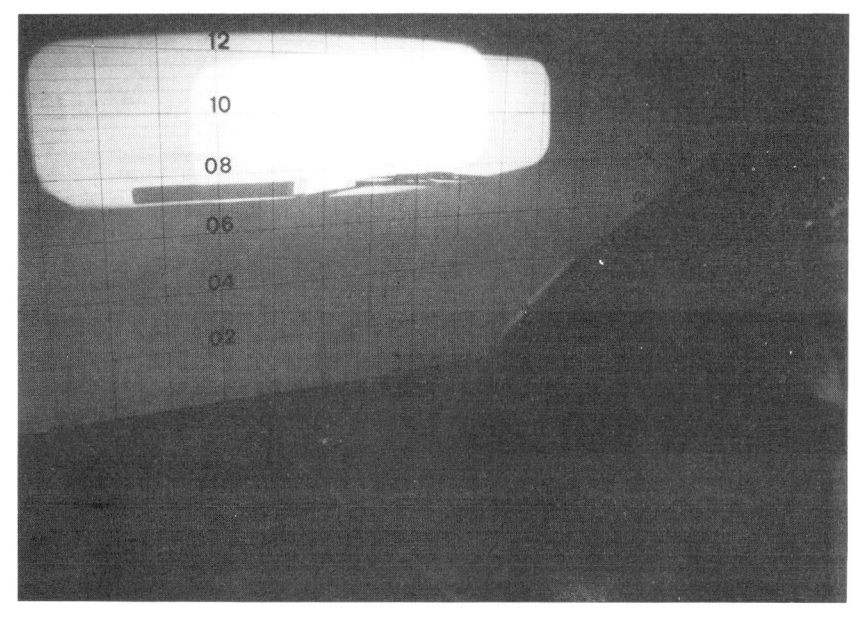

Figure 4.7. Measurement of field of view in the laboratory: (a) *Monocular view of forward field;* (b) *Binocular view in interior rear view mirror.*
(Reproduced with permission from the Motor Industry Research Association, Nuneaton, UK)

lasers or theodolites to scan the visual field. These optical methods avoid some of the previous problems and can be carried out with a high degree of repeatability.

Laboratory tests using the two light method

In a test facility using the light system, such as that illustrated in Figure 4.7, the light beams represent the sightlines collected by the eyes when viewing the surrounding world. The lights are chosen to have a very small filament size, and thus are effectively point sources of light. This size has advantages both in terms of representing the human eye relatively well, and in providing sharp shadow boundaries.

The effects of binocular vision can be seen very clearly in Figure 4.7b, since monocular obscuration casts lighter shadows than binocular obscuration. The technique can be used equally well for measuring the direct field of view and for measuring the view reflected in a mirror, when the light beam is reflected by the mirror onto screens to the rear of the vehicle, exactly reversing the collection of the road image in a driver's eye.

When a screen (marked with a reference grid) is placed outside the vehicle, the shadow boundary outlines the visible areas, effectively giving a vertical cross-section through the field of view. From the known position of the screen in relation to the eye point, the boundary can be translated by simple mathematical calculations into coordinate points within the reference frame, giving viewing angles or ground plans of the road view—or alternatively the view in some other horizontal plane, perhaps representing the height of a small child.

Figure 4.7a shows an analysis of the direct monocular view through the car's windscreen and side windows, which is recorded on a screen fixed in front of the car. The driver's eye (light) is positioned at a known point relative to the screen, in which case the ground plan can be calculated from the single screen record. Angles of obscuration of A-pillars or other objects within the vehicle (such as the interior mirror or windscreen wipers) can also easily be calculated. The use of a fixed screen is convenient for the analysis, but the positioning of the vehicle itself within the test facility can be very difficult. Not only does the eye point have to be located centrally but the vehicle axes must also be aligned with the reference axes of the screen.

A single screen cannot be used in this way to analyse the view in a mirror since the effective eye point (as reflected in the mirror) cannot be identified and located at the defined origin for the screen. In this case, as shown in Figure 4.7b, a mobile screen is used and positioned at two points to the rear of the car, with its surface perpendicular to the longitudinal axis of the vehicle. Markers need to be placed on the mirror or rear windscreen to identify a series of reference points around the shadow boundary (not shown in the photograph). The common reference points can then be identified for both screen positions and used to calculate the image projected onto a ground plane, through simple geometry. As an alternative to fixing the reference markers, it would be possible to insert a transparent reference grid at any convenient point in the light beam, but in practice analysts can find this confusing, particularly when a heater grid is also visible on the rear screen of the car.

Computer modeling of field of view

As an alternative to laboratory testing, various computer-based systems have been used within the vehicle industry for designing driver view packages. Ford started to use computer programs to analyse mirror views in the late 1960s (Devlin and Pajas, 1968, 1969) and since then many companies have developed their own analysis software. The geometry of the field of view can be defined by data input of the driver's eye position, mirror pivot, and the mirror dimensions and convexity.

The SAMMIE workspace modeling system incorporates viewing analysis, which can be used both in analyzing direct fields of view and in modeling mirror systems. Case *et al.* (1980) have described its use in a study of mirror systems for commercial vehicles, including their assessment for legislative requirements. Using SAMMIE, the view can be displayed either as a ground plan (or a vertical cross-section as on a test screen in laboratory trials) or as seen from the eyepoint of the driver him/herself, which enables designers to visualize the results easily.

Such packages are helpful in the design and development process, but detailed models of the vehicle bodywork are needed for accurate assessments of the fields of view. This is because critical parts of the view can be obscured by relatively small components or even by body trim. From experience in practical tests, the most critical areas are usually found at the A-pillars (which are close to the eyes so that any trim has a large influence on the binocular obscuration), at the offside rear wheel arch and edge of bumper (which may obscure the view of the ground close to the side of the vehicle), and at the rear screen surround (which may be critical when obtaining a view of the horizon in the interior mirror).

Developments in design and evaluation

Visual requirements obviously interact with many other design parameters, in particular with structural and styling features. For example, the positioning of the A-pillars is dictated by the need for mechanical strength and integrity as much as by the need for a clear field of view. The A-pillars and cant rail form part of the strong cage protecting the vehicle's occupants in the event of accident impact or rollover, and any move to reduce the thickness of the A-pillar to improve the view will have structural implications. In a similar way, the styling contours of the body shell and trim can obscure part of the view in the exterior mirror.

Vision therefore needs to be considered at an early stage in the design process, since any subsequent modifications can be difficult and expensive. The evaluation techniques which have been described are suited to the various stages in this design process, although for a full evaluation the body profiles need to be closely defined.

However, the existing criteria for such evaluations could be said to have been developed from 'good practice' within the industry, and for most road vehicles it seems to have been achieved relatively successfully. The legislative criteria are limited to minimum specifications for safe fields of view and very much simplify the complex needs of driver vision. Other ergonomic questions may arise during the design and development of a new model where no criteria are yet available.

Mirror vision

From the car driver's point of view, direct vision is usually good and mirror vision has improved noticeably over recent years. The remaining blind spot seems to be an off-side area close behind the car where passing vehicles are not seen, and the problem occurs mainly for small drivers who adjust the seat well forward. In this position, the mirror has to be turned further inwards to give the driver a view along the side of the vehicle and thus loses some field at the outer mirror edge.

The mirror view gives the driver information about other vehicles' speeds as well as about their positions on the road, which is important when considering appropriate mirror convexity. There has been a debate over the use of convex mirrors, since the more highly convex the mirror the smaller the image and the greater the risk that the driver will underestimate the distance or speed of the approaching vehicle. Drivers learn to interpret the image presented, although the period of adjustment may take some time when changing to a vehicle with a mirror of very different convexity, and it is advisable to standardize the convexity used for rear view mirrors, at least within limited ranges.

Manufacturing techniques now make it possible to design mirrors with varying surface curvatures. Such non-spherical curved mirrors are capable of displaying a wider view of the road, which eliminates the blind spot at the side of the vehicle, but the image size changes across the width of the mirror (Bockelmann, 1991). In view of the perceptual implications, it is obviously important to study the use of such mirrors and to evaluate their performance in terms of speed and distance judgments as well as for glance frequency, optical distortion of the image and glare potential. A procedure for evaluation has already been proposed by Burger and Ziedman (1987) which scores performance (as a Figure of Merit) on the basis of measures of driver errors in making position and motion judgments weighted according to their severity and of the effectiveness of the mirror in terms of glance frequency. Such evaluations are complex, since they need to take account of the use of the mirror under a variety of driving conditions, as well as for daytime, night or poor weather uses. Much further research is needed into the capabilities of both variable curvature mirrors and other devices which have been proposed for the development of complete rear view systems.

Buses and trucks

In buses and trucks, blind spots occur because the drivers are higher above the road, making it more difficult to see areas close to the vehicle without using larger mirrors (which increase the obscuration of the forward view, tend to be less aerodynamic and have greater problems of stability and vibration). In addition, the view to the rear is blocked by the long passenger compartment in the bus and by the load carried on the truck.

The problems of truck drivers trying to reverse were graphically described by Tait and Brennan (1980) who reported several accidents in which car drivers sat powerless to stop commercial vehicles reversing into them, while one refuse truck found maneuvering so difficult that the driver eventually ignored all obstacles in his way and drove directly out of the street, uprooting a lamp-post and damaging both a parked car and a garden wall in the process. Zlotnicki et al. (1980) in a survey of serious-injury HGV accidents found that the driver's view was a major factor in many accidents and that two blind spots occurred, one on the nearside just behind the front axle and the other at the rear of the vehicle.

Additional mirrors may be needed to provide views of these areas. An investigation of supplementary mirrors viewing the rear of an articulated truck during trailer swing is described by Case et al. (1980). Other solutions which have been suggested are telescopic devices or a camera mounted on the rear of the vehicle, as found on buses in Japan. Current techniques can easily be used to assess field of view in the additional mirrors, but new criteria are needed to define satisfactory performance.

Off-road vehicles

Off-road vehicles present more serious problems which have not been adequately covered by any existing standards. Hulbert (1980) describes a study of mine haulage trucks, and quotes the experience of some of the drivers:

> Driver confidence is initially rather low because they know that they cannot see in front, to the right and directly to the rear. However, drivers soon grow more confident of their truck handling and tend to forget the fact that they are often 'driving blind' ... .

As Hulbert reports, drivers could run over workers and equipment which were hidden in such 'blind' areas. A similar study was described by Chan et al. (1985). Their underground mobile roadheading (mining) machinery also worked in a very restricted space, and needed views of areas very close to the vehicle where other human operators might be working. They developed a technique for defining the available sightlines on an unrolled cylinder representing the 360 degree field of view, but further research on this type of equipment is needed to establish criteria for the minimum views needed in the different working zones.

Similar studies may well be needed for many types of industrial vehicles and they must take into account the operating conditions, which vary considerably, as illustrated by the work of Hella et al. (1988, 1991) in defining the visual needs of drivers of forklift trucks. They considered the various activities involved in driving, loading and unloading, and also found it necessary to study head movements in order to determine the degree to which a driver should be expected to turn to increase the field of view.

Improving vision for vehicle occupants

Such considerations are important for safe driving, but until now most emphasis has been placed on meeting minimum standards. Vision is the primary source of information in driving, and additional views obtained through either direct or mirror views could well assist by making the driver's task easier or less stressful (perhaps by reducing the need for concentration). A simple example is the question of what direct view is desirable through the side windows of a vehicle. There is no guidance on this either from standards or from the literature, but the side view probably supplements the mirror view of the rear quarter of the vehicle and certainly supplies part of the information on optical flow through peripheral vision. In addition to assessing requirements for direct side or rear view, there is a further problem in evaluating these requirements since there is very little information on head postures or eye locations of drivers when looking in these directions.

Passenger vision and comfort has scarcely been considered. There is some evidence that large windows giving good views for drivers may be less comfortable for front passengers sitting close to them, who may feel insecure or exposed, while headrests on front occupant seats severely restrict the view of rear seat passengers making them uncomfortable and claustrophobic. Several such visual aspects of vehicle design need further study.

One further consideration is that the most extensive studies of driver vision were made 20 years ago. Vehicle design, traffic density and behavior, and the road environment have all changed substantially in that time. The criteria used to assess the driver's field of view in current vehicles may well require some revision, and a review of the driver's task and relevant vehicle design parameters would be desirable.

Conclusions

This chapter has concentrated on the translation of drivers' visual needs into specifications for vehicle design. Evaluation techniques have been developed for the various stages of design and development, from theoretical mathematical analyses or computer evaluations on CAD systems to the initial model or prototype trials in the laboratory or design studio, and finally to standardized legislative tests for type approval or product compliance.

The geometrical relationships for direct and mirror-reflected fields of view are very well defined, but anyone engaged in such an evaluations quickly realizes that relatively small details of the vehicle structure and assembly can have a considerable effect on the visual field provided. As a result, simplified models of body structures are not adequate for accurate evaluation in computer-based studies—the full CAD definition of body panels would normally be required. Similarly, production compliance usually requires a laboratory measurement of the field of view on the vehicle. So, although theoretical analyses in CAD and computer facilities will considerably speed up the design process, they are unlikely to completely replace practical tests for evaluating visual requirements in vehicle design.

The test techniques are well developed, but evaluations should be based upon a thorough understanding of the task requirements and environmental influences. The use of visual information by the driver is a very complex human process, and much remains to be learned about the driving task. Current criteria appear to provide adequate standards for most aspects of the visual requirements for mass-produced road vehicles (some gaps having been noted above), but less is known about the needs of drivers of other more specialized types of vehicles, and all criteria may well develop further as research into driver vision progresses.

References

Bockelmann, W. D., 1991, Aspheric rear-view mirrors enhancing the safety of drivers, in Gale, A. G., Brown, I. D., Haslegrave, C. M., Moorhead, I. and Taylor, S. (Eds), *Vison in Vehicles*, Vol. III, pp. 399–408, North Holland: Amsterdam.

Burger, W. J. and Ziedman, D. A., 1987, Development of a uniform procedure for the evaluation of rearview systems, SAE Paper 870348, in *Automotive Crash Avoidance Research*, SAE SP-699, pp. 1–8, New York: Society of Automotive Engineers.

Case, K., Porter, J. M., Bonney, M. C. and Levis, J., 1980, Design of mirror systems for commercial vehicles, *Applied Ergonomics*, **11**, 4, 199–206.

Chan, W. L., Pethick, A. J., Collier, S. G., Mason, S., Graveling, R. A., Rushworth, A. M. and Simpson, G. C., 1985, Ergonomic principles in the design of underground development machines, IOM Report No. TM/85/11 on CEC Contract 7247/12/007, Edinburgh: Institute of Occupational Medicine.

Connolly, P. L. 1964, Human factors in rear vision, in *Design Aspects for Rear Vision in Motor Vehicles*, SAE SP-253, pp. 1–14, New York: Society of Automotive Engineers.

Devlin, W. A., and Pajas, M. R., 1968, Computer assisted packaging for driver's rear viewing, SAE Paper 680106, New York: Society of Automotive Engineers.

Devlin, W. A., and Pajas, M. R., 1969, Ford's computers aided design of driver's rear view package, *SAE Journal*, **77**, 6, 42–48.

Devlin, W. A. and Roe, R. W., 1968, The Eyellipse and considerations in the driver's forward field of view, SAE Paper 680105, New York: Society of Automotive Engineers.

Haslegrave, C. M., 1979, Measurement of eye heights of British car drivers above the road surface, TRRL Supplementary Report 464, Crowthorne: Transport and Road Research Laboratory.

Hella, F., Tisserand, M. and Schouller, J. F., 1988, Visibility requirements for the driver's stand of lift trucks, *Applied Ergonomics*, **19**, 3, 225–32.

Hella, F., Tisserand, M. and Schouller, J. F., 1991, Analysis of eye movements in different tasks related to the use of lift trucks, *Applied Ergonomics*, **22**, 2, 101–10.

Hulbert, S. F., 1980, A four-pronged human factors attack on mine truck accidents, in Oborne, D. J. and Levis, J. A. (Eds), *Human Factors in Transport Research*, pp. 178–84, London: Academic Press.

ISO, 1980, ISO 6549-1980E: Road vehicles—procedure for H-point determination, Geneva: International Standards Organization.

SAE, 1967, SAE Information Report J985, Vision factors considerations in rear view mirror design, New York: Society of Automotive Engineers.

SAE, 1969, SAE Eyellipse adapted for use when driver's head turns, SAE Paper 691000, *SAE Journal*, **77**, 5, 39–41.

SAE, 1973a, SAE Recommended Practice J182a, Motor vehicle fiducial marks, New York: Society of Automotive Engineers.

SAE, 1973b, Field of view from automotive vehicles, SAE SP-381, New York: Society of Automotive Engineers.

SAE, 1977, SAE Recommended practice J1050a, Describing and measuring the driver's field of view, New York: Society of Automotive Engineers.

SAE, 1985, SAE Recommended Practice J941 OCT85, Motor vehicle driver's eye range, New York: Society of Automotive Engineers.

SAE, 1987, SAE Recommended Practice J826 MAY87, Devices for use in defining and measuring vehicle seating accommodation, New York: Society of Automotive Engineers.

Tait, M. J. and Brennan, M., 1980, Characteristics of road accidents involving medium goods vehicles, in Oborne, D. J. and Levis, J. A. (Eds) *Human Factors in Transport Research*, pp. 160–8, London: Academic Press.

Zlotnicki, J., Hutchinson, T. P. and Kendall, D. L., 1980, Some problems and prospects with commercial vehicle safety, illustrated by case reports of accidents involving ergonomic factors, in Oborne, D. J. and Levis, J. A. (Eds), *Human Factors in Transport Research*, pp. 151–9, London: Academic Press.

5

Automotive seat design for sitting comfort

Herbert M. Reynolds

Introduction

The automobile's evolution as a universal means of personal transportation has focused seat design attention upon the occupant's comfort and health. However, the growth of an international automotive market has increased diversity in seat design. As a result, unique, but functionally equivalent, seats are built to satisfy similar design goals. Ideally, good seat design should apply knowledge of human sitting posture to accommodate occupant preference in vehicle seating. This chapter will, therefore, review current concepts of sitting posture for the design of comfortable automotive seats.

Ergonomic analysis of automotive seats

Much diversity in seat design is due to the need to match seat and vehicle purpose. There are three different motor vehicles used today for personal transportation:

1 family and personal business sedans,
2 minivan and off-road vehicles, and
3 sports cars.

Seats within these categories can be divided into performance or touring categories. In general, performance seats are stiffer with more contour (e.g., a bucket type design) than touring seats (e.g., 60/40 split seat design). Performance seats often have additional adjustable features such as lateral cushion and back bolsters. Thus, different seats are needed and the marketplace demands seats that vary between manufacturers, vehicle models, and vehicle function.

Independent of the vehicle, a task analysis reveals three different occupants in the vehicle: driver, front-seat passenger, and rear-seat passenger. There are four design criteria for a driver's seat.

1 The seat should *position the driver* with unobstructed vision and within reach of all vehicle controls;

2 The seat must *accommodate the driver's size and shape*;
3 The seat should be *comfortable for extended periods*;
4 The seat should provide a *safe zone* for the driver in a crash.

Passengers in the front and rear seats need comfortable supporting surfaces for a variety of postures unconstrained by vehicle operation. However, the front-seat passenger, unlike the back-seat passenger, may assist the driver in operating controls or providing visual assistance (Matasuoka and Hanai, 1988). Thus, design criteria for front and rear seat passengers should be carefully considered.

The seat, as an interface between occupant and vehicle, limits sitting postures available to each vehicle occupant. Adjustments to location and contour customize the seat for the occupant's size and preference. In general, these adjustments are for accommodation. To seat an occupant comfortably, however, demands correct position and support for a posture appropriate to the task.

Comfort describes an occupant's empirical perception of being at ease. Hertzberg (1972) defined comfort as the '... absence of discomfort'. Thus, the occupant is comfortable or does not perceive any discomfort. The perception of comfort may change over time. Reed *et al.* (1991) recently found that initial and final evaluations of seat comfort do not agree. The occupant has little control over seat parameters that contribute to comfort. If the seat is highly contoured or soft, changing position is difficult. Consequently, relief of discomfort from pressure or joint position may be impractical. Thus, comfortable seats require rigorous application of data on human geometry and long-term sitting effects on the occupant.

In addition to comfort, the health and safety of the occupant are also affected by seat design. Lower back pain and long periods of driving are associated (Kelsey *et al.*, 1984; Heliövaara, 1987). Previously, Troup (1978) identified postural stress, vibration, muscular effort, and impact and shock as the cause of backache and lower back pain in drivers. Postural stress is exposure to long-term sitting in the same position. Vibrations are transmitted from the car through the seat. Muscular effort contributes to occupant fatigue and discomfort through poor seat design and occupant packaging. Impact and shock are road hazards. In these factors, exposure is the critical parameter.

The seat is structurally an integral part of the occupant restraint system. Geometry for the head restraint is defined by a federal standard. The belts, shoulder and lap, have been attached to inertial reels that permit occupants to change their position. The evaluation of automotive seat comfort needs to consider the restraint system.

For example, the belt, crossing an occupant's body, follows a minimum path principle (Searle, 1974). Belt path is controlled by attachment locations and the shape of the seated occupant's body. Some occupants find the path across their neck and shoulder (States *et al.*, 1987) or over their pelvis (Sato, 1987) uncomfortable. A review of accident data shows that some occupants do not wear seat and shoulder-belts or wear them improperly. Thus, seat design includes restraint parameters and their effects on seat comfort should be evaluated. In summary, safety and comfort require appropriately designed environments, informed and cooperative occupants, and a systems approach to design and evaluation of seating comfort.

Human geometry for seat design

Static and dynamic geometries of the human body are used in seat design. Static geometry is obtained from anthropometric studies of the population. Dynamic geometry is obtained with a mechanical model of the human body. That is, mass-links are positioned to represent seated posture. Thus, static geometry describes the physical size to be accommodated in the seat. Dynamic geometry describes the functional position to be accommodated in the seat.

Body size

Seats are, theoretically, designed to fit at least 90 per cent of the population from small to large body sizes. A small female has some dimensions less than or equal to the fifth percentile. A large male has some dimensions larger than or equal to the 95th percentile. The range between the small female and large male approximates the adjustments needed in seating to accommodate anthropometric differences in body size.

Body size (see Kroemer, this volume) is defined primarily by the distributions of standing height and body weight. Stature and body weight are weakly correlated with coefficients that range from 0.246 to 0.460 for different age and sex groups (Abraham *et al.*, 1979). However, stature and weight are highly correlated with body height and breadth dimensions, respectively. Thus, distributions of stature and weight represent anthropometric variation within the population.

Since civilian, anthropometric data do not exist for ergonomical and biomechanical applications, military surveys are used. Military populations are more homogeneous than the general non-institutionalized civilian population. Comparison of stature, sitting height and body weight is made in Table 5.1 between the most recent United States Army (Gordon *et al.*, 1989) and civilian surveys (Najjar and Rowland, 1987). The fifth percentile female and 95th percentile male in the civilian data are larger and smaller respectively than the Army values by approximately 2 per cent. Censoring at the extremes reduces the range between fifth and 95th percentiles but increases correlations between dimensions. However, since the Army approximates the civilian population, anthropometric data from the Army will be used in this chapter.

Human linkage system

Design engineers utilize rigid body segments connected at joint centers to represent human body posture for equipment design. In a simple model of the body, size is described by link length. Mechanical links are straight lines between joint centers (Dempster, 1955). The locations and orientations of rigid bodies are specified by joint center position and the angle between adjacent links. A 22-link model of the body is defined in Table 5.2.

The torso is divided, typically, into five body segments: head, neck, chest, abdomen and pelvis (McConville *et al.*, 1980; Young *et al.*, 1983). However, the neck and abdomen are composed of multiple motion segments. These motion segments must be reduced for simple, rigid body models. Two lumbar

Table 5.1. Comparison of 1976–80 US general population and 1987–88 US army body sizes

	Female		Male		
	5th percentile	Avg. (SD)	Avg. (SD)	50th percentile	95th percentile
Stature (mm)					
USA	1509	1618 (66)	1755 (72)	1755	1870
Army (99)[1]	1528	1629 (64)	1756 (67)	1757	1867
Sitting Height (mm)					
USA	800	861 (37)	922 (38)	923	983
Army (93)	795	852 (35)	914 (37)	914	972
Weight (kg)					
USA	47.7	65.4 (15)	78.1 (13)	76.9	102.7
Army (124)	49.6	62.0 (08)	78.5 (11)	77.7	98.1

Note: [1] Numbers in parentheses are index to dimensions in Army Survey (Gordon *et al.*, 1989)

links, associated with the abdomen body segment, are sufficient to represent the curvature of the low back (Kohara and Sugi, 1972; Snyder *et al.*, 1972). The neck also may be reduced to two links. Therefore, segment volumes in Table 5.3 should be divided equally between the superior and inferior links of these bodies. With these modifications, the head, neck, chest, abdomen and pelvis can be positioned realistically to represent slumped and erect torso postures. Thus, the torso is composed of seven instead of five links.

Table 5.2. Linkage model for the seated human body

Body Segment	Definition	No. of Links
Head link	Eye–OC/C1	1
Neck links	OC/C1–C2/C3	1
	C2/C3–C7/T1	1
Chest links	C7/T1–Shoulder joint	2
	C7/T1–T12/L1	1
Lumbar links	T12/L1–L2/L3	1
	L2/L3–L5/S1	1
Pelvis links	L5/S1–Hip Joint	2
Arm links	Shoulder–Elbow	2
	Elbow–Wrist	2
	Wrist–Hand	2
Leg links	Hip–Knee	2
	Knee–Ankle	2
	Ankle–Heel Point	2
TOTAL		22

Table 5.3. Linkage anthropometry

Dimension	Lengths (mm) Female 5th percentile	Male 50th percentile	Male 95th percentile	Range (95th–5th percentile)	Volume (cc) Female Avg. (SD)	Volume (cc) Male Avg. (SD)
Head Link (D20)[1]	52.0	63.0	69.0	17.0	4369 (295)	3894 (267)
Neck Links:						
Total (D25)	80.0	106.0	123.0	43.0	1043 (182)	737 (122)
OC/C1–C3/C4	NDA[2]	NDA	NDA	NDA		
C3/C4–C7/T1	NDA	65.0	NDA			
Torso:						
Thorax Links:						
C7/T1–T12/L1 (D38)	330.0	398.0	434.0	104.0	24 385 (4749)	18 175 (3567)
C7–Shoulder Jt	181.0	212.0	227.0	46.0		
Abdomen Links:						
Total (D1)	32.0	47.0	75.0	43.0	2339 (693)	2817 (1465)
T12/L1–L2/L3[3]	NDA	74.0	NDA			
L2/L3–L5/S1[3]	NDA	79.0	NDA			
Pelvic Links:						
Ilium–H-point (D29)	102.0	145.0	171.0	69.0	11 390 (2565)	10 128 (3250)
L5/S1–H-point[4]	86.0	89.0	96.0			
Arm Links:						
Upper Arm (4)	285.0	341.0	369.0	84.0	1955 (389)	1557 (351)
Forearm (87)	219.0	269.0	297.0	7.8	1396 (247)	935 (194)
Hand (125)	59.0	69.0	78.0	1.9	512 (69)	344 (48)
Leg Links:						
Thigh (D37)	363.0	426.0	469.0	10.6	9884 (1690)	10 070 (2136)
Calf (D6)	363.0	433.0	476.0	11.3	3848 (629)	3111 (607)
Foot (75)	52.0	67.0	76.0	2.4	976 (155)	673 (103)

Notes: [1] Index to dimensions in Gordon et al., (1989).
[2] NDA = No Data Available.
[3] Estimated from Reynolds (1991).
[4] Estimated from Reynolds et al., (1981).

Links to attach shoulder and hip to the chest and pelvis are made from the spinal column. For convenience, the right and left shoulders are attached to the spinal column at C7/T1. Also, the right and left hip joints are attached at the lumbosacral joint (L5/S1). The segmental volumes, however, are reported for the pelvis as defined from the iliac crest to the hip joint. Thus, these data are inconsistent with the model at the junction of the spinal column and pelvis.

Approximate lengths and volumes of these links are given in Table 5.3. Arm and leg link lengths were measured in the Army anthropometric survey (Gordon et al., 1989). Torso lengths have been derived from numerous sources (Gordon et al., 1989; Snyder et al., 1972; Reynolds, 1991; Reynolds et al., 1981).

The ischial tuberosities at the base of the pelvis transmit most of the torso weight to the seat cushion. A line passing through the contact points between seat and ischial tuberosities defines the torso axis of rotation. This axis also locates the largest cushion deflection, D-point in the SAE literature.

Table 5.4 reports data from Reynolds et al., (1981) on the three-dimensional locations of the anterior superior iliac spine (ASIS), H-point (hip joint), ischial tuberosity (Ischiale) and posterior superior iliac spine (PSIS). Coordinates are given in a right-handed, anatomical axis system defined by the right and left anterior superior iliac spines and pubic symphysis. The origin lies between the right and left anterior superior iliac spines. The axis, normal to the plane defined by the three points, is $+x$. The axis through the left anterior superior iliac spine is $+y$. The superior direction is $+z$.

The anterior superior iliac spines (ASIS in Table 5.4) are structures at the front of the pelvis. The seat-belt should lie beneath these points so it holds the occupant's pelvis in position on impact.

H-point is the midpoint of the hip joint. Thus, it represents a center of rotation between the thigh and pelvis.

Ischiale approximates the midpoint of the ischial tuberosity of the pelvis. It is the center of rotation for the seated torso. Ischiale also approximates the positions of highest pressure and largest seat cushion deflection.

The posterior superior iliac spines (PSIS in Table 5.4) are two points at the back of the pelvis. These points contact the seat back. The lumbar support should not be located beneath these points because it would then be pushing on the sacrum.

Position of the body

The successful performance of any task is facilitated by optimal interaction between task geometry and operator position. The driver's seat position is constrained by vision and reach. The power of these constraints influences the human operator to select the same seat position in repeated trials (Verriest and Alonzo, 1986). Fore-aft and vertical adjustments in seat position are necessary for operators of different body sizes to accommodate a constant vision and reach geometry. With the increased use of split and bucket seats, back angle adjustments are also available in current vehicles. Thus, three adjustments should allow each occupant to define an optimal seat position.

Table 5.4. *Three dimensional coordinates in mm of pelvic geometry for small females (SF), average males (AM), and large males (LM)*

Landmark		X Avg. (SD)	Y Avg. (SD)	Z Avg. (SD)
ASIS—Right:	SF	0 (0)	−109 (12)	0 (0)
	AM	0 (0)	−113 (9)	0 (0)
	LM	0 (0)	−115 (8)	0 (0)
ASIS—Left:	SF	0 (0)	109 (12)	0 (0)
	AM	0 (0)	117 (8)	0 (0)
	LM	0 (0)	115 (8)	0 (0)
H-point—Right:	SF	−44 (5)	−76 (25)	−47 (65)
	AM	−48 (6)	−82 (5)	−66 (7)
	LM	−50 (6)	−86 (5)	−67 (7)
H-point—Left:	SF	−43 (5)	81 (4)	−59 (7)
	AM	−48 (5)	83 (5)	−65 (7)
	LM	−52 (10)	78 (39)	−69 (20)
Ischiale—Right:	SF	−87 (9)	−66 (7)	−102 (14)
	AM	−96 (9)	−61 (9)	−119 (11)
	LM	−100 (8)	−62 (7)	−120 (9)
Ischiale—Left:	SF	−87 (9)	66 (5)	−100 (12)
	AM	−96 (7)	60 (5)	−118 (9)
	LM	−99 (7)	63 (6)	−120 (9)
PSIS—Right:	SF	−120 (11)	−40 (8)	34 (17)
	AM	−134 (9)	−36 (8)	31 (16)
	LM	−137 (11)	−37 (7)	35 (13)
PSIS—Left:	SF	−119 (11)	38 (6)	32 (17)
	AM	−134 (10)	37 (5)	29 (16)
	LM	−139 (10)	38 (7)	32 (11)

(Source: Reynolds *et al.*, 1981)

Occupant position in the driving position is modeled by a set of design tools (see Roe, this volume) developed by the Society of Automotive Engineering through study of eye, hand and foot positions. These tools, however, encapsulated the occupant's response to old seat designs. When new seats are designed, the tools are inaccurate. For example, the addition of lumbar support in the seat back moved the eye outside the SAE eyellipse (Reynolds and Hubbard, 1986). The SAE drafting template and the H-point machine (SAE J826) are also based upon data obtained three decades ago. Tools must represent anatomical geometry in good sitting posture. Additionally, tools also should predict the postural response to a seat. The vehicle and seats should be designed for optimum occupant postures rather than positions of adaptable operators.

Posture of the body

In an automotive sitting posture, the seat cushion, back and floor board support 65, 10 and 25 per cent of the body weight respectively (Varterasian and Thompson, 1978). The pelvis and back transmit body weight to the seat through soft and hard tissues of the body. As a result, the arms and legs can move freely, and the head can be positioned for optimal vision and comfort. Sitting posture, however, is determined by joint angles that are constrained by seat–occupant interaction.

A mid-range position of joints optimizes stress of passive and active tissues about the joints (Keegan, 1953). Given appropriate seat position, the elbow, wrist, knee and ankle joints can be positioned optimally. The intervertebral joint angles in the neck and low back depend on head, chest and pelvis positions. Joint angles in the torso are affected by many seat parameters. However, the automotive seat may provide support that moves the body away from optimal joint positions. For example, a low seat height and reclined back angle flexes the hip joint and rotates the pelvis rearwards to flatten the lumbar spine. Thus, seat position must also incorporate appropriate seat support for optimal joint angles.

A lordotic torso posture subjects the intervertebral disc to lower pressure than a slumped, kyphotic torso posture. Nachemson and Morris (1964) measured a 30 per cent reduction in intervertebral disc pressure from slumped sitting without a backrest to erect standing. Disc pressure increases in slumped sitting, partially, because torso weight is transmitted through the intervertebral disc rather than distributed between disc and facets. As torso weight is increasingly supported by a seat back with lumbar support and an arm rest, disc pressure decreases (Andersson *et al.*, 1975).

Despite lower disc pressures in lordotic than kyphotic postures. Adams and Hutton (1985) identified three advantages to a flexed spine.

1 Stress is reduced on apophyseal joint ligaments and posterior annulus;
2 Disc metabolites are transported;
3 Compressive strength is high.

Therefore, sitting should be dynamic. Occupants should change their spinal posture rather than be fixed in a predetermined 'best' spinal geometry.

In addition to changes in torso posture while sitting, the seat must accommodate structural variation in lordosis. Variation in the population arises from differences in genes and age. The shape and size of the human back is highly variable (Branton, 1984). Flatter and more kyphotic spinal curves are observed in both sexes in the sixth decade of life (Milne and Lauder, 1974). This change in curvature with age includes osteoporosis in women. Thus, highly contoured seat backs will not 'fit' many people.

Furthermore, the lordotic curve decreases with short hamstring lengths. When the knee is extended from 90 to 135 degrees, the hamstrings are stretched. The hamstrings pull on the back of the pelvis, applying a moment that rotates the pelvis rearward. As the pelvis moves rearward, the coupled motions between sacrum and lumbar vertebrae flatten the lumbar curvature (Stokes and Abery, 1980).

Spinal curvature changes are voluntarily controlled by muscular effort. For example, a lordotic spine requires muscular effort to rotate the pelvis forward. Activity of the psoas muscle stabilizes the lumbar-pelvis-femur linkage system, increases lordosis, and flexes the thigh. However, when the psoas muscle contracts, disc pressure increases (Nachemson, 1968). Therefore, the seat should support the pelvis so minimal effort is needed by the iliopsoas to flex the hip and stabilize the pelvis.

In conclusion, the seat should reduce postural stress and optimize muscular effort. Comfortable support for many postures is essential (Åkerblom, 1948). However, the occupant can relieve postural stress and muscular effort by voluntarily changing position and limiting seated time to an hour (Bendix, 1986).

Vibration and ride comfort

The occupant's perception of ride comfort is based upon road shock, impact and vibration transmitted through the automotive seat. Ride comfort has been associated with vibration since the 1940s (Lay and Fisher, 1940; Paton et al., 1940). In the past five decades, exposure to vibration has been linked to low back pain (Troup, 1978). Thus, the seat design also must consider the vehicle suspension system and the vibration transmitted to the seated occupant.

In a laboratory investigation, Wilder et al. (1982) measured postural effects on the natural resonant frequencies of the torso. Of three resonant frequencies measured in males/females at 4.9/4.75, 9.5 and 12.7 Hz, the first frequency had the highest transmissibility. Although the body's response may be changed by posture, the highest frequencies and transmissibility remained within 4–8 Hz. Pope et al. (1987) measured differences between sinusoidal vibration and impact in erect and slumped seated postures. Again, effects remained within the 2–4 Hz, 4–8 Hz, and 8–16 Hz ranges.

In water and land vehicles, Wilder et al. (1982) reported peak frequencies to range from 3.25 to 7.50 Hz and accelerations from 0.88 to 25.64 m/sec^2. On a test track, Hosea et al. (1986) measured peak frequencies in an automobile at 6.8 and 12.8 Hz with accelerations at 0.011 and 0.012 g. Thus, the human body may be exposed to harmful vibrations in the automotive environment.

Geometric features of seat design

Seat design can be divided into accommodation and comfort requirements. Accommodation refers to seat size and adjustments for horizontal distance from controls, height and back angle. Comfort, however, refers to stiffness, contour, climate, memory and vehicle features that promote occupant comfort.

Accommodation

Cushion size accommodates the seated occupant's buttock and thigh dimensions. The distance from the buttock to popliteal region delimits the loaded cushion's length from seat back to the waterfall line. In Table 5.5, the length for the fifth percentile female is 440 mm which is comparable to previous recommendations of 432 mm (Keegan, 1964) and 440–550 mm (Grandjean, 1980). For cushions with adjustable thigh support, the length should not increase more than 105 mm.

Table 5.5. *5th percentile female, 50th and 95th percentile male values for selected anthropometric size dimensions (mm)*

Dimension	Female 5th percentile	Male 50th percentile	Male 95th percentile	Range (95th–5th)
Sitting Heights				
Top of Head (93)	795.0	914.0	972.0	177.0
Eye Height (49)	685.0	792.0	848.0	163.0
Body Breadths				
Shoulder (12)	397.0	491.0	535.0	138.0
Biacromial (10)	333.0	397.0	426.0	93.0
Chest (32)	250.0	319.0	367.0	117.0
Waist (112)	249.0	307.0	359.0	110.0
Hip (66)	343.0	365.0	432.0^2	89.0
Body Lengths				
Butt-Pop L (27)	440.0	500.0	545.0	105.0
Shoulder H (3)	509.0	598.0	646.0	137.0

Notes: [1] Numbers in parentheses are index to dimensions in Army Survey (Gordon et al., 1989).
[2] 95th percentile female value.

Lateral space is important for physical and psychological comfort. In a bench or split seat, breadth is primarily for psychological comfort. In a bucket seat, however, bolstering and lateral contours must accommodate the physical dimensions of the large torso.

Female hip breadth is greater than male hip breadth. Thus, the 95th percentile female hip breadth determines cushion breadth. In Table 5.5, the 95th percentile hip breadth is 432 mm. Grandjean's (1980) recommendation of 480 mm has been adjusted for clothing and leg splay.

The *seat back* supports the trunk while sitting, but it also must be considered a barrier to arm reach and vision. Seat back height is, therefore, determined by the small female, sitting shoulder height. Thus 509 mm is recommended. In comparison, Grandjean (1980) recommended 500 mm.

Seat back breadth may be divided into lower and upper regions. Large torso breadths at the hip, waist, and chest determine the lower space requirements. The lower region must accommodate a tapered shape from 432 mm at the hip to 367 mm at the chest (Table 5.5).

The upper region of the seat back should accommodate the static and dynamic breadth of the shoulder. Movement of the arm should not be constrained at the shoulder joint. The distance between the shoulder joints, biacromial breadth, is 426 mm. Total breadth is 515 mm when 89 mm is added for the deltoid muscle. Grandjean (1980), in contrast, recommends 480 mm for seat back breadth. For occupant lateral space, the larger dimension is appropriate. For mobility of the shoulder joint, a smaller breadth may be used.

Horizontal, vertical and *back angle adjustments* are needed in all seats. Other adjustments customize occupant comfort. The range of movement for each adjustment can be estimated from a theoretical linkage model of the human body. Experimental data on occupant preference is needed to confirm the theoretical estimates, however.

Horizontal adjustments accommodate differences in leg length that are associated with seat height and preferred knee angle. Grandjean (1980) recommends a minimum of 150 mm horizontal adjustment, and Rebiffé (1969) computed, from a biomechanical model of the body, the same distance. Comfortable joint angles in the automobile are typically between 95 and 120 degrees for the hip, and 95 and 135 degrees for the knee (Rebiffé, 1969).

Horizontal seat travel is a function of seat height and body size. Schneider *et al.* (1980) found greater fore-aft seat travel was needed after driving than expected from a linear model of stature and seat position. Average seat travel in six vehicles investigated was 148.6 mm. In five of six vehicles, additional travel was desired but the amount of travel depended on seat height.

Vertical adjustments accommodate differences in sitting eye height between the fifth percentile female and 95th percentile male. A simple trigonometric relationship can be established with link lengths and joint angles to compute the amount of seat adjustment needed in the vertical direction.

Adjusting a flat, non-deformable surface over a range of 163 mm (Table 5.5 sitting eye height range) maintains a constant eye height. However, small, lightweight females sit higher than large males. Seat cushion compression and suspension deflection are non-linear functions of the applied force. As a result, the vertical displacements needed in a soft seat are poorly calculated from anthropometric data. Adding to the vertical adjustment is the effect of an adjustable back angle. Thus, the range of adjustment in seat height must be determined empirically by seat cushion stiffness, back angle, and occupant body size. Additional adjustment is needed to accommodate the effects of seat age and wear.

Keegan (1964) recommended a maximum seat height of 406 mm but no minimum. Rebiffé (1969) defined seat height for vision requirements in the vehicle. Grandjean (1980), however, recommended a seat height between 250 and 300 mm. Thus, seat height is important for vision geometry and optimum seated posture in the vehicle.

Back angle adjustments accommodate differences in arm length and occupant preferred hip angle (Hallen, 1978). Typically, the seat back is reclined 15–25 degrees from vertical. Thus, the range of adjustment should be at least 10 degrees (Rebiffé, 1969). Increasing the seat angle decreases pressure on the disc (Andersson *et al.*, 1975). However, seat back angle has only a slight effect on lumbar lordosis (Andersson *et al.*, 1979). The seat back angle provides another accommodation parameter and changes the occupant's position.

The seat back pivot should be at the ischial tuberosities (Andersson *et al.*, 1979). Theoretically, this location also defines the largest deflection of the seat, D-point. If the motions of the seat and occupant's back are matched, shear forces between the occupant and back will be reduced. Thus, the seat back pivot point should be located approximately at D-point.

Grandjean (1980) recommends a seat cushion angle of 19 degrees with a range from 10 to 22 degrees. Recommendations for optimal seat back orientation range from 105 (Keegan, 1964) to 120 degrees (Hosea *et al.*, 1986) with a lumbar support. Experimental investigations indicate that the orientation of the cushion and back should be adjustable. The range of adjustment for driver and front-seat passenger should keep the occupant within safe boundaries. Thus, the front-seat passenger's back should not recline more than the driver's.

Seat comfort

Seat comfort is static and dynamic. Stiffness of the cushion and back, contour and climate (Diebschlag et al., 1988) primarily determine static seat comfort. Adjustments by the occupant to change position also contribute to comfort. Cushion and back stiffnesses primarily control dynamic seat comfort.

Stiffness

Cushion stiffness affects the vibrations transmitted to the occupant and pressure distribution in the seat. As stiffness of the seat increases, higher frequencies are transmitted. As stiffness of the seat decreases, pressure is distributed over a greater area. Thus, the occupant's evaluation of dynamic and static comfort is affected by cushion stiffness.

Neither the seat nor the body has a uniform, homogeneous stiffness coefficient. The body is composed of soft and hard tissues that have different load-deflection characteristics. The seat is also composed of different materials. For example, upholstery, foam, springs and internal metal structures contribute to seat stiffness. As a result, pressure distribution represents a complex interaction between occupant and seat.

Pressure distribution on hard, flat surfaces (Rosemeyer and Pförringer, 1979) clearly outlines anatomical structures. When the seat backrest angle is between 90 and 110 degrees, the highest pressure is at the ischial tuberosities (Rosemeyer and Pförringer, 1979). Thus, on a hard flat surface, pressure decreases from the ischial tuberosities. On a cushioned seat, pressure should be distributed in a similar pattern.

The physiological consequence of high pressure on a capillary bed is reduced blood flow. Thus, Diebschlag et al. (1988) recommended a maximum pressure under the tuberosities of 3 N/cm^2. However, absolute pressure maxima may be difficult to achieve. The vascularized tissues around the ischia have a non-linear elastic response to pressure that is related to age (Sacks et al. 1985). The relationship between comfort and pressure distribution remains a major research area.

Static comfort indicates that the highest pressures on the seat back should be in the lumbar region (Kamijo et al., 1982). However, long-term automotive sitting suggests that the stiffness of the lumbar support in the seat back must be carefully considered (Reed et al., 1991). Initially, occupants preferred a very stiff, well-defined lumbar support. After sitting for three hours, however, the initial evaluation was reversed.

Pressure distribution in the seat/occupant interface is affected by contour, upholstery seams, foam and suspension. As the contour changes, the distribution of pressure between the seat and body changes. Materials vary throughout today's vehicles, ranging from horsehair to latex foam and covered with vinyl, cloth and leather. Seat stiffness and materials should enhance the occupant's ability to change positions. Seats should be covered with material that provides friction between the occupant and seat. Thus, seats should support a variety of seated postures.

Contour

The contour of the seat cushion and back largely determines posture in the seat. The most important contours in the seat surfaces are defined in the sagittal and transverse planes. Loaded and unloaded contours also must be considered, but the occupant's perception of seated comfort is made while sitting in the seat.

The deflection of the seat surface from unloaded to loaded contour affects pressure distribution in the seat. Thus, the change in contour when the occupant sits in the seat is a measure of stiffness in automotive seating. The deflection of D-point describes how the seat contour changes to distribute the pressure of the cushion on the buttocks.

The loaded contour changes. When the occupant changes position, the contour changes. When the occupant changes an adjustment in the seat, the contour changes. For example, a lumbar support changes the sagittal curvature of the seat back. Bolsters in the seat cushion and back change the transverse curves. Thus, the contour should be considered a dynamic parameter in seat design.

The loaded sagittal curvature of the back defines the support provided to the *lumbar* region of the spine. Andersson et al. (1979) proposed that the seat back should have an adjustable lumbar support that extends 40 mm from the seat's back plane. Porter and Norris (1987) found in a laboratory investigation subjects who preferred a lumbar support of 20 mm. In both studies, seat back inclination had less effect on lumbar lordosis than the supporting surface.

Floyd and Roberts (1958) recommended a lumbar curve in office seats to have a radius not less than 300 mm but preferably 400–460 mm. The height of the lumbar support also should lie within a range of 200–330 mm above the seat surface. Grandjean et al. (1973) found experimentally that a radius of 450 mm located 100–140 mm above the loaded seat cushion was preferable in chairs (Grandjean, 1980). Andersson et al. (1979) did not find lumbar support height to affect the lordosis in the back, but the horizontal dimension was important. Porter and Norris (1987) recommended a height adjustment from 195–260 mm from D-point in the loaded seat cushion.

In an investigation of sitting posture for the United States Air Force (Reynolds, 1991), three-dimensional measurements of nine unembalmed cadavers were made. Subjects were seated in a wooden seat with a 105 degree back angle and a 6 degree seat angle. In erect sitting postures with and without lumbar support, the center of lumbar curvature lay between PSIS and T12, 156 to 219 mm above the seat-back intersection. The radius of curvature ranged from 206 to 348 mm. Thus, results from studies of living subjects and cadavers indicate that variation in preference and spinal structure requires an adjustable lumbar height support.

The loaded curve in the back for the *chest* should have a larger radius than the lumbar curve in the seat back. Grandjean (1980) proposed a slight curve with a maximum concavity 500 mm above the depressed seat surface. However, the upper surface should not counteract the occupant's desire to sit upright with a lordotic spine. Thus, support for the lower chest should be considered part of the upper lumbar support. The upper surface nearest the shoulder should be fairly flat. In particular, the design of the upper seat back needs to consider the motions of the shoulder blade. Thus, arm and torso motion should be free and unencumbered by seat back design.

The *cushion* must be contoured and soft at the 'waterfall' under the knee to avoid occlusion of fluids in the leg (Åkerblom, 1954). A flatter surface will raise pressure under the ischial tuberosities and a contoured surface will distribute pressure under the soft tissues. Keegan (1964) recommends a maximum deflection of 75 mm in the seat cushion. A slight deflection will distribute pressure under the soft tissues, and will provide a firm surface on which positions can be changed. Too much deflection restricts position changes and increases muscular effort to change to the new position. Thus, a firm cushion for touring and performance seating is recommended.

Climate

The transfer of water vapor from the skin through clothing and the seat is important for comfortable seating. If moisture accumulates between the occupant and the seat surface, the occupant quickly becomes uncomfortable. Discomfort arises from a) loss of evaporation as a heat transfer mechanism and b) increase in skin friction (Adams *et al.*, 1982). As a result, most seats should be constructed of materials that have high vapor permeability.

Very few studies of the effects of seating materials on vapor permeability have been conducted. The transport of vapor varies by foam and amount of compression. Water vapor permeability in foam increases up to 75–85 per cent compression (Diebschlag *et al.*, 1988). In general, the transfer of vapor through the seat increases as porosity increases. Porosity has two important components: a) diameter of the pore and b) distance between pores, or density. Either method of maximizing porosity will facilitate the transfer of water vapor through the seat. However, the optimal porosity in seating remains a subject for investigation.

In contrast to vapor transfer during the summer months, heated seats are comfortable in northern climates. It is typically a luxury item but it provides significant comfort to the occupant, especially in the winter.

Memory

Memory is a new feature that is usually only found in expensive cars. It is especially helpful when a seat has many adjustments. The occupant can become confused by numerous adjustments and use seat memory to return to an original position. It is sold to assist two drivers of different body sizes, e.g., husband and wife. Thus, if positions are programmed for each, different positions are easily obtained for seat accommodation and comfort.

Vehicle features

Any feature that assists the occupant change position, will increase comfort. Cruise control and tilt steering facilitate postural comfort of the seated occupant. Air conditioning does not affect the occupant's position, but clearly affects climatic comfort in the vehicle.

Conclusion

Predictive knowledge of the human body's linkage system is needed in an engineering design tool. However, the linkage system should be investigated in the context of human performance and preferred postures.

The seat functions as an interface between the occupant and the vehicle. As a result, each seat needs to fit the anthropometric and functional diversity in the population. Highly contoured seats are unable to accommodate differences between people or easily permit a change of position. As a result, firm seats should be designed with gentle contours.

Future investigations should consider the use of the clinical research methodology of controlled trials to evaluate seats. The seat should be studied in the moving vehicle to determine those features that contribute to comfort. In addition, occupant's behavior and personal preference need to be investigated using questionnaires and interview data. These data can assess bias toward the vehicle, i.e. the halo effect, as well as the occupant's knowledge of the seat and its controls. These factors, including weather and purpose of trip, influence the occupant's perception of the seat.

The seat must be considered a structure to support the body rather than a trimmed structure that fits the design of the vehicle. Thus, there needs to be a good balance between the design for task, occupant comfort and safety. In summary, good seat design can only be accomplished by collaboration between science, medicine and engineering.

References

Abraham, S., Johnson, C. L. and Najjar, M. F., 1979, Weight by height and age for adults 18–74 years: United States, 1971–74, Vital and health statistics, Series 11, Data from the National Health Survey, no. 208, Hyattsville, MD: DHEW Publication No. (PHS) 79–1656.

Adams, M. A. and Hutton, W. C., 1985, The effect of posture on the lumbar spine, *Journal of Bone and Joint Surgery*, **67-B**, 625–9.

Adams, T., Steinmetz, M. A., Heisey, S. R., Holmes, K. R. and Greenman, P. E., 1982, Physiologic basis for skin properties in palpatory physical diagnosis. *Journal of the American Osteopathic Association*, **81**, 366–77.

Åkerblom, B., 1948, *Standing and Sitting Posture with special reference to the construction of chairs*, Trans. by A. Synge, Stockholm, A.-B: Nordiska Bokhandeln.

Åkerblom, B., 1954, Chairs and sitting, in Floyd, W. F. and Welford, A. T. (Eds) *Human factors in equipment design*, pp. 29–35. London: H. K. Lewis & Co Ltd.

Andersson, G. B. J., Örtengren, R., Nachemson, A. L., Elfström, G. and Broman, H., 1975, The sitting posture: An electromyographic and discometric study, *Orthopedic Clinics of North America*, **6**, 105–20.

Andersson, G. B. J., Murphy, R. W., Örtengren, R. and Nachemson, A. L., 1979, The influence of backrest inclination and lumbar support on lumbar lordosis, *Spine*, **4**, 52–58.

Bendix, T., 1986, Sitting postures—A review of biomechanic and ergonomic aspects, *Manual Medicine*, **2**, 77–81.

Branton, P., 1984, Backshapes of seated persons—How close can the interface be designed? *Applied Ergonomics*, **15**, 105–107.

Dempster, W. T., 1955, Space requirements of the seated operator: Geometrical, kinematic, and mechanical aspects of the body with special reference to the limbs, WADC Technical Report 55–159, Wright Air Development Center, OH.

Dempster, W. T., Sherr, L. A. and Priest, J. G., 1964, Conversion scales for estimating humeral and femoral lengths and the lengths of functional segments in the limbs of American caucasoid males, *Human Biology*, **36**, 246–62.

Diebschlag, W., Heidinger, F., Kurz, B. and Heiberger, R., 1988, Recommendation for ergonomic and climatic physiological vehicle seat design, SAE 880055. Warrendale, PA: Society of Automotive Engineers, Inc.

Floyd, W. F. and Roberts, D. F., 1958, Anatomical and physiological principles in chair and table design, *Ergonomics*, **2**, 1–16.

Gordon, C. C., Churchill, T., Clauser, C. E., Bradtmitter, B., McConville, J. T., Tebbetts, I. and Walker, R. A., 1989, 1988 Anthropometric survey of US Army personnel: Methods and summary statistics. United States Army Natick Research Development and Engineering Center, Technical Report, NATICK/TR-89-044, Natick, MA.

Grandjean, E., 1980, Sitting posture of car drivers from the point of view of ergonomics, in *Proceedings of the International Conference on Ergonomics and Transportation*, pp. 205–13, London: Academic Press, 8–12 September.

Grandjean, E., Hunting, W., Wotzka, G. and Scharer, R., 1973, An ergonomic investigation of multipurpose chairs, *Human Factors*, **15**, 247–55.

Hallen, A., 1978, Comfortable hand control reach of passenger car drivers, SAE 770245, Warrendale, PA: Society of Automotive Engineers, Inc.

Hertzberg, H. T. E., 1972, The Human Buttocks in Sitting: Pressures, Patterns, and Palliatives, SAE 72005, New York, NY: Society of Automotive Engineers, Inc.

Heliövaara, M., 1987, Occupation and risk of herniated lumbar intervertebral disc or sciatica leading to hospitalization, *Journal of Chronic Diseases*, **40**, 259–64.

Hosea, T. M., Simon, S. R., Delatizky, J., Wong, M. A. and Hsieh, C.-C., 1986, Myoelectric analysis of the paraspinal musculature in relation to automobile driving, *Spine*, **11**, 928–36.

Kamijo, K., Tsujimura, H., Obara, H., and Katsumat, M., 1982, Evaluation of seating comfort, SAE 820761, Warrendale, PA: Society of Automotive Engineers, Inc.

Keegan, J. J., 1953, Alterations of the lumbar curve related to posture and seating, *Journal of Bone and Joint Surgery*, **53**, 589–603.

Keegan, J. J., 1964, Designing vehicle seats for greater comfort, *SAE Journal*, **72**, 50–55.

Kelsey, J. L., Githens, P. B., O'Conner, T., Weil, U., Calogero, J. A., Holford, T. R., White III, A. A., Walter, S. D., Ostfeld, A. M. and Southwick, W. O., 1984, Acute prolapsed lumbar intervertebral disc: An epidemiologic study with special reference to driving automobiles and cigarette smoking, *Spine*, **9**, 608–13.

Kohara, J. and Sugi, T., 1972, Development of biomechanical manikins for measuring seat comfort, SAE 720006, New York, NY: Society of Automotive Engineers, Inc.

Lay, W. E. and Fisher, L. C., 1940, Riding comfort and cushions, *SAE Journal*, **47**, 482–96.

Matsuoka, Y. and Hanai, T., 1988, Study of comfortable sitting posture, SAE 880054, Warrendale, PA: Society of Automotive Engineers, Inc.

McConville, J. T., Churchill, T. D., Kaleps, I., Clauser, C. E. and Cuzzi, J., 1980, Anthropometric relationships of body and body segment moments of inertia, AFAMRL-TR-80-119. Air Force Aerospace Medical Research Laboratory, Wright-Patterson AFB, OH.

Milne, J. S. and Lauder, I. J., 1974, Age effects in kyphosis and lordosis in adults, *Annals of Human Biology*, **1**, 327–37.

Nachemson, A., 1968, The possible importance of the psoas muscle for stabilization of the lumbar spine, *Acta Orthopadia Scandinavia*, **39**, 47–57.

Nachemson, A. and Morris, J. M., 1964, In vivo measurements of intradiscal pressure, *Journal of Bone and Joint Surgery*, A, **46**, 1077–92.

Najjar, M. F. and Rowland, M., 1987, Anthropometric Reference Data and Prevalence of Overweight, United States, 1976–80, US Department of Health and Human Services, Public Health Service, National Center for Health Statistics, Series 11, No. 238.

Paton, C. R., Pickard, E. C. and Hoehn, V. H., 1940, Seat cushions and the ride problem, *SAE Journal*, **47**, 273–83.

Pope, M. H., Wilder, D. G., Jorneus, L., Broman, H., Svensson, M. and Andersson, G., 1987, The response of the seated human to sinusoidal vibration and impact, *Journal of Biomechanical Engineering*, **109**, 279–84.

Porter, J. M. and Norris, B. J, 1987, The effects of posture and seat design on lumbar lordosis, in Megaw, E. D. (Ed.) *Contemporary Ergonomics, 1987*, pp. 191–6, New York: Taylor & Francis.

Rebiffé, R., 1969, The driving seat. Its adaptation to functional and anthropometric requirements, in Grandjean, E. (Ed.) *Proceedings of the Symposium on Sitting Posture*, London: Taylor & Francis, Ltd.

Reed, M. P., Saito, M., Kakishima, Y., Lee, N. S. and Schneider, L. W., 1991, An investigation of driver discomfort and related seat design factors in extended duration driving, SAE 910117, Warrendale, PA: Society of Automotive Engineers, Inc.

Reynolds, H. M., 1991, Erect, Neutral and Slump Sitting Postures: A study of the torso linkage system from shoulder to hip joint, Final Report, Air Force Aerospace Medical Research Laboratory, Wright-Patterson AFB, OH.

Reynolds, H. M. and Hubbard, R. P., 1986, Old problems and new approaches in seating biomechanics, in *Passenger comfort, convenience and safety: test tools and procedures*, SAE/P-86/174, Warrendale, PA: Society of Automotive Engineers.

Reynolds, H. M., Snow, C. C. and Young, J. W., 1981, Spatial geometry of the human pelvis. Federal Aviation Administration, Civil Aeromedical Institute, Memorandum Report AAC-119-81-5, Oklahoma City, OK.

Rosemeyer, B. and Pförringer, W., 1979, Measuring of pressure forces in sitting, *Archives of Orthopaedic and Traumatic Surgery*, **95**, 167–71.

Sacks, A. H., O'Neill, H. and Perkash, I., 1985, Skin blood flow changes and tissue deformations produced by cylindrical indentors, *Journal of Rehabilitation Research and Development*, **22**, 1–6.

Sato, T. B., 1987, Effects of seat belts and injuries resulting from improper use, *The Journal of Trauma*, **27**, 754–8.

Schneider, L. W., Anderson, C. K. and Olson, P. L., 1980, Driver anthropometry and vehicle design characteristics related to seat positions selected under driving and non-driving conditions, SAE 790384, Warrendale, PA: Society of Automotive Engineers, Inc.

Searle, J. A., 1974, The geometrical basis of seat-belt fit, *Ergonomics*, **17**, 401–16.

Snyder, R. G., Chaffin, D. B. and Schutz, R. K., 1972, Link system of the human torso, AMRL-TR-71-88. Aerospace Medical Research Laboratory, Wright-Patterson AFB, OH.

States, J. D., Huelke, D. F., Dance, M., Green, R. N., 1987, Fatal injuries caused by underarm use of shoulder belts, *Journal of Trauma*, **27**, 740–5.

Stokes, I. A. F. and Abery, J. M., 1980, Influence of the hamstring muscles on lumbar spine curvature in sitting, *Spine*, **5**, 525–8.

Troup, J. D. G., 1978, Driver's back pain and its prevention. A review of the postural, vibratory and muscular factors, together with the problem of transmitted road shock, *Applied Ergonomics*, **9**, 207–14.

Varterasian, J. H. and Thompson, R. R., 1978, The dynamic characteristics of automobile seats with human occupants, SAE 770249, Warrendale, PA: Society of Automotive Engineers, Inc.

Verriest, J. P. and Alonzo, F., 1986, A tool for the assessment of inter-segmental angular relationships defining the postural comfort of a seated operator, SAE 860057, *Passenger Comfort, Convenience and Safety: Test tools and procedures*, P-174, pp. 71–83, Warrendale, PA: Society of Automotive Engineers, Inc.

Wilder, D. G., Woodworth, B. B., Frymoyer, J. W. and Pope, M. H., 1982, Vibration and the human spine, *Spine*, 7, 243–54.

Young, J. W., Chandler, R. F., Snow, C. C., Robinette, K. M., Zehner, G. F. and Lofberg, M. S., 1983, Anthropometric and mass distribution characteristics of the adult female, FAA-AM-83-16. Washington, DC: Office of Aviation Medicine, Federal Aviation Administration.

6

Ergonomic guidelines for design of a passenger car trunk

W. Karwowski, J. W. Yates and N. Pongpatana

Abstract

This chapter outlines ergonomic guidelines for passenger car trunk design with respect to the maximum acceptable overlift height and rear body panel thickness. Such guidelines were based upon experimental studies which considered the biomechanical and psychophysical capabilities of the subjects in simulated loading and unloading of the car trunk. Human anthropometric dimensions and strength exertion capabilities required to place objects into and remove objects from the car trunk were included in the proposed guidelines.

Introduction

Consumer ergonomics is concerned with the application of human factors and industrial design principles in development of products used by consumers. According to the Institute for Consumer Ergonomics (1980), consumer ergonomists seek to provide information about the limitations and capacities of the user through scientific investigation in a way which is of practical benefit to those responsible for product design.

Recently, an increased attention has been given to the application of ergonomic principles in passenger car design. A computerized search of scientific reference data bases revealed the following areas of research into human factors in passenger vehicle design (Simmonds, 1983; Mortimer, 1978):

1. anthropometry (the driver's workspace, seat dimensions, seat comfort, health and performance),
2. information display (instrument legibility, panel layout, head-up displays),
3. vehicle controls (control location factors, biomechanical and informational properties of control, control coding, gain and feedback modes),
4. visibility (visual cues for steering control, obstructions in the forward field, windshield characteristics, rear visibility requirements),
5. vehicle markings and signaling, and
6. vehicle dynamics and driver control.

Unfortunately, no studies addressed the important problem of passenger car trunk accessibility, and its use in the transportation of typical household goods, which require loading and unloading of the trunk. It is well known that such tasks can contribute to an increased risk of lower back pain which is approaching epidemic proportions in the United States (Easterby et al., 1980; Chaffin and Andersson, 1986). In addition, the aging of the population calls for an increased attention to both physical and cognitive capacities and limitations of the older segment of society.

This chapter outlines ergonomic guidelines for passenger car trunk design with respect to the maximum acceptable overlift height (lip height) and rear body panel lip thickness. Such guidelines were based upon two experimental studies which considered the biomechanical and psychophysical capabilities of the subjects in simulated loading and unloading of the car trunk. Human anthropometric dimensions and strength exertion capabilities required to place objects into and remove objects from the car trunk were also addressed (Roebuck et al., 1975). In addition, this study assessed individual customer preferences with respect to trunk lip height, floor height and bumper extension.

Preliminary design

Background

Given good foot placement, individuals may be expected to exert their greatest lifting forces around knuckle height in the standing position (Pheasant, 1986). A lift which is initiated at knuckle height, provided the load is not excessive, can be continued comfortably to elbow height. In general, peak lifting (lowering) forces are exerted at 75 cm above the ground (see Table 6.1). When lifting box-like loads, heights between 65 cm and 100 cm are the most suitable. The optimal height for both pushing and pulling actions is around 100 cm for men and 90 cm for women (Pheasant, 1986).

It should also be noted that the maximum acceptable loads lifted occasionally at arm's length by individuals under 50 years old (two-handed lift in front of the body and in upright standing position) are 10 kg (23 lbs) for males and 6 kg (13.2 lbs) for females (Davis and Stubbs, 1978). These values should be reduced by about 60 per cent if reduced headroom used for exces-

Table 6.1. *Recommendations for the design of vertical dimensions for placement of objects storage (after Pheasant, 1986)*

Height (cm)	Comments
< 60.0	Fair accessibility for light objects, poor for heavy objects
60.0–80.0	Fair accessibility for heavy* items, good for light items
80.0–110.0	**Optimal zone for placement of objects**
110.0–140.0	Fair–good accessibility for light items, poor for heavy items
140.0–170.0	Limited visibility and accessibility
170.0–220.0	Very limited access, beyond reach for some people

Note: * Heavy items were defined as those greater than 22.1 lbs.

Table 6.2. Summary statistics for the trunk measures (cm) of 60 randomly selected passenger cars (Karwowski et al., 1987)

Car size	Head clearance	Lip height	Floor-lip height	Trunk depth
Large [n = 20]				
Mean	162.5	82.3	34.4	116.0
SD	6.7	7.2	7.6	17.6
Range	149.9–172.7	64.8–91.4	25.4–54.6	75.6–136.5
Medium [n = 20]				
Mean	154.5	83.7	32.8	97.2
SD	6.9	3.8	4.3	12.2
Range	129.5–167.6	78.7–91.4	26.7–61.3	80.0–121.9
Small [n = 20]				
Mean	138.9	79.3	30.4	60.8
SD	5.8	11.8	12.9	24.9
Range	129.5–147.3	53.3–86.4	6.4–48.9	33.0–90.2

Notes: Lip height = distance from ground to trunk lip
Floor-lip height = distance between trunk floor (bottom) and lip
Head clearance = distance from ground to trunk lid
Trunk depth = front-to-back, distance inside the trunk

sive reach prevents erect posture, as could be the case when loading or unloading objects from the trunk of a car (Pheasant, 1986).

Preliminary design data

Trunk dimensions of 60 passenger cars (20 large car trunks, 20 medium car trunks and 20 small car trunks) were measured in order to define parameters for the experimental trunk model to be used in the laboratory experiments. Four dimensions of each trunk were measured: front-to-back inside depth (ID), height of lip from trunk floor (HT/F), height of lip from ground (HT/G), and head clearance of trunk lid (HC). Table 6.2 shows those dimensions. For each trunk measured, the following data were also collected: the car manufacturer, the make, model, year and size of the car (large, medium, or small).

Experiment no. 1: Determination of the maximum acceptable trunk lip height

Objectives

The main goal of the first laboratory experiment was to study human preferences with regard to the maximum acceptable trunk lip heights. The subjects involved in the experiment, the methodology, and the elements of the actual experiment are described in the following sections.

Subjects

Thirty adult subjects participated in preference evaluation of the maximum acceptable overlift height as defined by the trunk's lip height. There were 15 males and 15 females. Ten of the males and 10 of the females were between the ages 18 and 35 years. This age group will be henceforth referred to as the *younger group*. Five of the males and five of the females were 45 years of age or older. This age group will be henceforth referred to as the *older group*. All subjects were required to sign consent forms required by the University of Louisville Institutional Review Board for the Protection of Human Subjects Committee. The anthropometric measurements and strengths of the subjects are shown in Table 6.3.

Methods

The subjects were asked to perform simulated loading (lifting into) and unloading (lifting out) of the car trunk, using six different items. The lifting

Table 6.3. Age, anthropometric and strength characteristics (cm) of the subject population involved with the determination of maximum acceptable trunk lip height

Variable	Males [n = 15]			Females [n = 15]		
	Mean	SD	Range	Mean	SD	Range
Age (years old)	33.5	15.9	20.0–67.0	34.2	15.0	18.0–62.0
Weight (kg)	78.9	9.3	56.3–95.9	60.8	7.4	46.7–73.0
Stature	175.2	8.3	161.0–195.4	162.8	8.2	148.0–175.3
Acromial height	138.3	8.1	125.0–162.7	134.5	7.4	120.1–145.8
Knee height (standing)	48.9	3.0	43.8–54.4	45.1	3.1	39.3–49.9
Iliac crest height	103.3	5.8	94.6–117.1	98.7	5.5	88.7–106.3
Arm length	78.0	4.2	71.6–88.6	71.6	4.7	62.5–78.8
Forearm length	47.1	2.6	42.9–52.5	42.6	2.6	36.6–47.2
Shoulder breadth	37.7	2.6	31.7–40.8	33.8	1.7	30.9–36.5
Chest breadth	32.1	3.0	27.6–38.3	27.5	2.2	23.6–32.2
Biceps circumference (flexed)	32.9	2.7	29.6–39.0	27.6	2.1	24.7–31.7
Biceps circumference (relaxed)	28.9	2.2	25.2–34.2	25.4	2.0	21.0–28.8
Lower thigh circumference	41.9	2.9	37.4–47.0	41.9	2.9	35.6–48.5
Abdominal circumference	90.6	9.4	69.9–101.0	88.6	9.6	69.9–102.8
Grip strength (kg)	43.5	9.8	22.0–56.0	24.6	6.9	15.0–34.0
Static (isometric) strengths						
Elbow flexion	75.1	18.6	36.0–104.0	40.5	11.5	21.0–56.0
Shoulder flexion	112.7	36.5	55.0–181.0	50.6	11.3	25.0–71.0
Stooped back	187.7	54.6	80.0–276.0	114.5	30.3	72.0–158.0
Leg	276.8	94.3	98.0–406.0	158.9	49.2	81.0–248.0
Composite	220.8	72.3	90.0–323.0	131.1	31.6	75.0–184.0
Dynamic (isokinetic) strengths						
Elbow flexion	42.5	12.5	21.0–65.0	24.9	8.0	11.0–43.0
Shoulder flexion	32.8	8.7	18.0–45.0	18.5	5.9	9.0–35.0
Dynamic lift	16.6	49.7	62.0–244.0	98.7	34.2	50.0–195.0
Back extension	119.1	50.8	51.0–242.0	79.0	29.1	32.0–131.0

Figure 6.1. Experimental car set-up

trials design method, based on the psychophysical methodology (Gescheider, 1976), was used to evaluate the subjects' preferences for the highest acceptable level of the trunk lip height. The adjustable trunk was built for this purpose using a Chevrolet Celebrity Eurosport car provided by General Motors Corporation (see Figure 6.1). The car was mounted on a lift table to allow its vertical movement. Also, the bottom of the trunk was cut out and and placed on another lift table allowing for its adjustability up and down. Both lift tables were provided by the Southworth Corporation.

Experimental design: Procedures

The subject was required to place six items in a specified order (one through six) into the car trunk, and then remove them in the reverse order (see Figure 6.2). The items lifted are listed in Table 6.4. Specific places were marked on the

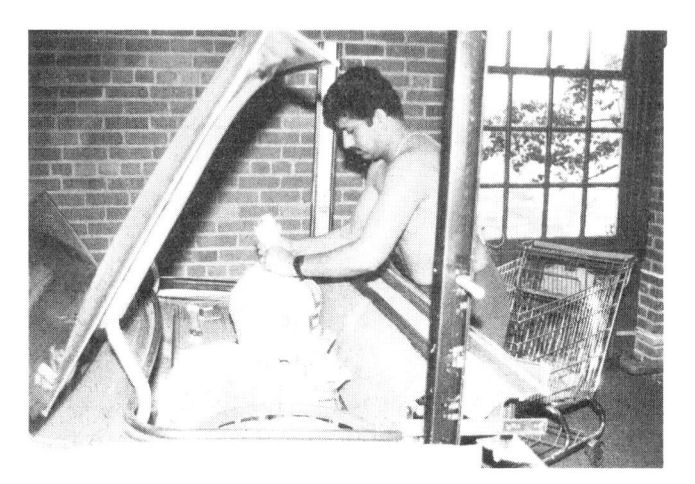

Figure 6.2. Illustration of lifting trials for determination of the maximum lip height

Table 6.4. Descriptions of items used in experiment no. 1

No.	Item	Weight (kg)	Height of lift
1	Laundry basket	13.7	From ground to trunk floor
2	Suitcase	15.3	From ground to trunk floor
3	Grocery bag 1	8.0	From cart (20 cm) to trunk floor
4	Grocery bag 2	10.0	From cart (20 cm) to trunk floor
5	Grocery bag 3	12.0	From cart (20 cm) to trunk floor
6	Bag of dog food	25.0	From cart (20 cm) to trunk floor

lab floor where items 1 and 2 were located, and specific places were marked in the grocery cart where the remaining items were located. Also specific areas where each item was to be placed were marked on the bottom of the trunk model. Each subject was instructed to lift at a pace with which he/she was comfortable, and would not become fatigued. However, if a subject did get tired, he/she sat down and rested as long as needed.

The subject could ask to have the car trunk lip adjusted 'a little', 'some', or 'a lot' until his/her highest tolerable lip height was reached (HTLH). Moving the lip height 'a little', 'some', and 'a lot' represented a movement of two cm, four cm, and six cm up/down, respectively. The subjects were encouraged to move the lip either up or down as many times as needed in order to determine the HTLH. They expressed their preference for the trunk lip height by requesting its adjustment either up or down to the highest level they could tolerate as users of the car.

If the subject could tell without lifting anything into or out of the trunk model that the lip height was too high or too low, he/she could request that the lip be moved in the appropriate direction until it was perceived as being near his/her HTLH. If, after lifting some but not all of the six items, the subject could tell that the lip was not at his/her HTLH (whether it was lower or higher than his/her HTLH), the subject could remove those items from the trunk model and return them to their original location. He/she could then request the desired movement of the lip, either up or down. The purpose for not requiring the subjects to lift all six items for each lip movement requested was to prevent them from getting fatigued and thus requiring the results to include a factor for fatigue. However, the subjects were required to put all six items into and remove them from the trunk model at a lip height they perceived as close to their HTLH.

No subject was allowed to select his/her HTLH without lifting all the items in and out at that lip height. This was done to ensure that the chosen lip height was indeed the subject's HTLH, because the removal of the items (particularly the dog food) was the deciding factor for many subjects. A total of twenty minutes was allowed for each adjustment session. An illustration of the experimental trunk adjustability is shown in Figure 6.3.

Experimental variables

The independent variables in this laboratory experiment were:

1   a lip height that could be adjusted from 68 to 150 centimeters from the ground,

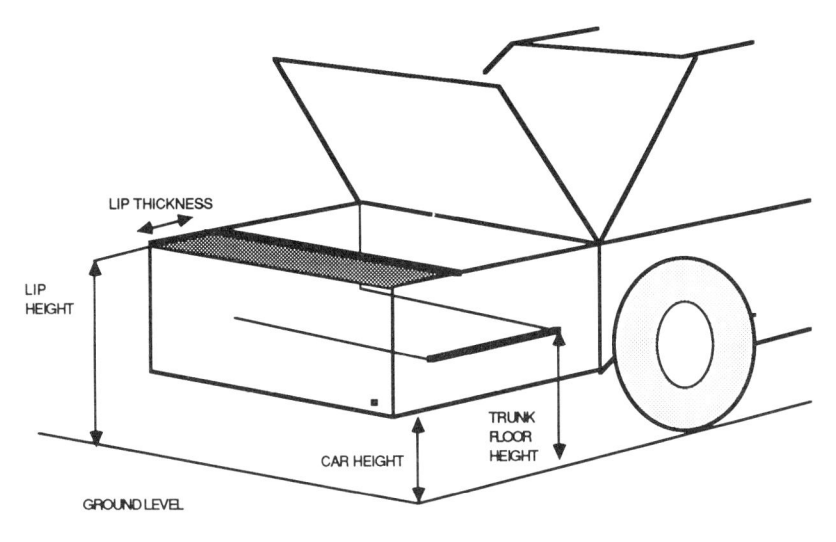

Figure 6.3. Illustration of investigated trunk dimensions

2 the height of the trunk floor (range from 51 to 89 centimeters from the ground),
3 the height (location) of the entire trunk, (the car could be raised 20 centimeters above the ground).

The bumper extension did not vary for this experiment; it was kept constant at 10 centimeters. The reason for this was to allow subjects to tolerate a higher lip height. For the purpose of ease of explanation, the customer preference experiment is explained in terms of six segments. Pairs {(1, 2); (3, 4); and (5, 6)} of these six segments (see Table 6.5) were performed in random order.

For segments 1 and 2, the trunk was in its original position with the bottom located 51 centimeters above the floor. The starting position of the lip height was either at 68 centimeters from the floor (low) at the subject's chest height (high). If the starting lip height was 68 cm from the floor in segment 1, then it

Table 6.5. Summary of experimental conditions for customer preference evaluation experiment

Segment	Car location	Trunk floor location (cm)	Option to adjust trunk height	Lip height above floor (cm)
1	At ground level	51	No	68
2	At ground level	51	No	Chest height
3	20 cm above ground	71	No	88
4	20 cm above ground	71	No	Chest height
5	At ground level	51	Yes	68 or chest height
6	20 cm above ground	71	Yes	88 or chest height

was set at the subject's chest height in segment 2, and vice versa. The reason for repeating the segment with a different starting lip height was to determine if starting at low or high levels of lip had any effect on the subjects' preference for the maximum acceptable height of the trunk lip.

For segments 3 and 4, the trunk was 20 centimeters above the floor, with the bottom located 71 centimeters above the floor. The starting lip height was either at 88 centimeters from the floor (low) or at the subject's chest height (high). Each subject performed his/her evaluation at both starting conditions. All subjects completed these segments of experimentation.

For segments 5 and 6, one of two beginning conditions was used: (1) car model on the floor, trunk floor 51 centimeters from the floor, and the lip height at 68 centimeters from the floor or at the subject's chest or (2) the car model 20 centimeters above the floor, trunk floor 71 centimeters above the floor, and the lip height at 88 centimeters from the floor or the subject's chest. For example, in condition 1, if the starting lip height was 68 cm from the floor in segment 5, then the starting lip height for segment 6 was equal to the subject's chest height, and vice versa. Condition 2, if the starting lip height was 88 cm from the floor in segment 5, the starting lip height for segment 6 was equal to the subject's chest height, and vice versa. Again, the reason for repeating the segment with a different starting lip height was to determine if starting high or low had any effect on the subject's choice. Fourteen of the 30 subjects completed segments 5 and 6 beginning with condition 1, and 16 subjects completed them beginning with condition 2.

Segments 5 and 6 differ from segments 1 through 4 in that for 5 and 6, the subject also had the option to adjust the trunk floor if he/she felt that moving the floor would help him/her tolerate a higher lip height. The subjects could adjust the floor in the same increments as the lip, i.e. 'a little', 'some', or 'a lot'. Thus during these two segments, the subject could adjust the lip height, the floor height, or both. The lip and the floor could be moved in the same or the opposite direction. In spite of the differences between what the subject could vary in segments 5 and 6, 'the goal was to determine the maximum acceptable height of the trunk lip he/she would tolerate as the user of the car'.

Results of experiment no. 1

The experimental data related to subject preferences for trunk lip height (LIPE), the associated values of the trunk floor height (FLR), and the extent of the bumper extension (BE) were analyzed using the Statistical Analysis System (SAS). The analysis of variance (see Table 6.6) revealed significant differences ($p < 0.05$) in the maximum acceptable lip height (LIPE) between males and females. The average selected values of LIPE were 98.7 cm and 107.9 cm for females and males, respectively. Although older subjects preferred lower height of the trunk lip (100.2 cm on the average) than the younger group (104.9 cm), the difference was not statistically significant.

With respect to the effect of car trunk variables on the preferences for LIPE, neither the initial trunk position (either low or high) at the start of the adjustment process, nor the option to adjust the trunk floor (FLR) during LIPE adjustments, had any significant effect on the LIPE value. However, when the whole car body was elevated by 20 cm above the normal vertical location (45.7 cm from the ground to the bottom edge of the car's body), the preferred

Table 6.6. Summary of the statistical analysis of data for subjectively chosen lip trunk (LIPE) height and trunk floor (FLR)

		Trunk lip height		Floor height	
Source	Values	p	Means* [cm]	p	Means* [cm]
Gender		0.0001		NS	
Males	0		107.9 A		62.9 A
Females	1		98.7 B		63.3 A
Age Group		NS		NS	
Younger (18–35 years old)	0		104.9 A		63.4 A
Older (>45 years old)	1		100.2 A		62.7 A
Car height from floor		0.05		0.0001	
Normal (25.7)	25.7		100.3 A		54.7 A
Elevated (45.7)	45.7		106.2 B		71.2 B
Trunk lip position at		NS		NS	
the start of experiment					
Low (68 cm)	0		101.8 A		62.9 A
High (chest height)	1		104.8 A		63.3 A
Option to adjust the trunk		NS			
floor during LIP adjustments					
No	0		102.2 A		
Yes	1		105.6 A		

Notes: NS = not significant at $p \leqslant 0.05$ level
       LIP = trunk lip height
       FLR = trunk floor height
* Means with the same letter are not significantly different (Duncan's Multiple Range Test)

LIPE increased significantly from 100.3 cm to 106.2 cm. This was due to the fact that at the same time the subjects had an option to adjust the trunk floor height. The FLR selected with the car body elevated, was 71.2 cm as compared to 54.7 cm in the normal car position. This was the only significant difference

Table 6.7. Summary of the results for lifting trials experiment to determine trunk lip and floor height values

Variable (cm)	Males (n = 15)			Females (n = 15)		
	Total	Younger	Older	Total	Younger	Older
Trunk lip height (LIPE)	n = 90	n = 60	n = 30	n = 90	n = 60	n = 30
Mean	107.9	111.0	101.0	98.7	98.4	99.4
SD	12.0	11.7	9.5	11.4	12.9	7.9
Range	88–142	90–142	88–126	78–130	78–130	82–112
Trunk floor height (FLRE)						
Mean	62.9	63.2	62.4	63.3	63.5	63.0
SD	11.0	11.1	11.0	9.8	9.8	10.0
Range	51–89	51–89	51–87	51–87	51–87	51–77

Notes: n = number of observations (divided by 6 to get the number of subjects in each category).
       Younger category: (18–35 years old), older category: (> = 45 years old).

Table 6.8. Maximum acceptable trunk lip heights (and associated floor heights) as a function of the initial lip position and option to adjust trunk floor height, where: LIPE = selected lip height, FLRE = selected trunk floor height associated with LIPE

Car body* height from the ground	Initial trunk floor height from the ground	Initial lip height from the ground	Option to adjust trunk floor	Variable	Males (n = 15)			Females (n = 15)		
					Mean	S.D.	Range	Mean	S.D.	Range
25.7	51	low	no	LIPE	99.7	9.0	88–122	91.5	10.9	82–122
				FLRE	—	—	—	—	—	—
25.7	51	low	yes	LIPE	105.8	6.5	96–116	104.4	15.6	82.126
				FLRE	61.4	10.7	51–87	63.4	12.5	51–87
25.7	51	high	no	LIPE	102.8	8.1	90–120	95.5	10.8	78–120
				FLRE	—	—	—	—	—	—
25.7	51	high	yes	LIPE	108.4	10.2	94–126	105.6	13.6	88–128
				FLRE	63.0	10.0	55–87	63.4	8.2	57–77
45.7	71	low	no	LIPE	111.6	11.2	96–134	98.5	11.1	88–128
				FLRE	—	—	—	—	—	—
45.7	71	low	yes	LIPE	114.0	17.9	90–142	96.2	6.7	88–110
				FLRE	73.7	9.1	59–89	70.0	3.5	65–77
45.7	71	high	no	LIPE	112.7	10.6	98–132	104.3	9.4	94–130
				FLRE	—	—	—	—	—	—
45.7	71	high	yes	LIPE	114.7	15.2	96–140	102.8	5.9	94–112
				FLRE	77.3	5.6	71–87	71.2	2.6	67–75

Notes: * All values in cm.
Initial lip height from the ground: low = 68 cm, high = subject's chest height.

in the trunk floor height with respect to all of the independent variables studied (see Table 6.6).

More specific data related to preferred values for the trunk lip and floor heights were also compiled. Summary of these results are given in Tables 6.7 and 6.8.

Experiment no. 2: Determination of the rear body panel (lip) thickness

Background information

In addition to such car trunk design parameters as ground to lip height (lip height), ground to trunk floor height (floor height) and bumper horizontal extension and vertical location, the thickness of the rear body panel (lip thickness) also plays an important role in the user acceptance of the trunk design. The objective of the second laboratory experiment was to determine the maximum acceptable lip thickness preferred by the subjects for simulated loading and unloading of the trunk under a set of experimental conditions described below.

The lip thickness was defined as the horizontal distance between the back end of the rear panel and the outer border of the bumper extension. Since the previous experiment showed no significant effect of the bumper extension on the preferred lip height, the bumper dimension was set at 10 cm from the front end of the rear panel, and remained constant throughout the current study.

Methods

A laboratory experiment was conducted in order to determine the preferred lip thickness. This experiment was based on the 'fitting trial method' (Gescheider, 1976; Pheasant, 1986) using the psychophysical approach. The adjustable car trunk used in the first experiment was modified in order to allow for changes in the lip thickness (horizontal extension of the rear panel). The adjustability range (25–50 cm) was determined based on analysis of the thickness values for the rear end panel representative for the twelve cars (see Table 6.9). Figure 6.4 shows the modified trunk with adjustable lip thickness. The items used during the trials involving simulated loading and unloading of the car trunk were the same as those used in the first experiment (see Table 6.4).

Methods: Subjects

A total of 24 subjects (12 males and 12 females) participated in the study. All subjects were asked to sign consent forms required by the University of Louisville Institutional Review Board for the Protection of Human Subjects Committee. The anthropometric and static strength characteristics of the subject population are shown in Table 6.10. The anthropometric measures were made in accordance with the NASA *Anthropometric Source Book* (1978). The instrumentation and procedures for these measures, as well as for the strength measures, were described by Karwowski *et al.* (1987). University of Louisville students, faculty and staff participated in the experiments. The subjects were

Table 6.9. Trunk lip thickness (cm) of various car models

Car Model	Lip thickness (cm)
Ford Taurus	31.25
Toyota Supra	32.50
Ford Scorpio	31.25
Honda Prelude	25.00
GM Pontiac 6000	33.75
Nissan 30	34.38
Ford Mustang	38.75
Ford Lincoln	43.75
Honda Legend	32.50
GM Camaro	45.00
Ford Avenue	35.00
Honda Accord	27.50
Mean	34.22
SD	5.89
Range	25.00–45.00

stratified according to their age into two groups: *younger* (18–25 years old) and *older* (>40 years old).

Experimental procedures

The subjects were asked to lift the six items listed in Table 6.4 into and out of the car trunk, and encouraged to adjust the horizontal lip thickness (either

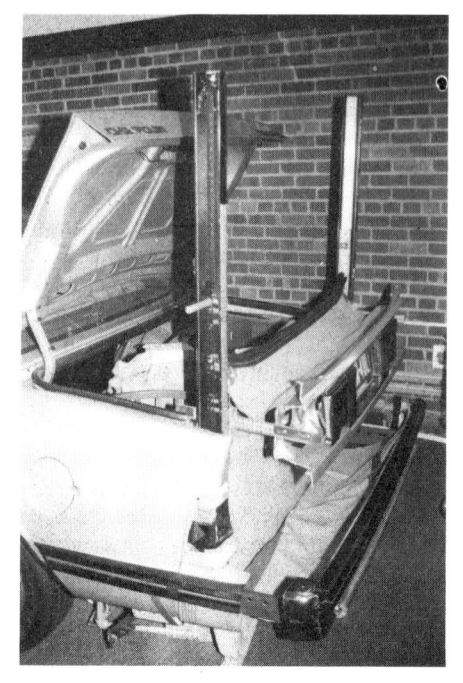

Figure 6.4. Experimental set-up for lip thickness study

Table 6.10. Anthropometric characteristics (cm) and static strengths of the subject population involved with determination of maximum acceptable lip thickness

Variable	Males (n = 12)			Females (n = 12)		
	Mean	SD	Range	Mean	SD	Range
Age (years)	39.1	15	22.0–64.0	34.8	15.7	18.0–55.0
younger	26.3	3.8	22–32	20.3	2.5	18–25
older	51.8	9.6	40–64	49.2	6.8	40–55
Weight (kg)	78.1	10.3	59.3–102.1	56.0	9.0	41.7–79.4
Stature	174.4	4.6	168.0–182.5	161.6	4.8	152.9–168.4
Acromial height	145.4	4.8	138.4–152.0	133.5	4.9	123.8–139.3
Knee height (standing)	49.8	2.2	46.0–54.0	45.8	3.4	38.6–51.3
Iliac crest height	102.7	3	97.6–107.9	96.4	5.7	86.7–104.2
Elbow height	112.5	4.2	105.5–118.2	101.8	4.1	94.0–107.1
Arm length*	76.9	3.1	73.5–82.7	70.1	3.9	63.3–77.4
Static strengths (Isometric)						
Arm	73.0	20.1	35.5–104.5	46.7	10.5	24.5–66.0
Stooped back	195.4	74.2	80.0–334.0	132.1	45.5	72.5–207.0
Shoulder	102.8	20.0	55.0–130.5	52.8	13.0	25.6–82.5
Leg	313.6	104.6	97.5–489.0	173.0	59.5	81.5–278.0

Note: * Arm length defined as forward grip reach

reduce it or extend it) by requesting to change it 'a little', 'some' or 'a lot' until the maximum acceptable lip thickness (LIPTH) was reached. Changing the LIPTH 'a little', 'some' and 'a lot' corresponded to adjustment by 5, 10 and 15 cm, respectively. The subject was asked to make as many adjustments as needed in order to determine the preferred value of LIPTH. A total of twenty minutes was allowed for each adjustment session.

Each subject was required to lift the six items in a specified order (first through sixth) into the trunk, and remove them in the reverse order. Each item was placed in a predetermined section which was marked on the trunk floor by numbers. The subjects were instructed to lift the objects at a pace which was comfortable and not tiring.

Experimental conditions

In order to utilize the information gathered in the previous experiments regarding the preferred values for lip height (LH), trunk floor height (FH) and bumper height (BH), the experimental conditions shown in Table 6.11 were chosen for this study. The conditions 1 and 3 for both males and females represent the combinations of the maximum acceptable lip height values given two pre-set bumper heights (50 and 68 cm) and two pre-set values for the trunk floor height (51 and 71 cm) used in the previous experiments. The lip height values represent the averages over the initial lip height positions (low and high) at the start of previous experiments. It should be recalled here that

Table 6.11. Summary of the experimental conditions (cm) for selecting the rear body panel thickness

Gender	Condition	Car height	Bumper height	Lip height	Floor height
M	1	25.7	48.0	101.0	51.0
	2	25.7	48.0	107.0	62.0
	3	45.7	66.0	112.0	71.0
	4	45.7	66.0	114.0	75.0
F	1	25.7	48.0	93.5	51.0
	2	25.7	48.0	105.0	63.0
	3	45.7	66.0	101.0	71.0
	4	45.7	66.0	99.0	70.5

Notes: Lip thickness: adjustable with minimum thickness of 25 cm
M—males, F—females
* The 51.0 and 71.0 cm values for the floor height, and all values of the bumper height were pre-set in the previous experiment. All other values were selected by the subjects

the effect of initial lip position, as shown by the previous study, was not significant. Conditions 2 and 4 represent similar combinations of independent variables, in these cases, both the lip height and trunk floor height values were selected by the subjects in the previous experiments. The subject repeated each of the experimental conditions twice, with the initial position of the lip thickness set at 25 or 50 cm.

The selection of these experimental conditions allowed for a comprehensive study of the interactions between all of the important trunk design variables as well as for the efficient utilization of laboratory resources. All experimental treatments were performed in a random order.

Results of experiment no. 2

Tables 6.12 and 6.13 show the summary of the lip thickness (LIPTH) values selected by subjects. The analysis of variance showed that no significant differences between males and females. There were, however, differences in LIPTH within the six groups due to age (see Table 6.14). While the younger females

Table 6.12. Values of the selected lip thickness (cm) for all subjects*

	Males (n = 12)			Females (n = 12)		
Statistics	Total n = 96	Younger n = 48	Older n = 48	Total n = 96	Younger n = 48	Older n = 48
Mean	39.0	40.3	37.7	37.8	36.6	39.1
SD	5.1	4.5	5.4	5.3	4.3	5.9
Range	25.0–50.0	30.0–50.0	25.0–45.0	25.0–45.0	30.0–45.0	25.0–45.0

Note: * The differences between males and females were not significant at $p \leqslant 0.05$ level.

Table 6.13. Summary of the selected maximum acceptable lip thickness (cm) as a function of the experimental condition

	Males			Females		
Condition	Mean	SD	Range	Mean	SD	Range
1	39.8	6.2	30–50	39.0	4.8	30–45
2	38.3	5.0	30–45	35.4	6.2	25–45
3	38.9	5.3	25–45	37.5	4.7	30–45
4	38.9	3.9	30–45	39.2	4.8	30–45
Total	38.9	5.3	25–50	37.8	5.2	25–45

selected on the average the LIPTH value of 36.6 cm, the older females chose 39.1 cm. The reverse trend occurred for males, with the younger group (average LIPTH = 40.3 cm) selecting greater values than the older group (average LIPTH = 37.7 cm). The effect of the initial LIPTH value (either 25 or 50 cm) was not significant for males or females.

Surprisingly, these subjective values of LIPTH chosen for different experimental conditions (Table 6.13) also did not differ significantly among each other. The above results could lead to the observation that the range of experimental conditions chosen for the study was sub-optimal with respect to the subjects' preferences for forward reach activity. Further analysis of experimental data included the regression analysis aimed to predict the preferred LIPTH based on the subject variables (see Table 6.14). For comparison, the prediction model for the maximum acceptable lip height based on previous experiment is also given. The results showed that the maximum acceptable lip thickness can be predicted based on the subject's forward arm reach (shoulder to grip length). Table 6.15 depicts the US population data that can be used for this purpose.

Table 6.14. Regression models for the preferred lip thickness and lip height values

(a) All subjects:
Lip thickness (cm) = 0.3185 + 0.5145 (forward arm reach)
$R^2 = 0.98$, $p \leqslant 0.0001$, MSE = 6.0
(b) Males:
Lip thickness (cm) = 0.5014 + 0.5115 (forward arm reach)
$R^2 = 0.97$, $p \leqslant 0.0001$, MSE = 10.8
(c) Females:
Lip thickness (cm) = 0.2414 + 0.5194 (forward arm reach)
$R^2 = 0.98$, $p \leqslant 0.0001$, MSE = 2.3
(d) All subjects:
Lip thickness (cm) = −32.9921 + 0.4838 (trunk floor height) + 0.6259 (stature)
$R^2 = 0.92$, $p \leqslant 0.0001$, MSE = 92.1

Table 6.15. Selected anthropometric data for subjects and corresponding US population (cm)

	Sample population		US population*			
	Mean	SD	Mean	SD	5 percentile	95 percentile
Stature						
Males	174.4	4.6	175.5	7.1	164.0	187.0
Females	161.6	4.8	162.5	6.4	152.0	173.0
Forward arm reach						
Males	70.1	3.9	71.0	3.2	65.5	76.5
Females	76.9	3.1	78.5	3.5	72.5	84.5

Note: * Based on Pheasant (1986)

Spatial design recommendation for trunk design

Background

The results of the two experimental studies, as well as the principles of engineering anthropometry and biomechanics, were used to develop a set of spatial trunk design recommendations. These recommendations are based on the following design variables:

(a) ground to trunk lip height (LIPH),
(b) ground to trunk floor height (FH),
(c) trunk floor to lip height distance (TF-LH),
(d) rear body panel thickness (including a 10 cm horizontal bumper extension),
(e) ground to rear bumper height (BH), and
(f) head clearance (ground to open trunk lid height).

In the sections that follow, the rationale and methods for selection of specific values for the design variables and their interaction are discussed. Also, the diagrams to be used as designers' aids are presented.

Relevant anthropometric and biomechanical data

Some of the concepts used to derive the design recommendations for trunk design are the concept of a percentile, and principles of design for extreme and range of population dimensions. In addition, relevant anthropometric and biomechanical data related to workspace design were utilized.

Data shown in Table 6.1 provided additional useful information regarding the recommendations for the design of vertical dimensions for placement of objects (Pheasant, 1986). It should be noticed that the suggested *optimal zone* for vertical placement of items is in the range 80–110 cm above the ground, while fair accessibility even for objects that weight more than 22 lbs can be achieved at the vertical height of 60–80 cm.

Design data for lip and trunk floor height dimensions

The means and standard deviations for the maximum acceptable lip heights and corresponding trunk floor heights were used to develop percentile values

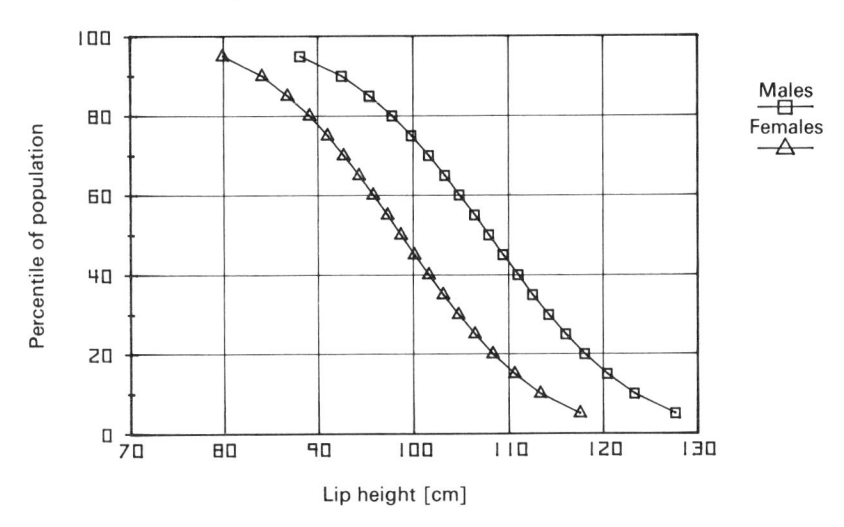

Ergonomic guidelines—passenger car trunk design

BASED ON ALL CONDITIONS

Figure 6.5. Design data: Percentile for lip height

for the trunk design data. The derived values were then plotted over the range from fifth percentile to 95th percentile in Figures 6.5 and 6.6.

Design data for trunk lip thickness dimension

The derived percentiles for maximum acceptable values of lip thickness for males and females are illustrated in Figure 6.7.

Trunk floor heights: 51 & 71 [cm]

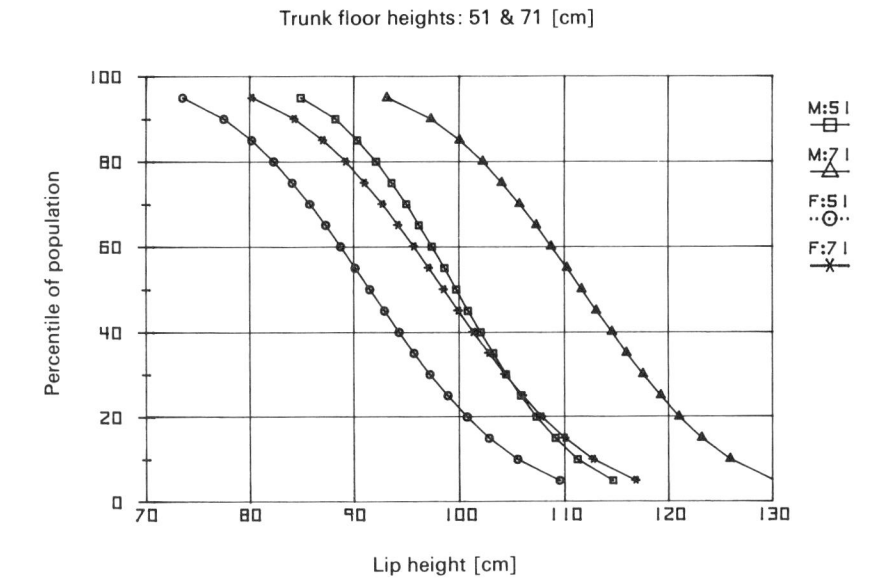

Figure 6.6. Design data: Percentile for lip height as a function of trunk floor height

BASED ON ALL CONDITIONS

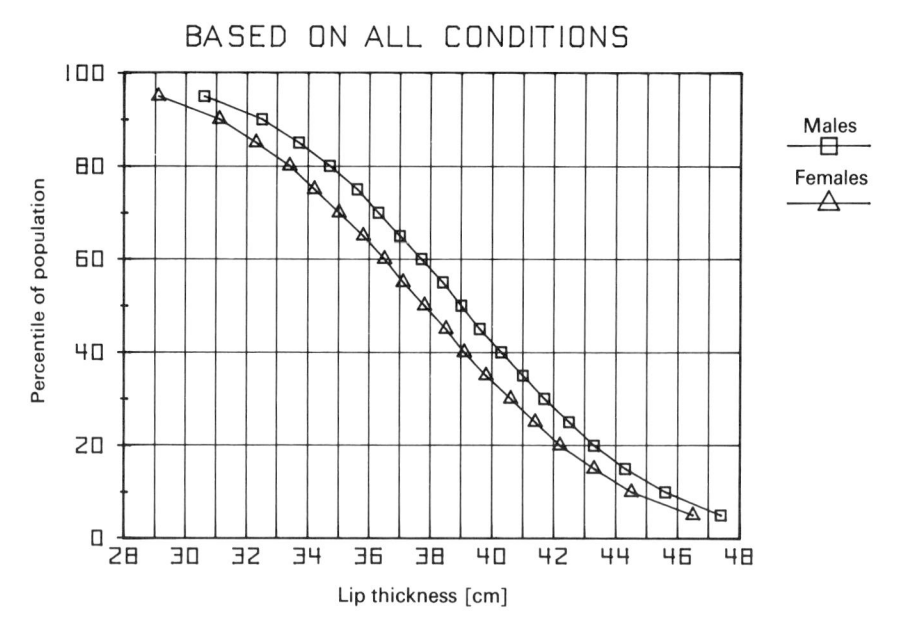

Figure 6.7. Design data: Percentiles for lip thickness

Design data for head clearance

The vertical and horizontal head clearance variables were not researched empirically in this study. Vertical head clearance was defined as the vertical height of the tip of the head from the ground, and corresponded to the height from the ground to tip of the open trunk lid. Preliminary information for this variable was gathered by Karwowski *et al.* (1987) and was shown in Table 6.2. Horizontal head clearance was defined as the horizontal distance between the end of the rear panel and the vertical plane from the tip of the open trunk lid.

In order to develop approximate percentile values for the vertical head clearance, the US population data for stature and hip height (shown in Table 6.16) was used (Pheasant, 1986). For that purpose, it was assumed that a person bends the upper torso at hip joint when getting access to the trunk, and that the standard deviations for the corresponding mean values of stature in the stooped posture remain the same as in the case of the erect posture. Pictorial representation of the derived approximate percentile values are shown in Figures 6.8, 6.9 and 6.10. The corresponding values for horizontal head clearance were expressed as a function of the angle between the upper body and vertical (frontal) plane in stooping posture. These values are shown in Table 6.16.

Interactions between design variables

It should be noted here that there exists a strong trade-off between the lip height and trunk floor height. In general, the higher the lip height, the higher the preferred floor height. This trade-off occurs because car users tend to minimize the amount of effort needed to place objects on the trunk floor after they

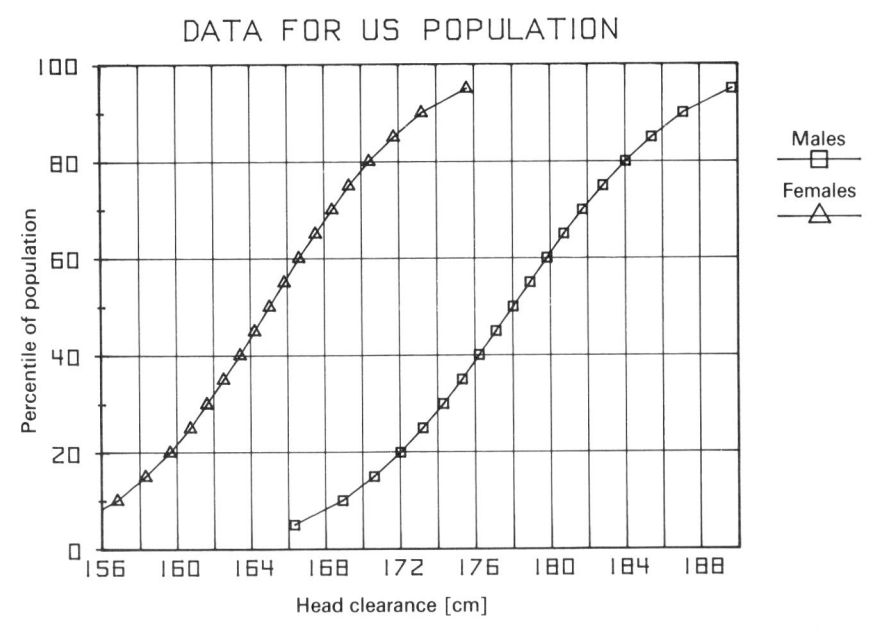

Ergonomic guidelines—passenger car trunk design 135

Figure 6.8. Design data: Percentile values for head clearance based on US population

have been lifted over the trunk lip. In the extreme case, this trade-off relationship leads to the *lip-floor paradox* (see Figures 6.11 and 6.12). This paradox illustrates the fact that in order to accommodate 90 per cent of population with respect to the lip height and floor height variables, the difference between

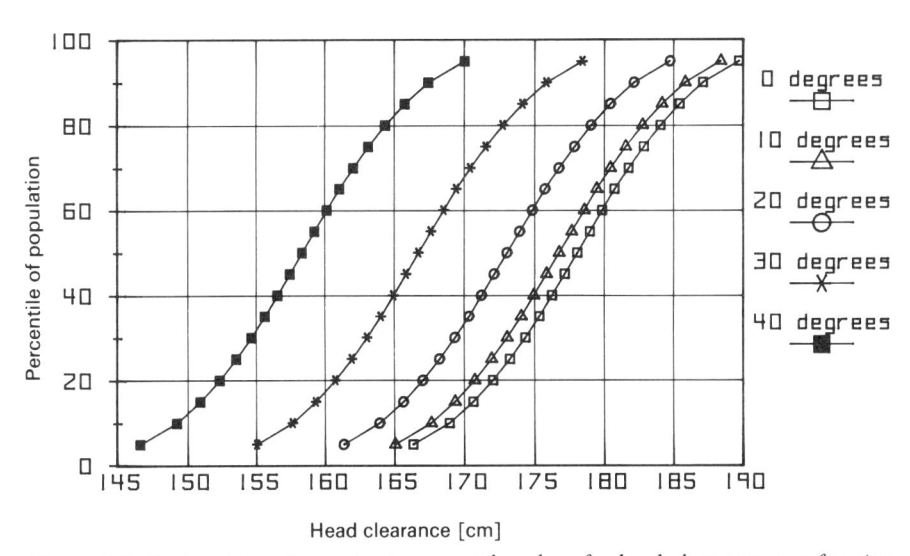

Figure 6.9. Design data: Approximate percentile values for head clearance as a function of upper body forward inclination at hip joint (males)

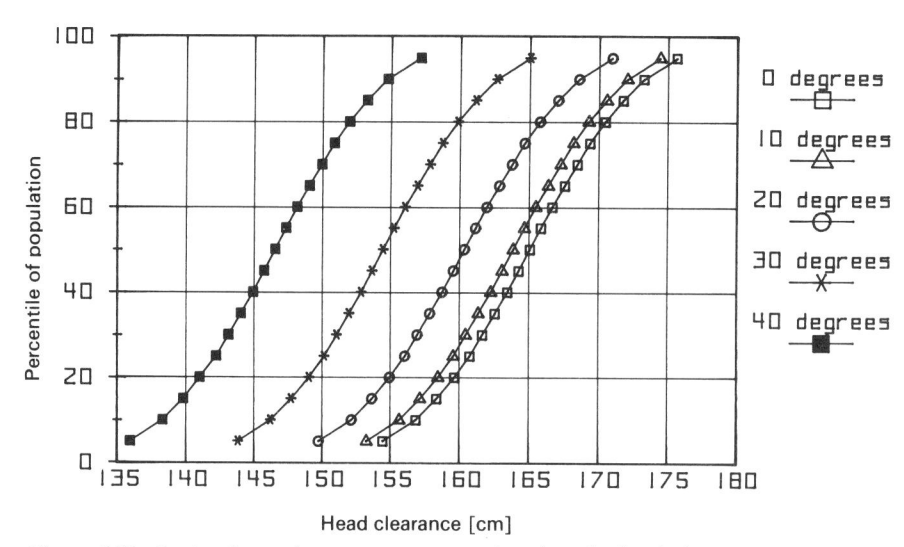

Figure 6.10. Design data: Approximate percentile values for head clearance as a function of upper body forward inclination at hip joint (females)

Table 6.16. Approximate of percentiles for vertical and horizontal head clearances as a function of forward upper body inclination at hip joint (including 2.5 cm for shoes)

Variable	Males		Females	
	Mean	SD	Mean	SD
Stature	175.5	7.1	162.5	6.4
Hip height	91.5	5.0	83.5	4.5

Degree of inclination	Horizontal head clearance[1]		Vertical head clearance[2] (cm)			
			Males		Females	
	Males	Females	Mean	SD	Mean	SD
0	0	0	178.0	7.1	165.0	6.4
10	15	14	176.7	7.1	163.8	6.4
20	29	27	173.0	7.1	160.3	6.4
30	42	40	166.7	7.1	154.4	6.4
40	54	51	158.3	7.1	146.5	6.4

Notes: Based on US population data (Pheasant, 1986). Horizontal head clearance defined as the horizontal distance from the end of rear panel towards the open area inside of the trunk.
[1] Horizontal clearance = [(stature − hip height) sin β]
[2] Head clearance = [(stature − hip height) cos β + hip height] + 2.5
β = degree of (forward) upper body inclination from vertical position at hip joint

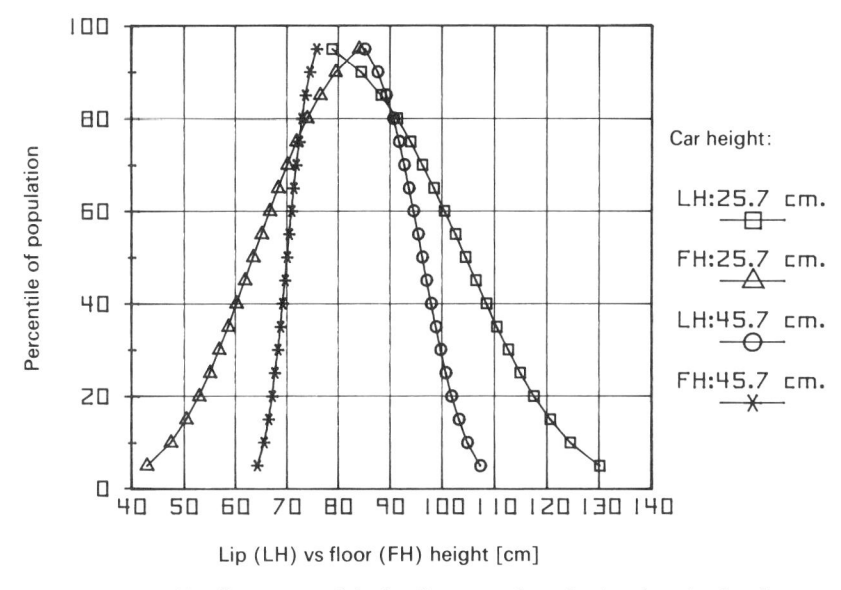

Figure 6.11. Illustration of the lip-floor paradox: Design data for females

the lip and floor heights need to be minimized, and consequently the trunk becomes very shallow. Figures 6.11 and 6.12 illustrate the general guidelines for acceptable ranges of both design variables (60–110 cm), based on the recommendations given by Pheasant (1986).

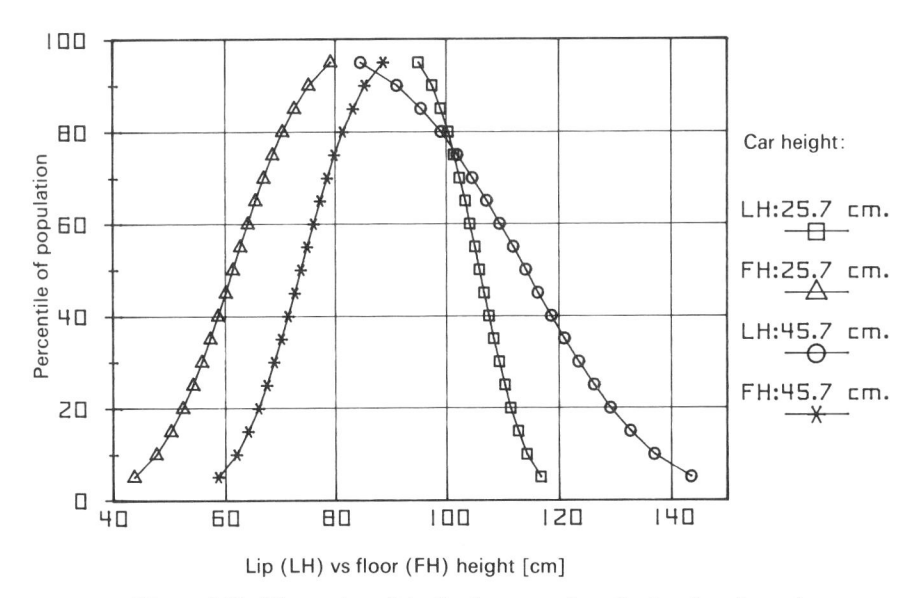

Figure 6.12. Illustration of the lip-floor paradox: Design data for males

Design data integration

The experimental data collected and other relevant information from the literature were integrated in order to develop the spatial recommendation for car trunk design. These recommendations are based on a range between 10th and

Table 6.17. Classification of the design criteria categories

Design category	Criterion based on the percentage of population accounted for by the design
Optimum	>90%
Acceptable	50–90%
Tolerable	10–50%
Unacceptable	<10%

Table 6.18. Car trunk design spatial recommendations (cm) based on percentile values

Design criteria	Optimum		Acceptable		Tolerable		Unacceptable	
Design variable	Males	Females	Males	Females	Males	Females	Males	Females
Trunk lip height (all conditions)[1]	60–92.5	60–84	93–108	84–99	108–123	99–113	>123	>113
Corresponding trunk floor height[2]	49–60	51–60	49–63	51–63	63–77	63–76	>77	>76
Trunk lip height as a function of the bumper & floor heights [cm]:								
48 & 51 (fixed)	51–88	51–77.5	88–100	78–91	100–111	91–106	>111	>106
66 & 71 (fixed)	71–97	71–84	97–112	84–98	112–126	98–113	>126	>113
48 (for selected trunk floor of)	(61.4) 62–97	(64) 64–84	97–106	84–104	106–114	104–113	>114	>113
66 (for selected trunk floor of)	(73.7) 74–91	(70.0) 70–85	91–114	85–96	114–173	96–105	>134	>105
Lip height minus floor height[3] as a function of bumper height								
48 cm	<31	<23	31–45	23–42	45–58	42–60	>58	>60
66 cm	<26	<22	26–39	22–29	39–52	29–36	>52	>36
Lip thickness (including 10 cm horizontal bumper extension)	<32	<31	32.5–39	31–38	40–46	38–44	>46	>45
Vertical head clearance[4]								
—for erect posture	>187	>173	178–187	165–137	169–178	157–165	<169	<157
—for 40 degrees forward bending of the upper body	>167	>155	158–167	146–155	149–158	138–146	<149	<138

Notes: [1] Lower limit (60 cm) based on the recommended height for the storage of heavy objects (Pheasant, 1986)
[2] The trunk floor should not be lower than 40 (cm) for males and 51 (cm) for females
[3] Based on the differences between means for experimental conditions

90th percentile values in order to accommodate approximately 90 per cent of the population. The respective percentile values were used to propose four categories for the trunk design criteria, i.e., *optimum, acceptable, undesirable,* and *unacceptable.* (see Table 6.17). All design variables were then pooled together and their respective values classified into the above design categories. The derived design recommendations are given in Table 6.18. When using presented data for car trunk design, all relevant interactions between design variables should be carefully considered.

Acknowledgments

This study was sponsored by a grant from the General Motors Corporation, CPC Engineering North, and partially supported by the Speed Scientific School of the University of Louisville. The authors express their appreciation to Dr B. Peacock and Mr David Houy, of the Human Factors Group, General Motors Corporation (CPC Engineering North), Pontiac, Michigan, for making this project possible. We also acknowledge the cooperation of Dr M. R. Wilhelm, Chairman, Department of Industrial Engineering, and Dr E. Gerhard, Dean Emeritus, Speed Scientific School, and logistic support provided by Mr John Jones, Senior Technician, Department of Industrial Engineering, University of Louisville.

References

Chaffin, D. B. and Andersson, G., 1986, *Occupational Biomechanics*, New York: John Wiley and Sons.
Davis, P. R. and Stubbs, S. T., 1978, Safe levels of manual forces for young males, *Applied Ergonomics*, **9**, 33–37.
Easterby, R., Kroemer, K. H. E. and Chaffin, D. (Eds), 1980, *Anthropometry and Biomechanics, Theory and Applications*, New York: Plenum Press.
Gescheider, G., 1976, *Psychophysics: Method and Theory*, New York: Lawrence Erlbaum Publishers.
Institute for Consumer Ergonomics, 1980, Leicestershire, United Kingdom.
Karwowski, W., Yates, J. W., Bryant, V. and Pongpatanasuegsa, N., 1987, *An Ergonomic Evaluation of Vehicle Trunk Dimensions*, Final Report to General Motors Corporation, Human Factors Group, Center for Industrial Ergonomics, Louisville, KY: University of Louisville.
Mortimer, R., 1978, *Human Factors in Vehicle Design*, Pontiac, MI: General Motors Corporation.
NASA, 1978, *Anthropometric Source Book, Volume I: Anthropometry for Designers*, edited by Webb Associates, NASA Publication 1024, Washington, D.C.: Scientific Technical Information Office.
Pheasant, S., 1986, *Bodyspace: Anthropometry, Ergonomics and Design*, London: Taylor & Francis.
Roebuck, J. A., Kroemer, K. H. E. and Thomson, W. G., 1975, *Engineering Anthropometry Methods*, New York: John Wiley and Sons.
Simmonds, G. R., 1983, Ergonomics standards and research for cars, *Applied Ergonomics*, **14**, 97–101.

7

Physical aspects of car design: occupant protection

Mark R. Lehto and James P. Foley

Introduction

Occupant protection focuses on the prevention of harm to automobile occupants. As such, occupant protection depends on the engineering of the vehicle, engineering of the protective device (e.g., seat belts and air bags), and on human biomechanics (Evans, 1987). Taming this potentially deadly interface of man and machine poses a big challenge to the field of human factors engineering. An environment must be created to work within the limitations of the human body to withstand the violent and rapid deceleration that occurs during a crash. The human factors analysis must also take into account the psychological environment generated by the specific countermeasures used to provide crash protection. Are there forms of protection that the occupant will find obtrusive and unacceptable? If the new, protected environment is too benign, will this cause an increase in risk taking that offsets any engineering gain?

Some perspective into the role of, and need for, occupant protection, is given by the statistics that in 1991 there were a total of 41 150 fatalities in 36 540 fatal crashes on public roads and highways in the United States (NHTSA, 1992). Among these fatal accidents, approximately 50 per cent of the impacts were classified as frontal, 40 per cent side, and 6 per cent rear (NHTSA, 1992). The preliminary fatality rate for 1991 was 1.9 deaths per 100 million vehicle miles travelled (VMT), the lowest rate ever recorded. The peak rate was 5.5 deaths per 100 million VMT in 1966. This significant reduction in fatality rates over the past 25 years can be traced, in part, to an increased focus on occupant protection. This trend is due to a more general focus on preventing and reducing the severity of crashes by designing both safer vehicles and roadways. The passage of the National Highway Traffic Safety Act 1966, in particular, focused attention in both the public and private sectors on the sharp increase in deaths and injuries that followed the rapid increase in the personal use of automobiles after World War II. Dr William Haddon, the first director of the National Transportation Safety Bureau (later to be the National Highway Traffic Safety Administration—NHTSA), is generally credited with changing the emphasis in traffic safety countermeasures from those dealing only with driver behavior to an engineering problem of properly packaging the occupants in a crash. This change in focus has led to significant advances in automobile safety based on engineering design changes of the vehicle itself to provide additional protection from the hazards encountered in a crash.

A significant body of mandatory and consensus standards have concurrently emerged. Among the mandatory standards, increased occupant protection is required by the Federal Motor Vehicle Safety Standards (FMVSS 208) that were part of the 1966 Traffic Safety Act. FMVSS 208 mandates certain levels of protection achieved and verified in standardized testing (i.e. predicted survivability in a 30 mph frontal collision into a fixed barrier based on three crash dummy injury criteria). It also mandates the provision of safety belts in all automobiles and requires the passive protection of both the driver and right front seat occupant of passenger cars and light trucks (pickups, vans and utilities) by 1994. Passive protection is defined within the standard as that which does not require any overt action on the part of the occupant (such as fastening a seat belt). To date this requirement has been met with either passive belts with knee bolsters or air bags. Other passive approaches to increasing occupant safety include energy absorbing steering columns, instrument panel and other interior padding, side door beams, and increased car mass or crush space.

Among consensus organizations, the Society of Automotive Engineers (SAE) has played a particularly prominent role. The SAE has developed numerous voluntary standards pertaining to crash performance, belt designs, testing procedures, use of anthropomorphic dummies, seat designs, energy absorbing steering columns and instrument panels, etc. These standards developed by the SAE have often been adopted for mandatory application by the federal government. For example, SAE standard J4 for seat belt assemblies served as the basis for FMVSS 209. The automobile manufacturers also set their own internal standards which are frequently more stringent that those mandated by the federal government or within the consensus standards. This is often done to ensure compliance with the mandatory rules in spite of test to test variations. With increased safety awareness on the part of consumers, doing so may also form the basis for an effective marketing strategy. Volvo and Mercedes-Benz have taken such an approach for years. More recently Chrysler has taken the lead among the domestic manufacturers by promoting air bags in all domestically produced (non-joint venture) automobiles. General Motors and Ford have also placed greater emphasis on vehicle safety in recent years.

Theoretical background

Analysis of the crash sequence from a theoretical perspective provides substantial insight into the incredibly difficult problem occupant protection poses, the generic modes of protection, and the numerous associated human factors issues. From this perspective we will first present the anatomy of a frontal collision, revealing the fundamental difficulties of crash protection and a taxonomy of generic countermeasures. The limits of protection will then be addressed, followed by a discussion of specific human factors issues related to biomechanics, anthropometrics, aging and risk taking.

Anatomy of a crash

An automobile crash is an extremely quick event. It is literally over in the blink of an eye. The duration of a frontal barrier crash, from first contact to maximum crush, is approximately one tenth of a second. Even more severe constraints are associated with side impacts. The change in the velocity of the vehicle's occupant compartment over the time ($\partial V(t)/\partial t$) during the collision phase of a motor vehicle crash (i.e., from the moment of initial contact until the moment of separation) defines the acceleration ($A(t)$) at any given moment. The crash pulse of a particular vehicle describes this relationship for a given crash. Figure 7.1 presents examples of representative crash pulses for vehicles in frontal barrier crashes at 30 mph. Note that the decelerations do not remain constant over time, but in fact have peaks and valleys reflecting the stiffness of particular sectors of the vehicle body as they deform over time. The vehicle with the 'hard' crash pulse has a rapid onset of a high initial pulse. For the vehicle with a 'soft' crash pulse, the peak decelerations are lower and are spread out over a longer time interval. For both types of crash pulses, the peaks will increase with the initial velocity of the vehicle.

The immediate human factors issue which becomes apparent after observing the crash pulse curves is one of determining how these decelerations translate to human injury. As a rough rule of thumb, some insight can be obtained from general guidelines (i.e. Damon, Stoudt and McFarland, 1966) which indicate whole body accelerations of 50 Gs or less are survivable when the durations are less than 0.1 seconds. Woodson (1981) concludes that head accelerations should not exceed 80 Gs for over 3 ms, chest accelerations should not exceed 60 Gs for 3 ms, and pelvic accelerations should be less than 50 to 80 Gs. (A significant collection of more sophisticated measures of human tolerance to injury have been collected within sources such as SAE J885, 1986, which more accurately reflect human biomechanics and variability as discussed further later in this chapter.) The important point is that comparison of the acceleration levels in Figure 7.1 to the general guidelines for survivability given above reveals that many crashes might be survivable if the occupant experienced the same crash pulse as the vehicle. Attaining this goal is referred to as riding

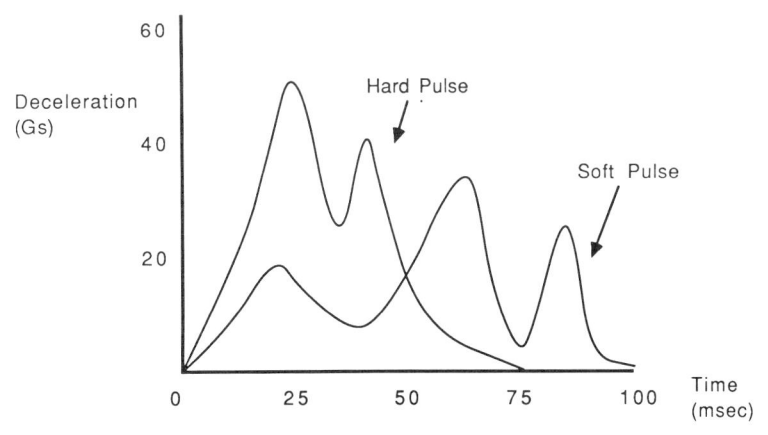

Figure 7.1. Comparison of generic hard and soft crash pulses

down the crash. Unfortunately, it is impossible to attain this goal for an unrestrained occupant and very difficult for a restrained occupant, as is shown in the following analysis.

There is a general relationship in the form of an S-shaped curve between the change in velocity of the vehicle (Delta-V) during a crash and injury severity. Fatalities can occur even at very low delta-Vs, e.g., 50 per cent occur at a velocity change of 33 mph or less. At a delta-V of 50 mph, a crash is virtually non-survivable. Because the injury-severity curve is steep for values of delta-V between approximately 25 and 45 mph, even a small shift in the threshold of contact forces that produce fatal injuries would result in a significant reduction in overall fatalities at this level (Mackay, 1988). The reason fatalities occur even at the lower levels of delta-V is that the actual accelerations incurred by the occupant are higher than those in the vehicle's crash pulse. Figure 7.2 depicts changes in occupant velocity that occur over time for an unrestrained occupant in an idealized 30 mph frontal collision. Figure 7.3 does the same for a restrained occupant.

Review of these two figures reveals three time intervals that are of particular interest. The first ($\Delta 1$) interval depicts the period (approximately 20 ms) from the initiation of the crash and ending when relative movement begins between the occupant and seat. This movement is entirely due to the vehicle and seat slowing down while the velocity of the occupant remaining unchanged. The second ($\Delta 2$) interval depicts the time between the beginning of relative movement by the occupant and the initial contact with some component of the vehicle interior (the windshield, instrument panel, or other object in Figure 7.2; a restraint in Figure 7.3). This latter interval (varying greatly between Figures 7.2 and 7.3) therefore describes a period of unrestrained movement culminating in a *second collision*, that between the occupant and the vehicle interior or restraint. The third ($\Delta 3$) interval depicts the duration (30–40 ms) of

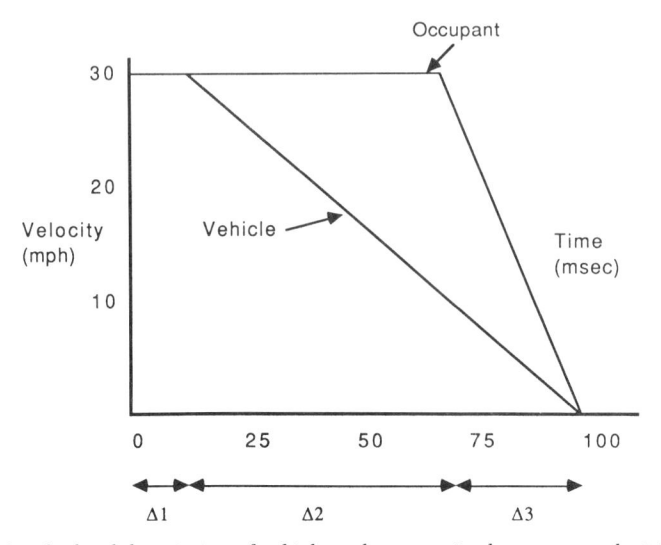

Figure 7.2. Idealized description of vehicle and unrestrained occupant velocities over time in a collision

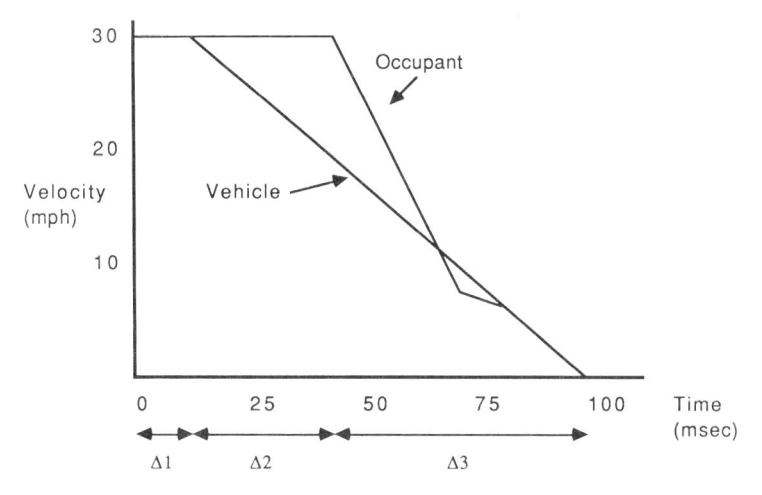

Figure 7.3. Idealized description of vehicle and restrained occupant velocities over time in a collision

the second collision. Note that a *third collision* also occurs during this latter interval when internal organs (such as the brain) contact rigid body structures (such as the inner surface of the skull). A *fourth collision* may also occur if the occupant rebounds, i.e., is thrown back after the second collision by a restraint system or other vehicle component which acts as a spring rather than dissipating all the energy.

The important point is that the velocity of the occupant remains unchanged until Δ3 begins. Consequently, the accelerations incurred by the unrestrained occupant are much higher than the peaks of the crash pulse. Furthermore, they are also significantly higher for the restrained occupant. These accelerations translate into amazingly high forces. For example, a 200 pound occupant decelerated at the relatively low maximum whole body level of 30 Gs will require a 6000 pound restraining force to prevent movement.

Generic countermeasures

A number of generic countermeasures become obvious from the preceding analysis. The first of these is to modify the crash pulse of the vehicle itself. Norbye (1984) notes that controlled crumpling of the vehicle structure is one way of attaining this goal. In general, high mass vehicles have softer crash pulses, because they have more crush space and more momentum to dissipate. A 1000 pound increase in car mass decreases the odds of driver injury in a crash by 34 per cent for an unrestrained driver and 25 per cent for a restrained driver (Jones and Whitfield, 1984). Increasing the car mass generates the largest reductions in occupant fatalities (Evans, 1989), but is generally unacceptable because of decreased fuel economy and increased consumption of resources in manufacture (Evans, 1989).

Other approaches can be classified in terms of the three time intervals within a crash described above. The first strategy is to reduce or eliminate the Δ2 interval during which the movement of the occupant is unrestrained.

Attaining this goal is extremely important, since the acceleration incurred in the second collision increases as a function of the $\Delta 2$ interval. (Minimizing the $\Delta 2$ interval maximizes the $\Delta 3$ interval). At the most general level, this simply corresponds to restraining the occupant with a belt and/or an air bag. More technologically advanced approaches make use of the $\Delta 1$ interval to couple the occupant more tightly to the vehicle prior to relative movement. Along these lines, TRW has developed a pyrotechnic belt pretensioner (Haland and Skanberg, 1989). By tightening the belts in the $\Delta 1$ interval, the amount of occupant movement prior to significant restraint by the belt system is significantly reduced. This of course increases the potential for riding down the crash at the same rate as the vehicle structure. Air bags also make partial use of the $\Delta 1$ interval, as a time window in which they can deploy.

Several other strategies are focused on minimizing the effects of the second collision occurring in the $\Delta 3$ interval. Among such strategies, early research on occupant protection focused on dissipating the forces transferred to the head, chest, and knees as the occupant impacted the interior. This research resulted in the development of energy-absorbing steering assemblies, high-penetration-resistant windshields, and slow-recovery padding. While these technologies provided improved protection, the concept of occupant restraint soon was recognized as the principal safety feature (Viano, 1988). Numerous approaches become viable for improving the performance of restraints in the $\Delta 3$ interval. In generic terms there are four objectives:

1 reduce the variance in the applied restraint force;
2 increase the excursion distance over which the restraint force is applied;
3 dissipate, do not store energy; and
4 spread the force over the greatest practical area.

The first three objectives are subject to the two constraints of: 1) keeping restraint forces within the tolerances of the human body; and 2) not exceeding the allowable excursion distance. Tolerances for restraint forces are of course determined on the basis of injury experience (i.e. SAE J885, 1986) and differ greatly depending upon the part of the body to which the force is applied. The allowable excursion distance is often defined in reference to the point where contact occurs between the head and windshield. These three objectives and two constraints interact in many complex ways and can be satisfied in a variety of ways using either air bags or belt systems. To begin with, maximizing the excursion distance over which the restraint force is applied, results in minimizing the overall restraint force. This follows because:

$$KE = (1/2)MV^2 = \int F\partial L$$

where KE refers to the kinetic energy
        M refers to the mass of the occupant
        V refers to the relative velocity of the occupant
        F refers to the restraint force
        L refers to the excursion distance over which the restraint force is applied.

Consequently, the restraint forces can be kept within reasonable bounds by allowing the restrained occupant to move within the passenger compartment in a controlled manner (i.e, by controlling the rate at which belts stretch or air bags deform). This becomes especially necessary when riding down the crash results in accelerations which exceed those tolerable or when the $\Delta 2$ interval is long enough for an appreciable relative velocity in the second collision.

At the ideal extreme, the criteria of minimizing the variance in the restraint force refers to the need for a fairly constant force which is below the maximum tolerable. However, the restraint force provided by a traditional belt or unvented air bag is generally a linear function of its stretch or deformation, respectively. A substantial body of research has been directed toward developing belt configurations which limit the forces applied (McElhaney et al., 1972; Viano, 1988; and others). Air bags are similarly designed to vent gases during contact to help limit the forces incurred. Both approaches also help attain the fourth criteria, that of dissipating energy rather than storing energy. This is important because, if a belt or bag stores rather than dissipates significant energy, a new problem emerges. Namely, if significant energy is stored the occupant will be thrown back violently after forward movement ceases with a force comparable to the original restraint force (i.e., a slingshot effect).

The final objective (spreading the force over the greatest practical area) is one of the primary justifications for the use of air bags. In fact, it has been shown that the broad area of an air bag reduces the severity of chest compression and consequently the severity of injury in frontal impacts (Groesch et al., 1986: Viano, 1988). Inflatable belts (Vrzal, 1975) also attempt to meet this objective by spreading the restraint force over a larger area than does a regular belt. From a more general perspective, inflatable belts have the potential of attaining the most comprehensive application of all the generic objectives discussed here. First, they reduce the $\Delta 2$ interval by inflating tightly around the occupant during the $\Delta 1$ interval. Second, as noted above, they distribute the restraint force over a larger area than a standard belt does. Finally, they keep the restraint forces at a fairly constant level and also dissipate energy by venting gas during their deformation. Consequently, vehicles equipped with air belts have been found to perform very well in frontal barrier tests, i.e., Fitzpatrick (1975) found 50 mph frontal barrier crashes in several experimental vehicles to be survivable when air belts were used to restrain the crash dummies.

Limits of protection

As desirable as it is to provide unbounded protection, that is not possible. Evans (1987) effectively points out that there are very definite limitations on occupant protection. An occupant protection device might be expected to have greater effectiveness (averaged over all severities) for some crash types and a lesser effectiveness for others. The overall effectiveness reflects a suitable weighted average over the complete universe of crash types. Simplifying Evans' formal analysis as that representing typical behavior averaged over all crash types, leads to an estimated overall effectiveness of 40 per cent in the field (40 per cent fewer fatalities for a protected driver than a driver without protection).

The reduction predicted by laboratory testing alone is not realized in field use because a surprisingly large number of fatal crashes are of an unusual and bizarre nature not anticipated in such testing to determine compliance to standards. Such circumstances include foreign objects entering the passenger compartment, cars being dragged for long distances along a railroad track, crashes involving greater mass and velocity than can be accommodated in the laboratory, etc. Crashes involving significant intrusion into the vehicle passenger compartment are particularly difficult to solve with restraint systems. Such crashes are associated with both high levels of energy transference and severe injury (SAE SP769, 1989).

Even though advances have been, and will continue to be, made in occupant protection, occupant protection alone will not be adequate for eliminating fatal or serious accidents. On such a basis, Viano (1988) proposes that other technologies are needed which focus on the human factors of accident prevention. Within this broader perspective the focus is placed on reducing human error via approaches such as computerized crash warning and avoidance systems, night vision enhancement systems for low visibility driving conditions, removal of high risk drivers from the road, intelligent guidance systems, etc.

Human factors issues

From the human factors perspective, a wide variety of highly pertinent design issues emerge. Among such issues, the role of, and means for, testing occupant protection is of great importance. Biomechanical modeling of human movement in general, and of injury mechanisms in particular, are of great concern. Human variability and behavior (e.g., seat belt usage, occupant positioning, etc.) also must be addressed, particularly in reference to their influences on the ultimate effectiveness of active versus passive restraint systems.

Testing methods

All major automobile companies have made significant commitments to testing occupant protection systems as part of the design process. Test methods can be separated into the categories of dynamic and static testing. Dynamic test methods are implemented in a wide variety of ways, including tests in which a vehicle is propelled into a fixed barrier (SAE J850, 1980; FMVSS 208) or has the roof crushed (SAE J996, 1980; FMVSS 216). Other forms of dynamic tests employ a wide variety of mockups in which parts of the vehicle are accelerated under carefully controlled conditions (i.e. pistons which push steering wheels rims at high speeds, sled bucks containing a seat and restraint system which are accelerated together with simulated crash pulses (SAE J117, 1970), etc.). Dynamic tests are particularly important for studies of occupant protection, as they provide a means for documenting the forces, accelerations, velocities, and movements which actually occur. Certain restraint systems such as air bags can only be tested dynamically. Furthermore, anthropomorphic dummies such as those used by automobile manufacturers and governmental agencies provide particularly sophisticated information about what happens to the occupant throughout a crash. The Hybrid II and Hybrid III dummies (Foster et al., 1977; Alderson et al., 1986) are used by most major manufacturers in their testing programs. EUROSID, BIOSID, and SID (Irwin et al. 1989) are also being introduced to measure forces in side impacts.

Static test methods focus on recording the stress/strain relationship during the slow application of carefully controlled forces to a structure. This relationship can then be used to infer the behavior of the structure under dynamic loading conditions. While much of the current focus is on dynamic testing, static tests such as currently required for FMVSS 214 (Side impact protection) do provide useful information (e.g, stress-strain curves documenting crush resistance).

Biomechanics

Understanding of the biomechanics of human movement at the high acceleration levels faced in a crash is essential to the development of modeling techniques capable of providing useful predictions and guiding the ultimate development of effective restraint systems. At the most basic level, such analysis takes the form of describing the human body as a set of linked masses, which has the obvious implication that the human body must be restrained at multiple locations to prevent relative movement between linkages.

Restraint systems have accordingly been developed in a wide variety of configurations. With regard to safety belts, several varieties of two-point and three-point systems have been developed. The earliest version in common use was the two-point lower torso (lap) belt. With increased recognition of the adverse effects of upper torso rotation around the lower torso belt, three-point systems were introduced to restrain the upper torso as well. Other variations of two-point belts have also been developed which restrain the upper torso with a shoulder belt and control movements of the lower torso with knee bolsters. These latter systems were developed as a method for passive restraint as discussed later. Another set of extensions have focused on head-neck injury mechanisms, and include head restraints and driver side air bags. There also has been a focus on reducing the forces applied to the sternum and pelvic region, as noted earlier, by providing restraint with force-limiting safety belts or air bags. Air bags, of course, require a knee bolster or seat belt to reduce submarining in frontal collisions (Klove and Oglesby, 1972; Kallieras et al., 1982) and offer little protection in side impacts or roll-overs.

Research in the biomechanics area is devoted to at least three issues: 1) defining the inertial properties of the human body, 2) determining the dynamic response of body subsystems to input forces, and 3) determining injury mechanisms. The inertial properties of the human body are of course determined by the distribution of mass over their physical dimensions (i.e. Damon, Stoudt and McFarland, 1966; provide a useful summary of such data obtained by analyzing cadavers). While static measures are easily obtained, their application in modeling has proved to be more difficult because the inertial properties of the human body change over time at the high acceleration levels experienced in a crash (i.e., as the body deforms its inertial properties change). Nevertheless, substantial progress has been made in this area, to the extent that current anthropomorphic dummies such as Hybrid III match 'average' human inertial qualities closely (Kaleps and Whitestone, 1988).

Other research has been directed toward measuring the dynamic response of body subsystems to input forces (e.g. SAE J1460, 1985) and determining the biomechanics underlying injury mechanisms (Sances et al., 1984; SAE J885, 1986). At the most basic level, the susceptibility to injury can be classified in terms of the directionality of the acceleration (i.e., Damon, Stoudt and McFarland, 1966, provide monograms defining tolerance limits for 1) forward acceleration, 2) backwards acceleration, 3) upwards acceleration and 4) downwards acceleration). At a more detailed level, substantial effort has been directed towards isolating separate systems for analysis, such as the head, head–neck, sternum, pelvic region, and lower torso. This latter approach focuses the analysis and helps guide the design of many specific aspects of restraint systems.

In conjunction with such efforts, a substantial body of work has been directed toward developing anthropomorphic dummies with high fidelity dynamic responses to localized loading (i.e. Irwin et al. (1989) compare the impact responses of the SID and EUROSID side impact dummies; Viano et al. (1986) evaluate the head dynamics and facial contact forces in the Hybrid III dummy). The availability of such information is clearly critical to determining the potential for various types of injuries that traditionally have been difficult to measure during testing, such as facial injuries due to contact with the steering wheel or chest injuries caused by localized distribution of forces by safety belts.

Anthropometrics

One of the major stumbling blocks to the successful design of restraint systems is how to accommodate the large range of potential occupants. The driver has been the focus of attention for the initial development of restraint systems, particularly passive restraints, because drivers are most at risk. Another consideration is that there is a smaller variance in size among the driving population than for all automobile occupants. The FMVSS 208 requires barrier testing to be done with a 50th percentile male dummy. However, the development of an effective occupant protection system must consider a much wider range of occupants.

Most automotive design standards attempt to accommodate occupants from at least the fifth percentile female to the 95th percentile male. Variation in occupant mass poses a particularly difficult problem, because the acceleration of the occupant is a function of both the restraining force and occupant mass [i.e. $A(t) = F/M$]. This means, for belt systems in particular, that the weight of the occupant will introduce a significant source of variance into the expected levels of deceleration. For the same restraint force, a 100 pound occupant will be decelerated at a level twice that experienced by a 200 pound occupant. This problem is potentially serious if a light occupant has a significant relative velocity prior to the initiation of restraint, since the force at which the belt begins to stretch may correspond to an unacceptably high deceleration. On the other hand, a heavy occupant may stretch the belt beyond the maximum desirable excursion distance. Both problems are obviously reduced if the restraint forces primarily reflect a riding down of the crash pulse, as becomes more likely when there is no slack present between the belt and the occupant prior to initiation of the crash pulse.

The problem above may be aggravated by efforts to provide increased comfort and/or to implement a passive belt system, since these approaches usually increase the amount of slack in the belt. Also, the difficulty in positioning a belt system effectively to restrain all sizes of potential drivers and occupants is formidable (Wells et al., 1986). This difficulty follows because the height and circumference of the driver, seat position and belt mount locations interactively determine the angle at which a belt crosses the occupant's upper torso. Because of such variability, it becomes difficult to place the belt in the most effective position for each possible occupant. More adjustability is being designed into new systems, such as by providing adjustable D-rings which allow the upper anchor point of a shoulder belt to be moved to multiple vertical locations on the B-pillar. By doing so, slack can be reduced and belt comfort significantly improved for shorter occupants. Directly coupling belt anchorage locations with seat position is an alternative approach which provides similar benefits and can be combined with the concept of adjustable D-rings (Harberl, Ritzl and Eichinger 1989). This latter approach of integrating the seat and the belt restraint is currently used in the 1991 BMW 850i—a vehicle that can absorb the high cost of such a system.

For air bag systems, there are difficulties both with the mass of the driver, and the distance from the driver to the point of bag deployment. As this distance decreases it becomes more difficult to deploy the bag without injuring the driver, because air bags possess considerable amounts of energy when deploying. Since small drivers tend to place themselves closer to the steering wheel, they are closer to the deploying air bag. This proximity combined with the fact they are lighter in mass makes it more probable that some injury might occur. On the other hand, as the distance increases, the occupant attains a greater relative velocity prior to contacting the deploying air bag. This latter result increases the required restraint force and the chance a heavy occupant will go through the air bag without adequate attenuation of their relative velocity.

A related problem that has delayed the development of effective passenger side air bag is the concern for the out-of-position child (Patrick and Nyquist, 1972). That is, an infant or small child who is not sitting normally in the seat, but might be lying on the seat, or even leaning against the portion of the instrument panel from which the device will deploy. The force of deployment might injure the child via 'bag slap'—where the leading mass of the deploying bag strikes the child with high velocity—or 'catapulting'—where the child is thrown violently into the seat back and rebounds into the bag (Biss *et al.*, 1980).

Aging

The average age of the driving population in the United States is increasing. From 1990 to the year 2000 the number of persons over age 55 will increase by some 6 million, or 11.5 per cent (Czaja, 1990). The greatest growth will be in the segment 75 years and older (increasing by nearly 5 million). It is well established that the aging process reduces tolerance to crash forces (Viano *et al.*, 1989; Verhaegen *et al.*, 1988). Not only are bones more brittle, but the recovery of normal functioning is more difficult for the older population than for younger age groups. The probability of death or serious injury is increased by about 70 per cent for those over 60 years of age in comparison with the 20 year-old group (Mackay, 1988).

Even when seat belts are used, an older occupant is at higher risk than a younger occupant. Restraint systems that are usable and effective for younger occupants may not be able to accommodate the reduced tolerance to restraint forces of the older occupants. Furthermore, for some older people the reaches required to grasp a belt, or the fastening and unfastening of buckles may be difficult or impossible, due in part to arthritic limitations (Mackay, 1988).

When an older person is injured, it is more likely to be a severe or fatal injury. The physiological changes that occur with aging make one more susceptible to injury and reduce the amount of force that can be successfully withstood in a crash. Persons over 75 are more frequently passengers than drivers (Smith, 1990), which has additional implications for the design of passenger restraints. Air bags have an advantage because they distribute the force over a larger surface area than belts. However, the problems of bag slap (the leading edge of the bag hitting the occupant) and other undesirable interactions may be greater with older occupants. The passive design of an air bag overcomes many of the difficulties experienced in finding and attaching seat belts. However, it does not overcome all the problems because a seat belt is still needed to provide protection in other than frontal crashes.

The increased vulnerability of older people and the difficulties they may have in using active restraint systems pose special concerns to vehicle designers. The generic approaches to increasing effectiveness mentioned earlier are, of course, desirable for all categories of occupants. However, it must be emphasized that because of the older occupant's increased vulnerability, systems optimal for younger drivers, or designed to meet government standards, may in fact induce injury to older occupants (i.e., increasing the maximum G-force exerted by a restraint system may increase effectiveness for young drivers in severe crashes, but at the same time be more likely to injure older drivers unable to tolerate forces at similar levels). The feasibility of restraint design specifically for the older population is therefore an option which should be considered.

Risk-taking

Driving, particularly crash avoidance, involves the assessment of risk. While for most drivers this decision-making process may be largely done at a subconscious level, the outcomes of the process can have life threatening results. Because driving is part of an everyday routine for most adults in the US, little thought is given to the true complexity of the task and its consequences. When confronted with a dangerous situation, the driver has to recognize the danger, perceive it, and respond in an appropriate manner. For example, some automobile occupants choose not to wear seat belts, even when they are aware that not wearing them increases the risk of injury. This follows because the cost of wearing a belt associated with a loss of comfort and freedom of movement is perceived to be high (Slovic et al., 1978; Bloomquist and Peltzman, 1981; Fhaner and Hane, 1974).

Risk compensation theory argues that if an individual is forced to 'consume' a higher level of safety than is desired in a free market, the individual will increase risk taking behaviors to return the system to a balanced state (Bloomquist and Peltzman, 1981). This theory has been promoted to explain why engineering changes in occupant protection have not been as successful as predicted. The theory has its critics (e.g. Joksch, 1976), but presents a view that is important to consider in designing effective occupant protection. Based on this theory, one might argue that safety enhancements which are not readily perceptible (such as crumple zones inherent to the vehicle's structure) will be more effective than those with high-levels of perceived effectiveness (such as air bags) because the latter forms of enhancement increase the possibility of offsetting behaviors traceable to risk compensation.

Active versus passive systems

From the preceding discussion, several means of occupant protection become apparent. These include improving the crashworthiness of the vehicle by the use of crumple zones to dissipate energy while maintaining the structural integrity of the passenger compartment; reducing the possibility of intrusion into the passenger space; and restraining the occupant to prevent ejection as well as allowing the occupant to 'ride down' the crash with the vehicle's acceleration.

These existing approaches fall into two general categories: active and passive. An active system requires an overt action on the part of the occupant to provide protection—such as buckling a seat belt. A passive system will provide protection in some types of crashes without any action on the part of the occupant. Methods for attaining this latter goal include improving the crashworthiness of the vehicle, maintaining the structural integrity of the passenger compartment, providing a friendly interior, and including restraint systems such as air bags or automatic belts which require no overt response by the occupant to be effective.

To be effectively implemented either approach must satisfy some minimum guidelines such as those proposed for effective restraint systems by Vrzal (1975). These guidelines suggest that any alternative system should:

1. be within a logical projection of the state of the art;
2. be practical to implement;
3. be configured for occupant comfort and convenience;
4. be available for all types and sizes of people; and
5. be designed for real world accidents and not just a frontal barrier crash.

The following discussion will first provide an overview of passive and active approaches. Tradeoffs between the two approaches in terms of the above guidelines, along with future directions for improving restraint systems will then be briefly addressed.

Passive systems

Improving the crashworthiness of the vehicle is a passive approach which despite its importance, is beyond the scope of this chapter in that it does not deal as directly with the interface of the human and the machine. However, it should be noted that minimization of intrusion takes on even more importance as the National Highway Traffic Safety Administration prepares to require manufacturers to meet injury criteria in a dynamic side impact test (FMVSS 214). This new procedure requires a 3000 pound wheeled impactor, known as a moving deformable barrier, to impact the side of the target vehicle at 33.5 mph. The specific requirements of the test are such that it is intended to simulate a crash between a vehicle moving at 30 mph into the side of a vehicle travelling at 15 mph. Acceptable criteria are determined by averaging the values from three accelerometers located in a Side Impact Dummy (SID). Particularly for smaller automobiles, this new standard will require major redesign of the vehicle structure and interior.

The concept of the 'friendly' interior is another passive approach which has been promoted for occupant safety. This approach focuses on the use of energy absorbing materials and careful design of interior surfaces and shapes that would negate the damage caused by impact with the interior of the car without the use of any add-on devices (such as seat belts or air bags). Thus far it has not proven possible to protect an unrestrained occupant in this manner for barrier type crashes at 30 mph or higher. However, this approach has clearly reduced the incidence of injuries associated with contact with solid protruding objects (such as switches or steering wheel rims) or lacerations associated with protrusion of the head through the windshield. Furthermore, work continues on developing interiors so as to minimize injury to unrestrained occupants. The NHTSA proposed regulation on interior head impact (FMVSS 214) has renewed the importance of making the interior more forgiving, even with restraint systems.

Current passive restraint designs fall into two categories, self-applying seat belts or air bags. Although there are differences in the details of design of passive or automatic belts, purchasers of vehicles with automatic belts favor them over manual belts by 3 to 1 (NHTSA 1988). There also is evidence that consumers will pay for the additional costs needed for the installation of air bags (Winston and Mannering, 1984). In general, consumers will favor a car they perceive as safer over those that are judged less safe (McCarthy, 1987). Manufacturers have recognized this shift in consumer behavior and have made it a keystone of their corporate advertising, e.g. Volvo, Mercedes-Benz, and Chrysler (Braunstein, 1990).

Air bags have recently soared in acceptance (*Automotive Electronics Journal*, 1990) while passive belts have been criticized because of their lack of acceptance (Johannessen and Yates, 1972; Snyder, 1969) and subsequent misuse or disuse. Passive belts can of course be easily disabled by occupants determined to do so, because all such systems provide a quick release mechanism intended for use under emergency conditions. Air bags are designed as a supplemental restraint, and thus require the occupant to use a seat belt for optimum protection. As noted earlier, they also require a knee bolster or seat belt to reduce submarining in frontal collisions (Klove and Oglesby, 1972; Kallieras et al., 1982) and offer little protection in side impacts or roll-overs. Although the precise benefit of providing air bags has yet to be determined empirically, due to the small proportion of the total fleet so equipped (TRB, 1989), public acceptance has made great strides. Much of the change in attitude may be due to anecdotal stories of air bag effectiveness, including those used in advertising.

Active systems

Active systems consist of seat belts. Lap/shoulder belts have been well documented as the single best protection against injury, given that a crash has occurred (Bohlin, 1967; Moreland, 1962; Grime, 1979; Huelke et al., 1985).

Estimates of effectiveness in injury reduction range from 40 to 90 per cent, depending on the speed of the crash and the type of injury. Evans (1986) found a 43 per cent reduction in fatal injury risk when the lap/shoulder belt system was used by front seat occupants.

The difficulty with seat belts is that even when mandated by law, general usage rates typically stabilize at around 50 to 60 per cent (Wagenaar, 1986). Those involved in fatal crashes have seat belt use rates substantially lower—about 25 per cent (NHTSA, 1989). In certain countries with long standing mandatory use laws and consistent enforcement, general use rates have been over 90 per cent. Efforts to promote similar use rates in the United States have been notoriously ineffective. As noted by Robertson (1983) extensive communication campaigns have failed dismally and the seat belt safety interlock system met with immense public opposition, to the extent that many such systems were disabled and quickly prohibited by an act of Congress. Other approaches such as buzzers and warning lights have been only moderately effective in motivating use (Westefeld and Phillips, 1976).

Tradeoffs and future trends

Because of the failure to obtain high use rates of active systems in the United States, NHTSA has mandated the installation of passive restraints in passenger cars. Effective for the 1990 model year, the driver must be provided with passive protection and by the 1994 model year both driver and passenger must have passive protection. Furthermore, the air bag is rapidly being phased in by all manufacturers, for both automobiles and light trucks (pickups, vans, and utilities) to the exclusion of passive belt designs. Light trucks are required to have passive restraints by 1998 model year.

This trend toward passive systems may be somewhat surprising, as the justification for moving toward air bags is somewhat suspect. Table 7.1 compares some of the advantages and disadvantages of the two types of passive restraints found in current production passenger cars. As is apparent from this summary, the current trend toward air bags reflects a large emphasis on their completely passive nature. It also reflects a lessened focus on their supplemental nature reflecting their inadequacy in crashes other than frontal collisions (only 50 per cent of fatalities occur in frontal collisions) and their dependence on safety belts and knee bolsters to provide optimal levels of protection.

Work continues on advanced and innovative restraint systems. Air bags are being proposed to provide protection in side impacts (Olsson, Skoette, and Svensson, 1989) and to protect rear seat occupants. GM has recently introduced a self-aligning steering wheel in 1987 model cars intended to reduce injury (Ealey, 1986). Additional features of belt systems such as pretensioning and webbing locks have shown promising results in the laboratory, but have to be verified empirically for final determination of their effectiveness (Viano, 1988).

Other innovative approaches include air belts (Vrzal, 1975). A deployable air belt was recently proposed by the authors as a means of combining some of the virtues of a traditional air belt and air bag. Preliminary testing of the concept has been accomplished (Lehto, Foley and Peacock, 1991) indicating that innovative approaches to occupant protection may eventually emerge, providing occupant protection superior to current systems.

Table 7.1. Airbags versus passive belts

System	Advantages	Disadvantages
Airbag	Completely passive No incentive to disconnect Disperses force over large area, reducing chest incursion Projected injury reduction in death and serious injury is 55% in frontal impacts Potential to protect head	Most expensive alternative Not effective in multiple impact crashes, roll-overs, or side-impacts May injure and doesn't protect out-of-position occupant Requires seat belt for maximum effectiveness, knee bolster for frontal collisions
Passive belts	Protects in nearly all types of crashes Relatively low cost	Low consumer acceptance Easily disconnected Interference with ingress and egress Difficulty in accommodating wide variations in occupant size

Conclusions

Consumer demand and the strong pressures of governmental agencies have made the need for improved occupant protection more critical than ever. Consumers have changed from being suspicious and unwilling to pay for air bags to virtually demanding that all vehicles, both cars and trucks, be equipped with them. Government regulation continues to press for additional advances in occupant protection and is calling for new test procedures as well as new levels of protection well beyond those currently feasible. In response to such demands, air bags are being installed in both cars and trucks as rapidly as manufacturers can effectively incorporate them in the vehicle designs. However, it must be emphasized that air bags alone do not provide a perfect solution, and may not provide the most cost effective approach. The three-point lap/shoulder belt restraint design is particularly effective, if users can be persuaded to wear them. Extensions in belt designs including the use of pretensioners, five-point harnesses, and energy absorbing belt materials seem particularly promising.

While there has been much improvement in occupant protection, the 'ultimate' system has yet to be developed. Much of the difficulty in the United States can be traced to the unwillingness of occupants to take active means of protecting themselves. While mandating such forms of behavior has been somewhat successful, a significant proportion of occupants continue to resist belt use. Consequently, additional innovation and fresh solutions may be needed to provide the maximum protection for all occupants.

Furthermore, there are still many aspects of the proposed solutions to occupant protection that are not adequately understood. Empirical verification of the designs being introduced will not be available for years until a substantial proportion of the vehicle fleet is so equipped. It will take nearly 15 years before 90 per cent of the fleet will be equipped with automatic restraints (TRB, 1989). In spite of these limitations, progress must continue to be made. Substantial research is obviously necessary to resolve some of the outstanding questions about the effectiveness of proposed solutions.

Simply put, design for occupant protection in car crashes is a difficult and critical task. Besides the incredible forces involved, human susceptibility to injury varies in complex ways. An additional problem is accommodating the wide variation in human bodily dimensions and weight. These factors combine to alter the positioning of occupants and accelerations they will experience for given restraint forces. They also make it difficult to manage the force distributions in optimal ways for preventing injury. Human factors practitioners can make a substantial contribution to solving such problems because of their familiarity with biomechanics and anthropometrics and their orientation towards engineering the system to overcome human limitations.

References

Alderson, S. W., Rozko, J. D., Stobbe, J. D. and Wright, S. G., 1986, Hybrid III testing: Problems and solutions, in *Passenger Comfort, Convenience and Safety: Test Tools and Procedures*, SAE paper 860202, Warrendale, PA: Society of Automotive Engineers.

Automotive Electronics Journal, 1990, Safety concerns to boost passive restraints market, January 1, 11.

Biss, D., Fitzpatrick, M., Zinke, D., Strother, C. and Kinchoff, G., 1980, A systems approach to airbag design and development, SAE paper 806031, Warrendale, PA: Society of Automotive Engineers.

Bloomquist, G. and Peltzman, S., 1981, An economist's view, in Crandell, R. and Lave, L. (Eds) *The Scientific Basis of Health and Safety Regulation*, pp. 37–52, Washington, DC: The Brookings Institute.

Bohlin, N. I., 1967, A statistical analysis of 28,000 accident cases with emphasis on occupant restraint value, in *Proceedings of the 11th Annual Stapp Car Crash Conference*, SAE Paper 670925, pp. 455–78, Warrendale, PA: Society of Automotive Engineers.

Braunstein, J., 1990, Convenience, simplicity may sell airbag as well or better than safety reasons, *Detroit Free Press*, 26 March.

Czaja, S. J., 1990, Special Issue Preface, *Human Factors*, **32**(5), 505–7.

Damon, A., Stoudt, H. and McFarland, R., (1986), *The Human Body in Equipment Design*, Cambridge, MA: Harvard University Press.

Ealey, L., 1986, No olds barred: Getting safe, *Automotive Industries*, October, 76–77.

Evans, L., 1986, Double pair comparison—A new method to determine how occupant characteristics affect fatality risk in traffic crashes, *Accident Analysis and Prevention*, **11**, 293–306.

Evans, L., 1987, Occupant protection device effectiveness—Some conceptual considerations, *Journal of Safety Research*, **18**, 137–44.

Evans, L., 1989, Passive compared to active approaches to reducing occupant fatalities, SAE Paper 896138, Warrendale, PA: Society of Automotive Engineers.

Fhaner, G. and Hane, M., 1974, Seat belts—Relations between beliefs, attitudes and use, *Journal of Applied Psychology*, **59**(4), 472–82.

Fitzpatrick, M., 1975, Inflatable belt development for subcompact car passengers, DOT HS-801 719, Washington, DC: NHTSA.
FMVSS 208, 1991, *Occupant crash protection*, 49 CFR 571.208, Standard No 208.
FMVSS 209, 1991, *Seat belt assemblies*, 49 CFR 571.209, Standard No. 209.
FMVSS 214, 1991, *Side impact protection*, 49 CFR 571.214, Standard No 208.
FMVSS 216, 1991, *Roof crush resistance—passenger cars*, 49 CFR 571.216, Standard No 216.
Foster, J. K., Kortge, J. D. and Wolanin, M. J., 1977, Hybrid III—A biomechanically-based crash test dummy, SAE Paper 770938, Warrendale, PA: Society of Automotive Engineers.
Grime, G., 1979, A review of research on the protection afforded to occupants of cars by seat belts which provide upper torso restraint, *Accident Analysis and Prevention*, **11**, 293–306.
Groesch, L., 1985, Chest injury criteria for combined restraint systems, SAE Paper 851247, Warrendale, PA: Society of Automotive Engineers.
Groesch, L., Katz, E., Marwitz, H. and Kassing, L., 1986, New measurement methods to assess the improved injury protection of airbag systems, *Proceedings of the 30th Annual Conference of the American Association for Automotive Medicine*, Montreal, Canada, 6–8 October.
Haland, Y. and Skanberg, T. 1989, A mechanical buckle pretensioner to improve a three point seat belt, SAE Paper 896134, Warrendale, PA: Society of Automotive Engineers.
Harberl, J., Ritzl, F. and Eichinger, S., 1989, The effect of fully seat-integrated front seat belt systems on vehicle occupants in frontal crashes, SAE 896136, Warrendale, PA: Society of Automotive Engineers.
Hobbs, C. A., 1989, The influence of car structures and padding on side impact injuries, SAE Paper 896115, Warrendale, PA: Society of Automotive Engineers.
Huelke, D., Compton, C. and Studer, R., 1985, Injury severity, ejection, and occupant contacts in passenger car rollover crashes, SAE Paper 850336, Warrendale, PA: Society of Automotive Engineers.
Irwin, A., Pricopio, L., Mertz, H., Balser, J. and Chkoreff, W., 1989, Comparison of the EUROSID and SID impact responses to the response corridors of the International Standards Organization, SAE Paper 890604, Warrendale, PA: Society of Automotive Engineers.
Johannessen, H. G. and Yates, G. A., 1972, Passive and semi-passive seat belts for increased occupant safety, SAE Paper 720438, Warrendale, PA: Society of Automotive Engineers.
Joksch, H., 1976, Critique of Sam Peltzman's study: The effects of automobile safety regulation, *Accident Analysis and Prevention*, **8**, 129–37.
Jones, I. and Whitfield, R., 1984, The effects of restraint use and mass in 'downsized' cars, SAE Paper 840199, Warrendale, PA: Society of Automotive Engineers.
Kaleps, I. and Whitestone, J., 1988, Hybrid III geometrical and inertial properties, SAE Paper 880638, Warrendale, PA: Society of Automotive Engineers.
Kallieras, D., Mattern, R., Schmidt, G. and Klause, G., 1982, Comparison of 3-point belt and air bag–knee bolster systems, *Proceedings of the 1982 IRCOBI Conference on the Biomechanics of Impacts*, pp. 166–38, Cologne, Germany.
Klove, E. H. and Oglesby, R. N, 1972, Special problems and considerations in the development of air cushion restraint systems, SAE Paper 720411, Warrendale, PA: Society of Automotive Engineers.
Lehto, M., Foley, J. and Peacock, B., 1991, Design and development of a concealed air belt system for protection of automobile occupants, working paper, West Lafayette, IN: School of Industrial Engineering, Purdue University.
Mackay, M., 1988, Crash Protection for Older Persons, in *Transportation in an Aging Society*, Washington, DC: National Research Council, Transportation Research Board.

Mannering, F. and Winston, C., 1987, Recent automobile occupant safety proposals, in *Blind Intersection? Policy and the Automobile Industry*. Washington, DC: Brookings Institution.

McCarthy, P., 1987, The effect of automobile safety on vehicle type choice: An empirical study, Falls Church, VA: AAA Foundation for Traffic Safety.

McElhaney, J., Roberts, V., Melvin, J., Shelton, W. and Hammond, A., 1972, Biomechanics of seat belt design, SAE technical paper No. 720972, Warrendale, PA: Society of Automotive Engineers.

Moreland, J., 1962, Safety belts in motor cars, an assessment of their effectiveness, *Ann. Occup. Hygiene*, **5**, 95–98.

NHTSA, 1988, Industry and consumer response to new federal motor vehicle safety requirements for automatic occupant protection, Phase II Report to Congress, Washington, DC: US Department of Transportation.

NHTSA, 1992, 1991 Traffic Fatalities Preliminary Report, Washington, DC: US Department of Transportation.

Norbye, J., 1984, *Car Design: Structure and Architecture*, Blue Ridge Summit, PA: Tab Books.

Olsson, J. A., Skoette, L. and Svensson, S., 1989, Air bag system for side impact protection, SAE Paper 896118, Warrendale, PA: Society of Automotive Engineers.

Patrick, L. and Nyquist, G., 1972, Airbag effects on the out-of-position child, SAE Paper 720442, Warrendale, PA: Society of Automotive Engineers.

Pletchsen, B., Herrmann, R., Kallina, I. and Zeudker, F., 1990, Essential requirements for an effective full scale frontal impact test, SAE Paper 900411, Warrendale, PA: Society of Automotive Engineers.

Robertson, L., 1983, *Injuries: Causes, Control Strategies, and Public Policy*, Insurance Institute for Highway Safety, Lexington, MA: Lexington Books.

SAE J4, 1955, *Motor vehicle seat belt assemblies.* Superseded by J114, J117, J140a, J141, J339a, J800c, Warrendale, PA: Society of Automotive Engineers.

SAE J117, 1970, *Dynamic test procedure—type 1 and type 2 seat belt assemblies*, SAE recommended practice J117, Warrendale, PA: Society of Automotive Engineers.

SAE J996, 1980, *Inverted vehicle drop test procedure*, SAE recommended practice J996, Warrendale, PA: Society of Automotive Engineers.

SAE J850, 1980, *Barrier collision tests*, SAE recommended practice J850, Warrendale, PA: Society of Automotive Engineers.

SAE J1460, 1985, *Human mechanical response characteristics*, SAE J1460, Warrendale, PA: Society of Automotive Engineers.

SAE J885, 1986, *Human tolerance to impact conditions as related to motor vehicle design*, SAE J885, Warrendale, PA: Society of Automotive Engineers.

SAE SP769, 1989, *Side impact: Injury causation and occupant protection*, SP769, Warrendale, PA: Society of Automotive Engineers.

Sances, A., Maiman, D., Myklebust, J., Larson, S. and Cusick, J., 1984, Biomechanics of vehicular injuries, in Peters, G. and Peters, B. (Eds) *Automotive Engineering and Litigation*, New York: Garland Law Publishing.

Slovic, P., Fischhoff, B. and Lictenstein, S., 1978, Accident probabilities and seat belt usage: A psychological perspective, *Accident Analysis and Prevention*, **10**(4), 281–5.

Smith, D. B. D., 1990, Human factors and aging: An overview of research needs and application opportunities, *Human Factors*, **32**(5), 509–26.

Snyder, R., 1969, A survey of automotive occupant restraint systems—Where we've been, where we are, and our current problems, SAE Paper 690243, Warrendale, PA: Society of Automotive Engineers.

Strother, C. E., Warner, C. Y., Woolley, R. L. and Wooley, M. B., 1990, The assessment of the societal benefit of side impact protection, SAE Paper 900379, Warrendale, PA: Society of Automotive Engineers.

Transportation Research Board (TRB), 1989, Safety belts, airbags and child restraints, Special Report 224 TRB, Washington, DC: National Research Council.

Verhaegen, P. K., Toebat, K. L. and Delbeke, L. L., 1988, Safety of older drivers: A study of the overinvolvement ratio, *Proceedings of the Human Factors Society*, 32nd annual meeting, Anaheim, CA.

Viano, D. C., Culver, C. C., Evans, L., Frick, M. C. and Scott, R., 1989, Involvement of older drivers in multi-vehicle side impact crashes, *33rd Annual Meeting of the Association for the Advancement of Automotive Medicine*, Baltimore, MD.

Viano, D. C., Melvin, J. W., McCleary, J. D., Madeira, R. G., Shee, T. R. and Horsch, J. D., 1986, Measurement of head dynamics and facial contact forces in the Hybrid III dummy, *Proceedings of the 30th Stapp Car Crash Conference*, Warrendale, PA: Society of Automotive Engineers.

Viano, D. C., 1988, Limits and Challenges of Crash Protection, *Accident Analysis and Prevention*, **20**(6), 421–9.

Vrzal, P., 1975, Occupant Protection . . . Back to the Basics, SAE Paper 750394, Warrendale, PA: Society of Automotive Engineers.

Wagenaar, A., 1986, Effects of mandating seatbelt use: A series of surveys on compliance in Michigan, *Public Health Report*, **101**(5), 505–13.

Warner, C. Y., James, M. B. and Strother, C. E., 1989, A perspective on side impact occupant crash protection, SAE 900373, Warrendale, PA: Society of Automotive Engineers.

Wells, R., Norman, R., Bishop, P. and Ranney, D., 1986, Assessment of the static fit of automobile lap-belt systems on front-seat passengers, *Ergonomics*, **29**(8), 955–76.

Westefeld, A. and Phillips, B., 1976, Effectiveness of various safety belt warning systems, DOT-HS-5-01154, Washington, DC: US Department of Transportation.

Winston, C. and Mannering, F., 1984, Consumer demand for automobile safety, *American Economic Review*, **74**(2), 316–9.

Woodson, W., 1981, *Human Factors Design Handbook*, New York: McGraw-Hill.

8

Vision and perception

Paul L. Olson

Introduction

Driving is thought of as being primarily a visual task, which, undoubtedly, it is. Some authors have estimated that 90 per cent or more of the information essential to the control of a motor vehicle is acquired visually, but this estimate may overstate the case. One characteristic of vision is the level of conscious awareness we have of its use. We are much less aware of information such as that provided by the kinesthetic senses (e.g., limb position, force feedback), without which it would be difficult or impossible to operate an automobile. Still, no one will seriously question that vision is of great importance in operating a motor vehicle. The most fundamental driving tasks, e.g., route selection, lane position, and the avoidance of conflicts rely entirely or primarily on vision. In addition, vision is the sense of choice in providing much other information, ranging from the very important (e.g., traffic control devices, engine status warnings, speed, fuel level) through the relatively unimportant (e.g., outside temperature, compass heading) to that which is not directly related to the driving task (e.g., billboards). In recent years the introduction of computer technology in cars has made it possible to provide additional information to drivers. Here again, the mode of presentation has generally been visual.

Regardless of the sense used, there are limits to how much information an individual can assimilate in a given period. Additionally, vision has strengths and weaknesses as a source of information. Some understanding of the basics of vision and visual perception is important in the design of information systems to minimize the likelihood that they will become part of the problem rather than part of the solution. The purpose of this chapter is to provide such an understanding.

This chapter is entitled 'Vision and Perception' and the terms 'vision' and 'visual perception' will appear throughout. It is not intended that the terms be used interchangeably. Vision will be used when discussing the mechanical aspects of seeing; i.e, how electromagnetic energy is converted into the neural impulses that we ultimately experience as vision. Visual perception is a complex and incompletely-understood process in which information in the form of neural impulses is processed into an integrated, meaningful experience.

This chapter consists of two main sections. The first is concerned with vision. It will discuss the organ of vision, the eye, its structure and general functioning. With that foundation in place, the second section of the chapter will deal with visual perception. Its purpose is to provide some understanding of what perception is and why it is important in the operation of a motor vehicle.

Vision

The visual stimulus

What we call 'light' is electromagnetic radiation. The electromagnetic spectrum, which is shown schematically in Figure 8.1, ranges from cosmic rays having wavelengths of 10^{-15} meters to the output of AC circuits having wavelengths of many kilometers. The portion of this spectrum to which the human eye is sensitive comprises a very small part of the total, ranging from wavelengths of approximately 400 to 700 nanometers. (A nanometer, nm, is one billionth, 10^{-9}, of a meter.) The eye is not equally sensitive to radiation within

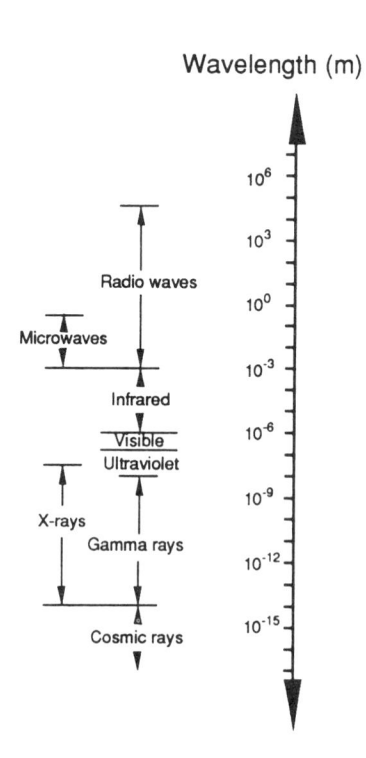

Figure 8.1. The electromagnetic spectrum, showing the position and relative size of the visible portion

the visible range. Peak sensitivity depends on the level of illumination, and is about 550 nm at high levels and about 500 nm at very low levels.

The Measurement of visual performance

The study of vision is complex. As a result, a great number of measures of visual performance have been developed. Of these, the one with which most people are most familiar is acuity. Acuity is an index of the ability to resolve fine detail. It is the principal interest in eye examinations conducted to determine the need for glasses, or a change in prescription for glasses. Poor acuity results in a relatively fuzzy image, which, while driving, may not be detected as readily as a clear image. Poor acuity also reduces the distance at which signs can be read, giving less time for the driver to respond once the message is legible.

Acuity is measured by a variety of standard tests. Examples of targets commonly used are shown in Figure 8.2. Each of these images represents a different way of measuring how well the subject can resolve fine detail. Acuity scores are normally given in comparative form, showing the performance of the subject relative to that of a 'standard observer'. An example of a score is 20/20, which is sometimes called 'perfect' vision. It is not perfect vision, but simply means that the subject can resolve detail as well at 20 feet as can the standard observer. A score of 20/40, which is a common upper limit for obtaining a driver's license, means that the subject can resolve detail as well at 20 feet as the standard observer can at 40 feet. A driver with 20/40 acuity would see objects (e.g., cars, pedestrians) less well than someone with 20/20 acuity, and would be able to read signs at about half the distance.

Acuity has an obvious relationship to the overall quality of vision and it is fairly simple to measure. For those reasons it is the main, and often the only criterion in the 'eye examination' given when applying for a driver's license. The wisdom of this approach was called into question in 1967, however, when Burg published the results of a very large scale study showing that static acuity scores do not correlate with traffic crashes. In a follow-up investigation Henderson and Burg (1974) attempted to develop a ranking of visual attributes. Their results showed that, for driving, the following were most important: dynamic visual acuity, saccadic fixation, static visual acuity, extent of the useful visual field, detection of movement in depth, and detection of angular movement.

Further review of the data from the original Burg study (Hills and Burg, 1977) indicated that there were no significant correlations between vision measures and crash data for participants under the age of 54. For those 54 and older, both static and dynamic acuity showed significant (although small) correlations with the crash data.

The investigations cited led to much research into the relationship between measures of visual performance and driving, and spurred work on the question of which aspects of vision should be measured in the driver's licensing examination. Recent reviews in these areas have been provided by Bailey and Sheedy (1988) and by Schieber (1988). Clearly, no substantial changes have come about in the licensing process as a result of this work to date, although some may appear soon.

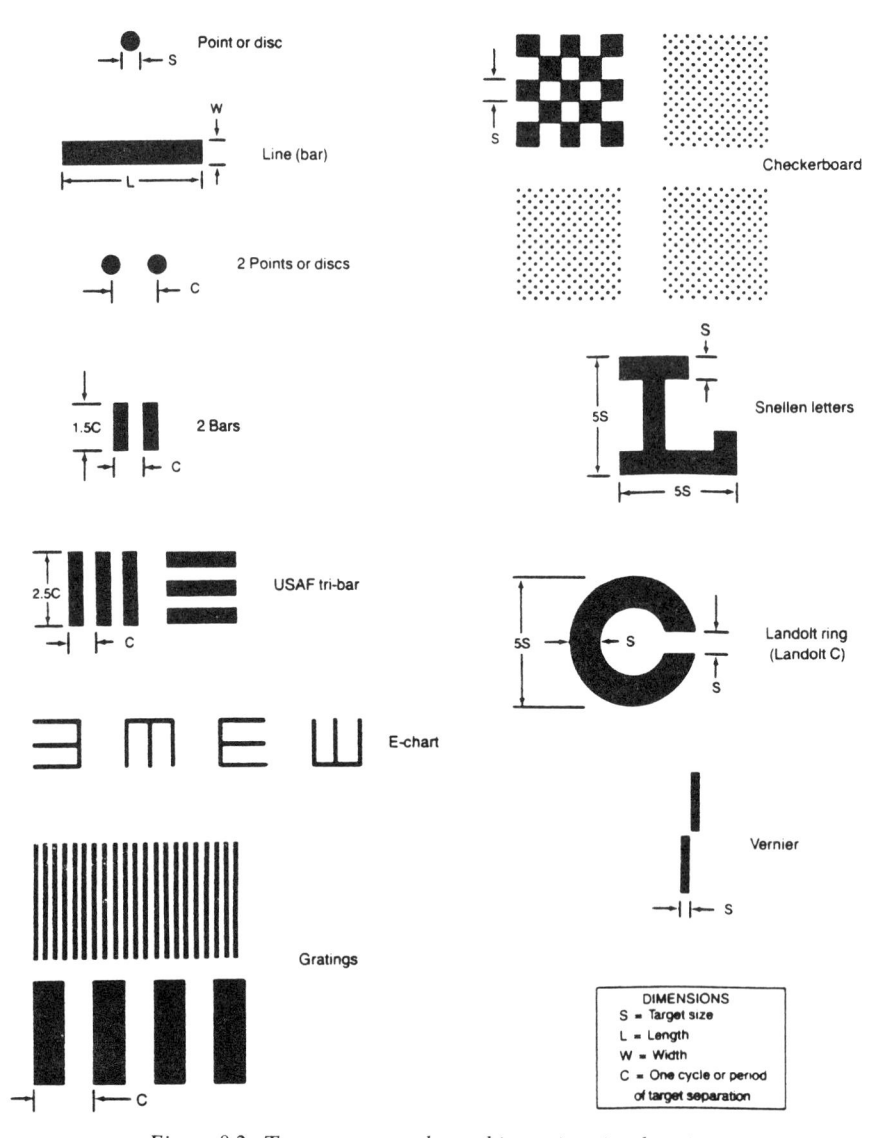

Figure 8.2. Targets commonly used in testing visual acuity

Structure and function of the eye

In terms of its gross structure, the eye is much like a simple camera. A camera has a lens to focus light, an aperture to control the amount of light entering, and a light-sensitive surface to record the image. In a camera the lens can be moved relative to the film plane to compensate for objects at different distances; the size of the aperture can be changed to adjust to different lighting conditions; and films of different sensitivity can be used to expand the range of lighting conditions in which photographs can be taken. The eye has all of these characteristics. Figure 8.3 is a schematic of the human eye with a number

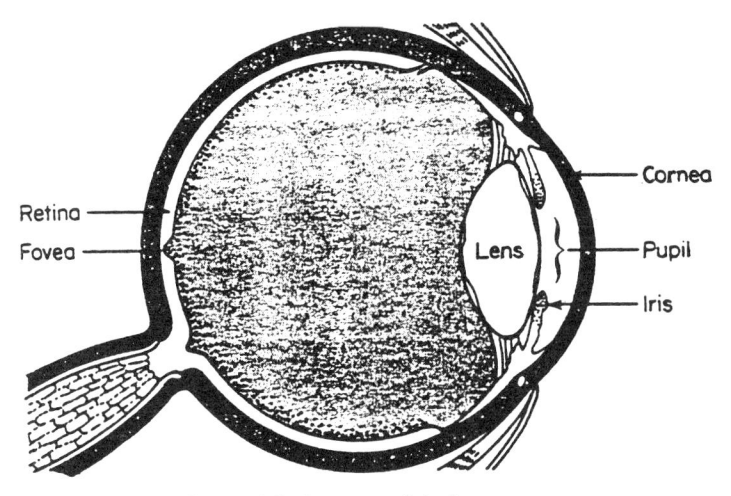

Figure 8.3. Diagram of the human eye

of the important features labelled. Each of these features will be discussed below.

Light entering the eye is brought to a focus by two elements. The first is the cornea, a portion of the outer surface of the eye that is both transparent and relatively highly curved. The cornea does most of the work of bending light rays to achieve a focus. The second element is the lens. The primary purpose of the lens is to provide a variable focal length so the eye can focus on objects that are at different distances, a process called accommodation. The lens does this by changing its shape, as shown schematically in Figure 8.4. As illustrated in the top part of Figure 8.4, when viewing relatively distant objects the entering rays of light are nearly parallel. Under this condition the network of muscles within which the lens is suspended causes it to become thinner, giving the optic system of the eye maximum focal length. When the eye is viewing

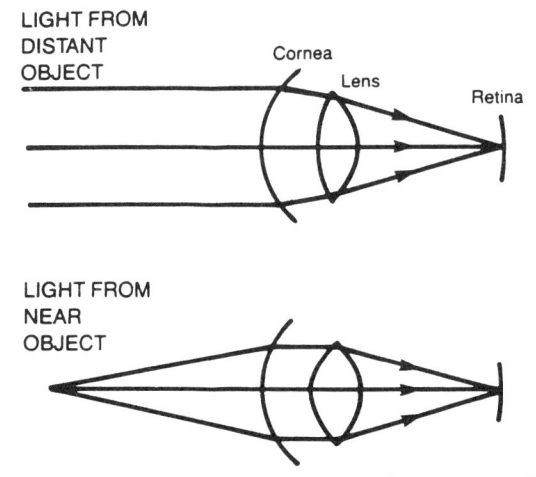

Figure 8.4. How the lens changes shape to accommodate to near and distant objects

objects that are close, the entering rays of light diverge. In this case the lens becomes fatter, reducing the system's focal length, as shown in the bottom part of Figure 8.4.

Accommodation is not accomplished instantaneously. In one study involving young to middle-aged adults (Campbell and Westheimer, 1960), the time of accommodation was measured for a change of stimulus distance requiring that the focal length of the eye be changed by a factor of two. On average, the accommodative response was initiated after a lag of about 0.3 second, and was complete in about 0.9 second. Thus, when a vehicle operator wishes to look at something in the instrument cluster, time is required for the eyes to accommodate from optical infinity to the panel and back again. Head-up displays, which project instrument images at optical infinity near the operator's normal field of view, were developed in part in recognition of the potential importance of the time lost in accommodation. Alternatively, key displays should be made large enough to be read while partially out of focus.

The flexibility of the lens decreases with age, which is a major reason middle-aged and older persons often have trouble bringing nearby objects into sharp focus (a condition known as presbyopia). The lens also acquires a yellow tint with age, which is a factor in the reduced night vision capability characteristic of older people. This phenomenon will be covered in more detail later in this chapter.

The opening in the eye through which light must pass is the pupil. Its size is controlled by a ring of muscle tissue called the iris. The iris constricts the size of the pupil at high levels of illumination, and causes it to open at lower levels. The iris thus plays a role in allowing the eyes to adapt to different levels of illumination. It is, however, a limited role. Because the iris responds quickly, it is of some help under conditions where an individual moves between areas having different illumination levels. The bulk of the adaptation that allows the eyes to function over a wide range of illumination levels is accomplished in the retina, as will be discussed later.

Images are brought to a focus on the retina, which covers about two-thirds of the interior of the eye. The construction of the retina is the reverse of what might reasonably be expected. The layers closest to the front of the eye are composed of a network of nerve fibers and blood vessels, through which the light must pass before reaching the light-sensitive receptors. At the optical center of the retina is a small depression called the fovea. Nerve fibers and blood vessels skirt the fovea, allowing light direct access to the receptors contained there.

There are two kinds of receptor cells in the retina, cones and rods. Some knowledge of their function and distribution in the retina is helpful in understanding certain characteristics of the visual system. Table 8.1 is a summary of some of the major characteristics of cones and rods. In brief, vision at high levels of illumination is by cones. Cones are found mainly in and near the fovea, enable us to see color, and have few neural interconnections, thus providing relatively good acuity. Rods are operational at low levels of illumination, are found throughout the retina with the exception of the fovea, do not provide color, and, because of a great number of neural interconnections, provide relatively poor acuity.

The operating characteristics of the two types of receptors, combined with their distribution in the retina, have important implications for what can be

Table 8.1: A comparison of the characteristics of cones and rods

Item	Cones	Rods
1 Approximate number in the eye	7 million	20 million
2 Operating illumination levels	High to medium	Medium to low
3 Provide color vision	Yes	No
4 Distribution in the retina	Concentrated in the fovea. Found in diminishing numbers as one moves away from the fovea.	None in the fovea. Distributed elsewhere throughout the retina, with the greatest concentration about 20° from the fovea.
5 Relative frequency of interconnections with other receptor cells	Low	High
6 Capability of resolving fine detail	Good	Poor

seen and how well it can be seen under different conditions. For example, vision is generally best in the small area of the fovea, due to the concentration of cones found there. Objects imaged outside the fovea are seen less well. The drop-off in acuity outside the fovea is substantial, as illustrated in Figure 8.5, which shows the relationship between acuity and angular distance from the fovea at different levels of illumination.

Figure 8.5 shows two things of importance. The primary point is the great reduction in acuity for objects imaged even a little way from the fovea. For example if foveal acuity is taken as 100 per cent, acuity for an object imaged 5° from the fovea will be about 33 per cent. The second point is the overall drop in maximum acuity as illumination is decreased. At very low levels of illumination the cones no longer function and the fovea is blind. Under those conditions acuity is better in the near periphery.

The peripheral area of the retina serves a valuable screening function. Given the relatively small area of the fovea, it is likely that unanticipated targets of importance will first appear in the peripheral retina. Once there the probability that it will capture the conscious attention of the observer is determined by certain characteristics. Research has shown that objects likely to be noticed first have high contrast, are flickering or flashing, are relatively large, or are moving (Notton and Start, 1971; Yarbus, 1967; Thomas, 1968). Once having gained the attention of the observer, the direction of gaze will generally be shifted so that the object is imaged in the fovea to facilitate subsequent identification and decision making.

The driver at night

Research has made it clear (e.g., Olson and Sivak, 1983) that visibility under night driving conditions is severely restricted, particularly when illumination is provided solely by the vehicle's headlamps. Yet vehicle speeds at night differ little from those measured during the day. The question is, if drivers cannot

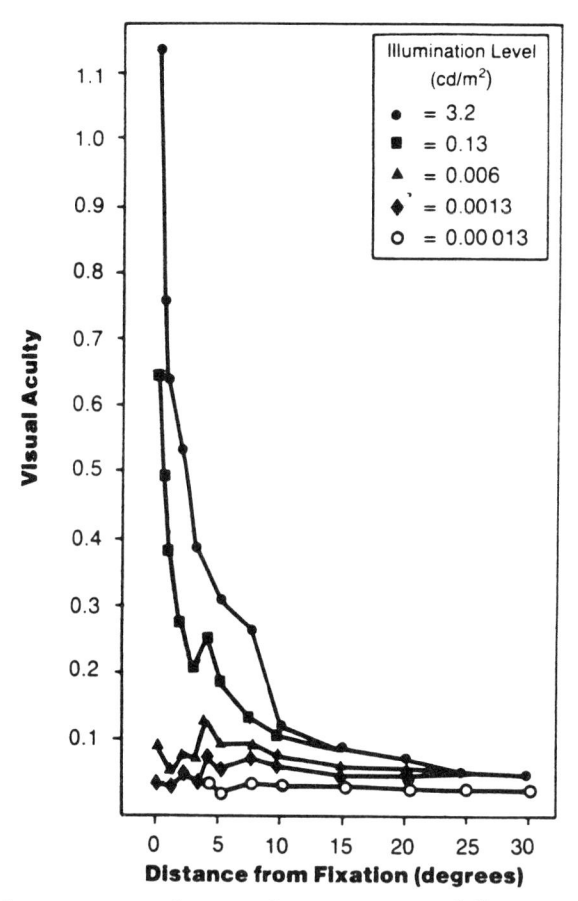

Figure 8.5. Relative acuity as a function of target position and illumination level. (Source: Mandelbaum and Sloan, 1947, *American Journal of Ophthalmology*, **30**, 581–588. Published with permission from The American Journal of Ophthalmology. Copyright by The Ophthalmic Publishing Company)

see nearly as well at night, why don't they slow down? The answer may be found in a theory advanced by Leibowitz *et al.* (1982). The theory rests on the fact that there are two independent modes of processing visual information. One is called the 'focal' mode, and is concerned with object discrimination and identification. Focal functions are optimal in the foveal area, and are affected by level of illumination and problems that affect acuity. The other mode is called 'ambient' and is concerned with spatial orientation. Spatial orientation can be accomplished in the foveal area, but, unlike the focal functions, it is adequate in the peripheral areas as well. In addition, ambient functions are much less sensitive to illumination levels and acuity than focal functions. Under night driving conditions there is a selective degradation of these two modes, with focal vision being much more affected. This means that we suffer relatively little loss of ambient vision, which is useful for maintaining lateral position on the road. The fact that focal vision is greatly reduced is less appreciated because the demands on it are intermittent. Since the driver can carry out the routine control functions about as well at night as during the day, overconfidence concerning the entire driving task may be generated.

Adaptation

'Adaptation' refers to changes in the sensitivity of the visual receptors that allow them to function at different levels of illumination. The range of illumination levels at which at least some visual functioning is possible spans about 12 log units, as illustrated in Figure 8.6. As noted earlier, vision is by means of the cones at high illumination levels, and rods at low levels, referred to as photopic and scotopic vision respectively. There is also a mid-range of illumination in which both receptors function, called mesopic vision.

Adaptation requires a certain amount of time. In general, adjusting to a higher illumination level (light adaptation) is accomplished more quickly than the opposite (dark adaptation). Adaptation to moderate increases in illumination may be accomplished in a minute or so. Extreme changes may require as long as ten minutes (Baker, 1949).

There are two phases to dark adaptation. There is an initial rapid stage of approximately five minutes due to cone adaptation, followed by a second stage, associated with rod adaptation, that may require as long as half an hour. The time course of dark adaptation, and the performance achieved at

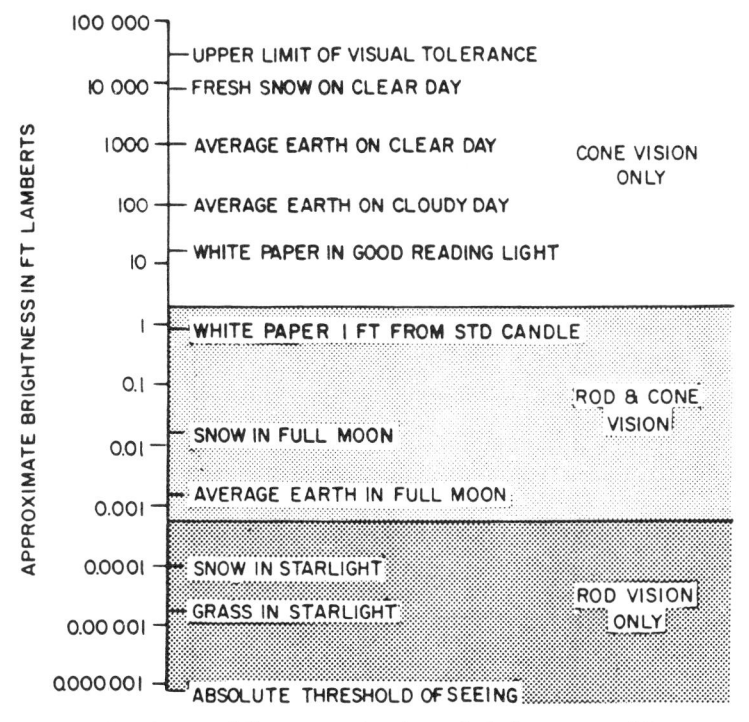

Figure 8.6. *Range of illumination levels at which there is visual function*
(Source: Grether and Baker, 1972)

the final level depends on a number of variables such as the intensity, size and wavelength of the adapting light, target size, and duration of the test target. See Boff and Lincoln, (1988) for a more complete discussion.

Recent research (Olson et al., 1990; Olson and Aoki, 1989) has provided some indication of the adaptation level of drivers under various conditions. For example, in a dark setting, using low-beam headlamps, a driver will be adapted to a level of about 0.3 ft Lamberts, which is in the upper third of the mesopic zone. Changing to high beams was found to increase adaptation level almost fivefold, to about 1.7 ft Lamberts, which is nearly at the top of the mesopic zone. These data were taken in the absence of glare from oncoming headlamps. Adding glare has a major effect on adaptation. Low beams about 100 feet distant on a two-lane highway produce an adaptation level of about five ft Lamberts. High beams under the same conditions produce an adaptation level of about 20 ft Lamberts. Both of these levels are in the photopic zone.

These results make it clear that dark adaptation changes greatly with driving conditions, and, under many of those conditions is in an almost constant state of flux. The problem is that these changes take time. It may take a minute or so for a driver to recover from exposure to high beam glare, for example. As a result visual performance at any given instant may be appreciably less than expected based on test data taken under controlled conditions.

Aging

Certain important changes in the vision system are associated with aging. From the perspective of this chapter the changes of interest are those that affect nighttime visual ability and susceptibility to glare.

Two changes that affect nighttime visual capability were mentioned in the section dealing with the structure of the eye, i.e, reduced pupil size and yellowing of the lens. The maximum diameter of the pupil decreases with age. According to Kornzweig, (1954) the area of the typical 20-year-old pupil under conditions of darkness is 12 times greater than that of an 85-year-old individual. Thus, when driving at night, the pupil of an older eye may admit only 10 per cent of the light that the pupil of a young eye would. The pupil condition, the accumulation of granular material and fibrillar protein (Dark et al., 1969), and the increased opacity of the lens (Wolf, 1960) combines to reduce greatly the amount of light reaching the retina of the older eye under conditions of low illumination

A consequence of reduced retinal illumination is that sources must be of higher intensity to be seen at night, as illustrated in Figure 8.7, taken from Domey et al., (1960). The figure shows the average minimum source intensity that could be detected by individuals in various age groups after 10 minutes in the dark (open circles) and 40 minutes in the dark (closed circles). The degree of gain in sensitivity over time seems not to change with age. The loss of retinal illumination, however, means that the stimulus, to be detectable, had to be about two log units (i.e., 100 times) brighter for subjects averaging 85 years old than it did for subjects averaging 15 years old.

In nighttime motor vehicle operation, objects are typically made brighter by approaching closer. Thus, the practical effect of lower retinal illumination is reduced visibility distance, as illustrated in Figure 8.8 (Olson, 1988). The figure

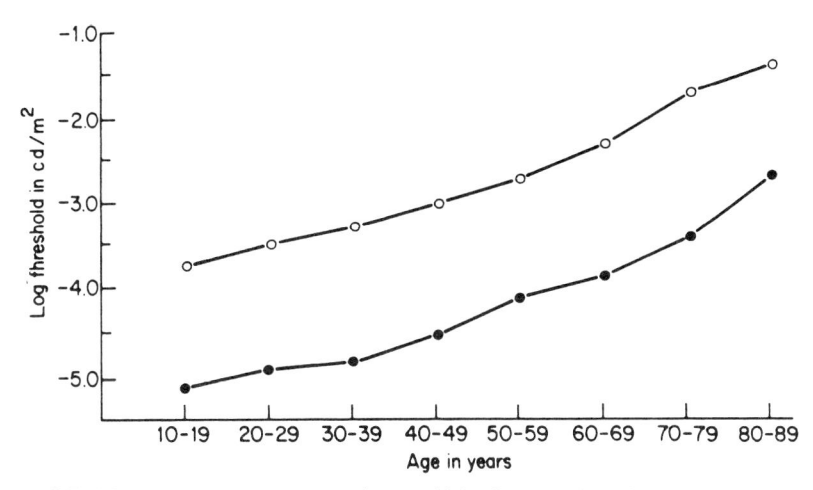

Figure 8.7. Minimum source intensity that could be detected by subjects of different ages after 10 minutes in the dark (open circles) and 40 minutes in the dark (closed circles). (Source: Domey, McFarland and Chadwick, 1960, *Journal of Gerontology*, 15, 267–279. Copyright © The Gerontological Society of America)

shows percentile distributions of detection distances of blank sign panels in environments having two levels of complexity for young and older subjects. For each complexity level the 50th percentile detection distances are about 200 feet less for the older subjects.

Light entering the eye is scattered by the optic media, with the result that some of it impinges on portions of the retina peripheral to the image. It is this scattering that causes bright light sources to appear larger than dim sources of the same physical size. It is also responsible for the loss of visibility that we suffer in the presence of glare (called disability glare), because the scattered illumination on a dark-adapted retina reduces the effective contrast between an object and its background (Fry, 1954).

One of the consequences of aging is that proportionally more scattering occurs, increasing the effects of glare, as illustrated in Figure 8.9. The figure shows the loss of visibility associated with the sudden onset of high beam glare, as well as recovery over time as the glare continues. There are three points to note in the figure. First, the older subjects were more affected by glare onset than were the young subjects; second, they took longer to reach a stable state; and third, their performance at the level of stability was substantially poorer than the young drivers.

Visual defects

There are a great number of vision problems that may affect driving capability. One of these, poor acuity, has aleady been discussed. Often poor acuity can be simply corrected, but many other visual defects do not yield to simple solutions. The listing of visual defects given here is not complete, and the descriptions are necessarily brief. A more comprehensive discussion can be found in Allen (1970).

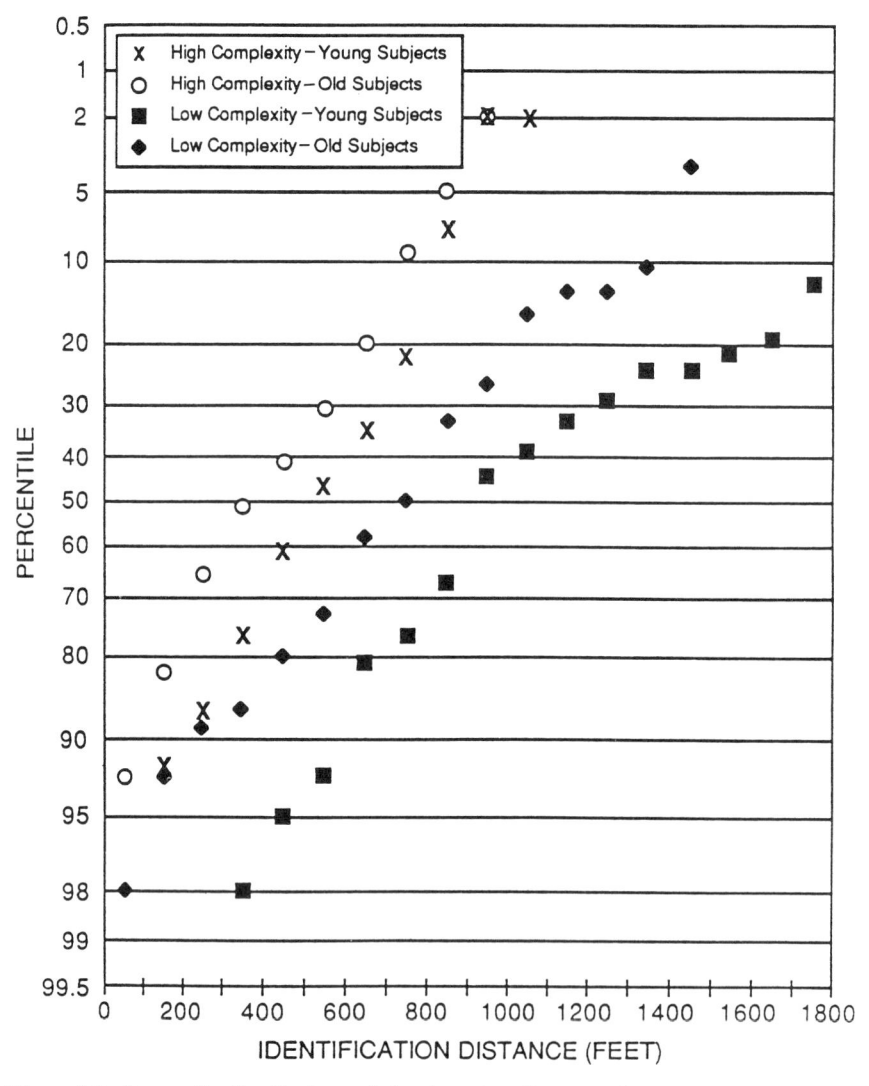

Figure 8.8. Percentile distributions of the detection distance of sign targets in areas of different complexity for subjects in two age groups
(Source: Olson, 1988)

Color blindness

Color blindness is a general term used to refer to a variety of conditions. Traditional classifications indicate that about 8 per cent of males and 0.5 per cent of females have some form of color blindness. Three levels of color deficiency are recognized:

Anomalous trichromats are 'color weak'. That is, their perception of one or more colors is less than normal, but not so severe as to fall into a more serious classification.

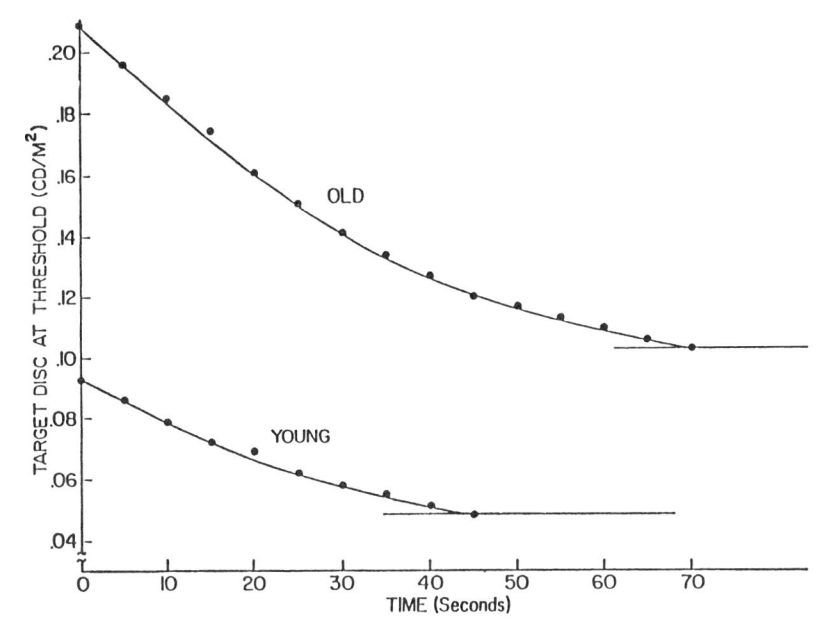

Figure 8.9. Recovery from sudden onset of glare as a function of age
(Source: Olson and Sivak, 1981)

Dichromats suffer from an inability to distinguish one of the primary colors. The two most common types are *protanopia*, an inability to distinguish red, and *deuteranopia*, an inability to distinguish green. These two conditions affect about 2 per cent of the population.

Monochromats are unable to distinguish any colors, seeing only in shades of gray. This condition is rare, afflicting about 0.003 per cent of the population.

Color deficiencies are of concern in the design and operation of traffic systems because of the wide use made of color coding. Probably the most obvious example is traffic signals. Shifting the transmissivity characteristics of the green lens more into the blue end of the spectrum has been of some help to persons with deuteranopia. Standardization of color position (i.e., red on top) has also been of benefit. Signals are, however, not the only color coded traffic indicators. Signs rely on color to indicate the class of message. Persons with a color deficiency will see certain signs less well than color normals, and may experience more problems in reading the message. Color coding can be a powerful tool in aiding information transmission. Care must be taken, however, to ensure that persons with color deficiencies are not severely disadvantaged, or the result may be a net loss in system efficiency.

Night blindness

Night blindness is a term used to refer to a condition in which the individual has a greatly reduced capability for vision at low levels of illumination. It is a consequence of a prolonged deficiency in vitamin A. Persons severely affected would probably not even attempt night driving. There are degrees of the condition, however, and those less affected may drive at night.

Night myopia

Accommodation was defined earlier as the ability of the eye to change its focal length so that both near and distant objects could be brought into sharp focus. In total darkness the eye accommodates to an intermediate state (called dark focus) that varies from person to person. Owens and Leibowitz (1976) have shown that the eye tends to accommodate to distances between infinity and that represented by dark focus as illumination levels are reduced. Because most events of consequence to a vehicle operator occur at visual infinity (i.e., beyond about 20 feet), an eye that is accommodated to an intermediate distance will see them less well. This condition is referred to as night myopia.

Available data indicate that a great number of people may have night myopia to some degree. There are large individual differences. In extreme cases a person may be focused on a point only a few feet away, with objects in the far field seriously blurred.

Theoretically, it should be possible to write a prescription for night driving glasses that would correct for night myopia. There are two factors that presently make this difficult. First, a person may have serious night myopia and not be aware of it. Thus, affected individuals are not likely to seek assistance voluntarily. Second, it is difficult to carry out conventional refractive measures at low levels of illumination, and night myopia cannot be predicted from measures taken at high levels. So, even if an individual felt the need, an ophthalmologist or optometrist could not conduct the necessary tests and write a prescription. Instruments have been developed that make such measurements practical (e.g., Leibowitz and Owens, 1975), but it may be some time before they are generally available.

Cataract

Cataract is a general term used to refer to any opacification of the lens. It may be manifest as a highly localized spot or a general loss of transmissivity in the lens. Cataracts have many causes, e.g., senile change, trauma, metabolic or nutritional defects, exposure to certain types of radiation (Davson, 1963). Cataracts reduce vision in the affected eye(s), and often increase the disabling effects of glare. They can be treated surgically, using a relatively simple procedure in which the lens is removed and replaced with an artificial one. There will always be some vision loss before surgery is recommended, and some people wait far too long before seeking medical assistance, with the result that a serious loss in vision results.

Aniseikonia

Aniseikonia is a condition in which the retinal image in one eye is significantly larger than in the other eye. This can result in a variety of symptoms, including spatial distortions and double vision. Many persons with this affliction can get along well enough during the day, due to the abundance of spatial cues, but they may experience problems at night.

Anisocoria

In anisocoria the pupils of each eye are of different size. The result is that the retina of one eye receives less illumination than the other, possibly producing a stereoscopic illusion known as the Pulfrich effect. This interesting phenomenon can be demonstrated to a person with normal vision by placing a dark lens over one eye and looking at something like a pendulum. The apparent path of the pendulum will be seen as elliptical, with significant movement toward and away from the observer. When driving, the Pulfrich effect can cause distortions in the perceived speed, distance, and trajectory of other vehicles.

Visual perception

To this point the discussion has centered on largely mechanical aspects of vision. The eye has been compared to a simple camera, which is a helpful way of studying its structure. The result of the process in the human visual system, however, can only loosely be compared to the film image from a camera. Certain processing and filtering of visual information affects what finally registers in our consciousness. The term 'perception' has been coined to refer to the result of this process.

Perception has been defined in various ways. For example, Gibson (1966) says, 'This conception and belief which nature produces by means of the senses, we call perception'. Bartley (1969) defines perception as the immediate discriminatory response of the organism to energy-activating sense organs. Both of these definitions start with sensation, and imply a processing of this information to produce the result, which is perception. Regardless of the nature of the sensation, however, the information it represents is dealt with selectively, processed, and compared with previous experience to arrive at perception.

The effects of the processing of visual information

It is easy to demonstrate that visual information undergoes some processing. For example, look at a pencil held up a few feet in front of your eyes. Since our eyes are separated by some distance, each eye obtains a somewhat different view of the pencil. It might reasonably be expected that these different images would yield a distortion or an impression of fuzziness. Instead, the pencil is perceived as standing out from its background. Similarly, due to the separation between the ears, a sound coming from one side reaches each ear at a different time. Again, one might expect a distortion or loss of clarity, but the perception is one of directionality. The processing and interpretation that these perceptions represent are clearly very useful in everyday life.

Optical illusions are visual situations that illustrate the processing of information in a dramatic way. Figures 8.10 and 8.11 show three relatively well-known examples. Figure 8.10 shows two pairs of lines labelled a and b, and a diagonal line labelled c. The pairs of lines, a and b, are equal in length and should be so perceived. Now, look at Figure 8.11. Nothing has changed about the original lines, but the upper line in a and b should appear longer than the lower line, and the top and bottom portions of the diagonal line in c do not appear to line up. The additional visual information presented in Figure 8.11

176 P. L. Olson

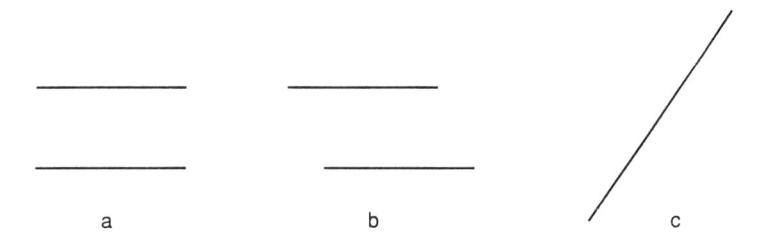

Figure 8.10. Illustration of some visual illusions (compare with Figure 8.11)

has the effect of distorting our perception of the information originally given in Figure 8.10. It is interesting that one can employ whatever means are necessary to satisfy oneself that the lines in Figure 8.11 are indeed as presented in Figure 8.10, but the perceptual system will stubbornly insist otherwise.

Visual illusions are two or three dimensional situations that cause something to appear different than it really is. Many such illusions appear to have no useful counterpart in the real world, but some do. An example is item a in Figure 8.11. If one imagines oneself standing in the middle of a sidewalk, the two lines leaning toward each other could represent the edges of the walk converging in the distance. If two rods were placed on the walk at different distances, clearly the further rod would be judged longer based on the simple fact that it comes closer to filling up the space between the edges. What this illusion suggests is that the judgment of relative length in such a situation is not so much an intellectual exercise as just described but something that the perceptual system does automatically.

Visual illusions play a greater part in everyday life than is generally recognized. The early Greeks are renowned for their skill in creating beautiful colonnaded structures. What is less well known is that the columns do not have parallel sides. Rather they are thicker at the top than the bottom to compensate for the effects of perspective. Arnheim (1965) quotes Vitruvius, a Greek architect, as follows:

> For the eye is always in search of beauty, and if we do not gratify its desire for pleasure by a proportionate enlargement in these measures and thus make com-

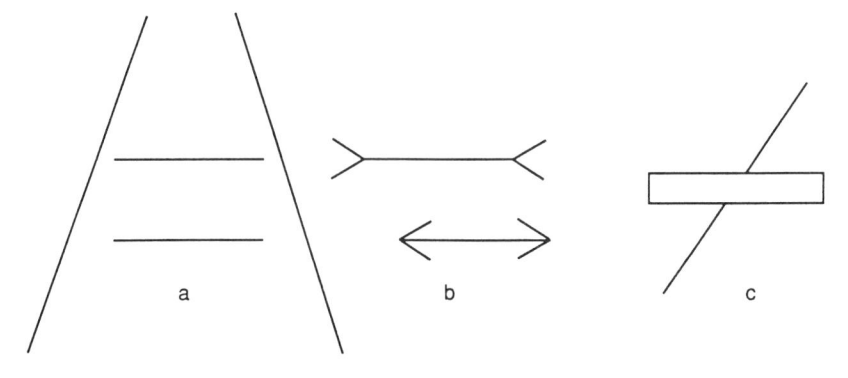

Figure 8.11. Illustration of some visual illusions (compare with Figure 8.10)

pensation for ocular deception, a clumsy and awkward appearance will be presented to the beholder.

Artists and architects have made frequent use of illusions to create a desired effect. One example cited by Arnheim is found in the Palazzo Spada in Rome. When Francesco Borromini was given this commission in 1635 it was his intention to have a deep architectural vista tapering off in a vaulted colonnade, despite having only a limited site at his disposal. The visitor stands in the courtyard and looks into the colonnade, seeing a long tunnel, flanked by columns and leading to an open space in which is located the large statue of a warrior. At least, so it appears. The colonnade is actually quite short, and Borromini distorted perspective to a degree that causes some visitors to become queasy when walking through it. The front arch is 19 feet high and 10 feet wide, the back arch, which is 28 feet away, is 10 feet high and three feet wide. The side walls converge, the floor rises, the ceiling slopes downward, and the intervals between the columns diminish. When the statue is reached the visitor is often surprised at how small it is. According to Arnheim these principles were typically employed, although to a lesser degree, by medieval architects to enhance the depth effect in churches.

The point is that perceptions can be manipulated through design. Very little use has been made of this fact in highway construction. One example is found in the case of toll booths. Persons who have driven at relatively high and constant speeds for a time will sometimes not slow sufficiently when entering the toll area. It has been found effective to paint stripes or install rumble strips across the road at ever closer intervals to exaggerate apparent speed.

Certain illusion situations occur on highways that may cause problems. One example is shown in Figure 8.12. The primary road curves sharply to the left in about the center of the photograph. However, the alignment of the connecting road, the line of utility poles and trees all combine to create the illusion

Figure 8.12. Photograph of a roadway containing a potential perceptual trap

that the road continues straight for some distance beyond the curve. Situations like the example shown pose a danger for unwary drivers, so much so that they are sometimes referred to as 'perceptual traps'. It would be desirable to eliminate such problem areas. Sometimes signing can be employed to good effect. In the case shown in Figure 8.12, the best solution would be to alter the alignment of the connecting road to destroy the illusion of continuity.

Some evidence suggests that illusions could be profitably employed in highway design. In one particularly interesting study subjects were shown paired photographs of curves (Shinar, 1977). The curves in each pair were similar in geometry, but differed in accident history. The subjects tended to judge the high-accident curves as closer, wider, and more visible than the low-accident curves. They did not perceive the high-accident curves as more dangerous, and they did not indicate that they would slow down any more for them. A related investigation showed that a driver's perception of curve radii could be altered through the placement of delineators (Shinar *et al.*, 1975).

The subject of illusions in driving would profit from further research. Clearly, illusions that lead to potentially dangerous situations should be avoided. Equally important, it might be possible to design roads in ways such that the resultant perceptions have a beneficial effect on driver behavior. For example, if people can be made to drive at desired speeds through curves and on entrance and exit ramps as a natural response to the roadway and its environment, this would be better than using signs. Such possibilities should be explored.

Stages in perception–response

There are stages in perception–response. Not all authors agree regarding the number of such stages or the appropriate terminology, but they are mainly differences in detail. The description that follows is adapted from Perchonok and Pollack (1981).

The first stage is detection. Detection occurs when the observer becomes consciously aware that 'something' is present. The detection interval starts when the relevant stimulus enters the observer's field of view, and ends when conscious awareness has been achieved. The ease with which an object in the field of view will capture a driver's attention depends in part on its characteristics, as noted earlier. Also of consequence are factors such as the driver's present focus of attention, where in the visual field the object occurs, and the number of other 'somethings' competing for the driver's attention.

Once attention is gained, the gaze is typically adjusted to bring the detected object or condition into focus in the foveal area to facilitate the next stage, identification. In identification sufficient information about the 'something' is obtained to permit the observer to decide what action, if any, is required. In the case of an object that is moving, or capable of movement, this must include information about its speed and trajectory.

Identification complete, the driver must next decide what, if any, action is required. Sometimes something as simple as blowing the horn or flashing the headlights is deemed sufficient. Where more drastic action is required, the choices come down to changes in speed and direction.

The final stage is response, in which the brain issues the necessary instructions to muscle groups to carry out the action decided upon. The sequence ends when an action such as turning the steering wheel or pressing on the brake pedal is initiated.

The first two stages, detection and identification, comprise the perceptual portion of the sequence. Although they are commonly thought of as the same, or at least rigidly linked, they are clearly different. A failure to detect precludes identification, obviously, but detection does not ensure timely and correct identification. A failure in the identification stage can lead to disaster just as surely as a failure in detection. Unfortunately, identification has largely been ignored as a causative factor in traffic crashes. The next section will deal with the issue of perceptual problems in driving.

Perceptual problems in driving

Expectancy

Many so-called perceptual problems in driving are attributable to conditions that do not match up to what the driver expected to encounter. Expectancy is a significant factor in driver performance, and therefore of vehicle and traffic system performance. Thus, it is appropriate to start any discussion of perceptual problems with the subject of expectancy.

Expectancy refers to a predisposition to believe that something will happen or be configured in a certain way. All drivers have certain expectations, which are based on their exposure to practices in road construction, traffic engineering, and observations of the behavior of other drivers.

Expectations influence human behavior in many areas other than driving. For example, the rule 'to increase, turn clockwise' applies to many devices. Its near universality makes it easier to do such things as tighten or loosen threaded fasteners, change the radio or TV volume, or adjust the toaster to a proper degree of darkness. Knowing that objects operate in certain ways makes life easier. In complex situations it helps to free the operator from some basic concerns so that attention can be focused on other matters. Conversely, when something does not work as expected, the result can be annoyance, distraction, errors, lost time, and/or damaged equipment.

The principle extends to such matters as the design of symbols, which are increasingly being used in preference to words on some highway signs and to label controls in vehicles, among other applications. Good symbols make intuitive sense to the user, and are easily learned and recognized. This topic is fully developed in Green's chapter on symbol design.

Conforming to expectancy facilitates performance in part by reducing the need for attention from the higher centers of the brain. If the radio is playing too loudly, we know that a counterclockwise twist on the knob will reduce the volume. Consequently, we do not have to think about how to accomplish the desired result. Of particular importance is the fact that expectancy affects the amount of information and/or the intensity of a stimulus that a driver needs to detect, identify, and reach a decision about a given situation. The amount of information, and the stimulus intensity required is minimal for conforming situations.

In traffic engineering two general types of expectations have been described, a priori and *ad hoc* (Alexander and Lunenfeld, 1986). A priori expectations are derived from general experience, and are the bases for assumptions about traffic operations that people bring with them whenever they take to the road. *Ad hoc* expectations are based on recent information. A driver noting a school zone sign, for example, should have his/her expectations heightened concerning the probability that children will be encountered.

One of the major efforts in US road construction and traffic engineering in the last few decades has been in the direction of establishing sound and uniform practices. Through publications such as the *Manual on Uniform Traffic Control Devices* (MUTCD), and concepts such as Positive Route Guidance (Alexander and Lunenfeld, 1975) considerable progress has been made toward achieving a nationwide highway system that is effective, uniform, and conforms to driver expectations. Because of this effort, non-conforming situations are encountered infrequently, and driver expectations are strengthened.

It is, however, an imperfect world, and situations that violate driver expectations can be found without much difficulty. A common example is left-hand exits on freeways. Most such exits are there for good reasons. They are contrary to expectations, however, and require more signing to alert drivers and give them time to change lanes. A general guideline is that expectancy violations should be avoided wherever possible; where they must exist, extra warnings are required.

Perceptual limitations

Most 'perceptual limitations' have already been discussed, e.g., the problems with vision at night, illusions of various kinds and violations of expectation that affect the ability to detect and identify some situations. The point to be made here is that the perceptual system is not equally effective in dealing with all stimuli. There are certain relatively common driving situations that present a higher likelihood of error than others due to a limited capability of the perceptual system to make sense out of the available cues.

Probably the most common example of a limitation of this type involves the assessment of relative speeds. Sometimes collisions involve slow-moving or stationary vehicles where there is no question of visibility. In the most extreme of these cases drivers will 'follow' a parked vehicle onto the shoulder before impacting it. Investigators are sometimes puzzled as to why the driver did not see the struck vehicle. The term 'moth effect' has been used to refer to this phenomenon, implying that some people are drawn to such hazards like a moth to a light source.

A more likely explanation can be found in the fact that the assessment of closing speeds is very difficult. A major cue to distance is image size. If the image is growing larger we know the object is coming closer. By the same logic the rate of change of image size should be a cue to the speed of approach. Unfortunately, the rate of change of image size depends both on the speed of approach, and on viewing distance.

Figure 8.13 illustrates the problem. It shows that the image size of an object doubles each time the viewing distance is halved. For example, suppose there were a stalled vehicle on the road ahead that first came into a driver's field of view at 1000 feet. Suppose further that the approaching car were travelling at 55 mph (about 80 ft/sec). A six-foot-wide car at 1000 feet subtends an angle of about 0.34 degrees. By the time the distance closes to 500 feet (which takes a

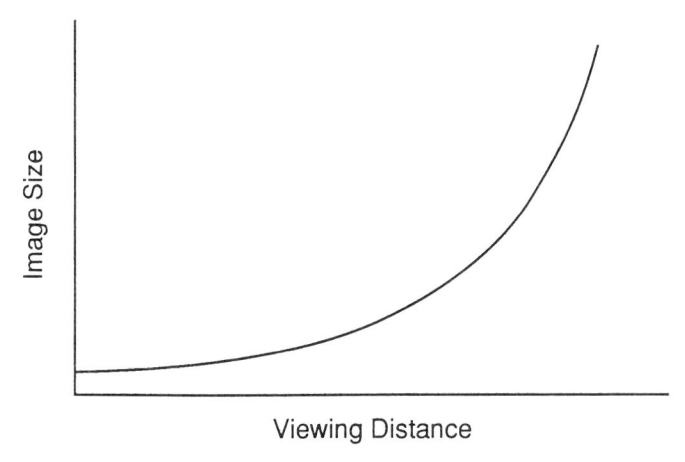

Figure 8.13. The relationship between target image size and separation distance

bit over six seconds) the image size doubles, to 0.69 degrees. At 250 feet, about three seconds later, the image size doubles again. It will double once again, one and a half second later, at 125 feet. Because of the non-linearity between image size and observation distance, the fact that the closing speed is very high may not become apparent until the separation distance has closed to a dangerous extent. Since such encounters are rare, drivers have little opportunity to become aware of the limitation. Instead, they apparently rely on an expectation that vehicles ahead of them are moving at speeds representative of other vehicles on the road in question. For this reason, special warnings, (e.g., flashing lights, flares) are very important for marking stopped or slow-moving vehicles.

The problem is particularly acute at night. With greatly diminished visual range, drivers tend to use whatever cues are available to give them advance information about where the road is going. One such source of information is the lights of vehicles ahead of them. While this is generally a reliable cue, it does bait the trap if the vehicle ahead is going very much slower than expected. In such a situation it becomes more understandable how a driver could follow the lights of a stopped or slow-moving vehicle even onto the shoulder before colliding with it.

Judgments of distance and closing speed can be further complicated by lighting arrays that differ from the usual arrangement. The laws of perspective dictate that more distant objects are not only smaller, but appear higher in the visual field. Thus, a lighting array that is relatively high mounted, or one in which the units are closer together than usual will look like one that is more distant and may lead to overestimations of separation distance. If, in addition, the output of the lamps is diminished for any reason the display can be seriously deceiving.

The perceptual system is a remarkably efficient means of acquiring information. It has proved adequate throughout human history. Recent technological advances have placed new demands on perception by requiring that more and different types of information be acquired. Clearly, there are limits to what the system can accomplish. An understanding of its operation and limitations will allow designers to avoid pitfalls that create unnecessary difficulties.

References

Alexander, G. J. and Lunenfeld, H., 1975, *Positive Guidance in Traffic Control*, Washington, DC: Federal Highway Administration, Office of Traffic Operations.

Alexander, G. J. and Lunenfeld, H., 1986, *Driver Expectancy in Highway Design and Traffic Operations*, Report No. FHWA-TO-86-1, Washington, DC: Federal Highway Administration, Office of Traffic Operations.

Allen, M. J., 1970, *Vision and Highway Safety*, New York: Chilton Book Company.

Arnheim, R., 1965, *Art and Visual Perception*, Berkeley: The University of California Press.

Bailey, I. L., and Sheedy, J. E., 1988, Vision screening for driver licensure, *Transportation in an Aging Society*, National Research Council, Transportation Research Board, Special Report 218, 294–324, Washington, DC.

Baker, H. D., 1949, The course of foveal light adaptation measured by the threshold intensity increment, *Journal of the Optical Society of America*, **39**, 172–9.

Bartley, S. H., 1969, *Principles of Perception*, New York: Harper and Row.

Boff, K. R. and Lincoln, J. E., 1988, *Engineering Data Compendium: Human Perception and Performance*, Wright-Patterson AFB, OH: AAMRL.

Burg, A., 1967, The relationship between vision test scores and driving record: General findings, Report No. 64-24, Los Angeles: Institute of Transportation and Traffic Engineering, UCLA.

Campbell, F. W. and Westheimer, G., 1960, Dynamics of accommodative responses of the human eye, *Journal of Physiology*, **151**, 285–95.

Dark, A. J., Streeten, B. W., and Jones, D., 1969, Accumulation of fibrillar protein in the aging human lens capsule, *Archives of Ophthalmology*, **82**, 815–21.

Davson, H., 1963, *The Physiology of the Eye*, Boston: Little, Brown and Company.

Domey, R. G., McFarland, R. A., and Chadwick, E., 1960, Dark adaptation as a function of age and time: II. A derivation, *Journal of Gerontology*, **15**, 267–79.

Fry, G. A., 1954, A re-evaluation of the scattering theory of glare, *Illumination Engineering*, **49**, 98–102.

Gibson, J. J., 1966, *The Senses Considered as Perceptual Systems*, Boston: Houghton Mifflin Company.

Grether, W. F. and Baker, C. A., 1972, Visual presentation of information, in Van Cott, H. P. and Kinkade, R. G., *Human Engineering Guide to Equipment Design*, Washington, DC: Joint Army–Navy–Air Force Steering Committee.

Henderson, R. L. and Burg, A., 1974, *Vision and audition in driving*, Santa Monica: Systems Development Corporation.

Hills, B. L. and Burg, A., 1977, *A reanalysis of California driver vision data: General findings*, Crowthorne, Berkshire: Transport and Road Research Laboratory.

Kornzweig, A. C., 1954, Physiological effects of age on the visual process, *Sight Saving Review*, **24**, 130–8.

Leibowitz, H. W. and Owens, D. A., 1975, Night myopia and the intermediate dark-focus of accommodation, *Journal of the Optical Society of America*, **65**, 1121–8.

Leibowitz, H. W., Owens, D. A., and Post, R. B., 1982, Nighttime driving and visual degradation, Report No. 820414, Warrendale, Pennsylvania: The Society of Automotive Engineers.

Mandelbaum, J. and Sloan, L. L., 1947, Peripheral visual acuity: With special reference to scotopic illumination, *Journal of Ophthalmology*, **30**, 581–8.

Manual on Uniform Traffic Control Devices for Streets and Highways (MUTCD), 1990, Washington, DC.: Federal Highway Administration.

Notton, D. and Start, L., 1971, Eye movements and visual perception, *Scientific American*, **224**(6), 34–43.
Olson, P. L., 1988, Minimum requirements for adequate nighttime conspicuity of highway signs, Report No. UMTRI-88-8, Ann Arbor: The University of Michigan Transportation Research Institute.
Olson, P. L. and Aoki, T., 1989, The measurement of dark adaptation level in the presence of glare, Report No. UMTRI-89-34, Ann Arbor: The University of Michigan Transportation Research Institute.
Olson, P. L., Aoki, T., Battle, D. S., and Flannagan, M. J., 1990, Development of a headlight system performance evaluation tool, Report No. UMTRI-90-41, Ann Arbor: The University of Michigan Transportation Research Institute.
Olson, P. L. and Sivak, M., 1981, Improved low-beam photometrics, Interim Report, Report No. UM-HSRI-81-4, Ann Arbor: The University of Michigan Highway Safety Research Institute.
Olson, P. L. and Sivak, M., 1983, Improved low-beam photometrics, Final Report, Report No. UMTRI-83-9, Ann Arbor: The University of Michigan Transportation Research Institute.
Owens, D. A. and Leibowitz, H. W., 1976, Night myopia: Cause and a possible basis for amelioration, *American Journal of Optometry and Physiological Optics*, **53**(11), 709–17.
Perchonok, K. and Pollack, L., 1981, *Luminous Requirements for Traffic Signs*, Washington, DC: Contract No. FHWA-RD-81-158, Federal Highway Administration.
Schieber, F., 1988, Vision assessment technology and screening older drivers: Past practices and emerging techniques, *Transportation in an Aging Society*, National Research Council, Transportation Research Board, Special Report 218, 325–78, Washington, DC.
Shinar, D., 1977, Driver visual limitations, diagnosis and treatment, Final Report No. DOT-HS-5-1275, Bloomington, IN: Indiana University Institute for Research in Public Safety.
Shinar, D., Rockwell, T. H., and Malecki, J., 1975, Rural curves: Designed for the birds? Or the effect of changes in driver perception on rural curve negotiation, Paper presented at the Eighth Summer meeting of the Transportation Research Board, Ann Arbor, MI.
Thomas, E. L., 1968, Movements of the eye, *Scientific American*, **219**(2), 88–95.
Wolf, E., 1960, Glare and age, *Archives of Ophthalmology*, **64**, 502–514.
Yarbus, A. L., 1967, *Eye movements and vision*, New York: Plenum Press.

9

Human factors considerations in the design of vehicle headlamps and signal lamps

Michael Sivak and Michael Flannagan

Introduction

This chapter will discuss two groups of vehicle components: headlamps and signal lamps. Both of these component groups are designed to enhance perceptual and psychomotor performance of motorists, and thereby decrease the likelihood of crashes and increase the efficiency of the transportation system.

The structure of this chapter is as follows. The next section will briefly define and discuss a range of basic terms that will be of relevance to the two main sections to follow. These two sections, dealing with headlamps and signal lamps, will briefly present some theoretical considerations, discuss a selection of studies that characterize past research issues, and present a sample of current research and future directions. The final section will discuss human-factors improvements of headlamps and signal lamps in the context of the natural strengths and weaknesses that humans bring to the driving task.

General considerations

This section will define and discuss several distinctions that have general importance in the evaluation of headlamps and signal lamps.

Detection vs. identification

Detection is the process of observing an object or event in the visual field, followed by a binary decision (yes or no), or a probabilistic decision about the presence of a stimulus. On the other hand, identification is the process of classifying an object or event. It may involve categorization into any number of potentially idiosyncratic categories.

Visibility vs. conspicuity

Visibility in the current context is the degree to which an object or event is above threshold for an observer who is primed to look for it. Conspicuity, on the other hand, involves the attention-getting properties of an object or event, given an observer who is not primed to look for it. Consequently, an object can be visible, but not conspicuous. On the other hand, if an object is conspicuous, it is also visible. In other words, visibility is not a sufficient condition for conspicuity, while conspicuity is sufficient for visibility. Visibility is governed exclusively by sensory limitations, while conspicuity is also affected by factors such as motivation, prior experience, and expectancy.

Disability glare vs. discomfort glare

The presence of a bright light in the visual field can result in a phenomenon called glare. The traditional view (e.g. Holladay, 1926) is that glare has two separate effects on the observer. The first aspect—disability glare—refers to an objective impairment in visual performance. Disability glare is thought to be primarily the consequence of veiling luminance resulting from light scattering in the optic media (Stiles, 1929). The other aspect of glare—discomfort glare—refers to a subjective impression of discomfort. Discomfort glare is thought to be related to the degree of brightness inhomogeneity between the glare source and its background (Schmidt-Clausen and Bindels, 1974). However, the physiological origin of this psychological phenomenon is not known. The nature of the relationship between discomfort and disability glare is not clear. Nevertheless, the concept of two separate effects of glare is dominant in the contemporary theory of headlamp performance (e.g. Bhise et al., 1977).

Alerted vs. unalerted observer

This distinction is often used to characterize subjects in experimental studies. While alertness can vary continuously, the two extremes are of primary interest here. A fully alerted observer is informed about the nature of the study. A fully unalerted subject is not aware that he or she is involved in the study at all. This distinction is important, because true conspicuity can be assessed only by using fully unalerted observers. On the other hand, visibility assessment is preferably studied by using fully alerted observers.

Laboratory vs. field studies

Empirical research in human factors can be classified into two broad categories. Laboratory studies are usually performed indoors, by trying to simulate the relevant aspect of a real-world situation. They employ, of necessity, subjects who are alerted to being in a study (although they might not be fully informed about the true purpose), and thus are suitable for visibility as opposed to conspicuity studies. The advantages of laboratory studies include good control over experimental conditions and subject population, time- and cost-efficiency, and subject safety. The primary disadvantage is the low level of face validity, although actual validity might be high if the relevant real-world conditions and tasks are well simulated. Field studies are performed outdoors, usually in actual traffic. They can employ alerted or unalerted subjects. The primary advantage of field studies is their high degree of face validity, although actual validity might be low if the relevant real-life conditions and tasks are not well represented. The disadvantages include relatively poor control over experimental conditions, poor time and cost-efficiency, and potential safety problems. (A full-scale simulation is a hybrid that attempts to combine the best of both approaches: the good experimental control characteristic of laboratory studies and the real-world conditions present in field studies.)

Sampling of subjects

The method for selecting experimental subjects will influence the generalizability of the findings. The ideal—a random sampling—is the exception, rather than the rule. One aspect of this issue—age of subjects—is of particular importance for human factors investigations involving vision. Most investigators recognize that the potential benefits of a treatment may vary with the age of subjects. The usual solution is to include subjects of two or more age groups.

Vehicle headlamps

Functions

Headlamps provide illumination for driving at low levels of ambient illumination. The primary function of headlamps is to assure efficient lane-keeping, detection of potential obstacles such as other vehicles and pedestrians, and detection and legibility of retroreflective traffic signs at night. While other traffic moving in the same direction might also benefit somewhat from this illumination, the primary beneficiary is the user. Furthermore, oncoming traffic is generally *impaired* by the use of this illumination because of disability and discomfort glare. The secondary function of headlamps is to increase the conspicuity of the user during low levels of ambient illumination. (An additional function—to increase the conspicuity of the user during high levels of ambient illumination—is discussed in the section on vehicle signal lamps under 'daytime running lights'.)

Theoretical considerations

The distinction between two modes of visual information processing that was proposed by Leibowitz and Owens (1977) provides one possible framework for nighttime (mesopic) driving, and thus for evaluating vehicle headlamps. They proposed a basic distinction between 'focal' and 'ambient' modes in vision. According to Leibowitz and Owens, focal vision deals with object discrimination and identification, or more generally, with the question of 'what'. It is mediated by the central visual field (cones), and involves analysis of higher spatial frequencies (fine detail). Consequently it is highly sensitive to illumination, with substantial degradation at low levels of illumination. On the other hand, ambient vision deals with spatial orientation, or more generally with the question of 'where'. Ambient vision is provided by the peripheral retina (rods), and relies on lower spatial frequencies (and the sensitivity to lower spatial frequencies is less dependent on illumination). Ambient vision can therefore function adequately at low levels of illumination.

The distinction between focal and ambient vision, and their selective degradations, provides a possible explanation for the higher crash rates at nighttime (NSC, 1990). In the driving context, focal vision (which is highly sensitive to illumination) is responsible for detecting and identifying relevant objects. On the other hand, ambient vision (which is less affected by low levels of illumination) provides sufficient information for lane-keeping. Leibowitz and Owens (1977) argued that 'since the major tasks of driving [lane maintenance and headway keeping] are relatively unimpaired by reduced illumination, the driver does not anticipate and is not prepared to deal with stimuli for which the focal system suffers a selective deficit' (p. 423). Consequently, this model suggests that relative effects on focal and ambient visual tasks should be of major concern in evaluating headlamp performance.

Past research

Where and how much light?

Vehicle headlamps have been around, in one form or another, for about 85 years (Perel *et al.*, 1983). While high-beam headlamps have been the focus of some studies (e.g. Helmers and Rumar, 1974), most studies have dealt either exclusively with low-beam headlamps (e.g., Olson and Sivak, 1983b; Rumar, 1970; Schmidt-Clausen, 1982), or with comparison of low- and high-beam headlamps (e.g., Graf and Krebs, 1976). The research issues have fallen into two basic categories: where should light be directed, and how much light is needed for safe and effective performance. The 'where' issue consists of lateral and vertical aspects. The lateral demands on light output stem from horizontal curvature characteristics of the roadway, and obstacles and retroreflective traffic signs that can appear to the left or right of the roadway. Vertical demands on the light output are influenced by traveling speed (in an inversely proportional manner), vertical curvature of the roadway, and vertical placement of retroreflective traffic signs. The issue of 'how much light', on the other hand, usually involves a trade-off between 'as much light as possible' for performance of visual tasks and 'as little as possible' to protect oncoming drivers from disability and discomfort glare.

An example of research that considered retroreflective-sign requirements is a study by Arens (1987). This study developed lighting intensity values required for minimum overhead sign visibility, reviewed the performance of US and European headlamps, and recommended some changes to the current US low beam standard. A related analytical study by Sivak, Gellatly and Flannagan (1991) developed data concerning relevant angles of the headlamp beam that are directed towards traffic signs in various positions, road profiles and viewing distances. By using the complex relationships between incident light and retroreflected light for different types of retroreflective sign materials, the required headlamp intensity levels were derived to yield a given level of sign luminance.

Over the years, different philosophies have emerged on the two sides of the Atlantic concerning the appropriate way to handle the conflict between visibility and glare. The European approach differs from the US approach primarily in a greater emphasis on protecting oncoming drivers (as well as drivers ahead) from glare, the ease of aiming the headlamp beam visually (relying on perceptual judgment of the lamp aimer), and aesthetic considerations. Consequently, the European low beam has 1) a sharper transition (cutoff) between where the light is needed for seeing and where it might impinge on the eyes of oncoming drivers, 2) less light above horizontal, and 3) a relatively bright and homogeneous foreground illumination. Each approach is superior to the other in certain traffic conditions (e.g., Rumar, Helmers and Thorell, 1973), but neither approach appears to be superior overall (Olson, 1977).

Computer models of headlamp performance

The dominant research effort during the last 20 years revolved around the development, validation and use of computer models of headlamp performance (Mortimer and Becker, 1973; Becker and Mortimer, 1974; Bhise *et al.*, 1977, 1984; Nakata, Ushida and Takeda, 1989; Olson *et al.*, 1990). The most influential headlamp-performance model has been CHESS (Comprehensive Headlamp Environment Systems Simulation). CHESS, developed by the Ford Motor Company (Bhise, Farber and McMahan, 1976; Bhise *et al.*, 1977), simulates thousands of nighttime encounters, and computes a Figure of Merit (FOM) as an index of headlamp performance. The FOM corresponds to the per cent of the simulated driving distance in which the following three conditions are simultaneously met: 1) the distance at which a pedestrian target is detected is equal to or greater than an appropriate critical distance; 2) the distance at which a road delineator is detected is equal to or greater than an appropriate critical distance; and 3) discomfort glare experienced by an opposing driver is less than a selected critical value. CHESS functions by exercising a core model called DETECT under a wide variety of encounters involving pedestrian and delineation targets with opposing vehicles on a three-dimensional roadway topography. The DETECT model (Matle and Bhise, 1984) sets up given encounters and predicts seeing distances and glare effects experienced by the observer and oncoming drivers.

The major contribution of CHESS/DETECT lies in simultaneously considering pedestrian visibility, road-delineation visibility and discomfort glare. Consequently, it provides a more comprehensive evaluation than models dealing with only one (or two) of these aspects of headlamp performance. (Additional advantages as well as some limitations of DETECT are discussed in more detail in Sivak and Olson, 1987.) CHESS has been used to evaluate relative merits of different headlamps, as well as effects on headlamp performance of variables such as mounting height and headlamp aim (Bhise, Matle and Hoffmeister, 1984).

Changeable beam pattern

The current low-beam patterns are the results of compromises between visibility demands and glare protection for the oncoming drivers, while assuming a fixed beam pattern regardless of the situation. However, a beam pattern that would be responsive to the nature of the situation would lead to improved performance. An early version of a headlamp that incorporates a sensor for detecting oncoming headlamps was described by Jones and Hicks (1970). If an oncoming headlamp was detected, the beam pattern was adjusted so as not to illuminate the trajectory of the eyes of the oncoming driver.

Polarized headlighting

Polarized headlighting was one of the most promising proposals for reconciling the conflict between maximizing visibility for the user of the headlamp and minimizing glare to the oncoming drivers (Helmers, 1972; Johansson and Rumar, 1970; Schwab and Hemion, 1971; Zirkle, Krebs and Curran, 1976). This system 'is based on the principle that polarized light is freely transmitted through a polarizing medium the axis of which is parallel to the direction of the light polarization, but is [virtually] extinguished when the axis is at 90 degrees' (Land, 1968, p. 334). One potential embodiment of this approach involves using in front of the headlamp a polarizing material (the polarizer) with its axis at, say, 45 degrees measured clockwise from the vertical. The driver views the road through polarizing material (the analyzer) whose axis is set in the *same* direction as the polarizer, and consequently all of the polarized light is available. However, the polarized light from oncoming headlamps is polarized at 45 degrees counterclockwise *from this head-on perspective*. Consequently, the light reaches the driver's analyzer at an angle of 90 degrees in respect to its axis, and is thus virtually extinguished. While in theory polarized headlighting appears to be a promising solution to the headlighting dilemma, no country has yet adopted this approach. Possible reasons for this include the need to compensate for the light loss in the polarizer and the analyzer by increasing the headlamp light output, increased problems for the drivers without analyzers during a transition period, and cost.

Current research and future directions

Basic research on discomfort glare

Traditional studies of discomfort glare in the driving context have been based on the assumption that discomfort is independent of visual task demands. There is reason to question this assumption. The discomfort that an observer feels when exposed to glare source may depend partly on the difficulty of a visual task in which the observer is concurrently engaged. This could be the case if the subject (incorrectly) attributes poorer performance on a more difficult task to the influence of the glare source.

This issue was studied by Sivak, Flannagan, Ensing and Simmons (1991), who evaluated the effect of task difficulty on discomfort glare. Subjects performed two tasks on each trial. The first was a gap-detection task. The difficulty of this task was manipulated by changing the size of the gap. The second task was a discomfort-glare rating, in which the subject gave a numerical rating of the discomfort experienced from a glare source that was presented simultaneously with the gap-detection stimulus. The hypothesis was that the resulting changes in the difficulty of the gap-detection task would influence discomfort glare. The results indicate that an increase in the difficulty of the gap-detection task resulted in an increase in discomfort glare. These results are consistent with the hypothesis that discomfort glare is related to task difficulty.

Consequently, a valid evaluation of discomfort glare in a given situation may require the presence of the relevant concurrent visual task.

International harmonization of low-beam photometrics

One of the main barriers to international harmonization of low beams is the greater emphasis in Europe on discomfort-glare protection. Sivak, Olson and Zeltner (1989) investigated the effects of prior headlighting experience on discomfort glare, as a potential explanation of this differential concern. Specifically, this field study compared discomfort-glare ratings provided by recently arrived West Germans in the US (who were presumably used to the relatively lower levels of glare associated with European headlamps) with glare ratings from age-matched US-born subjects. The West German subjects reported higher levels of discomfort glare than did the US subjects. This finding is in agreement with the so-called range effect (Lulla and Bennett, 1981), in which subjective judgments are affected by the range of stimuli experienced. This hypothesis suggests that if higher glare values were to be introduced in Europe, discomfort-glare reports of European drivers might initially increase, but would become, over time, comparable to those of US drivers.

An additional barrier to international harmonization of the low beam is the differential emphasis on aiming beams visually. Research is needed to determine how much the sharp European low-beam cutoff (the transition between the lighted and unlighted area) can be reduced without impairing visual aim. Some empirical research on the relation of several beam parameters to visual aim has recently been reported by Poynter, Plummer and Donohue (1989) and Sivak *et al.*, (1992a). The subjects in these studies were asked to adjust each of several beam patterns so that the visual cutoff overlaid a horizontal line drawn across a screen on which the beam pattern was projected. The location of the maximum intensity contrast between vertically adjacent points of the beam pattern proved to be a good predictor of the location of the cutoff. Furthermore, the variability of the performance was systematically related to the magnitude of the maximum contrast.

During the last few years, several new low-beam patterns have been proposed. Examples of these new developments include proposals from Europe (Padmos and Alfredinck, 1988), Japan (Taniguchi *et al.*, 1989), and the US (Burgett *et al.*, 1989; SAE, 1991). Sivak *et al.* (1992c) evaluated a total of eight proposed low-beam standards, as well as the current US, European and Japanese standards. The evaluation consisted of 1) developing a set of visual-performance functions for low-beam headlamps, 2) defining the representative geometry for these functions, 3) setting criterion illuminance values for the functions based on available empirical data and 4) evaluating the existing and proposed standards in relation to the criterion values by considering the worst allowed case.

High intensity discharge (HID) headlamps

During the 1980s several headlamp manufacturers began to consider the use of headlamps with HID light sources. The primary reasons for considering the use of HID headlamps are increased life, increased power efficiency and (because they can be made smaller than conventional lamps) increased stylistic freedom. The development of HID headlamps has raised questions regarding color rendition of objects they illuminate. The perceived color of an object depends on the spectral power distribution of the light source used to illuminate the object, and the spectral reflectance of the object. However, unlike the continuous spectral power distributions of daylight or tungsten/halogen lamps, HID lamps have high concentrations of energy at several narrow-band wavelength regions, while at other regions they have little or no energy. The primary concern has been with color perception of traffic signs, especially red signs (such as stop signs) because most HIDs are deficient in the long-wavelength end of the visible spectrum. In the extreme, such light sources may confuse the color-perception system. For example, consider the case of low pressure sodium light. Since virtually all of the emitted light is at one wavelength (589 nm), almost all the light reflected to the eye from an illuminated object will necessarily be of this wavelength. Thus, the appearance of colored surfaces under low pressure sodium is distorted (Jerome, 1977; Collins, 1988).

Color-naming empirical studies (Collins, 1988; Hussain, Arens and Parsonson, 1989; Arens, Saremi and Simmons, 1991) have indicated that color rendition of red might, indeed, be impaired under certain HID headlamps. In an analytical study, Sivak, Simmons and Flannagan (1991) differentiated between colorimetric *shift* of individual sign materials when illuminated by HID sources as opposed to tungsten/halogen, and colorimetric *separation* of red sign material from its nearest neighbors (yellow, orange and brown sign materials) when illuminated by HID sources. Sivak et al. argued that both of these aspects of color rendition are important. Their results indicate that the magnitude of the colorimetric shift generally increased with increasing correlated color temperature of the light source. However, the colorimetric separations of red from its nearest neighbors also tended to increase with increasing correlated color temperature, with the highest-temperature HID lamp tested yielding greater colorimetric separation than tungsten/halogen.

Flannagan et al. (1992) measured *in situ* chromaticities of 25 stop signs when illuminated by HID and tungsten/halogen headlamps. The results indicate that the magnitudes of the colorimetric shifts were moderate in relation to the range of chromaticities under tungsten/halogen, and the shifts were generally towards orange. In a related laboratory study on subjective preference for the color of stop signs, Sivak et al. (1992b) found that observers were differentially sensitive to colorimetric shifts from a baseline of a saturated red. They were most sensitive to hue shifts towards orange, and relatively insensitive to saturation shifts towards pink.

There are two fundamental issues which have not yet been addressed in the past research: 1) How important is color (in addition to other dimensions, such as shape and legend content) in achieving conspicuity and comprehension of traffic signs in general, and red signs in particular? In other words, how important is color rendition of traffic signs? 2) If color is important, how large a decrement in color rendition is acceptable from the safety point of view? Empirical evidence concerning these two issues is a prerequisite for arriving at rational recommendations in this area.

In addition to steady-state color rendition, other related color issues are of concern with HID headlamps. For example, there is a concern with color changes during the warm-up period (before a steady state is reached). Similarly, because of the sensitivity of the color output of HID lamps to manufacturing tolerances, different color output (even in the steady state) can be expected from two different headlamps. How much difference can drivers be safely subjected to, and how much difference are they willing to accept? Definitive answers to these and related color questions are not yet available.

It is also possible that the current concern about the potential color-rendition problems with HID headlamps has been somewhat misdirected. It is quite likely that, in comparison to tungsten/halogen headlamps, effects of HID lamps on the luminance of critical targets, such as stop signs, might be of more importance. The argument is similar to the above argument concerning color perception: tungsten/halogen headlamps provide most of their energy in the long-wavelength end of the visible spectrum, precisely in the area where HID lamps are deficient, and where red stop signs reflect most light. Consequently, the luminance of objects, such as stop signs, might be reduced when viewed under HID as opposed to tungsten/halogen illumination. An implication of this hypothesis is that HID headlamps might have more of a negative impact on visibility and conspicuity of traffic signs (and certain other targets) than on their identification. (The preceding argument holds only if the light output of HID headlamps were the same as of tungsten/halogen headlamps. To the extent that HID headlamps produce more light output, the potential problem might be reduced or eliminated.)

Ultraviolet (UV) headlamps

The problem of glare vs. visibility can, in principle, be solved by the use of UV headlamps. Since the UV radiation is not in the visible range, glare problems are eliminated. Visibility benefits may result from objects in the visual field with fluorescent properties—being able to convert the energy from the UV (invisible) part of the spectrum to the visible part of the spectrum. This approach has been most actively pursued in Sweden, but no widely-circulating research reports on this issue have been published thus far. Before UV headlamps can be considered as a viable supplement (or alternative) to current headlamps, two general issues need to be addressed: 1) How many of the relevant objects in nighttime driving (road delineation, pedestrians and pedestrian clothing, animals, and road debris) fluoresce? 2) Can potential health concerns about UV radiation (e.g., conjunctivitis and skin cancer) be successfully resolved?

Vehicle signal lamps (front, side and rear)

Functions

The primary components in this category are turn signals on the front, side marker lamps on the side, and presence, brake (stop), turn and back-up lamps on the rear of the vehicle. The functions of these components are as follows: Side marker lamps and presence lamps signal presence. Brake lamps indicate brake application, while back-up lamps indicate reverse direction of movement (and provide some limited illumination to the rear). Turn signals convey intention to turn. All of these components function properly only when they are detected and appropriately responded to by the relevant other motorists. (The illumination to the rear that is provided by back-up lamps does not fit this general pattern. It benefits the user directly.)

Theoretical considerations

Several theoretical models in perception, attention and human performance are relevant to vehicle signaling. Examples of models with direct relevance are those of Schneider and Shiffrin, and Norman and Bobrow. The basic distinction made by Schneider and Shiffrin (Schneider and Shiffrin, 1977; Shiffrin and Schneider, 1977) is between automatic and controlled processing. Automatic processing relies on learned associations between relevant stimuli and appropriate responses. It is initiated by the relevant stimuli and then proceeds automatically (without the person's control). It is generally parallel in nature (and thus it does not tax the information-processing capacity), and it does not require explicit attention. Controlled processing, on the other hand, is a temporary activation of a sequence of stimuli and responses, is frequently serial in nature and requires attention. The most direct relevance of this model to the topic at hand concerns the distinction between novice and experienced drivers. In the terminology of this model, responses to vehicle signals by novice drivers involve controlled (serial, attention-demanding, capacity-taxing) processing. Only when the associations between the stimuli and responses (e.g. between a brake signal and a release of the accelerator pedal) become well established by extended practice, does automatic processing become dominant. Consequently, this theory provides a conceptual framework for addressing one possible reason for the overinvolvement of young drivers in traffic crashes (Williams and Carsten, 1989). An implication is that research should attempt to identify stimuli (i.e., signals) that lead to a rapid establishment of stimulus-response associations needed for automatic processing. (It is fully recognized that there are other factors that appear to have critical roles in young driver crash etiology, including increased risk taking, Jonah, 1989, and alcohol consumption, Mayhew et al., 1986.)

Norman and Bobrow's (1975) distinction between data-limited and resource-limited processes is also relevant to the understanding of driver responses to vehicle signaling. As Norman and Bobrow define these processes, whenever an increase in the amount of processing resources (e.g. attention) can result in impoved performance, a task is resource-limited. On the other hand, if performance is independent of the amount of allocated resources, but is limited by the stimulus parameters, the task is data-limited. 'In general, most tasks will be resource-limited up to the point where all the processing that can be done has been done, and data-limited from there on' (Norman and Bobrow, 1975, p. 46). This model suggests that the efficiency of vehicle signaling can be influenced by motivational interventions only to a certain extent (defined by the information-processing capability of the driver). Additional improvements, however, can (in general) be obtained by data-related improvements (such as stimulus/background contrast).

Past research

Detailed reviews of past research in vehicle signaling are available in Cole, Dain and Fisher (1977), Sivak (1978), and Henderson et al. (1983). Past research in this area can be classified into two broad categories, depending on the basic issues investigated (Henderson *et al.*, 1983). One category of research dealt with the type of information needed to be presented, while the other category dealt with the optimal methods for displaying the desired information.

What information to present?

Automobile signal lamps currently display to other drivers four basic types of information: presence, application of brakes, intention to turn, and a reversed direction of movement. Traditionally, it has been considered important that all four of these types of information be presented when ambient illumination is low. On the other hand, until recently, presence information was not explicitly signalled during high levels of ambient illumination. An implicit assumption was that during such conditions the presence of a vehicle is sufficiently signalled by the vehicle itself. This assumption has been challenged in cases of both motorcycles and automobiles. For example, the data of Olson, Halstead-Nussloch and Sivak (1981) suggest that daytime conspicuity of motorcycles may be enhanced by the use of low-beam headlamps (cf. Olson, 1989b). Similarly, the argument for the use of daytime running lights (see below) is that the daytime presence information of automobiles should also be enhanced.

All four of the basic functions (presence, braking/deceleration, turn, and reverse) are presented currently only on a dichotomous yes/no basis. Voevodsky (1974) provided some evidence that signaling finer gradation of deceleration might be useful. Furthermore, there are other potential states or intentions that might be beneficial to convey to other motorists. These include speed (Jolliffe, Graf and Alden, 1971) and accelerator position (Mortimer and Sturgis, 1976; Liger, Cavallo and Peruch, 1991).

How to present the needed information?

Efficient detection and identification of signals is of primary concern. Signal detection can be influenced by lamp intensity (Sivak *et al.*, 1986; MIRA, 1988), lamp area (Lythgoe, 1973), and spatial separation between lamps signaling different functions (MIRA, 1988). Signal identification is affected by intensity difference (Sivak, Flannagan and Olson, 1987), spatial separation of functions (Attwood, Battiston and Madill, 1977), and degree of color coding (Mortimer, 1969; but cf. Projector and Cook, 1972).

Current research and future directions

Indicator location

While not a completely new topic, recent research in this area has led to the requirement that all automobiles sold in the US after 1 September 1985 have to be equipped with a supplemental center high-mounted brake lamp. US crash studies that employed fleets of automobiles (Malone *et al.*, 1978; Reilly, Kurke and Buckenmaier, 1980; Rausch, Wong and Kirkpatrick, 1982), have estimated that the use of such a lamp will reduce the rates of certain types of rear-end collisions by about 50 per cent. (A similar reduction in relevant rear-end collisions has recently been obtained also in a New Zealand study; McCormick and Allen, 1988.) The initial evaluations of actual benefits in the US after the introduction of the device have shown smaller, but still significant crash reductions (Kahane, 1987, 1989). There are several possible mechanisms for the benefits of center high-mounted brake lamps. These explanations include separation of function (since in the US the standard low-mounted brake lamps usually function as turn signals as well), greater total intensity, a more effective location, and a recognizable triangular pattern formed when the two standard brake lamps and the center high-mounted brake lamp are energized.

There is some empirical evidence supporting the location hypothesis. Sivak, Conn and Olson (1986) evaluated the distributions of driver eye fixations when following other cars in slow-moving urban traffic during daytime. The study was performed before the introduction of the US requirement for high-mounted brake lamps, and no vehicles in the study were equipped with such lamps. The results indicate that the eye fixations tended to concentrate on the rear-window of the lead car and not in the neighborhood of the standard low-mounted brake lights. Since brake lamps that are closer to eye fixations are likely to result in shorter driver reaction times than brake lamps farther away from the fixations (Cohen, 1983) and in fewer missed signals, this study provides some support for the location hypothesis.

Advanced indication of braking

Olson (1989a) evaluated a device that senses the rate at which the accelerator is released. If the rate is equal to or faster than a predetermined minimum, the device turns on the brake lights for one second. Consequently, this device is designed to give an advanced indication of a likely upcoming application of the brakes. The results indicate that when the device is utilized, the brake lights are illuminated 0.2 to 0.3 second sooner than they would be otherwise. However, there was a relatively large percentage of 'false alarms'—instances in which the device turned on the brake lamps, but no brake application followed. Out of all the instances in which the device turned on the brake lamps, 22 to 47 per cent (depending on the vehicle) were false alarms.

Daytime running lights

Daytime running lights are an example of presence signals designed to improve daytime conspicuity of automobiles. While a variety of different running lights have been tested (e.g. standard low-beam headlights, parking lights, add-on 'special' running lights), research tends to suggest that running lights reduce the number and severity of crashes which the lights were expected to affect, such as right-angle and head-on daytime crashes. The two most noteworthy of these studies were investigations of running-lights mandates in Finland and Sweden (Andersson, Nilsson and Salusjarvi, 1976; Andersson and Nilsson, 1981). Both of theses studies showed positive effects. However, a recent review (Theeuwes and Riemersma, 1990) is critical of the analyses that were performed in these two studies. (Theeuwes and Riemersma's study provides a discussion of theoretical issues related to daytime running lights and a comprehensive review of evaluation studies.)

An extensive study on this topic was recently completed in Canada (Sparks, Neudorf and Smith, 1989). In this study the crash experience of 4000 vehicles was examined by comparing a five-year pre-installation period with a two-year period in which the vehicles were equipped with a daytime-running-light system (low-beam headlamps and parking lights which turned on when the ignition was switched on). The results indicate a statistically significant 25 per cent reduction in the types of crashes which the lights were expected to affect, with no effect on other crashes.

Lamp rise time

An effective signaling system is frequently defined as one that leads to short reaction times by other traffic participants. However, the lamps that are used in vehicle signaling have a relatively slow rise time. Conventional tungsten-filament brake lamps require about 250 msec to reach 90 per cent of the eventual light output, thereby causing potentially important delays of warning information to following drivers. Consequently, several recent developments have attempted to modify the 'hardware' in order to reduce this built-in delay in driver reaction time. The most successful, although relatively expensive, solution thus far involves the use of LED light sources. Since these light sources have virtually instantaneous light onset, it is not surprising that they lead to shorter reaction times than standard tungsten lamps (Olson, 1987).

Another, less expensive, approach to reducing driver reaction times to brake signals was proposed by Flannagan and Sivak (1989). This approach involves a device that continuously applies low voltage to the tungsten filament and provides a brief overvoltage at the time of brake contact. Flannagan and Sivak (1989) have evaluated the benefits of such a device in a laboratory study that measured subjects' reaction times to the onset of brake lights in a simulated car-following situation. An additional manipulation involved the amount of simultaneous cognitive demand on the subject. The results indicate that the use of such a device would lead to a reduction in driver reaction time to brake signals by at least 115 msec. Furthermore, the data of Flannagan and Sivak indicate that the effect of the device is not simply to make the stimulus appear sooner, since the effect increased as concurrent cognitive demands on the subjects increased. This finding is consistent with some evidence in the basic human performance literature which suggests that abrupt-onset stimuli have special attention-getting properties (Yantis and Jonides, 1984).

International harmonization

The numerous differences between the US, Europe, and Japan in intensity and color specifications for automobile signaling have been comprehensively summarized in Hitzemeyer (1982). An example of such a difference is brake-lamp intensity. The increased concern with glare in Europe (see above) has resulted in European brake-lamp intensity specifications being lower than the US specifications. The European (EEC) requirements call for the intensity to be between 40 and 100 cd, while the US requirements specify the intensity to be between 80 and 300 cd. Consequently, there is only a small overlapping region in which lamps meet both sets of requirements. A recent study looked into the issue of reducing the US minimum intensity requirements in order to increase the overlap (Sivak, Flannagan and Olson, 1987). Specifically, this study evaluated the relationship between lamp intensity and differentiation between brake and presence signals. The findings indicate that reducing the current US minimum of 80 cd would lead to increased reaction time to brake signals and increased confusion about the nature of the signal. Consequently, the results of this study argue against a reduction in the US minimum specification.

High-contrast and body-color brake lamps

Traditional brake lamps appear red in both the 'off' and 'on' state. In recent years, spurred originally by stylistic considerations, some manufacturers have employed brake lamps that in the off state appear black (high-contrast lamps) or the same color as the surrounding car body (body-color lamps). In addition to stylistic advantages, there may be some behavioral benefits from using such lamps, including a reduction in driver reaction times to brake signals during high levels of ambient illumination. There are two possible mechanisms for such an effect. The first mechanism is based on the increased brightness difference between the off and on states. This would apply to both the high-contrast lamps and body-color lamps that appear darker in their off state than the current brake lamps. The second mechanism involves the increased color difference between the two states. While the standard brake lamp goes from darker red to brighter red, these novel lamps appear to change from black or body color to red.

Sivak, Flannagan and Gellatly (1991) evaluated the reaction-time effects of a high-contrast brake lamp. The study, performed in a laboratory, simulated daytime driving conditions with illumination from the sun being reflected by the lenses of the brake lamps. The results indicate that reaction times to the high-contrast lamp were shorter than reaction times to a standard brake lamp. The resultant difference (19 msec) was not large, but was highly significant statistically. In this study, brightness and color differences between the on and off states were confounded. However, a follow-up study on body-color lamps (Chandra et al., 1992) was designed to evaluate separately each type of difference. The results indicate that both have an effect on reaction time, with increased brightness difference and increased chromatic difference both leading to statistically shorter reaction times. However, the magnitudes of the effects were small and comparable to those found by Sivak, Flannagan and Gellatly (1991).

In real life, a high-contrast brake lamp (in the off state) might perceptually blend with the rest of the car body. Such a situation might lead to a certain degree of spatial uncertainty as to where brake signals might appear, although the uncertainty region would still be limited to the rear of the vehicle. Whether this somewhat increased location uncertainty would lead to an increase in reaction times is yet unknown.

Systems approach

Most past research has dealt with each type of signal separately. For example, a majority of studies on brake lamps did not deal with turn signals at the same time. However, the optimal parameters of a given type of signal depend on the values of parameters of other nearby signals on the vehicle. For example, while

Sivak, Flannagan and Olson (1987) found that reducing the current brake-lamp minimum of 80 cd is likely to lead to increased confusion between brake and presence signals, this finding is contingent on having presence lights of a given intensity—12 cd. Should the intensity maximum for presence lights (in the US) be reduced from the current 12 cd, the minimum for brake lamps could possibly be reduced from 80 cd without changing the current level of confusability of these signals. This example illustrates the need in the future to deal with all of the signals simultaneously. Examples of the few studies that have considered the possible interactions between different components include Cole, Dain and Fisher (1977), MIRA (1988), Mortimer (1970), and Mortimer (1989).

New technology—old issues

In the context of new technological developments, both sets of basic questions—what to present, and how to present it—are being re-examined. This research is being conducted primarily through ongoing programs such as the US Intelligent Vehicle Highway Systems (IVHS; Ervin and Chen, 1990), European PROMETHEUS (Glathe, 1991) and DRIVE (Karamitsos, 1991), and Japanese RACS and AMTICS (Yumoto, 1991). The 'what' issue needs to be studied in the context of the limitations of driver information processing (and consequently the proper priorities for various types of information). The 'how' issue needs to explore novel approaches to presenting visual information, such as head-up displays (Sojourner and Antin, 1990), and (because of the limitation of visual information processing) the possibility of presenting information in a nonvisual mode.

Concluding comments

Our evolutionary history has shaped us to deal simultaneously with only a limited number of objects and events, while moving at a leisurely pace of a few km/hr through an environment that is illuminated relatively uniformly over both time and space. We have not been prepared to move at speeds of over 100 km/hr, while other equally fast vehicles are lurking to enter our intended paths and passing next to us within a few centimeters. (A similar point has recently been made by Rumar, 1990.) The main practical effect of the increased speed has been the reduction in the available time for decision making and action. Evolution did not have a chance to help us to deal with visual information processing during constantly changing patterns of illumination. In many respects we are poorly prepared for driving. Probably the best example of this mismatch is the fact that at night we often drive at speeds for which the stopping distance is greater than the visibility distance to relevant objects, such as pedestrians (Olson and Sivak, 1983a).

The general role of human-factors research is to increase the compatibility of objects in our environment with our vast, but limited, capabilities. Cars (and other vehicles) are objects whose use leads frequently (from the societal point of view) to deadly consequences. (However, from an individual's point of view, driving is still a relatively safe activity, since an average person is likely to be involved in only one crash per 10 years of driving; Evans, 1991). Well-designed headlamps and signal lamps are some of the vehicle components that can make driving more user friendly, and thus safer and more efficient that it would be otherwise.

References

Andersson, K. and Nilsson, G., 1981, *The Effects on Accidents of Compulsory Use of Running Lights During Daylight in Sweden*, Report No. 208A, Linköping: Swedish Road and Traffic Research Institute.
Andersson, K., Nilsson, G. and Salusjarvi, M., 1976, *Effekt pa Trafikolyckor av Rekommenderad och Pakallad Användning av Varselljus i Finland*, Report No. 102, Linköping: Swedish Road and Traffic Research Institute.
Arens, J. B., 1987, The potential impact of automotive headlight changes on the visibility of reflectorized highway signs, in *Vehicle Highway Infrastructure: Safety Compatibility*, Special Publication P-194, Warrendale: Society of Automotive Engineers.
Arens, J. B., Saremi, A. R. and Simmons, C. J., 1991, Color recognition of retroreflective traffic signs under various lighting conditions, *Public Roads*, **55**(1), 1–7.
Attwood, D. A., Battiston, H. and Madill, M. D., 1977, *Automobile Rear Signal Research II: Effects of Functional Separation and Low Levels of Blood Alcohol on Laboratory Performance*, Technical Memo No. RSU 77/2, Ottawa: Transport Canada.
Becker, J. M. and Mortimer, R. G., 1974, *Further Development of a Computer Simulation to Predict Visibility Distance Provided by Headlamp Beams*, Report No. UM-HSRI-HF-74-26, Ann Arbor: Highway Safety Research Institute, The University of Michigan.
Bhise, V. D., Farber, E. I. and McMahan, P. B., 1976, Predicting target-detection distance with headlights, *Transportation Research Record*, **611**, 1–16.
Bhise, V. D., Farber, E. I., Saunby, C. S., Troell, G. M., Walunas, J. B. and Bernstein, A., 1977, *Modelling Vision with Headlights in Systems Context*, SAE Technical Paper Series No. 770238, Warrendale: Society of Automotive Engineers.
Bhise, V. D., Matle, C. C. and Hoffmeister, D. H., 1984, *Chess Model Applications in Headlamp Systems Evaluation*, SAE Technical Series Paper No. 840046, Warrendale: Society of Automotive Engineers.
Burgett, A., Matteson, L., Ulman, M. and Van Iderstine, R., 1989, *Relationship Between Visibility Needs and Vehicle-Based Roadway Illumination*, Paper presented at the 12th International Technical Conference on Experimental Safety Vehicles, Göteborg, Sweden.
Chandra, D., Sivak, M., Flannagan, M. J., Sato, T. and Traube, E., 1992, *Reaction Times to Body-Color Brake Lamps*, Report No. UMTRI-92-15, Ann Arbor: The University of Michigan Transportation Research Institute.
Cohen, A. S., 1983, *Einflussgrössen auf das nutzbare Sehfeld*, Research Project 8005, Zürich: ETH, Institut für Verhaltenswissenschaft.
Cole, B. L., Dain, S. J. and Fisher, A. J., 1977, *Study of Motor Vehicle Signal Systems*, Report No. 73/844, Parkville, Australia: Department of Optometry, Melbourne University.
Collins, B. L., 1988, *Evaluation of Colors for Use on Traffic Control Devices*, Report No. NISTIR 88-3894, Gaithersburg: National Institute of Standards and Technology.
Ervin, R. D. and Chen, K., 1990, Providing radical functionality to serve highway transportation: A 20-year vision for IVHS, in *Vehicle Electronics in the 90s*, Publication P-233, Warrendale: Society of Automotive Engineers.
Evans, L., 1991, *Traffic Safety and the Driver*, New York: Van Nostrand.

Flannagan, M. and Sivak, M., 1989, *An Improved Braking Indicator*, SAE Technical Paper Series, No. 890189, Warrendale: Society of Automotive Engineers.

Flannagan, M., Sivak, M., Gellatly, A. W. and Luoma, J., 1992, *Ranges of Stop Sign Chromaticity under Tungsten-Halogen and High-Intensity Discharge Illumination*, Report No. UMTRI-92-17, Ann Arbor: The University of Michigan Transportation Research Institute.

Glathe, H.-P., 1991, 'A PROMETHEUS Progress Report', a presentation at the 70th Annual Meeting of the Transportation Research Board, Washington, DC.

Graf, C. P. and Krebs, M. J., 1976, *Headlight Factors and Nighttime Vision* (Report No. 76SRC13), Minneapolis: Honeywell, Inc., Systems and Research Division.

Helmers, G., 1972, *Visible Distance and Visual Guidance as a Function of System Angle of Polarized Headlight Systems*, Report 126, Uppsala, Sweden: University of Uppsala, Department of Psychology.

Helmers, G. and Rumar, K. 1974, *High Beam Intensity and Obstacle Visibility*, Report 150, Uppsala, Sweden: University of Uppsala, Department of Psychology.

Henderson, R. L., Sivak, M., Olson, P. L. and Elliott, W. M., 1983, *Motor Vehicle Rear Lighting and Signaling*, SAE Technical Series Paper No. 830565, Warrendale: Society of Automotive Engineers.

Hitzemeyer, E., 1982, *Information Report: Lighting Devices—Compatibility of International Requirements*, SAE Technical Series Paper No. 820487, Warrendale: Society of Automotive Engineers.

Holladay, L. L., 1926, The fundamentals of glare and visibility, *Journal of the Optical Society of America and Review of Scientific Instruments*, **12**, 271–319.

Hussain, S. F., Arens, J. B. and Parsonson, P. S., 1989, Effects of light sources on highway sign color recognition, *Transportation Research Record*, **1213**, 27–34.

Jerome, C. W., 1977, The rendering of ANSI safety colors, *Journal of the Illuminating Engineering Society*, **6**, 180–3.

Johansson, G. and Rumar, K., 1970, A new polarized headlighting system, *Lighting Research and Technology*, **2**, 28–32.

Jolliffe, C., Graf, C. and Alden, D., 1971, An evaluation of a rear-mounted vehicle speed indicator, *Highway Research Record*, **366**, 130–2.

Jonah, B. A., 1986, Accident risk and risk-taking behavior among young drivers. *Accident Analysis & Prevention*, **18**, 255–71.

Jones, K. J. and Hicks, H. V., 1970, The Lucas 'Autosensa', *Technical Aspects of Road Safety*, **70**, 4.1–4.11.

Kahane, C. J., 1987, *The Effectiveness of Center High Mounted Stop Lamps—A Preliminary Evaluation*, Report No. DOT HS 807 076, Washington, DC: National Highway Traffic Safety Administration.

Kahane, C. J., 1989, *An Evaluation of Center High Mounted Stop Lamps Based on 1987 Data*, Report No. DOT HS 808 442, Washington, DC: National Highway Traffic Safety Administration.

Karamitsos, F., 1991, 'Future directions for DRIVE', a presentation at the 70th Annual Meeting of the Transportation Research Board, Washington, DC.

Land, E. H., 1968, The use of polarized headlights for safe night driving, *Traffic Quarterly*, **22**, 330–9.

Leibowitz, H. W. and Owens, D. A., 1977, Nighttime driving accidents and selective visual degradation, *Science*, **197**, 422–3.

Liger, M., Cavallo, V. and Peruch, P., 1991, Comparison of different brake light systems, in Gale, A. G., Brown, I. D., Haslegrave, C. M., Moorhead, I. and Taylor, S. (Eds), *Vision in Vehicles—III*, Amsterdam: North-Holland.

Lulla, A. B. and Bennett, C. A., 1981, Discomfort glare: Range effects, *Journal of the Illuminating Engineering Society*, **13**, 74–80.

Lythgoe, J. N., 1973, Visibility of vehicle rear lights in fog, *Nature*, **243**, 243–4.

Malone, T. B., Kirkpatrick, M., Kohl, J. S. and Baker, C., 1978, *Field Test Evaluation of Rear Lighting Systems*, Report No. DOT/HS 803 467, Alexandria, VA: Essex

Corp.

Matle, C. C. and Bhise, V. D., 1984, *User's Manual for the Visibility Distance and Glare Prediction Model (DETECT)*, Dearborn: Automotive Safety Office, Ford Motor Company.

Mayhew, D. R., Donelson, A. C., Beirness, D. J. and Simpson, H. M., 1986, Youth, alcohol and relative risk of crash involvement, *Accident Analysis & Prevention*, **18**, 273–87.

McCormick, I. A. and Allen, K., 1988, The evaluation of single centrally mounted auxiliary stop-lights: A New Zealand field test, *New Zealand Journal of Psychology*, **17**, 15–18.

MIRA (Motor Industry Research Association), 1988, *A Study of the Effectiveness of Rear Lighting Arrangements for Cars*, Report No. 92, Nuneaton, England: Author.

Mortimer, R. G., 1969, Dynamic evaluation of automobile rear lighting configurations, *Highway Research Record*, **275**, 12–22.

Mortimer, R. G., 1970, *Automobile Rear Lighting and Signaling Research*, Report No. HuF-5, Ann Arbor: University of Michigan, Highway Safety Research Institute.

Mortimer, R. G., 1989, The interaction of turn, hazard and stop signals—An examination of the SAE standards, in *Lighting Systems for Motor Vehicles*, Special Publication SP-786, Warrendale: Society of Automotive Engineers.

Mortimer, R. G. and Becker, J. M., 1973, *Development of a Computer Simulation to Predict the Visibility Distance Provided by Headlamp Beams*, Report No. UM-HSRI-HF-73-13, Ann Arbor: Highway Safety Research Institute, The University of Michigan.

Mortimer, R. G. and Sturgis, S. P., 1976, Evaluation of an accelerator position signal, *Transportation Research Record*, **600**, 33–35.

Nakata, Y., Ushida, T. and Takeda, T., 1989, Computerized graphics light distribution evaluation system for automobile headlighting using vehicle simulation, in *Vehicle Lighting and Driver Visibility for the 1990s*, Special Publication SP-813, Warrendale: Society of Automotive Engineers.

Norman, D. A. and Bobrow, D. G., 1975, On data-limited and resource-limited processes, *Cognitive Psychology*, **7**, 44–64.

NSC (National Safety Council), 1990, *Accident Facts*, 1990 edition, Chicago: Author.

Olson, P. L., 1977, *The Relative Merits of Different Low-Beam Headlighting Systems*, Report No. UM-HSRI-77-55, Ann Arbor: Highway Safety Research Institute, The University of Michigan.

Olson, P. L., 1987, Evaluation of a new LED high-mounted stop lamp, in *Vehicle Lighting Trends*, Special Publication SP-692, Warrendale: Society of Automotive Engineers.

Olson, P. L., 1989a, *An Evaluation of the Advance Brake Light Device*, SAE Technical Series Report No. 890190, Warrendale: Society of Automotive Engineers.

Olson, P. L., 1989b, Motorcycle conspicuity revisited, *Human Factors*, **31**, 141–6.

Olson, P. L., Aoki, T., Battle, D. S. and Flannagan, M. J., 1990, *Development of a Headlight System Performance Evaluation Tool*, Report No. UMTRI-90-41, Ann Arbor: The University of Michigan Transportation Research Institute.

Olson, P. L., Halstead-Nussloch, R. and Sivak, M., 1981, The effect of improvements in motorcycle/motorcyclist conspicuity on driver behavior, *Human Factors*, **23**, 237–48.

Olson, P. L. and Sivak, M., 1983a, Comparison of headlamp visibility distance and stopping distance, *Perceptual and Motor Skills*, **57**, 1177–8.

Olson, P. L. and Sivak, M., 1983b, *Improved Low-Beam Photometrics* (Report No. UMTRI-83-9), Ann Arbor: The University of Michigan Transportation Research Institute.

Padmos, P. and Alferdinck, J. W. A. M., 1988, *Optimal Light Intensity Distribution of the Low Beam of Car Headlamps* (Report 1988 C-9/E), Soesterberg, The Netherlands: TNO Institute for Perception.

Perel, M., Olson, P. L., Sivak, M. and Medlin, J. W., 1983, *Motor Vehicle Forward Lighting*, SAE Technical Report Series No. 830567, Warrendale: Society of Automotive Engineers.

Poynter, W. D., Plummer, R. D. and Donohue, R. J., 1989, *Vertical Alignment of Headlamps by Visual Aim*, Report No. GMR-6693, Warren: General Motors Research Laboratories.

Projector, T. H. and Cook, K. G., 1972, Should rear lights of motor vehicles be color coded?, *Journal of the Illuminating Engineering Society*, **1**, 135–42.

Rausch, A., Wong, J. and Kirkpatrick, M., 1982, A field test of two single center, high mounted brake light systems, *Accident Analysis and Prevention*, **14**, 287–91.

Reilly, R. E., Kurke, D. S. and Buckenmaier, C. C., 1980, *Validation of the Reduction of Rear-End Collisions by a High-Mounted Auxiliary Stoplamp*, Report No. DOT HS 805 360, Alexandria, VA: Allen Corporation of America.

Rumar, K., 1970, *Halogen and Conventional Continental European Headlights: A Comparison of Visibility Distances*, Uppsala, Sweden: Uppsala University, Department of Psychology.

Rumar, K., 1990, Driver requirements and road traffic informatics, *Transportation*, **17**, 215–29.

Rumar, K., Helmers, G. and Thorell, M., 1973, *Obstacle Visibility with European Halogen H_4 and American Sealed Beam Headlights*, Report 133, Uppsala, Sweden: University of Uppsala, Department of Psychology.

SAE (Society of Automotive Engineers), 1991, *Low Beam Design Guide, Proposal No. 7A*, Warrendale: SAE Headlamp Beam Pattern Task Force.

Schmidt-Clausen, H. J., 1982, *The Visibility Distance of a Car-Driver in Driving Situation*, SAE Technical Paper Series No. 820416, Warrendale: Society of Automotive Engineers.

Schmidt-Clausen, H. J. and Bindels, J. T. H., 1974, Assessment of discomfort glare in motor vehicle lighting, *Lighting Research and Technology*, **6**, 79–88.

Schneider, W. and Shiffrin, R. M., 1977, Controlled and automatic human information processing: I. Detection, search, and attention, *Psychological Review*, **84**, 1–66.

Schwab, R. N. and Hemion, R. H., 1971, Improvements of visibility for night driving, *Highway Research Record*, **377**, 1–23.

Shiffrin, R. M. and Schneider, W., 1977, Controlled and automatic human information processing: II. Perceptual learning, automatic attending and a general theory, *Psychological Review*, **84**, 127–90.

Sivak, M., 1978, *Motor Vehicle Rear Lighting and Signaling: Effects of Spacing, Position, Intensity, and Color—An Applied Literature Review*, Report No. UM-HSRI-78-8, Ann Arbor: Highway Safety Research Institute, The University of Michigan.

Sivak, M. and Olson, P. L., 1987, From headlamp illumination to headlamp performance: A critical review of issues relevant to the Ford Motor Company DETECT Model, *International Journal of Vehicle Design*, **8**, 271–81.

Sivak, M., Conn, L. S. and Olson, P. L., 1986, Driver eye fixations and the optimal locations for automobile brake lights, *Journal of Safety Research*, **17**, 13–22.

Sivak, M., Flannagan, M., Chandra, D. and Gellatly, A., 1992a, *Visual Aiming of European and U.S. Low-Beam Headlamps*, SAE Technical Series Report No. 920814, Warrendale: Society of Automotive Engineers.

Sivak, M., Flannagan, M., Ensing, M. and Simmons, C. J., 1991, Discomfort glare is task dependent, *International Journal of Vehicle Design* **12**, 152–9.

Sivak, M., Flannagan, M. and Gellatly, A. W., 1991, Reaction times to high-contrast brake lamps, in *Vehicle Lighting Design for Optimal Visibility and Performance*, Special Publication SP-857, Warrendale: Society of Automotive Engineers.

Sivak, M., Flannagan, M., Gellatly, A. W. and Luoma, J., 1992b, Subjective Preferences for the Red Color of Stop Signs: Implications for the Design of High-Intensity-Discharge Headlamps, *Color Research and Applications*, **17**, 356–360.

Sivak, M., Flannagan, M. and Olson, P. L., 1987, Brake lamp photometrics and automobile rear signaling, *Human Factors*, **29**, 533–40.

Sivak, M., Flannagan, M., Olson, P. L., Bender, M. and Conn, L. S., 1986, *Evaluation of Brake Lamp Photometric Requirements*, Report No. UMTRI-86-28, Ann Arbor: The University of Michigan Transportation Research Institute.

Sivak, M., Gellatly, A. W. and Flannagan, M., 1991, *Minimum Light above Horizontal of Low-Beam Headlamps for Nighttime Legibility of Traffic Signs*, Report No. UMTRI-91-3, Ann Arbor: The University of Michigan Transportation Research Institute.

Sivak, M., Helmers, G., Owens, D. A. and Flannagan, M., 1992c, *Evaluation of Proposed Low-Beam Headlighting Patterns*, Report No. UMTRI-92-14, Ann Arbor: The University of Michigan Transportation Research Institute.

Sivak, M., Olson, P. L. and Zeltner, K. A., 1989, Effect of prior headlighting experience on ratings of discomfort glare, *Human Factors*, **31**, 391–5.

Sivak, M., Simmons, C. J. and Flannagan, M., 1991, Colours of retroreflective traffic signs when illuminated by high-intensity-discharge headlights, *International Journal of Vehicle Design*, **12**, 284–95.

Sojourner, R. J. and Antin, J. F., 1990, The effects of a simulated head-up display speedometer on perceptual task performance, *Human Factors*, **32**, 329–39.

Sparks, G. A., Neudorf, R. D. and Smith, A. E., 1989, *An Analysis of the Use of Daytime Running Lights in the CVA Fleet in Saskatchewan*, Saskatoon, Canada: Clayton, Sparks & Associates Ltd.

Stiles, W. S., 1929, The scattering theory of the effects of glare on the brightness difference threshold, *Proceedings of the Royal Society (London)*, **105B**, 131–9.

Taniguchi, M., Kitagawa, M. and Jin, M., 1989, *Research in Japan on the Photometric Design Guidelines of Headlamp Passing Beams* (Report to GRE members), Tokyo: Japan Automobile Research Institute.

Theeuwes, J. and Riemersma, J. B. J., 1990, *Daytime Running Lights: A Review of Theoretical Issues and Evaluation Studies*, Report No. IZF 1990 A-28, Soesterberg, The Netherlands: TNO Institute for Perception.

Voevodsky, J., 1974, Evaluation of a deceleration warning light for reducing rear end automobile collisions, *Journal of Applied Psychology*, **59**, 270–3.

Williams, A. F. and Carsten, O., 1989, Driver age and crash involvement, *American Journal of Public Health*, **79**, 326–7.

Yantis, S. and Jonides, J., 1984, Abrupt onsets and selective attention: Evidence from visual search, *Journal of Experimental Psychology: Human Perception and Performance*, **10**, 601–21.

Yumoto, N., 1991, 'Status of advanced driver information systems in Japan', a presentation at the 70th Annual Meeting of the Transportation Research Board, Washington, DC.

Zirkle, R. E., Krebs, M. J. and Curran, R., 1976, *Advanced Headlighting Systems*, Report No. 76SRC12, Minneapolis: Honeywell Inc., Systems and Research Division.

10

Indirect vision systems

Michael Flannagan and Michael Sivak

The issue of indirect vision presents one of the most promising opportunities for improvements in the human factors design of vehicles. This is partly because of recent technical developments, including electronic control of reflectivity for control of glare from mirrors, heating of exterior mirrors, video systems, and technically advanced near-object warning systems as possible supplements to mirror systems. The current state of our knowledge, however, is such that even questions about the most essential aspects of the most conventional equipment—field of view and image quality of rearview mirrors—may benefit from further research and discussion. Much interesting work has been done, but even on fundamental issues the existing research has not built a satisfying level of confidence or consensus. A major illustration of this situation is the persistence of differing traditions concerning the use of convex mirrors in the United States versus Europe and Japan.

In this chapter we first review aspects of the human visual system that are relevant to the design of any indirect vision system. We then summarize previous human-performance research on indirect vision, which has been primarily on rearview mirrors, and conclude with some recommendations concerning future research.

Theoretical considerations

Relevant visual capacities

The field of view of the human visual system, including effects of movements of the eyes and head, is particularly relevant to the user of rearview mirrors and other indirect-vision displays. We will summarize the most basic aspects of the field of view here. Readers with specific interests in this area may find it useful to consult either or both of two publications of the Society of Automotive Engineers that discuss human visual capacities with specific reference to the use of rearview mirrors (SAE, 1964; 1967a). Both of these are clearly written and strongly oriented toward the applied problem of designing rearview mirror systems, but short on the technical details of vision science. More comprehensive information on eye movements is given by Alpern (1969); many special aspects of work on eye movements, encompassing basic research and applications, are covered in a volume edited by Senders, Fisher and Monty (1978).

Compared to the vision of many animals, human vision is strongly directional. Without head or eye movements, it provides basic awareness of the environment within a field that extends about 180 degrees horizontally. In contrast, according to Walls (1963) the right and left eye fields of some hares overlap both anteriorly and posteriorly, giving them visual awareness in all horizontal directions. Walls also notes that, across species, horizontal extent of vision is correlated with predacity. Hunters tend to have forward-directed vision in order to have the best vision of prey, and the hunted tend to have very wide fields of view in order to detect predators from any direction. However, he suggests that in the case of humans and other primates, visual guidance for the manipulation of objects is the most important function of forward-directed vision.

The Society of Automotive Engineers Information Report J985 (SAE, 1967a) gives the horizontal extent of human vision at 150 degrees for each eye, with 120 degrees of overlap, yielding 180 degrees of total horizontal sensitivity. The vertical limits of the field are given as 50 to 55 degrees above the primary line of sight and 60 to 70 degrees below. These numbers should be considered approximate; other sources give slightly different estimates. Also, the visual field seems to shrink with age. Burg (1968) reports an average total horizontal extent of 175 degrees for subjects 16 to 19 years old, and 139 degrees for subjects over 80.

The directional nature of human vision is even stronger than is suggested by the extent of the basic field of view described above. That field of view is defined by the ability simply to detect a stimulus. The ability to perform more demanding visual tasks, such as reading text, is much more spatially constrained. Highest visual acuity is concentrated in a small region around the point of fixation, corresponding to the area on the retina known as the fovea. Acuity decreases continuously toward the periphery of the retina. It is therefore impossible to specify the size of the useful field of view in general; it depends on the task. A good illustration of this variability is the recent work of Ball and her colleagues on the 'useful field of view' (UFOV) defined by performance on a task derived from the driving domain (Ball et al. 1988; Ball, Owsley and Beard, 1990). There is evidence that the UFOV defined in that way predicts accident involvement of older drivers better than the visual field measured by standard perimetry (Owsley et al., 1991).

One of the central puzzles of visual perception is that, in spite of a narrow concentration of visual resources, people have a sense of being able to see broad, complete, detailed scenes. In fact, people are often surprised when the narrowness of detailed vision is pointed out to them, as when they try to read text a few degrees of visual angle from the point they are fixating. The basic mechanism responsible for the expansiveness of perception is the ability to move the eyes and head to bring the high-acuity center of vision to bear on areas of interest. It is not very informative simply to define the maximum possible extents of eye and head movements. Unless highly motivated, people will not usually make the largest movements of which they are capable. In consideration of this, SAE Information Report J985 gives limits of various types for eye and head movements. For horizontal extent of eye movements, an 'optimal' range is given as 15 degrees left or right of the principal line of sight, and an 'acceptable' range is given as 30 degrees left or right. For vertical extent of eye movements, the 'optimal' range is from 15 degrees up or down, and the 'maximum' range is 45 degrees up to 65 degrees down. For horizontal extent of head movements, an 'easy' range is 45 degrees to either side, and a 'maximum' range is 60 degrees to either side. For vertical extent of head movements, the 'easy' range is 30 degrees up or down, and the 'maximum' range is 50 degrees up or down.

The extent of human visual awareness is thus determined by the distribution of sensitivity over the retina, and by head and eye movements. Sanders (1970) has treated these factors comprehensively in his work on the 'functional visual field'. He divided the functional visual field into three levels: 1) the stationary field, within which useful information can be gathered without eye or head movements, 2) the eye field, within which eye movements but not head movements are necessary, and 3) the head field, within which head movements are necessary. He explored the extents of these fields using a variety of tasks. His distinctions are conceptually useful, although field sizes cannot be specified universally because he found them to depend on the task used.

The need for indirect vision systems

Movements of the head and eyes can extend the effective visual field, but they take time. If vehicles moved at a walking pace or even a running pace, drivers would have enough time to move their eyes and heads to see the traffic around them. At high speed, the forward field must be monitored almost continuously, allowing little time for backward glances. Rearview mirrors, as well as more exotic rear vision systems such as periscopes and electronic video systems, address this problem by moving portions of the rearward scene into the forward field, thus reducing the extent and duration of the head and eye movements necessary to see the rear scene. The central importance of speed in motivating rearview mirror use is underscored by the fact that one of the earliest recorded uses of a rearview mirror on a motor vehicle was on a racing car at Indianapolis in 1911 (Connolly, 1964).

In the case of automobiles, almost all parts of the roadway environment to the rear are potentially in direct view from the driver's eye point. (The primary exception is usually obstruction by the rear roof pillar.) Therefore time constraints on head and eye movements are the primary reason for having rearview mirrors in automobiles. For other vehicles, such as large trucks and buses, vehicle geometry makes it impossible to see some surrounding areas from the driver's eye position. This results in a second important function for indirect vision aids: providing views of areas that would otherwise be blind spots, even if head and eye movements were used. That function is important even in low-speed maneuvers. Two important examples are the areas directly behind large vehicles during backing, and the areas to the sides when turning. The 'crossview' mirror for school buses is an example involving a blind spot in front of a vehicle. United States Federal Motor Vehicle Safety Standard (FMVSS) 111 (NHTSA, 1991), which nominally covers 'rearview' mirrors, requires each school bus to have a convex mirror that provides the driver with a view of the front bumper and the area in front of the bus.

Basic constraints on design of indirect vision systems

The nature of the problem of providing indirect vision to drivers imposes certain constraints on the design of any such system. The most basic of these is that the human field of view cannot truly be extended; mirrors and other devices merely rearrange the images available. Thus placing a rearview mirror in a driver's forward field of view necessarily obstructs part of that field. If a mirror or other device presents an image with unit magnification it will block exactly as much of the forward field as it presents of the rear field. Minification, perhaps by using a convex mirror, makes it possible to present a much wider field of view, but at a potential cost in problems with perception of detail, distance, or speed.

A related problem is the possibility that images in a rearview mirror may be misperceived as being actually present in the forward environment. This might be most important at intersections, where traffic might be present in several directions and decisions might have to be made quickly. Kelley and Prosin (1969) suggest that one effective way of dealing with this problem is to present rearward images so that they are not at the same elevation as the most critical forward images. This is illustrated by the typical arrangement in which automobile center mirrors are at a higher elevation and outside mirrors are at a lower elevation than most of the forward road scene.

Glare from the headlamps of other vehicles at night is particularly a problem with rearview mirrors because glare from following vehicles is normally present for much longer periods than glare from oncoming vehicles. As always, two aspects of the glare problem can be distinguished: 1) discomfort, which may be particularly important for mirror glare because it increases with duration of exposure (Olson and Sivak, 1984), and 2) disability. In the case of mirror glare, two types of disability can be distinguished. Glare from rear headlamps will decrease the visibility of rearward stimuli, seen in the mirror, as well as forward stimuli, seen directly. Because the design of rearview mirrors has little influence on the effects of glare on rearward visibility, disability in forward visibility will be of most concern. According to standard models of discomfort glare (Schmidt-Clausen and Bindels, 1974) and disability glare (Stiles, 1929), both discomfort and disability should decrease in predictable ways with decreases in the effective intensity of the glare source (such as might be accomplished by reducing the reflectivity of a mirror). Both problems should also decrease as the angle between a driver's normal fixation (which is approximately straight ahead) and the position of the rearview mirror (the effective position of the glare source) increases.

Past research

Field of view

There is reason to believe that rearward fields of view should be larger than those provided by most current passenger car rearview mirror systems in the United States. In Recommended Practice SAE J834a, which covers passenger car rear vision, the Society of Automotive Engineers recommends that the horizontal angle of binocular rear view vision for an inside rearview mirror be at least 20 degrees wide, and that the horizontal angle for a left-hand outside mirror be at least 10 degrees wide (SAE, 1967b). However, as demonstrated by Marcus (1964), even mirror fields substantially wider than these minimum recommendations will leave blind areas, large enough to hide small cars, just behind and on either side of an observer's car. Marcus assumed a basic visual field 180 degrees wide, and movements of the head and eyes that could extend an observer's direct field of view to 280 degrees, centered on the forward line of vision. He then determined that the minimum field width for a single, center rearview mirror that would make blind areas too small to conceal a small car in an adjacent lane was 38 degrees. Kelley and Prosin (1969) studied several extremely wide rear vision systems and recommended serious consideration of field widths of 90–100 degrees, adding that, 'the eventual goal should be to provide the motor vehicle driver with the entire 360 degrees field ...' (p. 137). Henderson and others summarized three independent engineering analyses of overall visibility requirements, and concluded that each of them yielded field of view requirements encompassing the entire 360 degrees around a passenger car (Henderson et al., 1983).

Although neither SAE J834a nor United States FMVSS 111 (NHTSA, 1991) requires a 360-degree field of view, there is a high degree of consensus in work on field of view requirements that there are simply no unimportant directions of sight from a passenger car. In light of this, designs that provide less than a full 360 degrees should be viewed as compromises between visibility and other factors. Because such compromises are being made, a valid procedure for quantifying the effectiveness of a rear vision system would be valuable. Some progress on that problem has been made (Burger and Ziedman, 1987).

If more is always better for rearward fields of view, it is encouraging that the history of rearview mirror use in the United States shows a trend toward greater coverage of the rear field. Almost 30 years ago, Marcus (1964) estimated that only 33 per cent of new cars sold in the United States were equipped with outside mirrors. Now all passenger cars are required to have outside mirrors at least on the driver's side. In addition, FMVSS 111 now requires cars that do not meet the field of view requirements for the inside mirror to have a plane or convex outside mirror on the passenger's side. Even when not required, such mirrors have become quite common. In the mid-1980s, passenger-side mirrors were found to be present on two-thirds of recently built cars, with plane and convex mirrors being about equally common in that position (Olson and Winkler, 1985).

Some progress has been made in understanding the special problems of indirect vision from large vehicles (e.g., Kelley and Prosin, 1969; Burger and Mulholland, 1982). The worst problem with indirect vision from large vehicles is probably the existence of blind spots directly to the rear. The geometries of most large vehicles prevent the use of interior rearview mirrors or direct looks to the rear. The rear blind spot is permitted by FMVSS 111, which requires only side mirrors for trucks and buses. There are also significant blind spots along the sides of and in front of many large vehicles (Burger, Mulholland, Smith and Bardales, 1980; Henderson et al., 1983). The variety of geometries involved in the cabs and driver locations of such vehicles has complicated the documentation of the visibility problems involved, but there is general recognition that they are substantial. One circumstance which promises some improvement is the availability of light, cheap, and reliable video systems. Such systems allow much greater flexibility in the selection of vantage points and display locations than is possible with mirrors or even periscope systems.

Convex mirrors

Convex mirrors offer one means of addressing the need for large fields of view because they present minified images, thus providing a driver with a wide indirect field of view without blocking a correspondingly large portion of the direct field of view. However, they may cause distortions of size, speed and distance perception. Also, convex mirrors present images that are optically much closer than the objects that they represent, possibly causing problems with drivers' visual accommodation.

The use of convex rearview mirrors varies considerably across nations. In the United States they were virtually absent from passenger cars until they recently became common as outside mirrors on the passenger's side. They are still prohibited as original equipment for the interior and driver-side locations (NHTSA, 1991). In contrast, they have been common in Japan, England, and the rest of Europe for many years. Morrow and Salik (1962) surveyed 200 cars in the London area and found that 75 per cent of them had convex rearview mirrors, with radii of curvature ranging from 16 to 82 inches (406 to 2083 mm).

A number of studies have investigated effects of convexity on perception of distance and/or speeed. Two general tasks have been used. One task requires observers to indicate the last possible moment at which they would change lanes in front of an overtaking car; thus indicating when they perceive the distance between themselves and the overtaking car to be just large enough to avoid a collision, given their perception of the closing speed involved. The overtaking car is observed through a rearview mirror that has one of the curvatures to be evaluated. The observer is typically in an automobile, and may be driving in actual traffic, or may be parked in a so-called 'semidynamic' setup.

Most studies that have used the lane-change task have found that observers accept smaller gaps (consistent with perception of distance as greater) as mirror radius of curvature decreases (e.g., Bowles, 1969; Mortimer, 1971; Mortimer and Jorgeson, 1974; Fisher and Galer, 1984). However, one study found little effect of convexity, even for subjects who were inexperienced with convex mirrors (Walraven and Michon, 1969). Also, if subjects have a plane mirror available in addition to a convex mirror (Mortimer, 1971; Mortimer and Jorgeson, 1974) the size of gaps that they accept is not affected by the radius of the convex mirror, perhaps because they rely completely on the plane mirror for distance and speed judgments. This finding supports the suggestion that people may use convex mirrors as 'go or no-go' indicators, proceeding with a maneuver if no vehicle is present in the convex mirror, but otherwise supplementing the convex-mirror view with a plane-mirror view or a direct look before making a decision (Mourant and Donohue, 1979).

The second type of task involves more direct judgments of distances observed through mirrors of various radii. The simplest version makes use of the psychophysical method of magnitude estimation, requiring an observer simply to estimate a perceived distance in feet, car lengths, or even a dimensionless number. Magnitude estimation is an appealing procedure because it is well established in psychological and human performance studies (e.g., Engen, 1971). However, it is probably not a good idea to assume that subjects' estimates of distance in feet or other units are meaningful in an absolute sense, as at least one study has done (Rowland et al., 1970). An alternative way of obtaining distance judgments requires an observer to make a 'null match' between a standard distance and a comparison distance. One way of implementing such a task is to have the observer indicate when a car seen approaching in a mirror appears to be the same distance away as a second car seen directly.

Studies that have employed more direct judgments of distance have also generally found an effect of convexity, with shorter radii of curvature leading to greater estimates of distance (Helmers, 1974; Rowland et al., 1970), even for Japanese observers with extensive experience using convex mirrors (Sugiura and Kimura, 1978). However there is some evidence that experience can decrease the effects of convexity on distance perception (Burger, Mulholland, Smith and Sharkey, 1980; Smith, Bardales and Burger, 1978).

As mentioned above, in addition to causing geometric distortions, convex mirrors also present images at optical distances much shorter than the distances to the actual objects. The image that an observer sees in a convex mirror is an erect, virtual image at a distance slightly further away than the mirror itself. The optical location of the image depends on the distance between the mirror and the actual object, and ranges from the position of the mirror surface to a position behind the mirror by one-half of the mirror's radius of curvature. As a practical example, if a convex mirror with a radius of curvature of 1 m is located 1.25 m from a driver's eye position, most of the image of the rearward scene will be at an optical distance of about 1.75 m.

If a driver's eyes are accommodated at virtual infinity while looking at objects several hundred feet in front of a car, and the driver then glances at a convex rearview mirror, he or she will have to change accommodation to see the mirror image clearly. This aspect of convex mirrors raises at least two potential problems. First, people may have to change accommodation quickly and often while driving. Second, some drivers, particularly older ones, may not be able to focus near enough to see the mirror image clearly. There is no direct evidence about the state of accommodation of a driver's eyes while viewing a roadway directly or while using a convex rearview mirror. Theoretical considerations suggest that loss of near accommodation with age would make it difficult for older drivers to use convex mirrors in near positions, such as anywhere on the driver's side or perhaps on the door of the passenger's side, but not in a far position such as the fender of the passenger's side (Seeser, 1974).

Placement of displays

Recommended Practice J834a (SAE, 1967b) advises that left-hand outside mirrors should be placed so that combined head and eye movement do not exceed 60 degrees. However, there may be an advantage in having mirrors (or other rearview displays) much closer than that to the forward line of sight. If the visual angle between a rearview display and the straight ahead is small enough, drivers will be able to maintain awareness of rear events without deflecting their gaze from the forward environment. When it is necessary to look directly at the display, to see details of rear stimuli, they will likewise be able to maintain some awareness of the road in front of them. People are good at monitoring the periphery of their visual field for potentially important events, even while their primary attention is focused on one location. However, that ability has spatial limits. In a laboratory study designed to simulate roughly the demands of driving, Morrow and Salik (1962) had subjects continuously perform a tracking task presented at the center of their visual field, and measured their ability to detect events at various angles in the periphery. They found a pattern, consistent across subjects, in which detection performance deteriorated rapidly when the detection stimuli were beyond 30 degrees from the line of sight.

Unfortunately there are difficulties with locating a display close to the forward line of sight. More central, and presumably more important, parts of the forward field may be occluded. Also, glare from reflected headlights will generally become worse as the angle between the forward line of sight and a mirror location is reduced. However, one evaluation of this issue suggests that within a reasonable range (between 35 and 55 degrees from the principal line of sight) location will have only a minor effect on the severity of glare (Olson and Sivak, 1984).

Reflectivity and glare

When mirrors are used to provide rear vision at night, drivers may experience harmful effects of glare from headlamps of vehicles behind them. Reducing mirror reflectivity will ameliorate the effects of glare, increasing the driver's comfort and ability to see the forward scene; but it will also decrease the driver's ability to see the rearward scene. A number of studies have addressed the resulting tradeoff between glare and rearward visibility (Mansour, 1971; Olson, Jorgeson and Mortimer, 1974; Ueno and Otsuka, 1988). All of these studies demonstrated the basic tradeoff: visibility improves with higher reflectivity but glare becomes worse. Unfortunately, none of them explicity attacked the technically difficult problem of measuring and properly weighting the utilities of visibility and glare protection so that an optimum tradeoff between them could be arrived at quantitatively.

Because lower reflectivity impairs visibility, and because glare protection is only required when ambient illumination is low and rear glare is present, it is desirable to vary mirror reflectivity in response to changing conditions. Currently, this is most often accomplished by the use of a prism mirror. Prism mirrors have two reflective surfaces: a front, unsilvered surface with a reflectance of about 4 per cent; and a rear, silvered surface with a reflectance of about 80 per cent. A two-position lever allows the surfaces to be moved into position to view the rearward scene alternately. Two of the studies mentioned above included explicit recommendations applicable to devices with two-state control of reflectivity, such as the prism mirror. Mansour (1971) concluded that an appropriate lower level of reflectivity, to be used at night when rear glare is present, would be between 10 and 20 per cent; and that a higher level, to be used at other times, should be at least 55 per cent. For a single level to be used at all times he recommended 30 to 40 per cent. Ueno and Otsuka (1988) specify a low-reflectivity level of 8 to 10 per cent, and a high-reflectivity level greater than 40 per cent. Although neither of these recommendations should be regarded as definitive, it is interesting that they agree in recommending less extreme reflectivities, for both low and high levels, than current prism mirrors provide.

Current and future research

Electronic control of reflectivity

Recent technical developments have made available rearview mirrors with reflectivities that can be varied continuously and automatically, under electronic control, over a wide range (e.g., Kato and Nakaho, 1986; Lynam, 1987). This development provides an alternative to the two-state prism mirrors that have been widely used to alleviate rearview mirror glare. The fact that control can be continuous in both time and degree of reflectivity, without adding to the driver's workload, suggests that a much better overall tradeoff between visibility and glare protection can be achieved than with prism mirrors. Electronically controlled rearview mirrors can be equipped with separate sensors to monitor the overall luminance of the forward scene, and the intensity of any light originating within the field of view of the rearview mirror. Generally, a combination of high intensity in that narrow region and a relatively dark forward field indicates that reflectivity should be lowered to protect the driver from glare. If a designer knows the reflectivity that achieves the optimum tradeoff between visibility and glare protection for any combination of overall luminance and glare intensity, the mirror can be made to track that best level continuously. This possibility provides stronger motivation than has previously existed to understand quantitatively the tradeoffs involved. It also raises questions about how drivers will react to automatic and possibly subtle changes in mirror reflectivity. If lower reflectivity reduces the visibility of the rearward scene, it is important that drivers recognize the degree to which they are impaired.

We recently performed a study to begin to address these concerns (Flannagan, Sivak and Gellatly, 1990). Subjects were seated in a laboratory mockup of a passenger car that was equipped with an electronically controlled mirror. They were exposed to varying intensities of glare from the rear while they performed three tasks: 1) a visual task in the rearview mirror that required high acuity, 2) rating of their confidence in their performance of the visual task, and 3) rating of the discomfort they experienced from the glare. The reflectivity of the rearview mirror was changed randomly between trials of the visual task without explicit notice to the subjects. Results confirmed the overall pattern of the tradeoff between visibility and glare protection that had been observed in previous studies. Both visual performance and discomfort were well described by linear functions of the log of reflectivity level, although the linearity of visual performance broke down when conditions led to either perfect or chance-level performance. Subjects' confidence ratings were generally conservative, but accurately reflected the effects on their performance due to changes in mirror reflectivity. This suggests that, at least for one type of task, subjects accurately perceive the impairment caused by changes in mirror reflectivity even when those changes are random and not explicitly identified.

Convexity

Existing research on the perceptual effects of convex mirrors is not adequate to make rational decisions about their use. It is clear that they can provide much wider fields of view than plane mirrors, and that may be of substantial benefit. There is credible evidence that they negatively affect perceptions of speed and distance when drivers are first exposed to them, but the effects of learning are not clear. The experimental evidence that does exist is inconsistent; although most of that evidence, as well as the long-term use of convex mirrors in many parts of the world, suggests that experience does allow drivers to overcome their initial problems. On the negative side, it is possible that even if drivers learn to use convex mirrors properly, they may revert to earlier and inappropriate use under the pressure of an emergency.

Most human performance data on convex rearview mirrors have been obtained using spherical mirrors. Mirrors with more complex curvature may be able to provide wide fields of view while minimizing problems with distance perception. One such design for a driver-side mirror combines a large area having a spherical, large-radius curvature with an outer edge of decreasing radius curvature (Pilhall, 1981). Overtaking vehicles are imaged in the spherical section until they are very close to the observer's vehicle. They then remain visible, although in more distorted form, in the outer section until they become visible to direct, peripheral vision. The mirror thus allows a wide total field of view with relatively little distortion of size and distance information until vehicles are close enough to preclude a lane change, thus rendering more exact perception of distance irrelevant. The promise that such 'aspheric' or 'complex surface' mirrors may provide a good tradeoff between field of view and image quality should be further investigated with human performance data.

Overlapping fields of view

With multiple-mirror systems (such as the two- or three-mirror systems currently common in the United States) there is typically some overlap of the fields of view of the individual mirrors. Considering effects on visibility alone, this redundancy may be beneficial, especially for integrating the information from multiple mirrors. At night, however, when following headlights are in regions of overlap, their glare effect is multiplied. The issues of how many mirrors should be used and how they should overlap involve many considerations, but evaluation of those issues would be helped by a quantitative treatment of the problem of multiplied glare.

Concluding comments

In the vast majority of vehicles, rear vision, and other indirect vision, is currently provided by mirrors. Over the years other, more elaborate options have been studied, including periscopes and electronic video systems, and recently there has been a growing interest in technically advanced detection systems that could provide a form of artificial indirect vision, such as radar or sonar near-object warning systems. Although these alternatives have some advantages over mirrors, their generally higher cost makes it unlikely that in the near future anything will supplant mirrors as the primary means of providing indirect vision. Although the more innovative technologies have given rise to interesting and important research questions, the most important research need is still to provide a thorough understanding of how to optimize the use of mirrors. Perhaps the most important unresolved issue concerning the design of mirror systems is the tradeoff between field of view and image quality involved in choosing among plane mirrors and mirrors with various curvatures. The current emphasis on international harmonization of government standards may provide the motivation finally to reach a true consensus on this issue.

References

Alpern, M., 1969, Movements of the eyes, in Davson, H. (Ed.) *The Eye*, Vol. 3, pp. 1–254, New York: Academic Press.
Ball, K., Beard, B., Roenker, D., Miller, R. and Griggs, D., 1988, Age and visual search: Expanding the useful field of view, *Journal of the Optical Society of America A*, 5, 2210–9.
Ball, K., Owsley, C. and Beard, B., 1990, Clinical visual perimetry underestimates peripheral field problems in older adults, *Clinical Vision Sciences*, 5, 113–25.
Bowles, T. S., 1969, Motorway overtaking with four types of exterior rearview mirror, in *International Symposium on Man-Machine Systems, Vol. 2: Transport Systems and Vehicle Control*, IEEE Conference Record No. 69C58-MMS, Man-Machine Systems group, Institute of Electrical and Electronics Engineers.
Burg, A., 1968, Lateral visual field as related to age and sex, *Journal of Applied Psychology*, 52, 10–15.
Burger, W. J. and Mulholland, M. U., 1982, *Plane and Convex Mirror Sizes for Small to Large Trucks: Predictions from Truck Characteristics*, Report No. DOT HS-806-677, Washington, DC: National Highway Traffic Safety Administration.
Burger, W., Mulholland, M., Smith, R. and Bardales, M., 1980, *Measurement of Blind Areas on the Right Front of Heavy Trucks*, Report No. DOT HS-805-636, Washington, DC: National Highway Traffic Safety Administration.
Burger, W., Mulholland, M., Smith, R. and Sharkey, T., 1980, *Passenger Vehicle, Light Truck and Van Convex Mirror Optimization and Evaluation Studies, Vol. 1, Convex Mirror Optimization*, Report No. DOT HS-805-695, Washington, DC: National Highway Traffic Safety Adminstration.
Burger, W. J. and Ziedman, D. A., *Development of a Uniform Procedure for the Evaluation of Rearview Systems*, SAE Technical Report Series No. 870348, Warrendale: Society of Automotive Engineers.
Connolly, P. L., 1964, Human factors in rear vision, in *Design Aspects for Rear Vision in Motor Vehicles*, Special Publication SP-253, Warrendale: Society of Automotive Engineers.
Engen, T., 1971, Psychophysics II: Scaling methods, in Kling, J. W. and Riggs, L. A. (Eds) *Experimental Psychology*, New York: Holt, Rinehart and Winston.
Fisher, J. A. and Galer, I. A. R., 1984, The effects of decreasing the radius of curvature of convex external rear view mirrors upon drivers' judgments of vehicles approaching in the rearward visual field, *Ergonomics*, 27, 1209–24.
Flannagan, M. J., Sivak, M. and Gellatly, A. W., 1990, *Rearward vision, Driver Confidence, and Discomfort Glare Using an Electrochromic Rearview Mirror*, Report No. UMTRI-90-27, Ann Arbor: The University of Michigan Transportation Research Institute.

Helmers, G., 1974, 'Judgements of distance and changing distance in flat and convex rear view mirrors', unpublished manuscript, University of Uppsala.

Henderson, R. L., Smith, R. L., Burger, W. J. and Stern, S. D., 1983, *Visibility from Motor Vehicles*, SAE Technical Report Series No. 830564, Warrendale: Society of Automotive Engineers.

Kato, S. and Nakaho, J., 1986, *Study of Liquid Crystal Antiglare Mirrors*, SAE Technical Report Series No. 860639, Warrendale: Society of Automotive Engineers.

Kelley, C. R. and Prosin, D. J., 1969, *Motor Vehicle Rear Vision*, Santa Monica, CA: Dunlap and Associates, Inc.

Lynam, N. R., 1987, *Electrochromic Automotive Day/Night Mirrors*, SAE Technical Report Series No. 870636, Warrendale: Society of Automotive Engineers.

Mansour, T. M., 1971, *Driver Evaluation Study of Rear View Mirror Reflectance Levels*, SAE Technical Report Series No. 710542, Warrendale: Society of Automotive Engineers.

Marcus, K. H., 1964, In the vehicle and on the road, in *Design Aspects for Rear Vision in Motor Vehicles*, Special Publication SP-253, Warrendale: Society of Automotive Engineers.

Morrow, I. R. V. and Salik, G., 1962, Vision in rear view mirrors (Part 1), *The Optician*, **144**, 314–318.

Mortimer, R. G., 1971, *The Effects of Convex Exterior Mirrors on Lane-Changing and Passing Performance of Drivers*, SAE Technical Report Series No. 710543, Warrendale: Society of Automotive Engineers.

Mortimer, R. G. and Jorgeson, C. M., 1974, *Drivers' Vision and Performance with Convex Exterior Rearview Mirrors*, SAE Technical Report Series No. 740961, Warrendale: Society of Automotive Engineers.

Mourant, R. R. and Donohue, R. J., 1979, Driver performance with right-side convex mirrors, *Transportation Research Record*, **737**, 95–104.

NHTSA (National Highway Traffic Safety Administration), 1991, FMVSS (Federal Motor Vehicle Safety Standard) 111: Rearview mirrors, in *Code of Federal Regulations [Title] 49*, Washington, DC: Office of the Federal Register.

Olson, P. L., Jorgeson, C. M. and Mortimer, R. G., 1974, *Effects of Rearview Mirror Reflectivity on Drivers' Comfort and Performance*, Report No. UM-HSRI-HF-74-22, Ann Arbor: The University of Michigan Highway Safety Research Institute.

Olson, P. L. and Sivak, M., 1984, Glare from automobile rear-vision mirrors, *Human Factors*, **26**, 269–82.

Olson, P. L. and Winkler, C. B., 1985, *Measurement of Crash Avoidance Characteristics of Vehicles in Use*, Report No. UMTRI-85-20, Ann Arbor: The University of Michigan Transportation Research Institute.

Owsley, C., Ball, K., Sloane, M. E., Roenker, D.L. and Bruni, J. R., 1991, Visual perceptual/cognitive correlates of vehicle accidents in older drivers, *Psychology and Aging*, **6**, 403–15.

Pilhall, S., 1981, *Improved Rearward View*, SAE Technical Paper Series No. 810759, Warrendale, PA: Society of Automotive Engineers.

Roscoe, S. N., 1984, Judgments of size and distance with imaging displays, *Human Factors*, **26**, 617–29.

Rowland, G. E., Silver, C. A., Volinsky, S. C., Behrman, J. S., Nichols, N. F. and Clisham, W. F., Jr., 1970, *A Comparison of Plane and Convex Rearview Mirrors for Passenger Automobiles*, Report No. FH-11-7382, Haddonfield, NJ: Rowland and Company, Inc.

SAE (Society of Automotive Engineers), 1964, *Design Aspects for Rear Vision in Motor Vehicles*, Special Publication SP-253, New York, New York: Author.

SAE (Society of Automotive Engineers), 1967a, Vision factors considerations in rear view mirror design, SAE Information Report J985, Warrendale, PA: Author.

SAE (Society of Automotive Engineers), 1967b, Passenger car rear vision, SAE Recommended Practice J834a, Warrendale, PA: Author.

Sanders, A. F., 1970, Some aspects of the selective process in the functional visual field, *Ergonomics*, **13**, 101–17.

Schmidt-Clausen, H. J. and Bindels, J. T. H., 1974, Assessment of discomfort glare in motor vehicle lighting, *Lighting Research and Technology*, **6**, 79–88.

Seeser, J., 1974, *Automotive convex mirrors—optical properties*, Report No. 201, Holland, MI: Donnelly Mirrors, Inc.

Senders, J. W., Fisher, D. F. and Monty, R. A., 1978, *Eye Movements and the Higher Psychological Functions*, Hillsdale, NJ: Erlbaum.

Smith, R. L., Bardales, M. C. and Burger, W. J., 1978, *Perceived Importance of Zones Surrounding a Vehicle and Learning to Use a Convex Mirror Effectively*, Report No. DOT HS-803-713, Washington, DC: National Highway Traffic Safety Administration.

Stiles, W. S., 1929, The scattering theory of the effects of glare on the brightness difference threshold, *Proceedings of the Royal Society (London)*, **105B**, 131–9.

Sugiura, S. and Kimura, K., 1978, *Outside Rearview Mirror Requirements for Passenger Cars—Curvature, Size and Location*, SAE Technical Report Series No. 780339, Warrendale: Society of Automotive Engineers.

Ueno, H., and Otsuka, Y., 1988, *Development of Liquid Crystal Day and Night Mirror for Automobiles*, SAE Technical Report Series No. 880053, Warrendale: Society for Automotive Engineers.

Walls, G. L., 1963, *The Vertebrate Eye and its Adaptive Radiation*, New York: Hafner. (Original work published 1942).

Walraven, P. L., and Michon, J. A. 1969, *The Influence of Some Side Mirror Parameters on the Decisions of Drivers*, SAE Technical Report Series No. 690270, Warrendale: Society of Automotive Engineers.

11

The effects of age on driving skill cognitive–motor capabilities

George E. Stelmach and Ariella Nahom

Adequate motor skills are important for a variety of daily tasks, one of which is the ability to drive a motor vehicle. Of course cognition and vision are essential for effective driving, however the end product and key to successful driving performance lies in the ability to control efficiently a vehicle with specific motor acts. In essence, driving requires the capability to act upon spatial and temporal information in a coordinated fashion, which involves movements of the upper and lower limbs as well as those of the head and neck.

There is abundant literature in the movement control area suggesting that the elderly exhibit deficits in various cognitive–motor tasks (Falduto and Baron, 1986; Goggin and Stelmach, 1990; Gottsdanker, 1982; Ponds, Brouwer and van Wolffelaar, 1988; Simon and Pouraghabagher, 1978; Stelmach, Goggin and Amrhein, 1988; Stelmach, Goggin and Garcia-Colera, 1987). Therefore, it is assumed that this population may also have some difficulty in operating a vehicle effectively. In this chapter some of the cognitive-motor impairments observed in older adults will be reviewed and these impairments will be discussed in the context of driving a motor vehicle. In the conclusion, some recommendations will then be suggested with regard to vehicle and road designs which may benefit those older adults who choose to drive. The overall purpose of these suggestions is to improve the general safety of all road users.

Movement initiation

One of the most prevalent conclusions in the study of aging and motor skill is that there is a slowing of motor performance with increasing age (Salthouse, 1985b). In addition, movement responses become more variable, relying less on motor programming and more on feedback processes (Stelmach and Goggin, 1988). Older adults therefore, may plan, prepare and organize their movements differently from young adults. Their action sequences may involve additional cognitive-motor processes and, as a result, become more computationally complex.

Reaction time (RT) is the most common measurement used to investigate the slowing in movement initiation of older adults (Welford, 1984b). It is defined as the time interval between the presentation of a stimulus and the initiation of a response and is believed to reflect cognitive-motor processes. This measurement can be used in the examination of driving performance, as response latency is an important part of this behavior. A driver must react with speed and precision to prevent an accident when an unexpected vehicle or pedestrian appears. As cognitive processing demands are added and response output remains the same, it is typically noted that the absolute difference in RT increases between the young and old (Birren, Woods and Williams, 1980; Cerella, 1985; Salthouse, 1985a). This observation demonstrates that the slower speed of response in the elderly is mediated, for the most part, by changes in the rate of mental processing (Bashore, Osman and Heffley, 1989).

In the context of driving, Olson and Sivak (1986) performed an experiment on brake reaction/movement time which they defined as the interval between the release of the accelerator and contact with the brake pedal. Actual driving situations were used and responses to roadway hazards were examined. These researchers found that the latencies measured were longer in older adults (50–84 years) than in younger adults (18–40 years).

Generally, the literature on movement slowing with age suggests some alteration in cognitive-motor processes. From neuropsychological studies, findings show in the elderly:

1 a failure to use advance preparatory information (Botwinick, 1965);
2 a difficulty in processing stimuli and making responses which are spatially incompatible (Rabbitt, 1968);
3 a deficit in initiation when dealing with increased task complexity (Jordan and Rabbitt, 1977); and
4 an inability to control performance speed (Rabbitt, 1979; Salthouse, 1979; Salthouse and Somberg, 1982).

Reaction time commonly reflects the speed and, by implication, the efficiency of the cognitive-motor system. Many investigators have attempted to address the question of what the reason(s) may be for the observed slowing reaction time. The following sections are organized around attempts by investigators to examine components which are thought to influence RT. It is thought that these functional manipulations may also be helpful in better understanding those vehicle-operating behaviors most affected by advancing age. Therefore, studies found in the driving literature as well as the general aging literature will be mentioned.

Response preparation

Response preparation is a subprocess whereby advance information is provided to determine how well older adults can use this task knowledge to prepare for a forthcoming movement. If given sufficient opportunity to prepare for a response, RT differences typically observed may be reduced between the young and elderly. To examine this suggestion, Gottsdanker (1980a, 1980b, 1982) performed a series of experiments which looked at how response preparation is affected by age. Gottsdanker (1980a) found only slight differences between young and old when using a simple RT task where preparation was easy. More recently, however, Gottsdanker (1982) found that RT increased significantly among older adults when preparation was difficult and/or hard to maintain.

In another study, Stelmach, Goggin and Garcia-Colera (1987) examined response preparation in an attempt to determine whether the processes involved in motor planning are responsible for the observed slowing in older adults. Subjects in three age groups (elderly, middle-aged and young) received complete, partial, or no information with regard to which arm, which direction and what extent to move. When all the advance information conditions were collapsed, they translated into 0, 1, 2 or 3 levels (choices) of uncertainty. As the response uncertainty level increased, the RT results showed that older (60–66 years) adults were disproportionately slower than young (18–25 years) and middle-aged (40–47 years) adults. In addition, these averages indicated that the elderly were particularly slower when preparation was not possible (uncertainty levels 2 and 3). RTs were faster in all age groups, however, when subjects obtained complete information about an upcoming response (see Figure 11.1). To explore age group differences further, Stelmach et al. (1987) estimated the time taken to specify a dimension of movement (see Rosenbaum, 1980 for methods). They found that it took 18, 34 and 23 milliseconds respectively on average for extent, direction and arm dimensions in the young. In contrast, the elderly took much longer on average: 60, 77 and 75 milliseconds, repsectively, to specify the same movement dimensions. They concluded that the slowing seen in older adults is in part due to an inability not only to choose efficiently among response alternatives but also to specify dimensions of movement efficiently.

Many motor control studies have specifically manipulated the temporal parameters in movement preparation. When preparatory intervals were employed such that older subjects were exposed to a stimulus for a longer

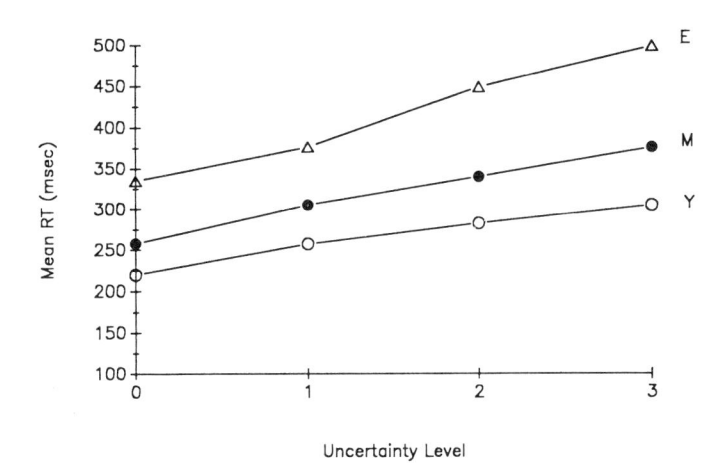

Figure 11.1. Reaction time as a function of uncertainty level in young (18–25 yrs), middle-aged (40–47 yrs) and elderly subjects (60–66 yrs). Level 1 refers to having to select only one dimension, either arm direction or extent. Level 2 refers to having to select two dimensions, either arm and direction, arm and extent, or extent and direction. Level 3 refers to having to select three dimensions: arm, direction and extent.
Source: adapted from Stelmach, Goggin and Garcia-Colera, 1987, p. 42.

period of time or had longer or more predictable intervals between stimuli, the subjects benefited from the higher inspection times by performing better and showing an increased speed of movement (Eisdorfer, 1975; Goggin, Stelmach and Amrhein, 1989). Thus, it appears that preparatory intervals and length of stimulus viewing time are essential in age-related studies on the preparation and planning of movement (Goggin et al., 1989).

If one takes a step further and applies these findings to the older driving population, it might be implied that older drivers may be particularly disadvantaged when they are required to initiate a movement while in traffic in which there is little opportunity to prepare a response. In relation to the studies mentioned above, suggestions have been made in the driving literature regarding the placement of road signs on streets. Since older adults benefit from longer exposure to stimuli, Winter (1985) proposed that signs should be spaced further apart or more in advance of a hazard to allow the driver enough time to process the information and decide on the action to be taken.

Response selection and complexity

Another manipulation used to examine slowed initiation responses in older adults and is thought to reflect decision-making processes is response selection. In studies of this type, RT data are compared as the amount of response choices increase. According to Fozard (1981), older adults may require a longer period of time to identify the stimulus signal, and/or distinguish between or among stimuli in order to choose a suitable response. This decision may involve both short term and long term memory. According to Schonfield and Wenger (1975), sensory memory appears to decline with age. Continual memory scanning and identifying stimuli, and subsequent response selection is required when driving. It appears then, that older drivers may be more at risk than younger drivers when faced with more than one alternative of action (Vegega, 1989).

Using choice RT tasks in which subjects are required to decide among response alternatives, Simon and Pouraghabagher (1978) found that older adults were considerably slower in making movement responses. Further, they allege that response encoding processes are primarily responsible for the observed slowing in older adults. Therefore, in addition to a generalized slowing of the central nervous system, aging appears to disrupt decision-making processes and other higher cortical functions through difficulty in response encoding.

Researchers studying the elderly generally find that as the complexity of the response processing increases, so does the prevalence of slower movements in older adults compared to the young (Griew, 1959; Falduto and Baron, 1986). Results from a study by Falduto and Baron (1986), in which complexity was manipulated by increasing the number of stimuli, indicated that the older subjects were considerably slower in their initiation times and they took longer to produce movement responses as complexity increased. In a meta-analysis, Cerella (1985) evaluated a variety of studies dealing with psychomotor abilities throughout the life span. On the generally simple RT tasks, middle-aged (40–50 years) and older adults (65 years and above) showed only a small increase in latency. On tasks of greater complexity requiring higher cognitive ability, the oldest group of adults showed noticeably slower responses, thus far more decrement was exhibited as difficulty increased. As a result, Cerella concluded that as processing complexity increases, the absolute difference in RT between young and older subjects also increases. This statement supports the inference that age-related slowing in RT is caused primarily by declines in central processing speed (Bashore, Osman and Heffley, 1989).

There are few studies cited in the driving literature which focus directly on this issue. Winter (1985) has suggested that road signs, namely the amount of information presented on them, represent a type of stimulus which may influence driving performance. Also, if signs require too much symbolic motor translation, in general, they may cause confusion, hesitation and may even lead to an accident. In an experimental study, Halpern (1984) found that symbolic signs, as opposed to verbal signs, seem to be more difficult to process centrally, as RTs were slower for older adults in a condition using signs with symbolic messages compared to those with verbal messages written on them. Therefore, traffic signs should be simple, succinct and easily understood to reduce possible confusion and avoid unnecessary delays in processing information.

If these data are correct, then the elderly driving population may be at a disadvantage compared to the younger population of drivers, when faced with numerous response alternatives and general increased complexity. Some specific accident examples illustrating the inefficiency of the older driver in processing information and making decisions include changing lanes carelessly, inappropriate reversing and faulty turning (Planek, 1974).

Response programming

A third subcomponent of RT which has been investigated is the programming and/or restructuring of planned motor actions. Stelmach, Goggin, and Amrhein (1988) performed an experiment to determine whether younger and older adults plan and reprogram movement in a comparable fashion. Subjects received advanced information prior to the time the actual movement was to be executed. In one condition, the information was fairly accurate, with a high (75 per cent) probability level (i.e., programming situation). When the signal to respond was displayed, movement could be prepared in advance, however, it was sometimes necessary to alter the intended response. On these incorrect trials, with a low (25 per cent) probability level (i.e., reprogramming situation) the subjects had to restructure the motor plan. The results of this study showed that older subjects, as expected, were slower to initiate the response, especially when the reprogramming task was to be executed. The authors concluded that there are age-related deficits in ability to restructure a planned movement.

With respect to driving behavior, Staplin, Janoff and Decina (1985) conducted an experiment to assess the driving skills of young and older adults when planning time was allowed. They found that, with increasing age, movement is performed only slightly slower when responses are preplanned. In terms of specific driving tasks, when activities such as brake, accelerator and steer control are highly anticipated, limited age differences may occur. However, in this experiment there were no attempts made to study the effects of restructuring planned motor responses in driving. For example, if a situation suddenly requires an altered response, little is known from the driving literature about the effects these changes may produce in older drivers. An assumption based on the results of Stelmach et al. (1988) predicts that older drivers may have greater difficulty in those situations when quick changes must be made to unanticipated driving maneuvers.

Perceptual style may also be a significant factor in the success of reprogramming situations. Studies have shown that people who are field dependent (i.e., less sensitive to body cues and more sensitive to environmental cues) show more difficulty in these types of situations in driving simulations (Barrett and Thornton, 1968a, 1968b). Older adults have been found to possess this characteristic more so than young adults (Schwartz and Karp, 1967). In a later study, however, Karp (1967) found that more active subjects were less field dependent than age-matched subjects who were inactive.

Explanations of slowing movement initiation

Speed/accuracy trade-off

Several studies have been published which examine and discuss the relationship between advancing age and the speed/accuracy trade-off. This terminology refers to the strategy of whether to move swiftly and compromise accuracy, or to move with precision and risk a slower speed. It is typically found that higher error rates coincide with faster RTs (Wickelgren, 1977), no matter which age group is observed. Many studies have demonstrated that older adults are more cautious in their response strategies than are young persons and that this conservatism may be increased by higher difficulty (e.g., Rabbitt, 1979; Salthouse and Somberg, 1982). Scientists generally agree that part of the slowing which occurs with increased age is related to the view older adults have that maintaining response accuracy is more commendable than speeded responses (Welford, Norris and Shock, 1969; Birren, Woods and Williams, 1980; Botwinick, 1984; Smith and Brewer, 1985; Ponds, Brouwer and van Wolffelaar, 1988).

To examine the speed/accuracy trade-off seen in older adults, Welford et al. (1969) conducted an experiment using a spatial aiming movement task. Results indicated that those characteristics related to the sense of position and distance covered worsen with advancing age, more so than the precision of actual target contact. Ponds et al. (1988) also found that older adults were more accurate than younger adults in an experiment utilizing a self-paced task. Salthouse (1985a) suggests that older adults are more apprehensive of making mistakes and, for that reason, prefer slower to faster movements.

While the studies mentioned above seem to show that the elderly utilize different response strategies from the young, there are clear methodological confounds in such research. Investigations specifically aimed at examining changes throughout the life span may be obscured by shifts in response strategies. Cerella (1985) and Salthouse (1985b) have suggested that scientists involved in cross-sectional age-group comparisons use measurements that systematically diversify response sets. In other words, subjects in an experiment should be encouraged to be quick in one condition, accurate in another, and to balance the two in a third (Bashore et al., 1989). Using such procedures will make the observed differences between age groups more straightforward. This approach may be most revealing and may aid in explaining response strategy differences between age groups. Explaining speed/accuracy trade-offs is essential for driving research since the older adult typically drives considerably slower than the middle-aged and young adult. This slowness often produces dangerous circumstances on highways and city streets. However, this lack of speed very well may be their strategy for avoiding these types of circumstances.

Neuropsychological hypotheses

Salthouse (1985a) has proposed three hypotheses for the slowing of RT in older adults which are analagous to the operation of the computer. The first, 'input or output rate', suggests that the transmission of information from the sensory receptors to the central nervous system (CNS) and between the CNS and the musculature is impaired. Prolongations in transmission are thought to occur because of changes in the central/peripheral nervous system with increasing age. The second hypothesis, 'hardware differences', deals with CNS deficits in which the internal mechanisms responsible for controlling the cognitive–motor system are impaired. Salthouse and Somberg (1982) conducted experiments to examine the age-related slowing in temporal tasks and determined that the observed slowing is general in nature, rather than isolated in one processing stage. They concluded that the entire CNS was involved in the RT slowing of older adults. In connection with this proposal, Welford (1981, 1982, 1984a) proposed the neural noise hypothesis, which implies that the signal-to-noise ratio in the transmission path of older adults is much smaller than that of young people. Explanations include weaker signals, increased noise, or both (Welford, 1981). According to this hypothesis, RT becomes slower mainly because of central transmission decline. The third hypothesis proposed by Salthouse (1985a) is named 'software differences'. This explanation points to deficits in the sequences of control processes, differences in strategy, poor preparation, task complexity, and speed/accuracy trade-offs. In this hypothesis, all of these aspects contribute to the observed slowing with advancing age and are processing operations.

Reduction of age differences in movement initiation

The foregoing three hypotheses attempt to explain why RT slowing occurs, however, some research suggests that response slowing can be minimized. Spirduso (1982) has shown that physical activity can modify the rate of decline, by providing evidence that RTs are faster in physically trained versus physically untrained older adults of the same age. Practice has also been found to minimize and/or even erase observed age differences (Spirduso, 1982; Murrell, 1970; Baron and Mattila, 1989). In their study on cognitive–motor skills, Baron and Mattila (1989) found that when subjects were given specific time limits to respond, the RTs of the elderly decreased with practice. Murrell (1970) also found age differences to be significantly reduced with practice, and noted that response selection plays an important role in the amount of practice needed. Older subjects required a greater amount of rehearsal than younger subjects, since they began to improve at later stages' of the task compared to the young. It may be that instead of interpreting the results as 'age' effects in these experiments, the findings are simply the product of the effects of disuse or less practice at these particular tasks.

In the context of operating a vehicle, these implications of the above experiments may be very important because 1) people who have driven more frequently or for a longer period of time have more experience (i.e., rehearsal), and are therefore, at an advantage, and 2) practice in more variable environments, might result in faster responses to particular problems on the road (e.g., skid recovery, braking, etc.), especially for those at a higher level of ability. Schmidt (1988) explains that when learners are involved in practices which are more varied in nature, greater learning and generalizability will result. This explanation may be applied to driving situations as well. However, as previously mentioned, the assertion of differences between young and old should not be entirely rejected since subjects in all of the experiments done at least by Stelmach et al., 1987 were well-practiced in their corresponding tasks. Age differences in these experiments were still manifested. Thus, it is our belief that although experience and practice can minimize these effects, age differences are substantial and contribute to the decline in driving ability.

Movement execution

Reaction time has been the main measurement discussed up until this point, and age differences observed with this dependent measure reflect central processing deficits. While central processing is always involved in driving behavior and certainly plays a part in the presumed decline of driving ability in older adults, motor factors (i.e., variables associated with the execution of movement) also contribute to the decline of driving performance. Age-related motor decrements have been linked to known changes in loss of sensory receptibility, decreases in muscle mass and elasticity, decreases in bone density and a reduction of central and peripheral nerve fibers (Black, 1977; Larsson, 1982; Kenshalo, 1977; Scheibel, 1979; Welford, 1982). These changes with advancing age affect the ability of the older adult to control movement rapidly as well as accurately.

For illustration purposes, driving skill may be analyzed in terms of neck, leg, arm and hand movements. Observing the ability of the elderly to coordinate actions of these body parts provides insight into the type of coordination problems they may encounter when operating a vehicle. However, since there is a lack of data in the driving literature on movement execution characteristics, this section again focuses mainly on research evidence from laboratory studies concerned with the way in which these characteristics are altered by age.

Movement time

A measure commonly used to assess movement deficits in the elderly is movement time (MT), defined as the time period between the initiation of movement and its termination. Welford (1984b) states that MT is related to muscle control, rather than the time needed for muscle contraction. This measurement has not been abundant in the experimental aging literature compared to RT, probably because applicable manipulation of variables is difficult. Nevertheless, there are several studies which show evidence of slower movement execution in the elderly versus the young (see Stelmach and Goggin, 1988; and Welford 1984b for the reviews). In a driving example, older adults tend to use the brakes for longer durations than young drivers. Corso (1981) attributes this behavior to an attempt to decrease the pace of the movement tasks related to driving or to compensate for the decline in motor reaction functions. It is typically reported that movement time age differences are in the range of 20–30 per cent. While these data and those of others (Welford, 1984b) have shown that older adults are slower in executing movement than the young, they have not been very informative as to why such slowing occurs.

Movement trajectories

It may be that specific movement trajectories produced by older adults are different from those of younger adults and thus elderly driving performance is affected. Murrell and Entwisle (1960) proposed that there are three phases of a typical movement pattern: acceleration, steady speed and deceleration. They conducted an experiment comparing young and old in these three dimensions. Results revealed somewhat similar MTs between the two groups, however, examination of the trajectories showed that older subjects accelerate more slowly and show a smaller negative acceleration phase than the young. So, even though general timing was not affected, more specific aspects of the movement time were different.

In another study examining movement trajectories in the aged, Goggin and Stelmach (1990) tried to answer the question of whether or not movement reprogramming occurred while movement was being executed. Subjects performed aimed movements to the right or left, while a digitizing tablet recorded the coordinates of the horizontal and vertical components. From the results, it was found that the elderly show more asymmetrical velocity profiles than the young. These data reveal that the elderly spent much more time in the decelerating phase of the movement which suggests a greater amount of feedback control during the closing phase of the profile as opposed to ballistic control (see EMG evidence). In addition, results showed that RT, MT and time to peak velocity was slower in the older than in the younger adults. The elderly subjects found it more difficult to attain peak velocity quickly, compared to the young subjects, and they displayed a considerably smaller change in the rate of velocity (i.e., acceleration). Since this type of finding was observed in all of the movement amplitudes studied, the authors further deduced from the data that the elderly may have difficulty in scaling velocity to match movement amplitude, especially when amplitude was the sole criterion of the movement. It was reported that this result may suggest a decrement of muscular-force control in the elderly.

Force

Vrtunski, Patterson, and Hill (1984) performed an experiment which specifically investigated muscular force in older adults. They found that braking ability, while releasing a button after pressing, was weaker in the elderly than in younger adults. They also observed that the agonist and antagonist muscles were less coordinated in the older adults compared to the young. In an experiment conducted more recently, Stelmach, Teasdale, Phillips and Worringham

(1989) examined the magnitude and variability of force production in young and elderly subjects. The subjects' tasks were to make isometric contractions against a strain gauge with the wrist and forearm. Maximum force data for every subject was collected, and then subsequently each subject was asked to generate a force equivalent to 15, 30, 45 and 65 per cent of their maximum (see Figure 11.2). The data shown in the figure are maximums in absolute terms for comparison purposes. Note that the older adults produced less peak force than the young and that the force pulses they generated demonstrate some irregularity and more variability.

The ability to drive safely can be influenced by a deficit in force generation and velocity scaling. If, for example, abrupt braking is required to stop an automobile, the amount of speed and force produced per time unit (i.e., impulse) must be sufficient to stop it quickly. This observation may also be correct for other driving maneuvers such as making turns, shifting gears and skid recovery. Therefore, it is crucial to be able to generate the force needed swiftly enough to avoid a road hazard.

Coordination

Operating a vehicle often requires complex, coordinated movements to be made by various parts of the body. It has been shown that bimanual coordination is reduced in older adults (Stelmach, Amrhein and Goggin, 1988). These authors report that bimanual movement initiations and terminations are not as tightly coupled compared to the young, and these differences are magnified

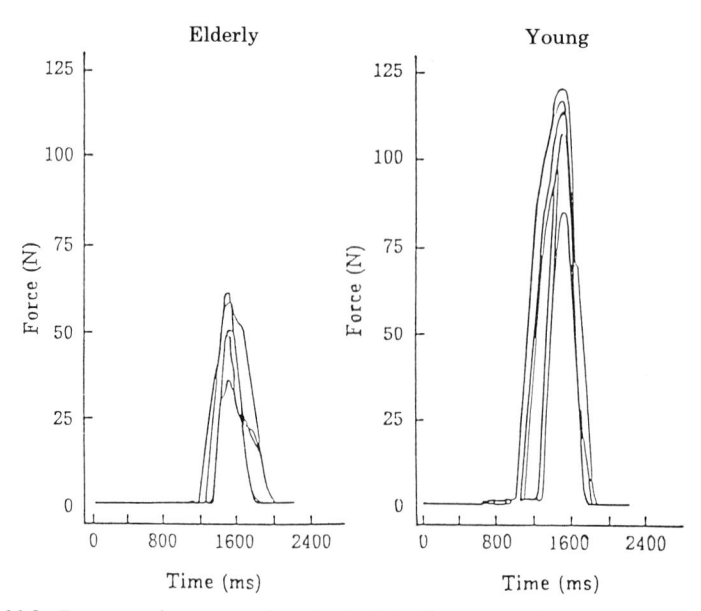

Figure 11.2. Force production in the elderly (60–67 yrs) and the young (18–27 yrs) as a function of time Note: Individual lines are from trials of typical subjects in each age group.
Source: reprinted with permission from Stelmach, Teasdale, Phillips and Worringham, 1989.

when their subjects performed asymmetrical movements with both limbs. Coordination does not only involve limb movements, but it also consists of trunk and head movements as well. Regardless of whether a person is standing or sitting, the trunk must be somewhat stabilized before movement initiation, which requires the production of rapid coordinated actions where force is applied (i.e., turning, lifting). If this is not achieved, the trunk will be destabilized and the coordination and control of voluntary movement will be negatively affected (Stelmach, Populin and Müeller, 1990).

Belinkii, Gurfinkel and Pal'tsev (1967) conducted an experiment to examine preparatory processes associated with voluntary movement execution. When subjects were required to raise their arms, it was found that patterns of muscle activation could be broken into two parts: a preparatory phase, in which the postural muscles were activated before the prime mover muscles in order to compensate for the destabilizing effects of the movement, and a compensatory phase, in which the postural muscles were exerted following the prime mover muscle activation. Applying this type of methodology, Man'kovskii, Mints and Lysenyuk (1980) showed that the elderly exhibit fewer anticipatory postural adjustments just before the initiation of voluntary movement. Man'kovskii et al. (1980) compared the features of anticipatory postural muscle onsets and voluntary muscle onsets for the young (20–29), old (60–69) and very old (90–99) years and observed a slowing of both the preparatory and prime mover phases. The preparatory phase, however, was affected the most on a relative basis. The authors noted that when the elderly were required to move as quickly as possible, they activated the postural and prime movers simultaneously, instead of using an anticipatory activation sequence to stabilize the body (Woollacott, 1990).

Additional credence for a decrement in voluntary movement and body stability as reported by Stelmach, Phillips, DiFabio and Teasdale (1989). As voluntary movements were made, functional postural muscles on both sides of the body were activated relatively simultaneously in the young but they were not as synchronous in the elderly. Moreover, the young adults swiftly suppressed postural responses which were inappropriate when carrying out a voluntary movement, whereas the elderly showed less inhibition.

Ponds et al. (1988) explains that control of coordinated movement may consist of three distinct tasks: controlling the main movements at hand (i.e., the two tasks which are executed simultaneously, in this case from the prime mover and the antagonist muscles) and a third supervisory task that joins the two main tasks into one whole activity. They contend that older adults utilize the same control strategy as the young, however, they do so in a less efficient manner. This is why age differences may be observed. In the section that follows, attempts will be made to explain further these and other movement execution differences seen in older adults.

Explanations of impaired movement execution

Proprioception

The awareness and monitoring of muscular control in limb and trunk movement may play an important underlying role in driving ability as well as other motor skills in general. This sensory–motor integration requires feedback to the brain from kinesthetic receptors in the muscles and joints, in order for skilled movements to occur. In driving, skilled coordinated actions of both the upper and lower body parts are very important.

It has been shown that older adults have more difficulty in performing complex movements. Research also indicates that the elderly show a decrease in kinesthetic sensitivity (Stelmach and Worringham, 1985). Older adults seem to have a higher proprioceptive threshold for passive motion (Birren, 1947; Skinner, Barrack and Cook, 1984), and are less accurate than younger adults when reproducing and matching joint angles (Kokmen, Bossemeyer and Williams, 1978; Stelmach and Sirica, 1987). Woollacott, Shumway-Cook and Nashner (1986) manipulated proprioceptive and visual information in a study which examined the effects of unreliable sensory information in the young and in elderly. The results showed that the elderly were more negatively influenced by the incongruent proprioceptive information. Anatomically, the elderly have fewer hair cells in the semicircular canals (Rosenhall and Rubin, 1975), and a reduction of axons on the optic nerve (Johnson, Miao and Sadum, 1987) which may contribute to their poorer performances.

These declines may place the elderly at a disadvantage when the integration of changing sensory information is necessary. For example, since older adults have more difficulty monitoring proprioception, they may be less aware of limb positions at the start of a driving maneuver and less aware during the execution of those maneuvers as well.

Muscle mass

Another explanation for the deficits seen in older adults with respect to motor skills might be changes in the muscular system (see Larsson, 1982 for review). There has been evidence to show that the amount of muscle fibers and the size of these fibers decline with advancing age. Fast-twitch fibers, especially those found in the lower limbs, are more affected by the aging process (Larsson, Grimby and Karlsson, 1979; Asmussen and Heebøll-Nielsen, 1961). In addition to physiological research, behavioral studies also have shown that muscle mass plays a role in the movement impairment observed in older adults. For instance, muscular strength has been shown to decline in the lower extremities (Asmussen and Heebøll-Nielsen, 1961; Kroll and Clarkson, 1978) as well as in the hands and arms (Montoye and Lamphier, 1977). Some factors influencing these deficits include denervation at the neuromuscular junction, nutrition and disuse (Larsson, 1982). As noted previously, physical activity may slow the aging process down, particularly at the level of muscle composition. Additionally, the impairments found in the muscular system may, in fact, be related to those seen in force production in the elderly. A certain amount of motor units obviously need to be recruited in order to produce a corresponding amount of force.

Joint flexibility

Joint flexibility is also a necessary component in operating an automobile efficiently. If upper extremity range of movement, using the upper limbs and spine, is limited in the older driver, mobility and coordination may be seriously impaired. Scanning the rear through mirrors and turning the head to view blind spots are just some of the driving tasks in which joint flexibility is necessary (Yee, 1985). Smith and Sethi (1975) have estimated that range of motion declines by approximately 25 per cent in the older adult. This downgrade is a result of joint deterioration, arthritis, and greater calcification of cartilage. McPherson et al. (1988) deduced from their study that those older adults with less joint flexibility performed worse using on-road performance testing measures than those with larger ranges of motion. Specifically, older male drivers exhibited less flexibility than their younger counterparts in shoulder, torso and neck joints, with 13, 13 and 11 per cent declines respectively. Walker et al. (1984) have shown that the elderly are seriously impaired in range of limb movement and head rotation. A comprehensive survey of elderly drivers conducted by Yee (1985) indicated that 35 per cent of older drivers reported problems with arthritis, and 21 per cent claimed that it was difficult to turn their head and look to the rear when driving. Thus, joint flexibility is another important part of driving skill which seems to be compromised in older adults.

Summary and conclusions

The existing literature pertaining to motor control and aging was summarized in this chapter. Movement initiation and execution were examined in detail and found to be somewhat impaired in the elderly compared to young adults. The main conclusion drawn from this review is that motor performance is generally slowed as age advances, and is noted in all sub-processes of movement initiation including the preparation, selection and programming of outward responses. These phenomena may result simply because of a different strategy taken by older adults in which accuracy is esteemed over speed, or they possibly may result from actual neuropsychological impairments in the aged. Nevertheless, it has been suggested that physical activity and practice may play a role in retarding the aging processes so that observed changes may not be so pronounced.

Movement execution was also found to be inferior in older compared to young adults, as seen in movement time, and specifically in movement acceleration, deceleration, force production and in movement coordination. Explanations of impaired movement execution were discussed in terms of proprioception, muscle mass, and joint flexibility. Here again, training has been suggested as a way to limit the motor declines observed.

Throughout this chapter, implications of motor performance were applied to driving situations and street designs. In reality, road designs which have been long established are based on performance characteristics of a younger driver population (Transportation Research Board, 1988). At the time these designs were created, not only were there fewer older people, but also, of these people, fewer of them drove. For example, the Transportation Research Board report showed that in 1965, just 40 per cent of those aged 65 or older had drivers' licenses. By 1985, almost 65 per cent of this population were licensed to drive. As the number of older drivers will be increasing substantially within the next few years, it is important that writers of highway and vehicle design manuals, and others who influence public policy, consider the cognitive–motor abilities of older adults when designing or planning for the driving population to insure safety for all on the road.

Some specific suggestions for road and highway engineers might include devising effective street signs which do not require extensive attentional resources (too much information), are simple to read, and are spaced far enough apart to allow for better response preparation. This way, movement reactions to handle vehicle controls might not only be faster but more accurate. For those who design vehicles, it would be advised to make vehicle controls as simple and as easy to learn as possible (i.e., requiring simple limb movements) to allow for both effective reactions and movement capability. Since older adults are generally slower in executing movements, another suggestion may be to design easy access to vehicle controls, so that less time may be spent reaching for them. This modification would also reduce the problem of declining flexibility. Force production and strength have been shown to decline in the elderly. Therefore, these factors should be taken into consideration as well. Anti-lock brake systems and power assisted controls may be useful in this respect.

The foregoing suggestions are based on limited experimental evidence, therefore it is imperative that we continue our research on aging and motor performance in the context of driving. Of course, not all older drivers perform poorly. Still, others may have weakened abilities but compensate for them by driving with more caution. However, there are those who are not able to compensate and put themselves and others at risk on the road. It is the scientist's responsibility to distinguish among those groups and explain the reasons for these different behaviors and performances among older adults. Those endeavoring in the experimental pursuit must make sure that future studies include large and representative sample sizes, a comprehensive set of test trials and adequately controlled methodology to insure that the results and conclusions are reliable as well as convincing. Of course, researchers must make sure that their tests assess the skills that are relevant to the driving task. Designing studies in this manner should help provide a better understanding of driving behavior in the elderly, and in turn, might lead to better ideas of what may be done to retrain lost skills and reduce accidents among the elderly.

References

Asmussen, E. and Heebøll-Nielsen, K., 1961, Isometric muscle strength of adult men and women, in Asmussen, E., Fredsted, A. and Ryge, E. (Eds) *Community Testing Observation Institution Danish National Association Infantile Paralysis*, **11**, pp. 1–43.

Baron, A. and Mattila, W. R., 1989, Response slowing of older adults: Effects of time-limit contingencies on single- and dual-task performances, *Psychology and Aging*, **4**, 66–72.

Barrett, G. V. and Thornton, C. L., 1968a, Relationship between perceptual style and driver reaction to an emergency situation, *Journal of Applied Psychology*, **52**, 169–76.

Barrett, G. V. and Thornton, C. L., 1968b, Relationship between perceptual style and simulator sickness, *Journal of Applied Psychology*, **52**, 304–8.

Bashore, T., Osman, A. and Heffley, E., 1989, Mental slowing in elderly persons: A cognitive psychophysiological analysis, *Psychology and Aging*, **4**, 235–44.

Belinkii, V. Y., Gurfinkel, V. S. and Pal'tsev, Y. I., 1967, Elements of control of voluntary movements, *Biofizika*, **12**, 135–41.

Birren, J. E., 1947, Vibratory sensitivity in the aged, *Journal of Gerontology*, **2**, 267–8.

Birren, J. E., Woods, A. M. and Williams, M. V., 1980, Behavioral slowing with age: Causes, organization, and consequences, in Poon, L. W. (Ed.) *Aging in the 1980s: Psychological Issues*, Washington, DC: American Psychological Association.

Black, O., 1977, The aging vestibular system, in Han, S. S. and Coons, D. H. (Eds) *Special Senses in Aging*, pp. 178–85, Ann Arbor: University of Michigan.

Botwinick, J. E., 1965, Theories of antecedent conditions of speed of response, in Welford, A. T. and Birren, J. E. (Eds) *Behavior, Aging and the Nervous System*, Springfield, IL: Charles C. Thomas.

Botwinick, J. E., 1984, *Aging and Behavior*, New York: Springer Publishing Co.

Cerella, J., 1985, Information processing rates in the elderly, *Psychological Bulletin*, **98**, 67–83.

Corso, J. F., 1981, *Aging Sensory Systems and Perception*, New York: Praeger.

Eisdorfer, C., 1975, Verbal learning and response time in the aged, *Journal of Genetic Psychology*, **109**, 15–22.

Falduto, L. L. and Baron, A., 1986, Age-related effects of practice and task complexity on card sorting, *Journal of Gerontology*, **41**, 659–61.

Fozard, J. L., 1981, Speed of mental performance and aging: Costs of age and benefits of wisdom, in Pirozzolo, F. J. and Maletta, G. J. (Eds) *Behavioral Assessment and Psychopharmacology*, pp. 59–94, New York: Praeger Publishers.

Goggin, N. L. and Stelmach, G. E., 1990, Age-related differences in a kinematic analysis of precued movements, *Canadian Journal on Aging*, **9**(4), 371–85.

Goggin, N. L., Stelmach, G. E. and Amrhein, P. C., 1989, Effects of age on motor preparation and restructuring, *Bulletin of the Psychonomic Society*, **27**, 199–202.

Gottsdanker, R., 1980a, Aging and the use of advance probability information, *Journal of Motor Behavior*, **12**, 133–43.

Gottsdanker, R., 1980b, Aging and the maintenance of preparation, *Experimental Aging Research*, **6**, 13–27.

Gottsdanker, R., 1982, Age and simple reaction time, *Journal of Gerontology*, **37**, 342–8.

Griew, S., 1959, Complexity of response and time of initiating responses in relation to age, *American Journal of Psychology*, **72**, 83–88.

Halpern, D. F., 1984, Age differences in response time to verbal and symbolic traffic signs, *Experimental Aging Research*, **10**, 201–4.

Johnson, B. M., Miao, M. and Sadum, A. A., 1987, Age-related decline of human optic nerve, *Age*, **10**, 5–9.

Jordan, T. C. and Rabbitt, P. M. A., 1977, Response times to stimuli of increasing complexity as a function of ageing, *British Journal of Psychology*, **68**, 189–201.

Karp, S. A., 1967, Field dependence and occupational activity in the aged, *Perceptual and Motor Skills*, **24**, 603–9.

Kenshalo, D. R., 1977, Age changes in touch, temperature, kinesthesis and pain sensitivity, in Birren, J. E. and Schaie, K. W. (Eds) *Handbook of the Psychology of Aging*, pp. 562–79, New York: Van Nostrand Reinhold.

Kokmen, E., Bossemeyer, R. W. and Williams, W. T., 1978, Quantitative evaluation of joint motion sensation in an aging population, *Journal of Gerontology*, **33**, 62–67.

Kroll, W. and Clarkson, P. M., 1978, Age, isometric knee extension strength, and fractionated resisted response time, *Experimental Aging Research*, **4**, 389–409.

Larsson, L., 1982, Aging in mammalian skeletal muscle, in Mortimer, J., Pirozzolo, F. and Maletta, G. (Eds) *Aging Motor System*, pp. 60–97, New York: Praeger.

Larsson, L., Grimby, G. and Karlsson, J., 1979, Muscle strength and speed of movement in relation to age and muscle morphology, *Journal of Applied Physiology*, **46**, 451–4.

Man'kovskii, N. B., Mints, A. Y. and Lysenyuk, V. P., 1980, Regulation of the preparatory period for complex voluntary movement in old and extreme old age, *Human Physiology* (Moscow), **6**, 46–50.

McPherson, K., Ostrow, A., Shaffron, P. and Yeater, R., 1988, *Physical fitness and the aging driver*, Washington, DC: AAA Foundation of Traffic Safety.

Montoye, H. J. and Lamphier, D. E., 1977, Grip and arm strength in males and females, age 10 to 69, *Research Quarterly*, **48**, 109–20.

Murrell, F. H., 1970, The effect of extensive practice on age differences in reaction time, *Journal of Gerontology*, **25**, 268–74.

Murrell, K. F. and Entwisle, D. G., 1960, Age differences in movement pattern, *Nature*, **185**, 948–9.

Olson, P. L. and Sivak, M., 1986, Perception-response time to unexpected roadway hazards, *Human Factors*, **28**, 91–96.

Planek, T. W., 1974, Factors influencing the adaptation of the aging driver to today's traffic, *Clinical Medicine*, **81**, 36–43.

Ponds, R. W. M., Brouwer, W. H. and van Wolffelaar, P. C., 1988, Age differences in divided attention in simulated driving task, *Journal of Gerontology*, **43**, 151–6.

Rabbitt, P. M. A., 1968, Age and the use of structure in transmitted information, in Talland, G. A., (Ed.) *Human Aging and Behavior*, New York: Academic Press.

Rabbitt, P. M. A., 1979, How old and young subjects monitor and control responses for accuracy and speed, *British Journal of Psychology*, **70**, 305–11.

Rosenbaum, D. A., 1980, Human movement initiation: Specification of arm direction and extent, *Journal of Experimental Psychology: General*, **109**, 444–74.

Rosenhall, U. and Rubin, W., 1975, Degenerative changes in the human vestibular sensory epithalia, *Acta Oto-Laryngologica*, **79**, 67–81.

Salthouse, T. A., 1979, Adult age and the speed-accuracy tradeoff. *Ergonomics*, **22**, 811–21.

Salthouse, T. A., 1985a, Speed of behavior and its implications for cognition, in Birren, J. E. and Schaie, K. W. (Eds) *Handbook of the Psychology of Aging*, pp. 400–26, New York: Van Nostrand Reinhold.

Salthouse, T. A., 1985b, *A Theory of Cognitive Aging*, Amsterdam: Elsevier.

Salthouse, T. A. and Somberg, B. L., 1982, Isolating the age deficit in speeded performance, *Journal of Gerontology*, **37**, 59–63.

Scheibel, A. B., 1979, Aging in human motor control systems, *Sensory Systems and Communication in the Elderly*, **10**, 297–310.

Schmidt, R. A., 1988, *Motor Control and Learning: A Behavioral Emphasis*, 2nd Edn, Champaign, IL: Human Kinetic Publishers, Inc.

Schonfield, D. and Wenger, L., 1975, Age limitation of perceptual span, *Nature*, **253**, 377–8.

Schwartz, D. W. and Karp, S. A., 1967, Field dependence in a geriatric population, *Perceptual and Motor Skills*, **24**, 495–504.

Simon, J. R. and Pouraghabagher, A. R., 1978, The effect of aging on the stages of processing in a choice reaction time task, *Journal of Gerontology*, **33**, 553–61.

Skinner, H. B., Barrack, R. L. and Cook, S. D., 1984, Age-related decline in proprioception, *Clinical Orthopaedics and Related Research*, **184**, 208–11.

Smith, B. H. and Sethi, P. K., 1975, Aging and the nervous system, *Geriatrics*, **30**, 109–15.

Smith, G. A. and Brewer, N., 1985, Age and individual differences in correct and error reaction times, *British Journal of Psychology*, **76**, 199–203.

Spirduso, W. W., 1982, Physical fitness in relation to motor aging, in Mortimer, J., Pirozzolo, F. and Maletta, G. (Eds) *Aging Motor System*, pp. 120–51, New York: Praeger.

Staplin, L., Janoff, M. and Decina, L., 1985, *Reduced Lighting During Periods of Low Traffic Density*, Final Report: FWHA Contract DTFH 61-83-C-00056, August. McLean, VA, Turner Fairbank Highway Centre.

Stelmach, G. E., Amrhein, P. C. and Goggin, N. L., 1988, Age differences in bimanual coordination, *Journal of Gerontology*, **43**, 18–23.
Stelmach, G. E. and Goggin, N. L., 1988, Psychomotor decline with age, in Spirduso, W. W. and Eckert, H. M. (Eds) *Physical Activity and Aging*, pp. 6–18, American Academy of Physical Education Papers, No. 22, Champaign, IL: Human Kinetics Books.
Stelmach, G. E., Goggin, N. L. and Amrhein, P. C., 1988, Aging and restructuring of precued movements, *Psychology and Aging*, **3**, 151–7.
Stelmach, G. E., Goggin, N. L. and Garcia-Colera, A., 1987, Movement specification time with age, *Experimental Aging Research*, **13**, 39–46.
Stelmach, G. E., Phillips, J., DiFabio, R. P. and Teasdale, N., 1989, Age, functional postural reflexes and voluntary sway, *Journal of Gerontology*, **44**, 101–6.
Stelmach, G. E., Populin, L. and Müeller, F., 1990, Postural muscle onset and voluntary movement in the elderly, *Neuroscience Letters*, **117**, 118–94.
Stelmach, G. E. and Sirica, A., 1987, Aging and proprioception, *Age*, **9**, 99–103.
Stelmach, G. E., Teasdale, N., Phillips, J. and Worringham, C. J., 1989, Force production characteristics in Parkinson's disease, *Experimental Brain Research*, **76**, 165–72.
Stelmach, G. E. and Worringham, C. J., 1985, Sensorimotor deficits related to postural stability: Implications for falling in the elderly, *Clinics in Geriatric Medicine*, **1**, 679–94.
Transportation Research Board, 1988, *Transportation in an aging society*, Vol. I, Special report 218, Washington, DC: National Research Council.
Vegega, M. E., 1989, *The effects of aging on the cognitive and psychomotor abilities of older drivers: A review of the research*, Unpublished document.
Vrtunski, P. B., Patterson, M. B. and Hill, G. O., 1984, Factor analysis of choice reaction time in young and elderly subjects, *Perceptual and Motor Skills*, **59**, 659–76.
Walker, J. N., Miles-Elkousy, N., Ford, G. and Trevelyan, H., 1984, Active mobility of the extremities in older subjects, *Physical Therapy*, **64**, 919–23.
Welford, A. T., 1981, Signal, noise, performance and age, *Human Factors*, **23**, 91–109.
Welford, A. T., 1982, Motor skills and aging, in Mortimer, J., Pirozzolo, F. and Maletta, G. (Eds) *Aging Motor System*, pp. 152–87, New York, Praeger.
Welford, A. T., 1984a, Between bodily changes and performance: Some possible reasons for slowing with age, *Experimental Aging Research*, **10**, 73–88.
Welford, A. T., 1984b, Psychomotor performance, in Eisdorfer, C. (Ed.) *Annual Review of Gerontology and Geriatrics*, pp. 237–73, New York: Springer Publishing Co.
Welford, A. T., Norris, A. H. and Shock, N. W., 1969, Speed and accuracy of movement and their changes with age, *Acta Psychologica*, **30**, 3–15.
Wickelgren, W. A., 1977, Speed-accuracy trade-off and information processing dynamics, *Acta Psychologica*, **41**, 67–85.
Winter, D. J., 1985, Learning and motivational characteristics of older people pertaining to traffic safety, in *Proceedings of the Older Driver Colloquium*, pp. 77–86, Orlando, FL: AAA Foundation for Traffic Safety.
Woollacott, M. H., 1990, Changes in postural control and the integration of postural responses into voluntary movements with aging: Is borderline pathology a contributor? in Brandt, T., Paulus, W., Blis, W., Dietirich, M., Krafczyck, S. and Straube, A. (Eds) *Disorders of Posture and Gait 1990*, Stuttgart: Verlag.
Woollacott, M. H., Shumway-Cook, A. and Nashner, L. M., 1986, Sensory integration in aging, *International Journal of Aging and Human Development*, **23**, 97–114.
Yee, D., 1985, A survey of the traffic safety needs and problems of drivers age 55 and over, in Malfetti, J. W. (Ed.) *Needs and problems of older drivers: Survey results and recommendations*, Falls Church, VA: AAA Foundation for Traffic Safety.

12

Design and evaluation of symbols for automobile controls and displays

Paul Green

Introduction

This chapter is an exhaustive review of the research on symbols for labeling automobile controls and displays. It covers 1) the development of symbols in the international standard, 2) how symbols should be developed and tested, 3) size and other display requirements and 4) perceptual characteristics of easily identified symbols. Given space constraints, this chapter focuses only on automotive applications, a portion of the literature on symbols. Readers with broader interests in symbols should examine Easterby's research (1970), the Zwaga and Boersema study (1983) of the reactions of 11 600 railway passengers to public information signs, the Zwaga review (1974) of methodology, Verplank's work on icons for the Xerox Star (1986), NBS research on workplace symbols (Lerner and Collins, 1980) and Stauffer's 1987 book.

What is a symbol?

The terms symbols, pictograms, pictographs, pictographic symbols, glyphs and icons often appear interchangeably in the literature, although they have slightly different meanings. This chapter covers symbols—small, simple images used to represent an object, function, system or the associated state. The concept the symbol represents is called the referent. Symbols are usually monochromatic and static, and usually do not contain text.

Why use symbols?

Automobile controls (e.g., horn, fan) and displays (e.g., fuel, engine temperature) can be labeled with symbols, words or abbreviations. Symbols are generally more compact than either the associated words or abbreviations. In their classic study, Jacobs, Johnston and Cole (1975) found that people recognized symbolic highway signs (e.g., road narrows) at a smaller visual angle (equivalent to double the viewing distance) than text signs. (Paniati,

1988, who conducted a similar study, reported a legibility ratio of 2.8. Similar results were obtained by Babbitt Kline et al., 1990.) For automobile instrument panels, limited panel space makes small labels desirable. Symbols are likely to be most beneficial to those whose vision is closest to size-related legibility thresholds, namely older drivers.

It is commonly believed that symbols can be a means of communication that overcomes natural language barriers. Travelers in a foreign country must be able to operate cars safely, even if they cannot read the native language. Having a standard set of symbols can make language-specific versions of a car for each national market unnecessary. This uniformity reduces world-wide product cost. While some US engineers have advocated using English ('everyone should know English'), that ethnocentric view is politically unacceptable.

There is little empirical research that addresses the understandability of foreign words and abbreviations. Table 12.1 demonstrates the problem, showing the English name for several functions with French and German equivalents.

Related research is presented in Elsholz and Bortfeld (1978). German visitors to a BMW museum were shown 28 German words and 35 symbols and were asked what they represented. Interpretation of the words was usually superior, though neither the words tested nor how the results were scored was given.

A second group was shown 30 words and 34 symbols. After responding to them, participants were given their meanings. Ten or more days later, the process was repeated. Correct identification of the 'functions' associated with words went from 81 to 93 per cent. For symbols the results were 56 and 83 per cent, respectively. Even after training, the identification of symbols was only slightly better than the initial response to words—an argument against using symbols.

Whether symbols are responded to more quickly than text depends on how the question is asked. Dewar and Ells (1974) found verbal (text) highway signs elicited briefer response times than symbolic signs (823 ms vs. 937 ms) because the response (saying them) was more compatible with the sign format. Dewar, Ells and Mundy (1976) reported similar results. In Ells and Dewar (1979), sign names were read and then slides shown, to which people responded as either 'same' or 'different'. In this study, the order was reversed. Response times to symbolic signs were about 100 ms less than those to words.

To some extent, symbols are used because they are fashionable and fit in with product aesthetic themes. They have been popularized by the user inter-

Table 12.1. Automotive terms in several languages

English name	French Word	French Abbrev.	German Word	German Abbrev.
light	phare	phare	licht	licht
wiper	essuie-glace	essui	wischer	wisch
defroster	dégivreur	dgivr	entfroster	entfr
heater	chauffeur	chauf	heizung	heiz

face for Xerox Star and Apple Macintosh computers, and now are a central part of most graphical user interfaces.

The use of specific symbols may be required. For cars sold in the US, Federal Motor Vehicle Safety Standard 101 (US Department of Transportation, 1990) stipulates which symbols may appear on the instrument panel. Many countries call for compliance with ISO Standard 2575 (International Standards Organization, 1982), and 2575 has become a *de facto* world-wide requirement (see Figure 12.1). All of the symbols in SAE J1138, a voluntary construction standard, also appear in ISO 2575. Other relevant symbols appear in ISO Standard 7000 (International Standards Organization, 1989), over which 2575 has precedence for road vehicles.

Summary

While there is no empirical evidence, it is widely believed that words and abbreviations in foreign languages are not as well understood as the symbols for them. This is particularly important for internationally marketed products such as cars, where native language barriers can decrease usability. Also several studies have shown that symbols are at least two times more legible than their text equivalents—important for elderly drivers or in cars where panel space is limited. Finally, symbols are often required because of the performance advantages cited above. For cars, the key standard is ISO 2575.

How did the current international automotive set emerge?

Most of the early research on symbols for ISO 2575 concerned interpretability (see Kyropoulos, 1972; Frank, Koenig and Lendholt, 1973; Hoffman, 1976; and Wiegand and Glumm, 1979). The initial research (Jack, Heard and Pew, 1970) was carried out under the auspices of the Society of Automotive Engineers (SAE). Subsequent studies, mostly of national differences, were collaborative efforts carried out by members of Working Group 5 (Symbols) of ISO Technical Committee 22, Subcommittee 13 (Ergonomics of Road Vehicles). Since the mid-70s, the focus has shifted toward how well symbols in 2575 are understood. Recent proposals for symbols have not been accompanied by data.

Jack, Heard and Pew (1970)

Jack, Heard and Pew (1970) (see also Control Symbols Compared, 1970) tested 202 US visitors to a Ford plant and 77 foreign nationals at the International Institute in Detroit. They were shown symbols and wrote down the name of the control for each. There were no differences in the per cent correctly identified between US and foreign nationals, or between men and women. Only three symbols were somewhat well identified (above 60 per cent correct; see Figure 12.2).

Jack (1972)

Using a similar method, Jack (1972) had 1187 visitors to a Ford plant write down the name of symbols shown, a free response task. The per cent correct data is in Figure 12.3. At least one candidate symbol was identified correctly 75 per cent of the time (Jack's upper category) for seven referents. The per cent correct data reported are much higher than those of Jack, Heard and Pew (1970). The Jack (1972) paper is the first one to mention 75 per cent correct or better, as a category, a partitioning that has appeared in other SAE-related papers as well.

Heard (1974)

Heard (1974) is the most extensive study of automotive symbols in the literature. A total of 2593 people from four countries (France, Germany, UK and US) participated. While seated in a mockup, scenarios were read to people in their native language. (Multiple mockups were used in each country and no two countries used the same mockup.) The experimenter then removed the cardboard blocking the subject's view and timed how long it took to touch the

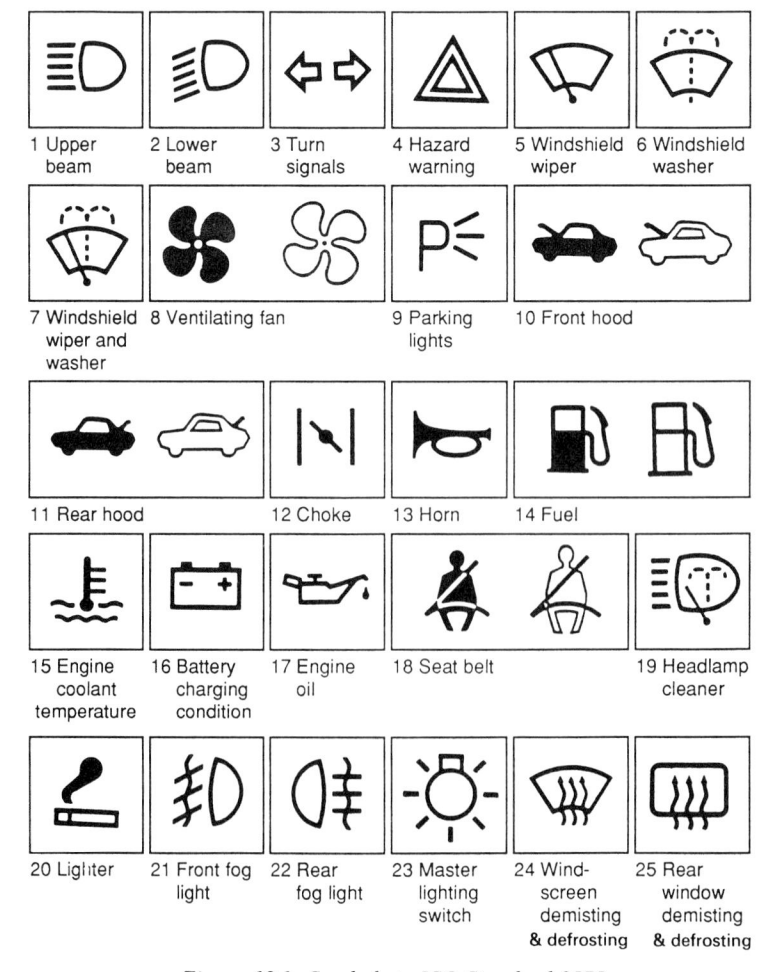

Figure 12.1. Symbols in ISO Standard 2575

Design of control and display symbols 241

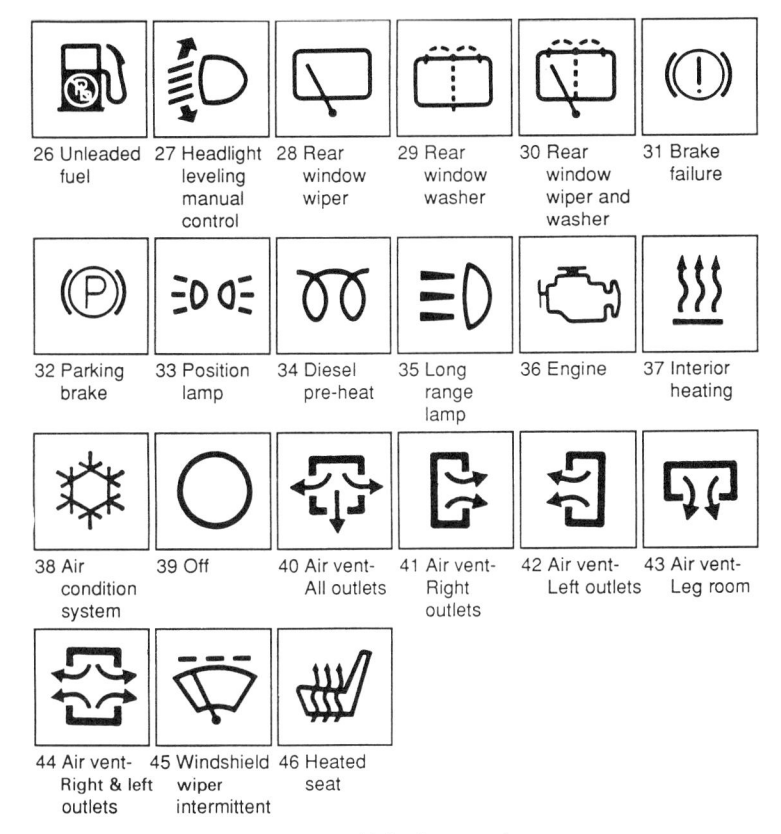

26 Unleaded fuel	27 Headlight leveling manual control	28 Rear window wiper	29 Rear window washer	30 Rear window wiper and washer	31 Brake failure
32 Parking brake	33 Position lamp	34 Diesel pre-heat	35 Long range lamp	36 Engine	37 Interior heating
38 Air condition system	39 Off	40 Air vent- All outlets	41 Air vent- Right outlets	42 Air vent- Left outlets	43 Air vent- Leg room
44 Air vent- Right & left outlets	45 Windshield wiper intermittent	46 Heated seat			

Figure 12.1. Continued

symbol cued. On the panel were nine symbols from the existing ISO standard plus one of three sets of 15 symbols (See Figure 12.4).

Figure 12.4 shows the per cent correct data. For the 15 referents there were significant differences between countries (12), and candidates for each referent (8), as well as significant candidate-country interactions (10). Thus, tests of symbol meaningfulness need to be carried out in multiple countries. While this would seem to be obvious for political reasons, Heard's study provides empirical evidence.

Table 12.2 shows the mean response time for each candidate. There was a reasonable correlation between the time and error data ($r = 0.78$ based on further analysis). Note that while the constrained nature of per cent correct suggests that rank correlations (e.g., Spearman's Rho) should be used instead of Pearson's r, differences between the two values were less than 1 per cent. Here r is used to be consistent with the literature.

Heard's acceptance criteria (for addition to the standard set) was 75 per cent correct with no more than 5 per cent confusions of symbols in the set. Twelve candidates, some of which did not meet the 75 per cent criterion, were recommended to be added to the International Standard—parking lights, front and rear hood symbols, horn, fuel, oil, seat belt, charging condition, temperature,

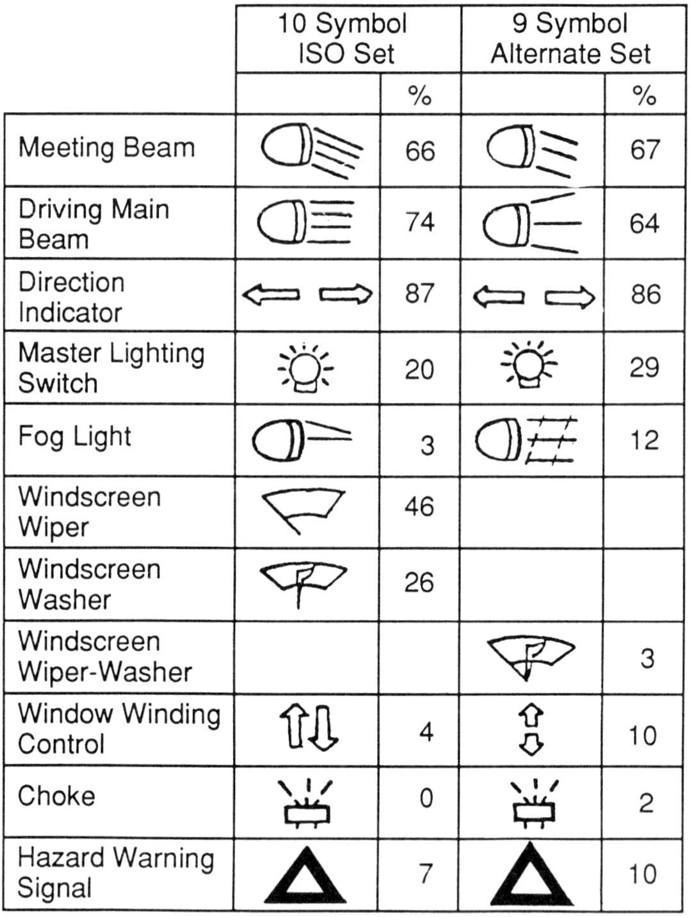

	10 Symbol ISO Set	%	9 Symbol Alternate Set	%
Meeting Beam		66		67
Driving Main Beam		74		64
Direction Indicator		87		86
Master Lighting Switch		20		29
Fog Light		3		12
Windscreen Wiper		46		
Windscreen Washer		26		
Windscreen Wiper-Washer				3
Window Winding Control		4		10
Choke		0		2
Hazard Warning Signal		7		10

Figure 12.2. Results from Jack, Heard and Pew (1970)

choke, and front and rear fog lights. In several cases the recommended symbol was not the best recognized candidate for a referent (either overall or when the per cent correct scores were weighted by the populations, number of drivers or number of vehicles in the responding countries).

The Heard (1974) work is noteworthy for its large sample size, attention to national differences, and impact on the international standard. No other study of automotive symbols has had a comparable influence. However, the statistics should be viewed with some caution. (In one ANOVA, 88 per cent of the data was discarded to balance cell sizes.)

McCormack (1974)

This replication of Heard (1974) (same method, etc.) used 150 Canadians as subjects. Significant differences due to gender and symbol sets were reported. Table 12.3 shows the per cent correct data for each of the candidates tested and the mean response times. Further, there was a strong relationship between

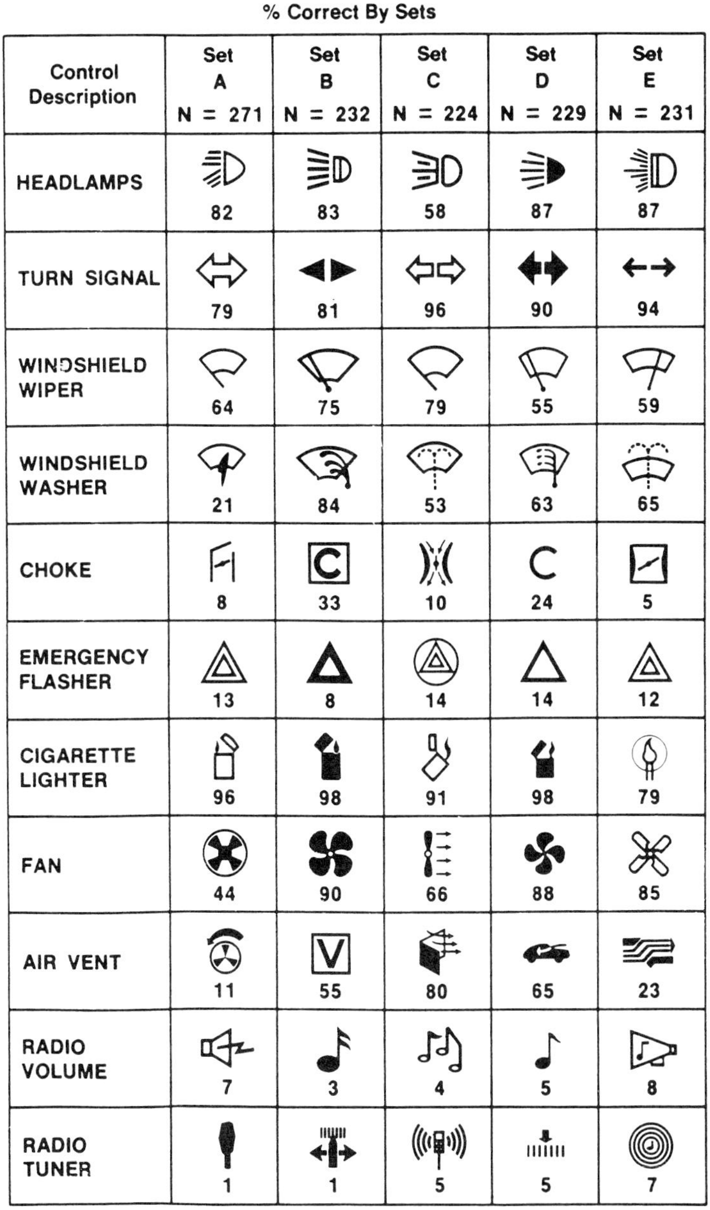

% Correct By Sets

Control Description	Set A N = 271	Set B N = 232	Set C N = 224	Set D N = 229	Set E N = 231
HEADLAMPS	82	83	58	87	87
TURN SIGNAL	79	81	96	90	94
WINDSHIELD WIPER	64	75	79	55	59
WINDSHIELD WASHER	21	84	53	63	65
CHOKE	8	33	10	24	5
EMERGENCY FLASHER	13	8	14	14	12
CIGARETTE LIGHTER	96	98	91	98	79
FAN	44	90	66	88	85
AIR VENT	11	55	80	65	23
RADIO VOLUME	7	3	4	5	8
RADIO TUNER	1	1	5	5	7

Figure 12.3. Results from Jack (1972)

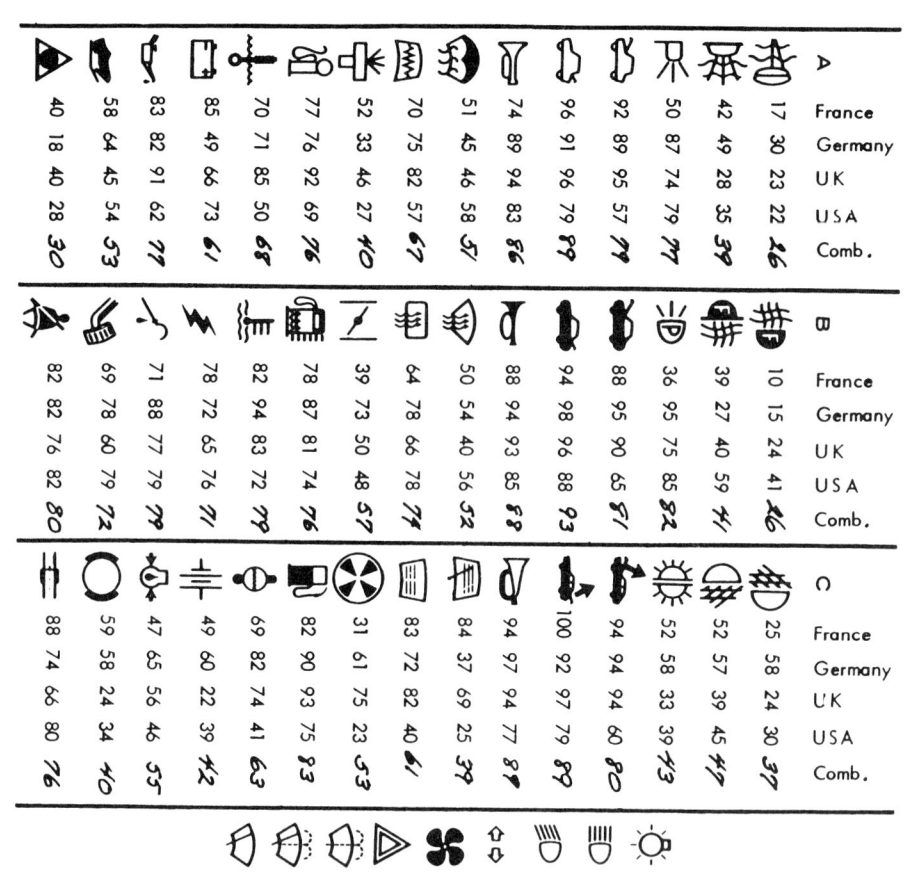

40	58	83	85	70	77	52	70	51	74	96	92	50	42	17		France
18	64	82	49	71	76	33	75	45	89	91	89	87	49	30		Germany
40	45	91	86	85	92	46	82	46	94	96	95	74	28	23		UK
28	54	62	73	50	69	27	57	58	83	79	57	79	35	22		USA
30	53	77	61	68	76	40	67	51	86	89	77	77	39	16		Comb.

82	69	71	78	82	78	39	64	50	88	94	88	36	39	10		France
82	78	88	72	94	87	73	78	54	94	98	95	95	27	15		Germany
76	80	77	65	83	81	50	66	40	93	96	90	75	40	24		UK
82	79	79	76	83	72	74	48	78	56	85	88	65	59	41		USA
80	72	79	71	79	76	57	71	52	88	93	91	82	41	16		Comb.

88	59	47	49	69	82	31	83	84	94	100	94	52	52	25		France
74	58	65	60	82	90	61	72	37	97	92	94	58	57	58		Germany
86	24	56	22	74	93	75	82	69	94	97	94	33	39	24		UK
80	34	46	39	41	75	23	40	25	77	79	80	39	45	30		USA
76	40	55	42	63	83	53	61	39	81	89	80	43	47	37		Comb.

Figure 12.4. Symbols examined by Heard (1974) and per cent correct for candidates tested

time and errors (r = 0.82 here versus r = 0.78 in Heard), with RT = −11.22(per cent correct) + 13.68. This result argues for using either time or errors (or both) as dependent measures in initial identification experiments.

As expected, the data for Canadians correlates very well with Heard's data for the US (r = 0.81 based on further analysis). However, for two referents there were considerable differences in the per cent correct, often over 30 per cent. US drivers were better at identifying charging circuit symbols while the Canadians were considerably better with front hood. Removing these data points increases the correlation to 0.92.

Saunby, Farber and DeMello (1988)

This study was conducted to see how the results of free response and matching tasks compared, and if driver understanding of symbols had improved since the early SAE studies. At a driver licensing office in the US, 505 drivers completed a survey. In part 1, people wrote in the names of 25 symbols. In part 2, drivers matched symbols with a list of 25 control/display names.

Design of control and display symbols

Table 12.2. Mean response time (seconds) for each candidate

	Symbol	Set A	Set B	Set C
1	Front fog lights	12	13	13
2	Rear fog lights	9	7	6
3	Parking lights	5	4	10
4	Front hood	6	7	6
5	Rear hood	3	4	3
6	Horn	3	3	3
7	Front defogger	8	9	11
8	Rear defogger	7	6	7
9	Choke	15	14	9
10	Fuel	6	8	6
11	Coolant temperature	8	6	9
12	Charging circuit	6	10	8
13	Engine oil	5	7	9
14	Brake	11	9	13
15	Seat belt	18	9	10

Only nine of the 25 exceeded a 75 per cent criterion in the free response task, as opposed to 16 of the 25 in the matching task (see Figure 12.5). The correlation of the two data sets was 0.82, quite good. From regression analysis, the equivalent of 75 per cent in the free response task is 86 per cent for the matching task. There were two symbols that were not readily identified but could be easily matched (washer, unleaded fuel), and several that were easily recognized (fuel, high beam, oil) but did not do as well in the matching task.

Table 12.3. Per cent correct and response time (secs) from McCormack (1974)

	Symbol	Per cent Correct Set			Response Time (Sec) Set		
		A	B	C	A	B	C
1	Front fog lights	27	45	32	11.0	7.0	11.5
2	Rear fog lights	40	51	49	9.0	5.0	3.5
3	Parking lights	77	96	45	5.5	5.0	14.0
4	Front hood	92	94	91	3.5	3.0	3.0
5	Rear hood	100	100	94	2.5	2.5	2.5
6	Horn	92	92	94	3.0	2.5	2.5
7	Front defogger	58	71	28	6.0	7.0	10.5
8	Rear defogger	48	75	49	6.5	4.0	6.0
9	Choke	15	45	15	7.0	6.5	7.0
10	Fuel	58	59	62	5.0	6.5	4.5
11	Coolant temperature	31	71	40	12.5	6.5	10.0
12	Charging circuit	31	37	26	11.5	10.0	9.0
13	Engine oil	63	75	43	7.0	7.5	8.5
14	Brake	62	80	43	8.5	5.0	11.0
15	Seat belt	17	86	70	14.0	6.5	6.5

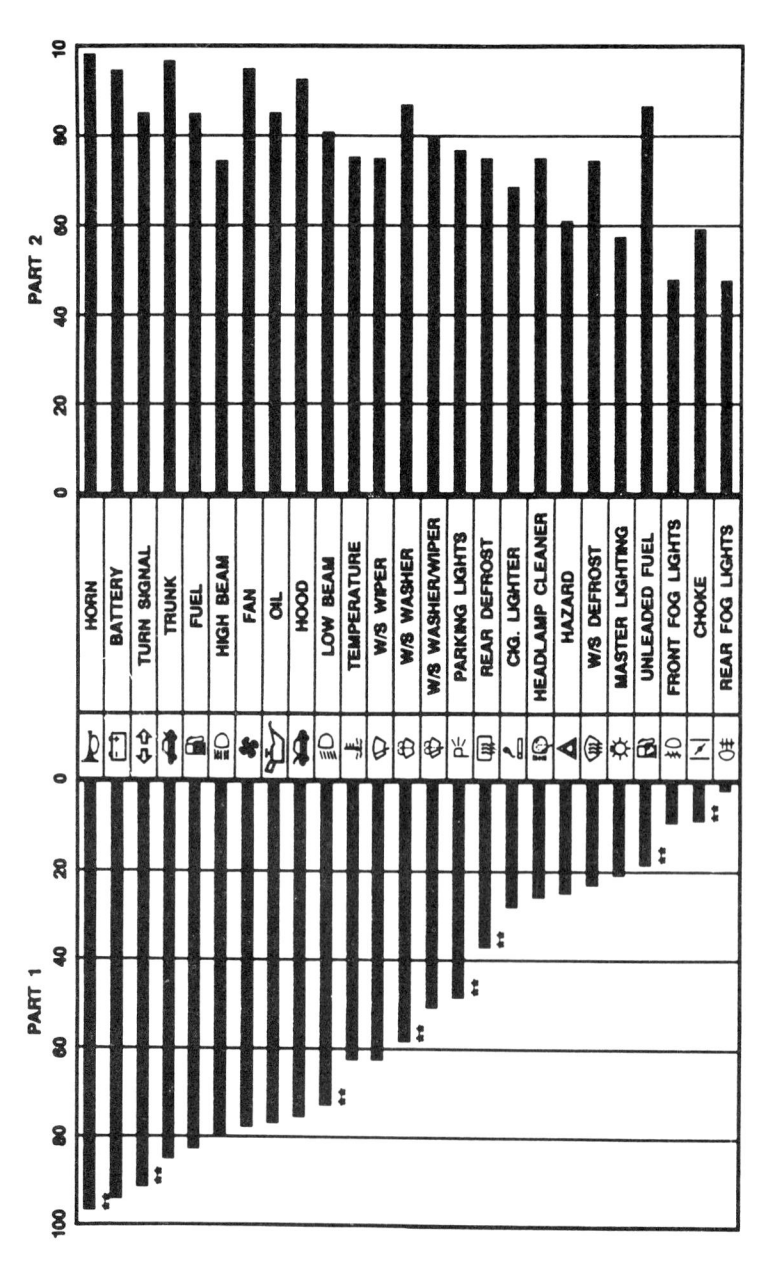

Figure 12.5. Results from Saunby, Farber and DeMello (1988)

Men did better than women (by 8 per cent in the free response task and 5 per cent in the matching task) and younger people did better than those over 50. A significant difference of 19 per cent was observed in both cases.

Saunby *et al.* did not compare their results with previous studies. The most appropriate comparison group is Heard's US drivers. Heard's task was similar to part 2 (matching), though data were collected in a mockup (see Table 12.4). The correlation was only 0.40, not very high. It is noteworthy that over the intervening 14 years, driver understanding of several symbols (parking lights, temperature, rear defrost) has remained at the limit of acceptability. Some consideration should be given to new images for these referents.

Summary

Several studies, mostly from the early 1970s, have examined the meaningfulness of candidate symbols for ISO 2575. Popular methods include free response naming and matching tasks. Where naming is used, the procedure for scoring the data is often not described in detail, making replication difficult. Since absolute identification is important, this is a significant problem.

Much of the matching work has involved paper and pencil (not in-vehicle) data collection procedures. If the goals of a study are to determine which of several candidates is best, paper and pencil work fine. Absolute levels should be determined in a vehicle.

Research (Saunby *et al.*, 1988) has shown that performance on the two tasks is well correlated, but not perfect. Also, response time in matching tasks (in-vehicle) is well correlated with error rates. However, the method that is most appropriate depends upon the type of symbol. For a control, drivers have a referent in mind and search for the appropriate image (matching task). For a display, drivers see an image and need to identify it (naming task).

The literature also emphasizes the importance of demographic factors. In general, people over age 50 and women have somewhat greater difficulty in identifying symbols. There are also country-specific differences in identifying individual symbols. Thus age, sex and country must be considered when absolute levels of symbol identification are important (such as deciding if a symbol should be added to ISO 2575).

Table 12.4. *Comparison of Saunby et al. (1988) with previous research*

Symbol	% Correct		Difference
	Heard (1974)	Saunby *et al.* (1988)	
Horn	85	98	13
Trunk	88	97	9
Battery	73	95	22
Hood	65	93	28
Fuel	75	87	12
Oil	62	86	24
Parking lights	79	77	−2
Temperature	72	75	3
Rear defrost	78	75	−3
Front defrost	56	75	19

Finally, recent research (Saunby et al., 1988) indicates that many symbols (e.g., fog lights) in the ISO standard are still not well identified, even though they have been in the international standard for over a decade. This does not reflect favorably on the work conducted to support symbol development.

How has the ease of learning symbols been measured?

The focus of the learning studies has been on the development of simple techniques to assess the learnability of symbols. Two major studies have been carried out, one by Frank, Koenig and Lendhold (1973); see also Lendholt, 1974) and one by Simmonds (1974b, 1976). The first study involved 100 visitors to a VW plant in Germany, the second involved 600 visitors to a Ford plant in the UK and 300 to a Ford plant in the US.

Participants were shown a small number of symbols (eight or nine), wrote what they meant, were told what they meant, and then after a tour, were shown the symbols and asked what they meant. In addition to the usual gender differences (men as a group usually doing better than women) and age differences (people over 45 doing worse), both studies reported improvements in identification after the tour, though identification was far from 100 per cent correct for all symbols. While factory visitors can be a steady stream of potential subjects with broad demographics, it is not clear if the measure of learnability from these types of studies reflects in-vehicle performance. As a practical matter, the Ford Rouge facility tour (which was the source of people for many studies) is no longer given. (In fact, most automotive factories no longer offer tours.)

Another approach considered was to include the mention of symbols in lectures on car safety and to test recall at the conclusion of the sessions. 'Recall was very high and of little use in discriminating between designs' (Simmonds, 1976, p. 3). Thus, for the most part, there has been little success in developing methods for assessing symbol learnability and interest in developing such methods has been low.

How should symbols be presented?

Symbol legibility is affected by the image size, resolution, contrast, luminance, orientation, and other factors. Contrast and other lighting requirements are similar to those for text and are described in the chapter on the fundamental characteristics of controls and displays. (See also Imbeau, Wierwille, Wolf, and Chun, 1989 for related work on the legibility of chromatic text displays.)

Simmonds (1974a)

This is the only study in the literature that considers the size of automotive symbols. All 16 participants had visual acuity better than 6/14 (about 20/50). Older drivers and those with very poor visual acuity were not in the sample.

Drivers were presented with a single easy-to-see symbol (eight possibilities) at the standard panel viewing distance. Subsequently, drivers identified the matching symbol in an array of nine symbols (from a set of 24; see Figure

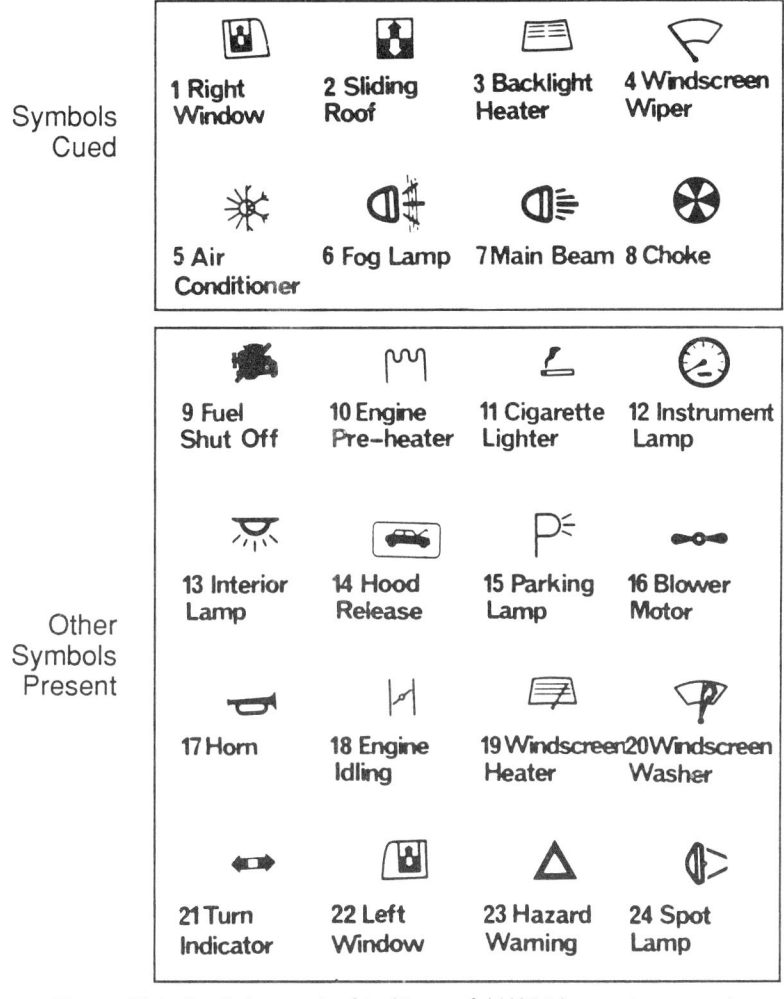

Figure 12.6. Symbols examined in Simmonds' (1974a) experiment on size

12.6). The size of the symbols in the array varied. Lighting conditions were typical for a car interior at night.

Simmonds reported no overall differences due to contrast direction, but for symbols more than 10 mm in diameter, error rates for white on black were about 2 per cent less than black on white. Further analysis based on these data shows both the expected inverse relationship between per cent correct and size, and superimposed on it, a curvilinear trend. Here, one should design for the worst case (the main beam symbol) whose relationship is:

% Correct = $111.7 - (346.1/\text{diameter}) + (292.5/(\text{diameter})^2)$

where the diameter is in mm

Accordingly, a diameter of 20 mm (about 0.8 inches) will lead to correct identification 95 per cent of the time. In contrast, Simmonds recommends ('arbitrarily') 14.4 mm (about 0.6 inches) diameter, the 50 per cent confidence point for 90 per cent correct (based on the mean for *all* per cent symbols). Since symbols vary considerably in their discriminability, some symbols will be identified far less accurately than 90 per cent at 14.4 mm. Discriminability problems result from a few symbols being confused with each other, rather than slight confusions between all symbols. In general, confusability (and size requirements) therefore depends on the specific symbols in the set though required sizes generally increase as the set size increases. Furthermore, symbols should not just be barely legible, but easy to read for everyone, especially older drivers (not included in his sample).

Green and Davis (1976)

This research shows that drivers could have difficulty recognizing symbols that are not upright, such as those on rotary knobs. Ten students were given forms on which pairs of ISO symbols (parking light, windshield wiper, lower beam) appeared, differing in orientation. Participants responded 'same' or 'different' to each pair (different if one symbol was a rotated mirror image of the other). Decision time, which included the time to write 's' or 'd' on the response sheets, is:

Decision Time (sec) = 1.35 + 0.0077 (difference in degrees)

Galer and Spicer (1986)

The required size for a symbol should vary with display resolution. Galer and Spicer (1986); (see also Galer, Spicer and Holtum, 1984) examined the legibility of 32 × 32 dot-matrix symbols. The test involved 195 drivers whose eyesight was typical of the adult population. The viewing distance and symbol height were the same as for normal driving. There were no differences shown between the two display formats (standard vs. dot matrix). This research must be applied with caution as the display conditions (symbol luminance, contrast ratio, size) are not given.

While of great interest, the relationship between these display parameters, and symbol size, display resolution and per cent correct has not been explored for automotive symbols. As use of pixel-based displays (e.g., CRTs) expands, this information will become more important.

What constitutes a good symbol?

Green (1977)

This study concerned the relationship between the psychological dimensions of symbols and response time to them. Students made magnitude estimates of symbols, identical to those in Green and Pew (1978) on nine dimensions with the front hood symbol serving as the anchor (rating = 1). (A symbol twice as complex would have a complexity rating of 2.) In a second experiment other students were shown a symbol name, then an array of 16 symbols, and subsequently pressed a button for the matching symbol.

Based on correlations, the test sequence was partitioned into four periods—response 1 (to each slide), response 2, responses 3–10, and responses 11–60.

Table 12.5. *Correlation of estimates with mean response times (Green, 1977)*

	Response			
Estimate	1	2	3–10	11–60
RT (early)	0.33	0.56*	0.28	0.20
RT (late)	0.27	0.40	0.31	0.30
Complexity	0.42	0.69**	0.41	0.31
Detectability	−0.25	−0.49*	−0.65**	−0.68**
Discriminability				
Most like it	0.45	0.23	−0.17	−0.39
from set	0.31	0.11	−0.25	−0.40
Communicativeness				
Label-symbol	−0.25	−0.52*	−0.36	−0.32
Referent-symbol	−0.30	−0.55*	−0.30	−0.24
Referent-label	−0.13	−0.37	−0.13	−0.18

Notes:
1 RT (early) and RT (late) are estimates of response time early and late in practice.
* = significant, $p < 0.05$
** = significant, $p < 0.01$

The correlations of the various estimates with the mean response times are shown in Table 12.5. Detectability became more important with practice. (More detectable symbols had briefer response times.) Surprisingly, discriminability became less important with practice, while communicativeness (meaningfulness) did not change.

Green (1979b)

This dissertation examines measures to predict the discriminability of symbols. People were shown 372 pairs of symbols (the ISO symbols for low beam, high beam, wiper, washer and modifications of them). Symbols varied in terms of the angle of one line (light ray or wiper blade), the number of such lines (for the lights only), and the extent to which the image was filled in (see Figure 12.7). Pairs were shown one at a time and people rated their dissimilarity. The front-rear hood pair (discriminability = 10) served as the standard.

All three variations (fill, line angle, number of lines) were rated as equally effective when by themselves. Fill, when combined with other dimensions, led to greater rated discriminability than other combinations of manipulations. Adding more dimensions on which symbols varied did not add equal amounts to rated discriminability. So, symbols that differed on two dimensions were not rated twice as discriminable as those that differed on one, though discriminability was enhanced. (See Figure 12.8).

In part 2, people were shown 471 pairs of symbols. On each trial a cue symbol appeared followed by two symbols flanking the cue's previous location. The subject indicated which symbol matched the cue by pressing a key. The response time data mirrored the ratings data in many ways, though there

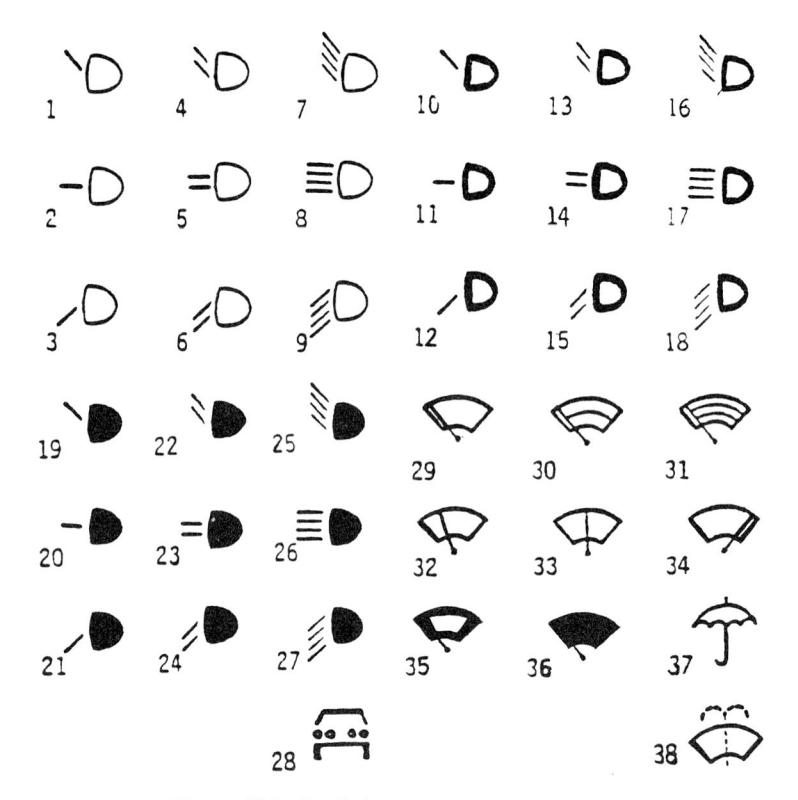

Figure 12.7. Symbols examined by Green (1979b)

were some interesting differences (r = 0.74). Figure 12.9 shows the response times for the headlight symbols. Again, increasing the number of dimensions on which symbols differed led to decreases in response time, but less for each added dimension. With regard to the individual dimensions, fill and number of lines tended to result in more salient differences than line angle.

The results relating response times to models of visual discrimination provided the most insight. Models fall into three classes—feature-based, overlap and spatial frequency. In feature-based models discrimination is based on the ratio of the number of common versus distinctive elements. For example, in upper case E and F, both have one vertical member and two horizontal members in common, as well as one distinguishing horizontal member. Identification of the feature elements requires human interpretation. These models do not deal with filled-in objects (such as symbols).

In overlap models, image similarity is derived from superimposing the two images (see Figure 12.10). There are many potential measures of similarity (the areas where both are the same polarity, that number divided by the area of differing polarity, etc.). Among them is r, the Pearson moment correlation of the images, in which each pixel is assigned a binary value.

Spatial frequency models are the least well known of these three types, but are well supported by physiological evidence (Hubel and Wiesel, 1962). In the

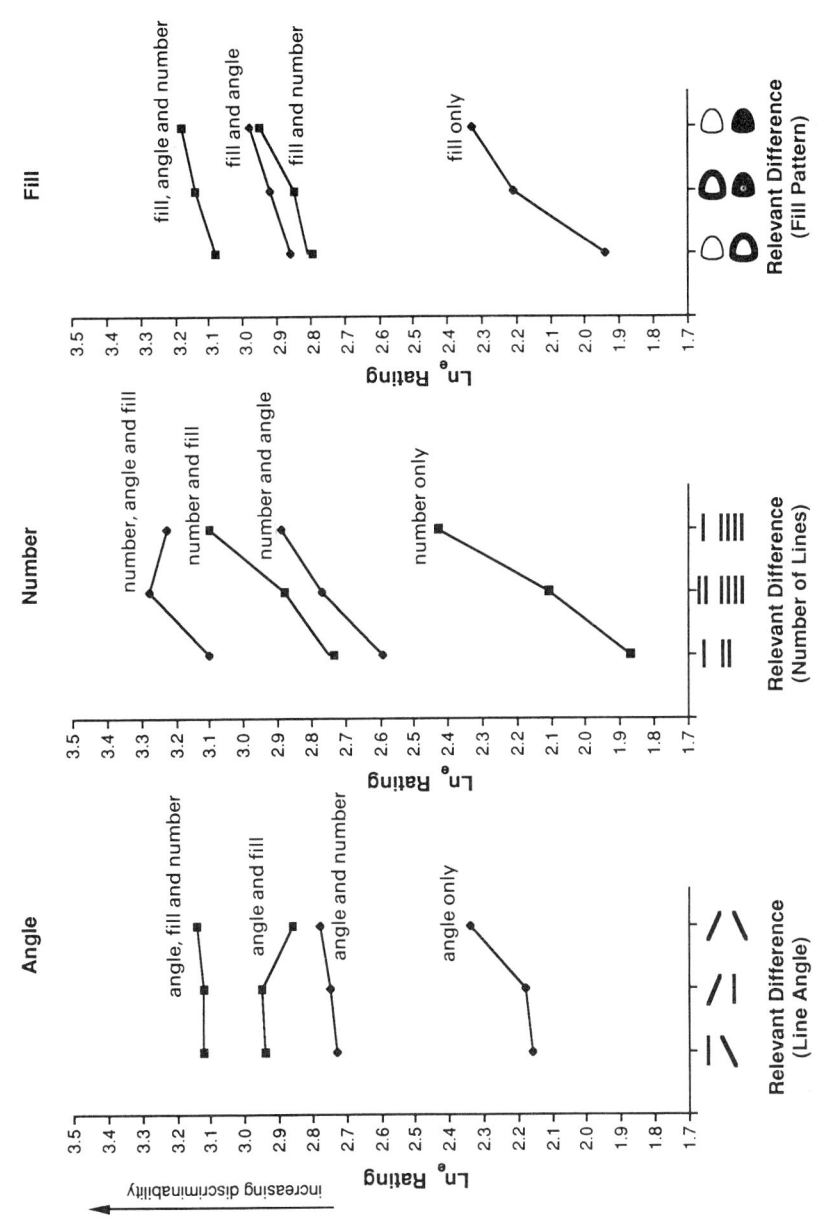

Figure 12.8. Rated discriminability of headlight symbols

254 P. Green

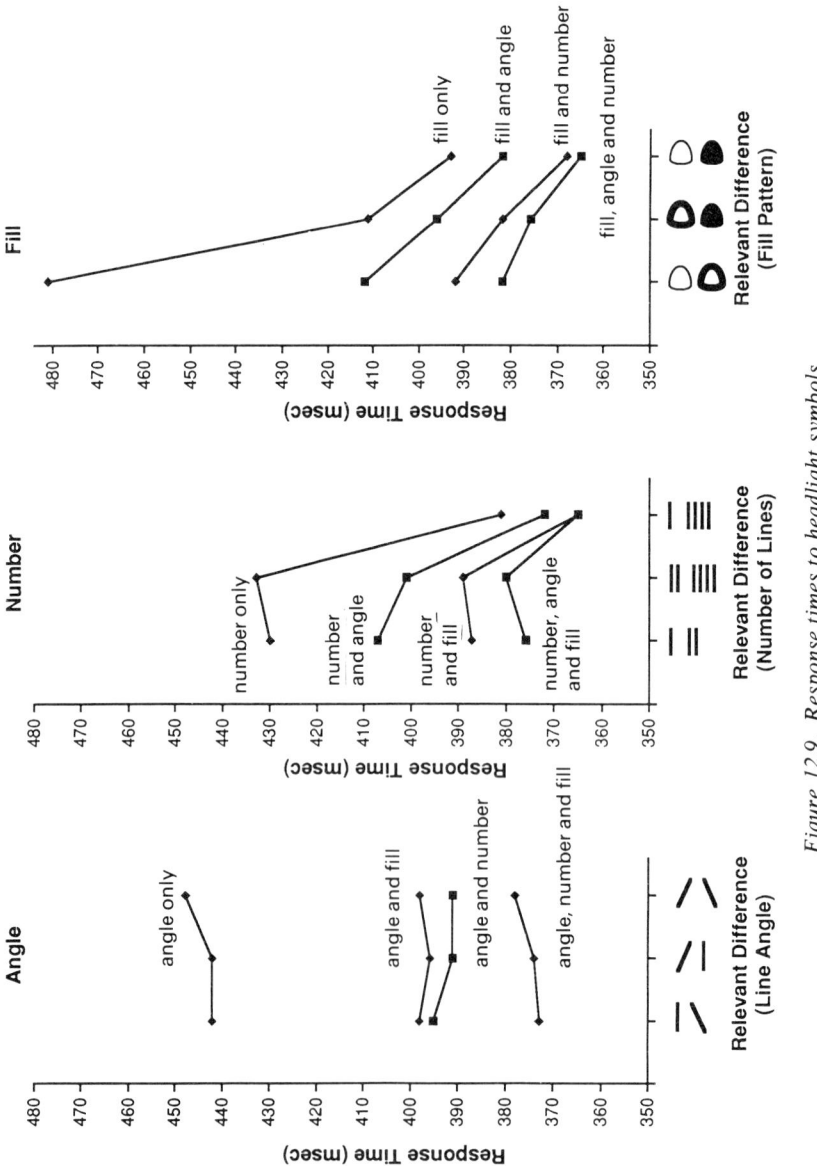

Figure 12.9. Response times to headlight symbols

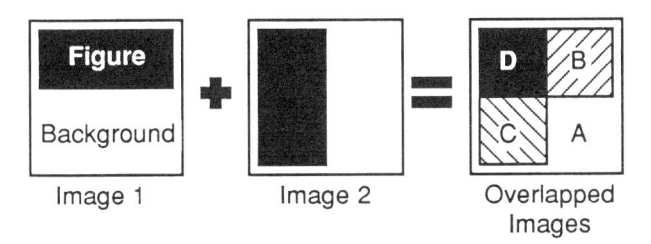

Figure 12.10. Image overlap representation

time domain, frequency analysis of a pure sine wave would result in the line spectrum shown in Figure 12.11. Similarly, in the spatial domain, a light whose intensity varies sinusoidally would have the spatial frequency pattern shown in Figure 12.11. But real images vary both in the x and in y dimensions, and hence real images have two-dimensional spatial frequency patterns. In spatial frequency analysis, the measure of similarity is the correlation of the power spectra of the images.

Table 12.6 shows the correlations of the various measures for headlight symbols. Correlations were smaller for the wiper symbols because there were fewer data points. The spatial frequency measures are as well correlated as any of the overlap measures, and far more overlap measures (actually 11 total) were examined. (The table refers to a filter. When people make visual discriminations, they pay the least attention to fine details. In the spatial frequency analysis such details were eliminated by a low-pass filter.)

Four measures were found to be reasonably well correlated with two-choice response times and ratings of discriminability—unfiltered spatial frequency, filtered spatial frequency, ratio of similar to dissimilar areas $((A + D)/(B + C))$, and the correlation coefficient (r). A subsequent experiment examined these predictors for set sizes of greater than two. The experiment involved six drivers. The procedure was the same as the first experiment; there were always two confusable lights symbols shown. Depending on the condition, there were

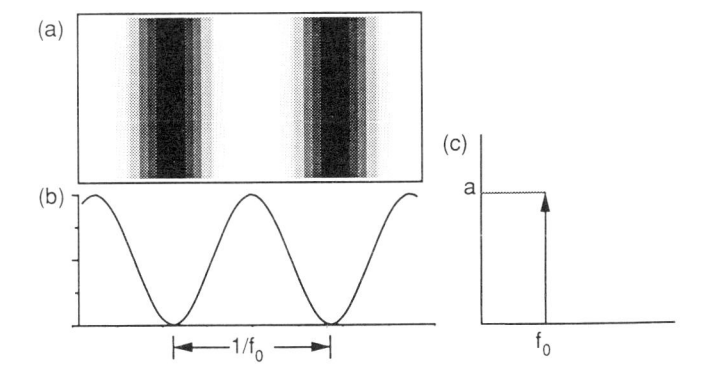

Figure 12.11. Line spectrum and spatial frequency of a sine wave

Table 12.6. Correlations of discriminability predictors for headlight symbols (Green, 1979b)

Model	RT	No. errors	Rating
Spatial frequency			
filtered	0.51	0.38	−0.75
no filter	0.48	0.31	−0.58
Overlap (template)			
(A + D)/(B + C)	0.51	0.34	−0.64
r (correlation)	0.50	0.33	−0.61
D/(B + C + D)	0.30	0.20	−0.35
D/(B + C)	0.29	0.18	−0.36
D	0.25	0.13	−0.27
(A ∗ D)/(B ∗ C)	0.21	0.18	−0.37

either none, three or eight symbols for other controls surrounding the cue location.

The effectiveness of various coding schemes (line angle, line number and fill) was similar to those in the previous experiment for the two-choice condition. For the five and ten-choice conditions, the relative effects were less. Table 12.7 shows the correlation of the various models with performance. As the number of choices increases, pairwise discriminability became less significant. This work suggests that some predictions of discriminability (and response time) can be made based on image characteristics alone.

Green and Burgess (1980) and Green (1981)

This experiment concerned both the selection of desired symbols and the development of a symbol language. In the first experiment, students drew pictures to symbolize various vehicle functions. Drawings were obtained for seven

Table 12.7. Correlation of models with performance in the second experiment (Green, 1979b)

Model	2-choice		5-choice		10-choice	
	RT	No. Err	RT	No. Err	RT	No. Err
Spatial frequency						
filtered	0.36	0.37	0.31	0.36	0.23	0.34
no filter	0.36	0.34	0.28	0.30	0.23	0.30
Overlap (template)						
(A + D)/(B + C)	0.43	0.54	0.42	0.54	0.40	0.53
r (correlation)	0.40	0.33	0.19	0.28	0.03	0.27
D/(B + C + D)	0.40	0.30	0.16	0.25	0.01	0.24
D/(B + C)	0.42	0.47	0.28	0.45	0.23	0.46
D	0.25	0.21	0.06	0.13	−0.01	0.18
(A ∗ D)/(B ∗ C)	0.36	0.53	0.29	0.52	0.27	0.53

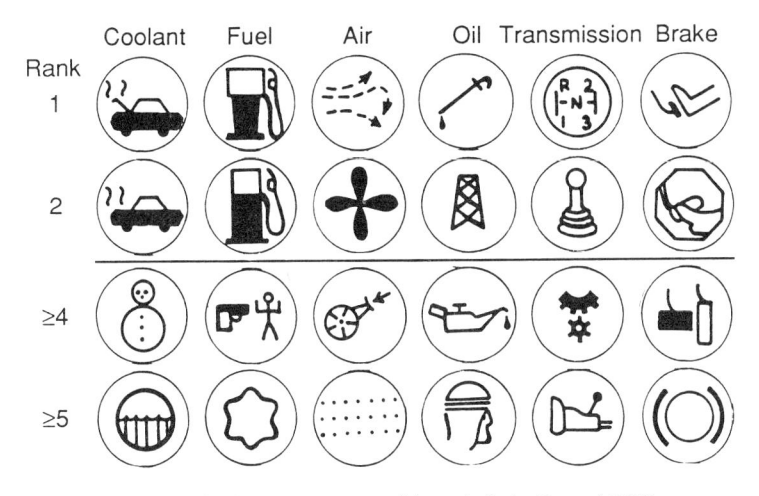

Figure 12.12. Most meaningful symbols in Green (1981)

'systems' (air, brake, coolant, fuel, hydraulic, oil, transmission), four system attributes (filter, fluid level, pressure, temperature), and 20 of the 28 system-attribute combinations. From those drawings and various standards, 216 candidates were developed for 26 functions.

In the second experiment, 26 people made magnitude estimates of the meaningfulness of each candidate. Based on those data, candidates and modifications of them were recommended for further testing. For each function, the top three candidates are shown in Figure 12.12. In a few cases, where a standard symbol was not rated as meaningful, its rank is shown as well.

Ratings for symbol elements (for systems such as 'air', and attributes such as 'filter') could be used to predict combined symbols (air filter) to a limited degree. Table 12.8 shows the predictions for the mean logarithmic ratings. Predictions were best when the symbol elements were adjacent to each other

Table 12.8. Regression equations predicting mean natural logarithmic rating (Green and Burgess, 1980)

				Prediction	
	R^2	n	Constant	System Coefficient	Attribute Coefficient
Overall	0.52	83	0.36	0.43	0.29
By combining rule					
Adjacent	0.69	34	0.151	0.59	0.27
Superimposed	0.44	49	0.37	0.36	0.34
By rule departure					
None	0.78	30	−0.085	0.58	0.41
Slight	0.25	17	1.03	0.17	0.13
Some	0.45	36	0.44	0.50	0.15

rather than superimposed. Typically, ratings for combined symbols were lower than those for the elements, with the system symbol contributing more to the combined rating than the attribute.

Summary

Good symbols are detectable, discriminable from others, and meaningful to drivers, although the correlations of such ratings with response time are not high. When symbols are well learned, detectability was the most important of these attributes.

In general, adding features to enhance discriminability was underadditive. To predict discriminability, several models have been developed. Of them, simple overlap models seem to be best at predicting response time and discriminability ratings followed closely by correlations of filtered image power spectra (from spatial frequency analysis).

Finally, the meaningfulness of combined symbols depends much more on the meaningfulness of the system symbol than the meaningfulness of the attribute. Ratings for combined symbols are usually less than ratings for the elements.

How should symbols be produced?

Engineers usually seek suggestions for symbols from either standards (such as ISO 7000) or symbol compendia (usually Dreyfuss, 1967, though sometimes Modley, 1976, is used). Most of these symbols have never been tested for their meaningfulness (see Chong, Clauer and Green, 1990 for recent examples). Automobile symbols, however, are often based on automotive designers' ideas (see Gingold, Shteingart and Green, 1981 for information on trucks). The literature indicates that there are better sources and processes for identifying candidate symbols.

Howell and Fuchs (1968)

Howell and Fuchs was the first study to examine use of the symbol production method. Students drew pictures for 52 military referents (e.g., missile site under construction). A second set of 20 people ranked symbols (based on the drawings) in terms of their applicability to the referents. The correlation between drawing frequency (stereotype scores) and applicability ranking was small ($r = 0.37$), but significant. In four other experiments Howell and Fuchs showed that high applicability (stereotypical) symbols were more readily matched with their intended name, learned in fewer trials, and more likely to be recalled after a day, than low applicability symbols.

Mudd and Karsh (1961) and Karsh and Mudd (1962)

More than in any other study, this pair clearly demonstrates that symbol production is the preferred method for creating candidates. Mudd and Karsh had 125 soldiers draw pictures for the 34 referents (spark advance, choke, winch, etc.). In a second experiment, 94 other soldiers were shown 90 symbols (34 based on the drawings, 34 existing symbols, and 22 others) along with the names of all 34 referents. (Each referent appeared at least twice.) For 23 of the 34 referents, soldiers did significantly better in matching the production-based symbol than the existing symbol with its referent.

Green (1979a)

In this study 43 people from a college campus drew pictures for seven functions. Figure 12.13 shows the drawings for the heater control, one of the functions considered. Some of the images shown included campfires, wavy arrows, the sun, and thermometers with arrows pointing up. Candidate symbols were developed from ideas represented and tested in a second experiment. In that experiment 62 members of a swim club (while poolside) rated the meaningfulness of each candidate relative to the ISO front hood symbol (rating = 10). Figure 12.14 shows the symbols evaluated and their geometric mean rating.

Sayer and Green (1988a, b)

This experiment examined whether drivers preferred ISO symbols to alternatives. To develop alternatives, 32 people from a college campus drew pictures

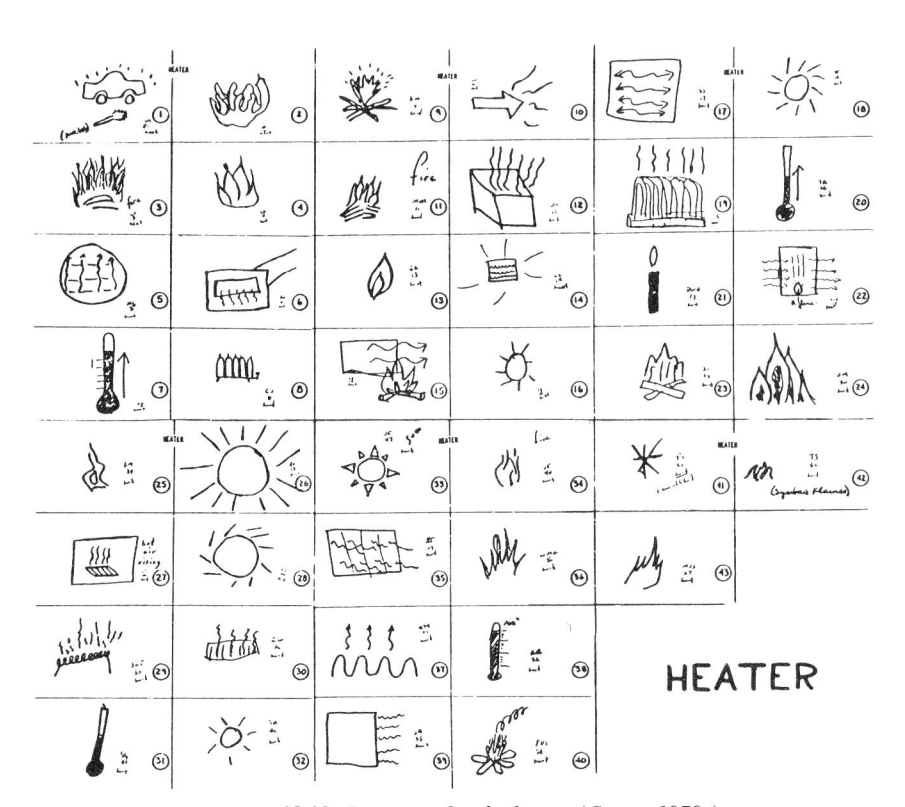

Figure 12.13. Drawings for the heater (Green, 1979a)

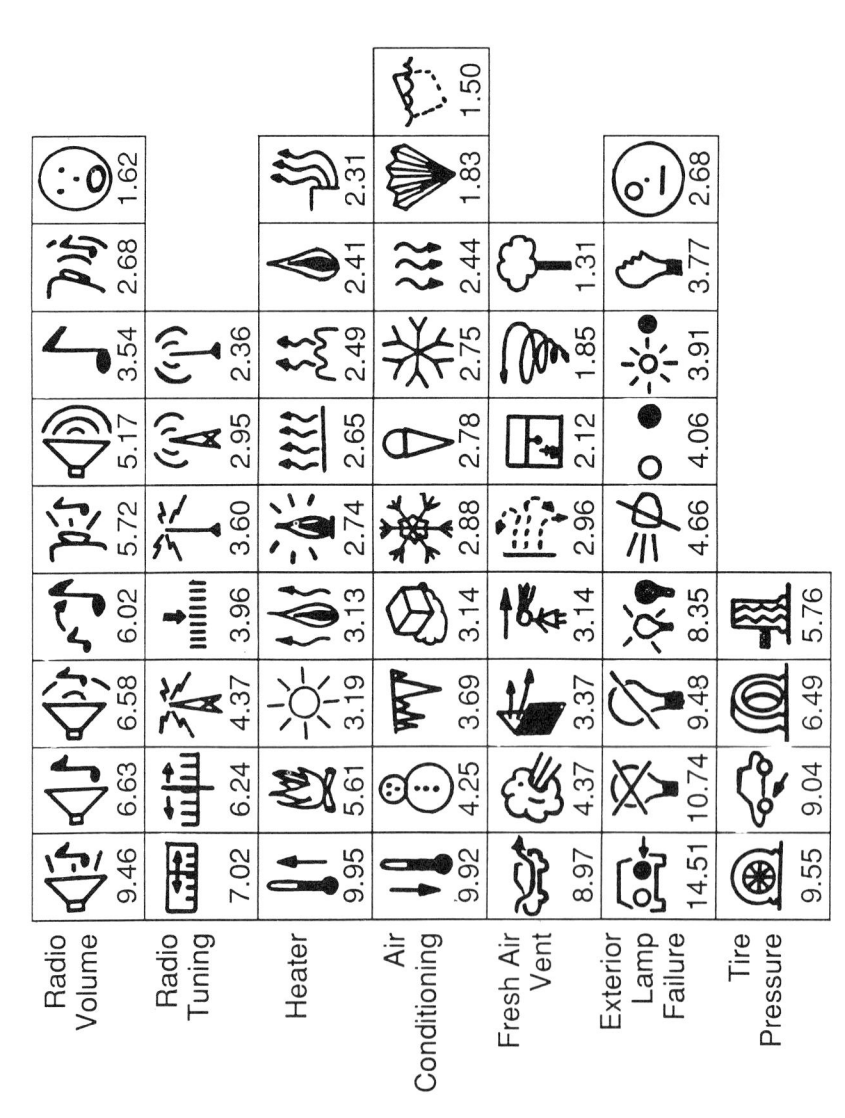

Figure 12.14. Symbol ratings from Green (1979a)

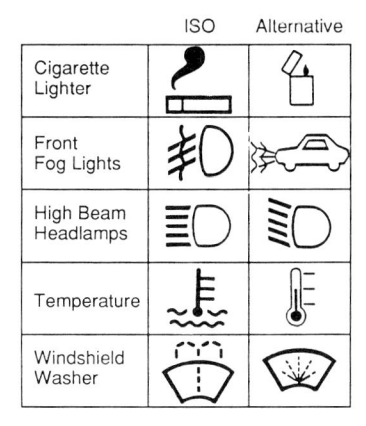

	ISO	Alternative
Cigarette Lighter		
Front Fog Lights		
High Beam Headlamps		
Temperature		
Windshield Washer		

Figure 12.15. ISO symbols and alternatives from Sayer and Green (1988a, 1988b)

to represent 25 functions. Subsequently, 104 people waiting to have their licenses processed ranked candidates (based on the drawings) from best to worst. The ISO symbol was preferred for only seven of the 25 referents. For the 18 other referents, a production-based candidate was significantly preferred for 15. Figure 12.15 shows the ISO symbol and the best alternative. Thus, even though the symbols have been in use for many years, there were better alternatives—at least for US drivers.

Summary

Symbol production studies show that symbols developed from user drawings are more likely to be understood, are easier to learn, and are more likely to be preferred than symbols developed without user input. This is true even if the comparison symbols have been in widespread use for several years. Hence, it is essential that any effort to develop new symbol sets should begin with a production experiment.

How should symbols be tested?

Testing requirements have been debated at length within International Standards Organization Technical Committee 22, Subcommittee 13, Working Group 5 (ISO TC 22/SC 13/WG 5). There were no requirements until the mid-1980s, when a resolution was passed stating that no symbol could be added to ISO 2575 unless it was tested in at least two countries. Recently, this requirement for supporting data was dropped. Tests were not being conducted, and symbols were not being added to the standard. Some observers believed that 'progress' was not being made. Currently, countries objecting to a proposed symbol are given two years to provide test results supporting their position. As is documented in this chapter, symbols that are not developed from production data, or tested, are generally poor. Adding poor symbols to a set is not progress. In contrast, ISO 7000 requires 85 per cent correct in matching tests where the risk in using a symbol is small, and 95 per cent when it is large.

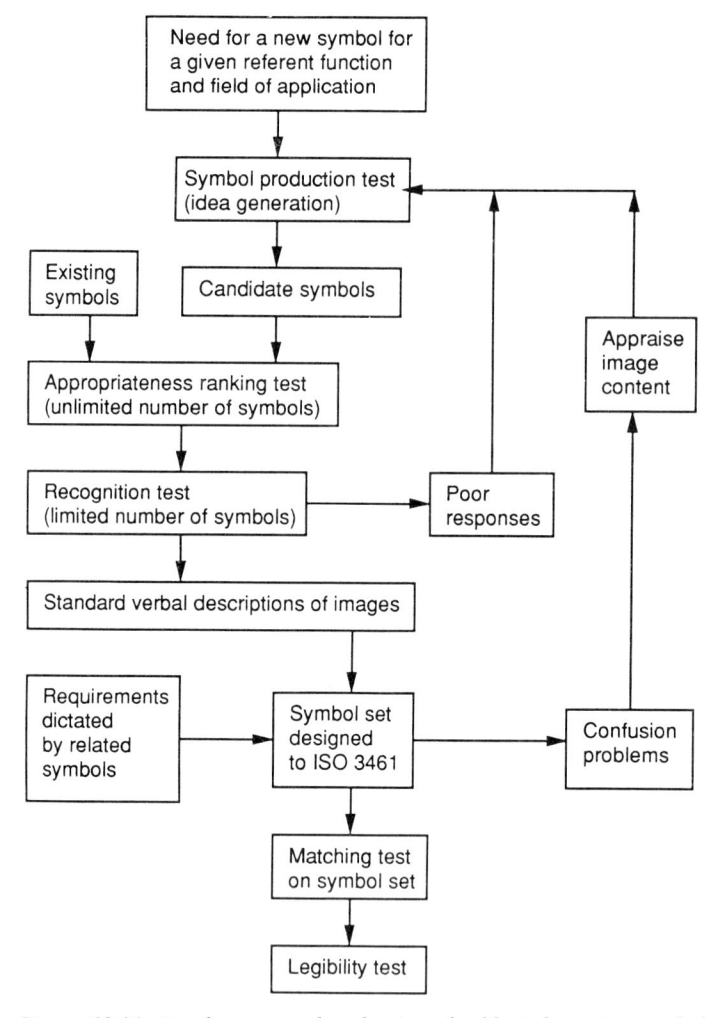

Figure 12.16. Development and evaluation of public information symbols
(Source: adapted from Zwaga and Easterby, 1984)

Several symbol acceptance protocols are proposed in the literature. Jack (1972) describes a three-step process. Jack's first step involves determining the natural association between a referent and a control (or display) using any one of several naming or matching tasks—such as those reported by Jack, Heard and Pew (1970) and Jack (1972). The dependent measure is the per cent correct. In a second step, the ease of recognition and learning of symbols should be assessed, with response time (over practice) being the prime candidate. Third, performance of a symbol in an actual driving situation should be assessed—with the response time to operate a control (from verbal prompts) being the suggested performance measure. The process was envisioned as sequential, with only symbols that passed each step receiving further testing.

Design of control and display symbols 263

The required protocol for public information symbols developed by ISO Technical Committee 145, Ergonomics (ISO, 1979; Zwaga and Easterby, 1984) is in direct contrast (see Figure 12.16). The first phase of symbol development involves a production test, an appropriateness test, and a comprehension test. The production test involves representative users drawing symbols for referents (e.g., Green, 1979a). It is intended for instances where no symbols exist, to generate ideas for symbols. In the appropriateness test, users are shown many candidates for each referent and rank them from best to worst. Candidates receiving low ranks are eliminated from further tests. In the comprehension test, people are shown symbols and the context in which they are used, and then indicate what they represent. In the second phase of testing, standard image descriptions are developed and the images are modified to meet ISO graphic requirements. Next, a matching test is conducted in which all of the symbols in a set are shown and respondents match them with referent names. This test examines issues relating to both recognition and likely confusions. The final test involves assessing symbol legibility. There is no agreed-upon procedure for the legibility test. This protocol is noteworthy because of the extent to which it specifies the development and evaluation process.

Green and Pew (1978)

Green and Pew is the only study to systematically explore alternative methods of assessing automotive symbols. Fifty students responded to 19 symbols included, or likely to be included, in ISO 2575. There were five tasks. In the first, people indicated with which symbols they were *familiar*. In the second,

Table 12.9. Results from Green and Pew (1978)

Symbol	Familiar n = 50 people	Meaningfulness	RT (ms)
Ventilating fan	2	7.41	501
Horn	0	19.23	504
Fuel	0	15.99	505
Hazard warning	3	2.62	507
Charging circuit	1	7.51	518
Parking lights	6	6.86	518
Seat belt	1	17.75	519
Turn signal	37	16.39	523
Choke	2	0.48	526
Front hood	3	10.00	535
Engine oil	0	4.24	535
Coolant temperature	0	1.63	539
Upper beam	26	7.48	551
Rear hood	1	8.99	565
Windshield washer	8	8.95	567
Windshield wiper	23	5.55	570
Lower beam	15	7.87	593
Front fog lights	1	0.40	667
Rear fog lights	0	0.40	677

Table 12.10. *Intertask correlations from Green and Pew (1978)*

Variable	6	5	4	3	2
1. No. familiar symbols	0.26	0.04	0.03	0.03	0.15
2. % correct (matching)	0.22	0.48*	−0.05	0.62**	
3. Meaningfulness rating	−0.08	−0.52*	−0.15		
4. % correct (learning)	−0.27	−0.19			
5. Response time	0.26				
6. No. RT errors					

* = p < 0.05, ** = p < 0.01

the experimenter read scenarios similar to those in Heard (1974) and participants pointed to the *matching* symbol in an array. In the third, magnitude estimates of how well each symbol *communicated* its intended meaning were obtained. Task four was a paired-associate *learning* task. Participants were shown symbols one at a time, named them, and then were told what each was. The performance measure was the number of trials required to name each symbol correctly. In Task five, people pressed buttons to indicate if an image was the same or different from a previously read symbol name. *Response time* was the primary dependent measure.

Table 12.9 shows some of the results. Most remarkable was the lack of correlations between the various performance measures (see Table 12.10). However, per cent correct in matching symbols with scenarios (association norms) was significantly correlated with meaningfulness ratings ($r = 0.62$). Response time was also correlated with association norms (0.48) and with meaningfulness ratings (-0.52). Thus, the qualities of knowing what symbols mean and how easy they are to learn tend to be independent and require separate tests to assess them. These qualities are independent of how familiar the symbols being tested are to drivers.

Conclusions

- Symbol development should begin with a symbol production task in which a sample of users draw suggestions for each symbol. There is considerable scientific evidence that symbols developed that way are much more likely to be understood by users. Production tasks are very easy to carry out.
- There is little information on the efficacy of various test methods. The author's research suggests there are several independent characteristics of good symbols (detectable, identified at first glance, easily learned, responded to quickly after practice, recognizable when they are small, not confused with other symbols, etc.). Hence, multiple tests are required to evaluate a symbol set.
- Symbol meaningfulness was the most commonly evaluated characteristic. Typically, such studies are conducted using matching or free response tasks. Such tasks have been conducted using paper and pencil methods, although in-vehicle tests have also been described. Performance on the matching and free response tasks is fairly well correlated. It seems reasonable to use matching tasks for controls ('I am looking for the _____. Which symbol is it?'), and free response tasks for displays ('This image has just appeared. It is a warning for _____').

- Several studies of meaningfulness were carried out in the early 1970s providing a basis for selecting symbols to include in ISO 2575. In recent years tests have been fewer and smaller. Recent studies in the US show that some of the symbols in the set are still poorly understood (e.g., hazard), despite many years of exposure. It has been suggested that any new automotive symbols should exceed 75 per cent correct on a matching task and have less than 5 per cent confusions with any symbol in the set. For ISO 7000 the requirements are 85 per cent for small risk symbols and 95 per cent for high risk. Resolutions relating to ISO 2575 require only that tests be conducted to block the addition of symbols to the standard, but neither the type of test nor the level of performance is specified. This is in stark contrast to the detailed protocol required for public information symbols by International Standards Organization Technical Committee 145 (Ergonomics) (ISO TC 145).
- Demographic characteristics affect how well people do in tests of symbol meaningfulness. In general, men do better than women, and those under about age 50 do better than those who are older. There are also differences between countries in terms of how well their citizens identify individual symbols, but citizens from one country do not do better than another, overall. Hence, tests of symbols' meaningfulness in which absolute levels are important (should specific symbols be added to a standard) must consider age, sex and the nationality of the participants.
- Only a few experiments on symbol learning have been conducted—typically testing factory visitors before, and after, a tour. Results have not shown the method to be favorable, and since the tours are no longer given, this method is not likely to be used in the future.
- There are some concerns about the quality of many studies reported in the literature. Many are not replicable either because the words or symbols tested were not given, the display conditions were not specified (size, color, character and background luminance, etc), or because procedural details were missing. With surprising consistency, scoring procedures were only generally described for free response tasks. Since decisions about which symbols should be added to a set are based on independent studies carried out in different countries, it is essential that the research be reported so it can be replicated.
- Research on automotive symbols has been very atheoretic, although there are some exceptions. While most of the research has been directed toward specific applied questions (e.g., which of several candidates should be added to the standard), the lack of theories has meant that each new effort starts from the beginning. A major roadblock to expanding the international standard has been the cost of driver tests. Theory-based predictions of detectability and confusability can potentially estimate real-world performance, thereby reducing the scope and cost of such tests. This aspect should be the focus of further research.
- The lack of research on symbol display characteristics is somewhat surprising. Symbols should be at least 20 mm in diameter to assure recognition by older drivers under poor lighting conditions. This topic will become more important as the use of CRTs increases.

Research on automotive symbols has addressed many applied questions. That research has been used to decide which symbols should be added to the international standard. However, the quality of a few studies is poor and the approach atheoretic. As a consequence, little progress has been made in advancing the state of the art. There have been discussions on vastly increasing the number of symbols in ISO 2575 and not supporting those additions with any empirical data. In recent years the automobile industry has been concerned with trying to be more customer-oriented, providing cars that are safer and easier to use. Providing obscure and arcane graphics to identify new technology hardware runs counter to that trend.

Acknowledgments

I would like to thank Marie Williams (UMTRI), Wendy Barhydt (UMTRI), Dan Jack (Ford), Dick Pew (Bolt, Beranek and Newman), Ken Socks (Chrysler), and Don Mitchell (GM) for comments which significantly improved the quality of this chapter.

References

Babbit Kline, T. J., Ghali, L. M., Kline, E. W. and Brown, S., 1990, Visibility distance of highway signs among young, middle-aged, and older observers: Icons are better than text, *Human Factors*, **32**(5), 609–20.
Chong, M., Clauer, T. and Green, P., 1990, *Development of Candidate Symbols for Automobile Functions* (Technical Report No. UMTRI-90-25). Ann Arbor: The University of Michigan Transportation Research Institute.
Control Symbols Compared, 1970, *SAE Journal of Automotive Engineering*, **78**(9), 11.
Dewar, R. E. and Ells, J. G., 1974, Comparison of three methods for evaluating traffic signs, *Transportation Research Record*, **503**, 38–47.
Dewar, R. E., Ells, J. G., and Mundy, G. 1976, Reaction time as an index of traffic sign perception, *Human Factors*, **18**(4), 381–92.
Dreyfuss, H., 1967, *Symbol Sourcebook*, New York: McGraw-Hill.
Easterby, R. S., 1970, The perception of symbols for machine displays, *Ergonomics*, **13**(1), 149–58.
Ells, J. G. and Dewar, R. E., 1979, Rapid comprehension of verbal and symbolic traffic sign messages, *Human Factors*, **21**(2), 161–8.
Elsholz, J. and Bortfeld, M., 1978, *Investigation into the Identification and Interpretation of Automotive Indicators and Controls* (SAE paper 780340), Warrendale, PA: Society of Automotive Engineers.
Frank, D., Koenig, N. and Lendhold, R., 1973, *Identification of Symbols for Motor Vehicle Controls* (SAE paper 730611), Warrendale, PA: Society of Automotive Engineers.
Galer, M. and Spicer, J. (1986). *The Recognition and Readability of Dot Matrix Warning Symbols in Cars* (SAE paper 860180), Warrendale, PA: Society of Automotive Engineers.
Galer, M., Spicer, J. and Holtum, C., 1984, The Readability of Dot-Matrix Warning Symbols in Cars, *Proceedings of the Ergonomics Society Annual Conference*, Loughborough, UK: Ergonomics Society, pp. 257–60.
Gingold, M., Shteingart, S. and Green, P., 1981, *Truck Drivers' Suggestions and Preferences for Instrument Panel Symbols* (Technical Report No. UM-HSRI-81-30), Ann Arbor, Michigan: The University of Michigan Highway Safety Research Institute, (NTIS No. PB 82 224279).
Green, P., 1977, *The Prediction of Choice Response Times for Pictographic Symbols*, Department of Industrial and Operations Engineering technical report, University of Michigan, *JSAS Catalog of Selected Documents in Psychology*, August 1980, p. 77 (manuscript 2102).
Green, P., 1979a, *Development of Pictographic Symbols for Vehicle Controls and Displays* (SAE paper 790383), Warrendale, PA: Society of Automotive Engineers.
Green, P., 1979b, *Rational Ways to Increase Pictographic Symbol Discriminability*, Unpublished PhD dissertation, Department of Industrial and Operations Engineering and Department of Psychology, University of Michigan, *Dissertation Abstracts International* (University Microfilms No. 79-25, 156).
Green, P. 1981, Displays for automotive instrument panels: Production and rating of symbols, *The HSRI Research Review*, July–August, **12**(1), 1–12.

Green, P. and Burgess, W. T., 1980, *Debugging a Symbol Set for Identifying Displays: Production and Screening Studies* (Technical Report No. UM-HSRI-80-64), Ann Arbor: The University of Michigan Highway Safety Research Institute.

Green, P. and Davis, G., 1976, The recognition time of rotated pictographic symbols for automobile controls. *Journal of Safety Research*, **8**(4), 180–3.

Green, P. and Pew, R. W., 1978, Evaluating pictographic symbols: An automotive application, *Human Factors*, **20**(1), 103–14.

Heard, E. A., 1974, *Symbol Study—1972* (SAE paper 740304), Warrendale, PA: Society of Automotive Engineers.

Hoffman, E. R., 1976, Symbols used for identification of motor vehicle controls, tell-tales and indicators (memorandum), Melbourne, Australia: University of Melbourne, Department of Mechanical Engineering.

Holmes, N., 1985, *Designing Pictorial Symbols*, New York: Watson-Guptill.

Howell, W. C. and Fuchs, A. F., 1968, Population stereotype in code design, *Organizational Behavior and Human Performance*, **3**, 310–39.

Hubel, D. H. and Wiesel, T. N., 1962, Receptive fields, binocular interaction, and functional architecture in the cat's visual cortex, *Journal of Physiology*, **160**, 106–23.

Imbeau, D., Wierwille, W. W., Wolf, L. D. and Chun, G. A., 1989, Effects of instrument panel luminance and chromaticity on reading performance and preference in simulated driving, *Human Factors*, April, **31**(2), 147–60.

International Standards Organization, 1979, *Public Information Symbols—Index, Survey and Compilation of Single Sheets* (ISO Draft International Standard 7001), Geneva, Switzerland: International Standards Organization.

International Standards Organization, 1982, *Road Vehicles—Symbols for Controls, Indicators and Tell-tales*, 4th Edn, (ISO Standard 2575), Geneva, Switzerland: International Standards Organization.

International Standards Organization, 1989, *Graphical Symbols for Use on Equipment—Index and Synopsis*, 2nd Edn, (ISO Standard 7000), Geneva, Switzerland: International Standards Organization.

Jack, D. D., 1972, *Identification of Controls, a Study of Symbols* (SAE paper 720203), Warrendale, PA: Society of Automotive Engineers.

Jack, D. D., Heard, E. A. and Pew, R. W., 1970, SAE research study on the interpretability of ISO vehicle controls symbols (memorandum of 9 June 1970), Detroit, MI: Society of Automotive Engineers.

Jacobs, R. J., Johnston, A. W. and Cole, B. L., 1975, The visibility of alphabetic and symbolic traffic signs, *Australian Road Research*, May, **5**(7), 68–87.

Karsh, R. and Mudd, S. A., 1962, *Design of a Picture Language to Identify Vehicle Controls III. A Comparative Evaluation of Selected Picture-Symbols Designs* (Technical Memorandum 15-62), Aberdeen Proving Ground, MD: US Army Human Engineering Laboratory, August.

Kyropoulos, P., 1972, *Identification of Controls: Background and Approach* (SAE paper 720202), Warrendale, PA: Society of Automotive Engineers.

Lendholt, R., 1974, *Identification of Symbols and Its Influence on Training for Motor Vehicle Controls*, (SAE paper 740995), Warrendale, PA: Society of Automotive Engineers.

Lerner, N. D. and Collins, B. L., 1980, *Workplace Safety Symbols: Current Status and Research Needs* (Technical Report NBSIR 80-2003), Washington, DC: National Bureau of Standards, US Department of Commerce.

McCormack, P. D., 1974, *Identification of Vehicle Instrument-Panel Controls* (SAE paper 740996), Warrendale, PA: Society of Automotive Engineers.

Modley, R., 1976, *Handbook of Pictorial Symbols*, New York: Dover.

Mudd, S. A. and Karsh, R., 1961, *Design of a Picture Language to Identify Vehicle Controls. I. General Method II. Investigation of Population Stereotypes* (Technical Memorandum 22-61), Aberdeen Proving Ground, MD: US Army Human Engineering Laboratory, December.

Paniati, J. F., 1988, Legibility and comprehension of traffic sign symbols, *Proceedings of the Human Factors Society—32nd Annual Meeting*, 568–72, Santa Monica, CA: Human Factors Society.

Saunby, C. S., Farber, E. I. and DeMello, J., 1988, *Driver Understanding and Recognition of Automotive ISO Symbols* (SAE paper 880056), Warrendale, PA: Society of Automotive Engineers.

Sayer, J. R. and Green, P., 1988a, *Automobile Instrument Panel Symbols: Do Drivers Prefer Alternatives Over Those in the ISO Standard?* (Technical report No. UMTRI-88-10), Ann Arbor: The University of Michigan Transportation Research Institute, (NTIS No. PB 88 181911/AS).

Sayer, J. R. and Green, P., 1988b, *Current ISO Automotive Symbols vs. Alternatives: A Preference Study*. (Society of Automotive Engineers, paper no. 880057, SAE Special Publication SP-752), Warrendale, PA: Society of Automotive Engineers.

Simmonds, G. R. W., 1974a, *The Influence of Size on the Recognition of Symbols for Motor Vehicle Controls* (SAE 740997), Warrendale, PA: Society of Automotive Engineers.

Simmonds, G. R. W., 1974b, Symbol Learning Tests (document ISO/TC 22/SC13 (WG5-15) 212, Geneva, Switzerland: International Standards Organization, Technical Committee Subcommittee 13, Working Group 5 (Ergonomics of Road Vehicles-Symbols).

Simmonds, G. R. W., 1976, 'Recall as a Basis for Selecting Symbol Designs', paper presented at the Ergonomics Research Society Annual Conference, 6–9 April, Edinburgh, Scotland.

Society of Automotive Engineers, 1991, *Symbols for Motor Vehicle Controls, Indicators, and Tell-tales* (SAE Standard J 1048 Mar 80), *SAE Handbook*, Vol. 4, 34.214–34.217, Warrendale, PA: Society of Automotive Engineers.

Stauffer, M., 1987, *Piktogramme fur Computer*, Berlin, Germany: Walter de Gruyter.

US Department of Transportation, 1990, *Controls and Displays* (Federal Motor Vehicle Safety Standard 101), 49 CFR Chapter V, 571.101, 188–195.

Verplank, W., 1986, *Designing Graphical User Interfaces* (Tutorial 1, CHI'86 Conference on Human Factors in Computing Systems), New York: Association for Computing Machinery.

Wiegand, D. and Glumm, M. M., 1979, *An Evaluation of Pictographic Symbols for Controls and Displays in Road Vehicles* (US Army Technical Memorandum 1–79), Aberdeen Proving Ground, MD: US Army Human Engineering Laboratory.

Zwaga, H. J. G., 1974, *Research on Graphic Symbols: An Attempt to a Methodological Review* (Technical Report 74-4), Utrecht, Netherlands: Psychological Laboratory, University of Utrecht.

Zwaga, H. J. and Boersema, T., 1983, Evaluation of a Set of Graphic Symbols, *Applied Ergonomics*, **14**(1), 43–54.

Zwaga, H. J. and Easterby, R. 1984, Developing Effective Symbols for Public Information, in Easterby, R. and Zwaga, J. (Eds), *Information Design*, pp. 277–97, Chichester, UK: Wiley.

13

Role of expectancy and supplementary cues for control operation

Walter W. Wierwille and John McFarlane

Introduction

One of the most fundamental human factors principles is that of recognizing and designing to include population stereotypes. Such stereotypes indicate the way in which users expect controls to operate. In an automobile, direction-of-motion stereotypes are very important because when not taken into account they lead to control actuation errors, confusion and increased driver workload. Stereotypes are not determined by edict. They must be determined experimentally using unbiased population samples and good experimental methodology. Stereotypes may vary with cultural or other demographic differences and they may vary with exposure. In addition, stereotypes may be strong (having nearly complete agreement within the population sample) or weak (having substantial ambivalence or inconsistency within the population sample). At present, however, there are no agreed upon definitions of what constitutes a strong or a weak stereotype.

If a direction-of-motion stereotype is strong and the designer heeds the direction, it can be expected that the resulting design can be used with a low error rate by drivers possessing the same cultural/demographic makeup as the test sample. If a direction-of-motion stereotype is weak, the designer can expect some errors and confusion, at least initially, by many drivers. Obviously, a designer should try to avoid any design that conflicts with either a strong or even a weak direction-of-motion stereotype. Such a design will cause error rates in all cases to be greater than 50 percent in initial attempts to use the control. Using a design that conflicts with a *strong* stereotype will result in initial error rates much greater than 50 percent.

When a direction-of-motion stereotype is found to be weak, it is important to ascertain the causes. Ideally, a designer should, insofar as possible, use controls that have strong direction-of-motion stereotypes. However, when no known controls have a strong stereotype for a given application, questions should be asked regarding what is causing the weak stereotype and how a design change might be developed that would cause a strengthening of the stereotype.

Angle of presentation, location

It is believed that the angle of presentation (angle of inclination, for example) of a control can often affect the strength of a stereotype. Roughly speaking, as plane of motion of a control more closely mimics (in some way) that of the actual or visualized display, stereotype can be expected to strengthen. Thus, angle of presentation should be included in any experimental investigation on stereotypes.

Similarly, it is important to recognize that location may effect stereotype direction and strength. As an example, consider the use of toggle switches on the center console or on the header. Quite likely the console switches would have a strong stereotype of forward—on. However, depending on the angle of presentation, the same switches on the header might have a stereotype that agrees, might agree somewhat, or might even disagree with that of the console.

How stereotypes occur: Their relationship to workload

In the general discussion of direction-of-motion stereotypes, Gibbs (1951) outlines two sources of stereotypes. Some stereotypes stem from normal spatial relationships and, thus, may be called *natural*. For example, when drivers turn a steering wheel to the right, it is expected that the vehicle will move to the right. Gibbs points out that this type of stereotype is related to eye-hand coordination practiced from birth. Other stereotypes may stem from manifestations of customs inherent to different cultures. Gibbs terms these latter stereotypes *expected*, *preferred* or *dominant*. The example of down-for-on light switch in the United Kingdom and up-for-on in the United States illustrates this stereotype.

When a driver encounters an automotive control that is to be activated for the first time, the driver will usually try moving the control in the direction thought most likely to produce the desired result. (We will assume for the moment that the control is not coded.) The driver will then sense whether the actuation has produced the desired result, and will then correct the direction of usage if necessary. Gradually, after some number of repeated actuations (anywhere from two or three to perhaps 1000) at various times, the driver becomes accustomed to the direction of motion used and seldom makes errors in further usage. The driver then memorizes the expectancy and would probably expect similar controls in other automobiles to operate in the same direction.

This typical scenario has certain ramifications associated with it. If the direction of actuation is different from expected, the effect will be to increase driver workload. There are several types of increased loading that can occur, namely visual, auditory, cognitive and manual. If the control is moved in the wrong direction, the resulting system response must be sensed, usually visually or auditorily. Thereafter, cognitive processing must be used to interpret what is sensed, to determine what is wrong, and to effect a new response. This response would usually then be executed manually. In effect, a closed-loop process takes place between the driver and the system in which the correct or desired outcome is eventually obtained.

It can be hypothesized that the control direction that minimizes driver workload is the one that matches expectancy. While loop closure still occurs when activation direction is correctly chosen (with the driver sensing and processing the results of the actuation), there are fewer steps in the procedure when actuation matches expectancy.

It is important to recognize that drivers have limited visual, auditory, cognitive and manual resources, and that 'messing around' with controls diverts the driver from the task of driving. In particular, visual diversion and attention distraction can pose problems. For normal driving there is spare capacity in vision (the driver can look away from the forward scene for a brief period), audition (the driver can listen to the radio without creating a hazard), cognitive loading (the driver can think about things other than driving), and manual capability (the driver can drive with one hand on the wheel). Nevertheless, if diverting visual and cognitive loads are placed on the driver, the driver may miss an important driving cue and may, in a very small percentage of cases, have a near-miss or accident. This is not speculation. Perel (1976) has shown using accident records that some accidents are indeed caused by drivers' diversion of attention to instrument panel controls. If so, then anything that can be done to minimize workload caused by controls should reduce accident rates.

It should also be mentioned that using a control that has a direction-of-motion opposite to a stereotype can be viewed by the driver as an annoyance. Such a control could potentially reduce the driver's perception of attention to detail in design and corresponding product quality. In today's competitive environment, such impressions could affect outcomes of first or second purchases of a specific make of vehicle.

Relationship to coding

Another important consideration in direction-of-motion stereotype work is control coding, namely labeling, shape coding, and color coding. Labeling can take the form of printed words or abbreviations with additional graphics such as arrows, or the use of icons. However, icons are ordinarily used to identify controls as opposed to aiding with direction of activation. (Certain power door lock controls have, however, used icons to indicate locking and unlocking direction.) The most prevalent form of shape coding today for direction of activation cueing is 'bump and dimple'. The concept is to use the bump side for up, on or raise and the dimple side for down, off or lower. This shape coding is most often used with rocker switches, which may not possess strong direction-of-motion stereotypes without coding of some kind. Color coding has also been used in some circumstances. Colors such as red, green or yellow have been used for on and black or no color for off. Often color coding is in the form of an illuminated indicator which may be integrated with the control or placed near it on the appropriate side for activation or deactivation.

The use of coding for direction of activation must be carefully considered. This consideration is more a matter of design philosophy than it is a matter of research. Nevertheless, improper use of coding can certainly result in confusion and can conceivably result in lower direction-of-motion stereotype strengths. The philosophy that should be used is as follows:

1 If possible a control should be selected, located and oriented so that it has a strong direction-of-motion stereotype without any form of coding.

2 If control coding is to be added to a control that already has a moderately strong or strong stereotype (without coding) such coding should be carefully tested to ensure that at least it does not reduce the corresponding uncoded stereotype strength.
3 If there is no known control that has a strong stereotype for the intended application (without coding), then control coding should be added to increase stereotype strength. Such coding should be carefully tested to ensure that stereotype strength is in fact increased

Finally, in regard to direction-of-motion coding, it should be recognized that there may be a price to be paid for coding. The price is possibly increased visual, tactual and cognitive demand. When a control is coded, the driver must sense the code and interpret it. Additional problems can occur when labels or colors are not illuminated for nighttime driving, and when labels are unclear or in a language the driver cannot read. Thus, the advantages and disadvantages of direction-of-motion coding should be carefully weighed against one another.

Plasticity of stereotypes

It has already been stated that stereotype strength can be modified by driver exposure. Another way of saying this is that direction-of-motion stereotypes have a degree of plasticity associated with them; they can be modified.

There are many influences at work in such stereotypes. A few are listed here:

Automobile control designs evolve. Drivers adapt to the cars they drive and then come to expect the controls to respond in a certain manner, usually the way their last or present car responds.
Drivers come from diverse backgrounds, ethnic groups and cultures. In some societies, for example, left, down and back represent activation, while in others right, up and forward represent activation.
When automotive controls are mounted near one another or are shared (such as a set of four power window controls), stereotype direction and strength for any one of them may be influenced by the other controls. This is a result of the need for cognitive *assignment* of each control to its function and the possible implication for consistency.
Stereotype conflicts can occur naturally within the context of a given control. For example, using a toggle switch with vertical actuation on the driver's door for power door locks can cause such a conflict. If the driver envisions the traditional post lock, he or she would push the toggle switch down to lock. On the other hand, engaging or *turning on* the power door locks would ordinarily be associated with an upward movement of the switch. Another example is the windshield wiper knob which would rotate clockwise to turn on, while the blade itself initially rotates counterclockwise from the rest position.
Drivers themselves may, as a result of aptitudes, adapt more quickly or more slowly to new or unusual control uses. If, for example, a driver temporarily rents a vehicle, he or she may adapt rapidly, may adapt slowly, or may never adapt to the rented vehicle, in terms of certain direction-of-motion stereotypes. Furthermore, on returning to the vehicle normally driven, the driver may readapt rapidly, may readapt slowly, or may not need to readapt as a result of never having adapted to the rental vehicle.

These influences on the driver can be viewed from a slightly different perspective, that of the driver's learning process. To sketch this, consider a hypothetical novice driver who comes to a given control situation having experienced various cultural and educational situations:

> The novice attempts to use a control under short-term exposure. The experience results in either enhancement of the driver's expectancy or conflict, depending on whether the activation direction is as assumed or in conflict. If there is no further exposure for a long period of time, the novice is likely to revert to the innate expectancy, because of forgetting. On the other hand, if the novice is repeatedly re-exposed to the control, over a long period of time, whether or not it enhances or conflicts with an innate expectancy is of little consequence. The novice gains experience and eventually learns to use the control without error. In the process the driver's choice of direction of motion for activation is likely to shift to correspond to the exposure. This is a relatively permanent change and is likely to remain so until modified by a conflicting control. If the exposure to a new, conflicting control is short-term, it probably will not affect the long-term acquired expectancy. If, however, there is repeated long-term exposure to the conflicting control, the driver will slowly modify his or her expectancy.

Simply stated, driver expectancies are a product of cultural expectancies, modified by combinations of short-term and long-term exposure. Long-term exposure will in most cases have the dominant influence. A clear-cut ramification of this hypothetical model is that direction-of-motion stereotypes change with time for the individual and for a given driver population. The instrument panels of today are quite different from those of 20 years ago. Furthermore, there are differences in other consumer products, and also in cross-cultural communications. Thus, direction-of-motion stereotype data can become dated for two reasons. First, the types of controls in current use can change, leaving the designer with new controls for which there are no data. Second, because exposure causes changes in stereotypes, population stereotypes will themselves slowly change. Notwithstanding this problem, proper data gathering and use of direction-of-motion stereotypes should increase driver acceptance, reduce workload, and possibly result in *de facto* standardization, because drivers will prefer those controls which they perceive as easy to use.

Brief literature review

Automotive direction of motion stereotypes

The amount of research that *directly* addresses direction-of-motion stereotypes with regard to automobile drivers is limited. The most recent work appears to be that carried out by Jack (1985). In this research, a series of direction-of-motion stereotype studies was conducted to determine how tactile coding schemes, switch orientation, and labeling influenced drivers' choices of initial control movement direction with rocker switches. The results of Jack's studies indicate that tactile coding such as bump/dimples and serrations influences which side of a rocker switch is pushed to obtain a desired action. By adding on/off labeling appropriately the likelihood of a driver's success in actuating the correct side of the switch is increased.

An investigation by Black, Woodson and Selby (1977) collected driver expectancy data with regard to control location and operability for ten functions. Functions analyzed were: windshield washer, windshield wiper, cruise control, headlights on/off, headlights high/low beam, hazard, interior fan, temperature controls, defrost/defog, and radio volume. Approximately 900 drivers were shown a two-sided board containing 20 panel-mounted controls on one side, and 10 steering column mounted controls, or stalk controls, on the other side. Drivers were asked to select a control which they felt matched a function. Once selected, subjects were then asked to indicate perceived control location and method of operation. Accompanying the hardware and instructions was an exterior picture of a car in which the hypothesized controls were to be found. This picture was provided to determine if the drivers were influenced in their expectancies by the kind of car they were visualizing. Subjects tended to associate chromed controls with domestic (US) vehicles and dull black controls with foreign vehicles. Location and operational expectancies were considerably stronger for panel mounted controls than for stalk controls. The salient results indicate that subjects expected the headlights on/off and windshield wiper/washer controls to be panel mounted (approximately 3 : 1). When mounted on a panel, a round knob that is pulled was expected for the headlight on/off switch. When the headlight on/off switch was located on a stalk control there was no strong agreement on the location (left or right column mounted) and mode of operation. For both the windshield wiper and washer, use of a panel mounted round knob had the highest expectancy. The knob was to be turned clockwise for the wiper and pushed in for the washer.

In analyzing high/low beam controls, only stalk controls were considered. However, it was indicated that most subjects expected the beam control to be located on the floor left of the brake pedal. When using a stalk for the high/low beam control, subjects preferred the left over the right, and pulled the stalk toward the driver for high-beam activation.

In general for the climate functions such as fan, temperature and defog, slide controls were expected by most subjects. To increase or turn on the climate controls, subjects preferred moving the slide either up or to the right. The radio volume was expected to be controlled by a round or irregular round knob, and it was rotated clockwise to increase the volume.

An advantage of the Black *et al.*, study is that expectancy feedback is obtained on three criteria: type of control, location and method of operation. However, Turner and Green (1987) have pointed out that a weakness in the research was that final assignment of the control type, control location and method of operation did not involve the instrument panel as a whole. This point is best illustrated by the fact that a button on the left stalk had the highest expectancy percentages for both washer and cruise controls, even though a combination of these controls on a single stalk is probably not feasible.

Other studies on driver expectancies is for controls do exist. However, the primary concern for each is *where* drivers expect controls to be located, with little or no data on control operability. The most comprehensive research in this area was conducted by Anacapa. The results are detailed in three technical reports. The first is an early progress report (Anacapa, 1974) that indicates how expectancy data should be collected. Two methods were tested. In one condition, subjects placed adhesive backed controls in expected locations on a blank instrument panel mounted in an unfamiliar full-size American car. In the second condition, subjects marked a sketch indicating their location expectancies with the same controls used during the in-car phase. The results from the survey showed that the expectancy distributions obtained from both methods (paper and pencil, in-car) were similar, thus, the simpler and less expensive paper and pencil method could be used for test purposes.

The study conducted by McGrath (1974) for Anacapa, reports control expectancies from a survey of 219 European and Japanese drivers. In general, McGrath found that European expectancies did not depend on whether the vehicle was European or American made. In addition, while most differences between left-hand and right-hand drive cars showed mirror image location reversal, some control locations were not reversed. For example, drivers expected the cigarette lighter on the right panel of a left-hand drive car and the left panel of a right-hand drive car, while they expected to find the defroster on the right panel of both types of cars. Lastly, McGrath showed that each nationality had its own set of distinct expectancies concerning locations of certain controls. Thus, no single design would be perfectly suited for world wide sales.

The last report by Anacapa (1976) was a continuation of the progress report mentioned earlier. Expected locations were marked for 14 controls on a mail-back questionnaire by 1708 drivers. Five versions of the questionnaire were distributed. Two were for full-size American sedans, one for compacts, one was for light trucks and vans, and one was for smaller foreign cars. An important finding from the surveys was that the vehicle a person drives does indeed have an influence on expectancies.

A recent report by Green et al. (1987) takes preferred direction of motion into account. In their study, 103 participants designed instrument panels by placing the controls they preferred for 24 functions where they wanted them, not expected them. There were 255 control designs (i.e., stalks, pushbuttons, switches, etc.) from which to choose. In addition to control selection and placement, drivers identified control motion. When the design was completed, they reached for each control while operating an elementary driving simulator. The Green et al. (1987) study is reasonably comprehensive in its examination of driver preferences for control type and location. Graphical illustrations and response percentages make the document easy for designers to read and use. In addition, participants in the study were told to design a dashboard for cars of the 1990s. However, the data for preferred motion are limited. For any given function, the subjects chose both control type and location before indicating the method of operation. This technique reduces the significance of the directional percentages because the sample size for each configuration is reduced as more control type and control location choices are made. In addition, for some functions such as power door locks, power seat, power windows, and climate controls, data regarding method of operation were not collected.

General stereotypes

Despite the limited volume of direction-of-motion stereotype data concerning automotive controls, ample literature exists on the effects of directional relationships between common controls and displays. Loveless (1962) conducted a review of the literature to assess the relative strength of various stereotypes. One effect considered was body position. According to Loveless, when direction-of-motion is described it may be related to some standard frame of reference or to the orientation of an operator's body, and the two do not necessarily coincide. If a discrepancy exists between these two frames of reference it may affect direction-of-motion stereotypes. Humphries (1958) illustrates this point with an experiment in which a vertical joystick, mounted in a horizontal plane, is used to control vertical and horizontal movements of a display mounted in a vertical plane. Results of the experiment showed that stereotypes were altered when operators moved from a normal frontal position to positions which required viewing the display over the right or left shoulder. It should be noted that the control and display movement reference planes do not coincide in the above example. In a second part of his study, Humphries changed the presentation of the joystick so that both the control and the display plane were vertical. For this case, stereotypes were not altered for different operator positions (i.e., frontal view, side view).

The compatibility between controls and displays has also been documented in many classical human factors texts geared towards equipment design. Salvendy (1987), for example, points out that if movement and perception stereotypes are taken into account, decoding steps and mental processing work will be minimized for the operator. Thus the safety of the system is increased due to a reduction of the response time and learning phase. Control movement recommendations have also been documented for specific types of controls. The SAE (1977) has provided guidelines for on or increase. The recommendations are similar to those outlined by other authors (McCormick and Sanders, 1982; Woodson and Conover, 1971).

Handedness

Another issue of concern in directional stereotypes is that of handedness. Depending on the situation, such as operator position or control location, operators may not be able to use their preferred or dominant hand for control operation. This is especially true in automotive design where controls are located on both sides of the driver. Chapanis and Gropper (1968) comment that most population stereotypes reported are right-handed relationshps. With this preference in mind, is there an effect on direction-of-motion stereotypes between use of the preferred or non-preferred hand? The review by Loveless (1962) does not disclose concrete evidence documenting there is an effect. In fact, Loveless states, 'on the present evidence, it would seem that as long as well marked stereotypes are used, there is little risk in assigning controls to the left hand'. Since left-handed individuals live in a predominantly right-handed world they are subject to right-handed biases, and therefore respond in a right-handed manner. This conclusion is supported in a study by Boles and Dewar (1986). In their experiment, individual differences in movement stereotypes for controls on common household appliances were examined. Samples of both right- and left-handed subjects from three countries (Canada, United States, Australia) were used. In general, results showed that directional stereotypes were stronger for right-handers. However, in no instance were stereotypes for left-handed subjects in the opposite direction of those for right-handers. Boles and Dewar concluded that there was no need for special consideration of left-handed people in the design of appliances tested.

The research by Chapanis and Gropper (1968) is not consistent with the findings above. Their study utilized an apparatus with a scale that could be oriented horizontally or vertically with scale values that could be made to increase in either direction for both the horizontal and vertical layouts. In addition, the linkage between the control knob and hairline was reversible so that a clockwise rotation of the knobs could be made to move the hairline toward either end of the scale. Results showed that performance of right-handed and left-handed operators was significantly different for some control-display (C-D) arrangements. The results also suggest that the effects of a given C-D arrangement on response time, reversal movements and initial direction of movement should be considered independently. A possible shortcoming of the Chapanis and Gropper study is that the significant arrangements were far-fetched, in that they violated commonly accepted C-D relationships. This may have had an adverse effect on subject performance, thus reducing the applicability of the results.

Unanswered questions

One of the most basic questions unanswered in the literature is whether automotive-related usage stereotypes differ from general usage stereotypes, and if so, how. In addition, there are many controls in current use for which there is no stereotype strength and direction information available. Typical of these is power window controls of which there are many varieties in many planes of presentation. While the literature is helpful, it is far from conclusive for many automotive stereotype applications.

Stereotype determination methodology

Caveats

When researchers set about the task of determining direction-of-motion stereotype strengths experimentally, they must proceed with caution. The concept of a direction-of-motion stereotype is certainly easily understood. However, there are many subtleties in the concept. Foremost among the subtleties is *transfer*. It cannot be assumed that a subject exposed to a first set of window controls, for example, and then to a second different set of such controls has responses in the second set that are independent of the first. It is generally assumed that the stereotype strength measurement should be based on first exposure to a set of controls. The reason for this assumption is that in practice a driver is only exposed to a specific set of controls (as opposed to two or more sets) in any given vehicle. Thus, control usage should be customary to the driver on first use.

One way to control for transfer is to use a between-subjects experimental design. If such a design is used, then transfer is eliminated by the fact that each subject is exposed to only one set of controls. While the between-subjects design is certainly rigorous, it is also usually unworkable. To give an idea of why such designs are impractical, consider that stereotype strengths for say eight different power window controls are to be determined. It will be shown later that at least 50 data must be obtained to assess stereotype strength with reasonable confidence. Accordingly, 400 subjects would be required for the experiment. (Many more would be required if transfer among individual window controls were to be eliminated.) While it is not inconceivable that such an experiment could be run, it is unlikely that anyone would do so because of the costs and data gathering time compared with the yield in terms of data and information. If data about other controls is also required, additional subjects would have to be run for those controls.

A more practical alternative is only to expose a given subject to widely different controls. For example, stalk control movements would probably not interfere in any substantive way with window controls. Furthermore, if experimenter commands associated with a given control or control type are separated from one another by commands associated with other controls that are quite different, there is then less likelihood of influence of commands associated with a given control type on one another. This method implies that a given subject can be used for a wide variety of controls as long as they differ substantially from one another in terms of design and purpose. If an experimenter then randomizes or counterbalances the order of presentation with the constraint that commands that may be related are separated from one another, an acceptable experimental design can be achieved. Of course, unless the researchers run a full between-subjects design, they can never be absolutely sure that transfer effects have not biased the data. However, conventional within-subjects designs of any kind run similar risks and are routinely accepted in human factors research.

An additional precaution that must be taken involves removing any coding cues. If, for example, a rocker switch uses bump and dimple tactual coding or color coding, it cannot be assumed that the same results would be obtained without such coding, as already discussed. While it may be desired to study the effects of coding approaches as a means of increasing stereotype strengths, it is usually first desired to determine stereotype strength of the control itself. The reason for placing emphasis on innate stereotype strength is that if strong innate stereotypes can be found, coding is unnecessary. Coding need only be added when innate stereotype strength is insufficient and needs to be augmented, as has already been discussed.

Yet another caveat involves the environment in which the data are gathered. For automotive stereotype work, the appropriate environment would appear to be an automotive interior. The concept is to give the driver/subjects sufficient environmental fidelity to allow them to visualize how the given control should be actuated. However, experimenters must be careful not to make the interior look too much like any single make or model of car, because this might bias the results or make them unrepresentative of the current population of vehicles. Similarly, the apparatus should not be too *laboratory* in its looks. Such an approach may cause difficulties in having subjects visualize how the controls are used in an automobile. The results would then be representative of general direction-of-motion stereotypes, instead of automotive direction-of-motion stereotypes.

Sample size and other statistical considerations

It must be recognized that the gathering of stereotype strength data is a sampling process in which individuals are polled. The results of such a poll are subject to all of the pitfalls of polling, namely, instability in the estimate resulting from small sample size, inaccuracy in the estimate resulting from biased or nonrepresentative sampling, and inaccuracy in the estimate resulting from inappropriate procedures, such as lack of control of transfer or improper subject instructions or conditions, as already discussed. Nothwithstanding these pitfalls, there are also certain fortuitous circumstances associated with and unique to polling. These circumstances make it possible to estimate sample size accurately, without a preliminary estimate of variance.

Assume for the moment that research subjects are requested to rotate a control in the direction they prefer to turn on, say, the windshield wipers. Assume that 50 representative subjects are used and that 41 choose the clockwise direction. The sample proportion, $\hat{p}$, then equals $41/50 = 0.82$, a sample probability. This sample probability will ordinarily differ somewhat from the population probability. If additional samples of 50 subjects are used, each sample probability would likely differ from the population probability, and the variance of the sample probabilities would represent a measure of instability in the estimates (small variance indicating low instability). It is well known that the sample probability $\hat{p}$ can be expected to become more stable (closer on the average to the population probability) as the number of subjects in the sample, N, increases, and one way to assess the stability is by means of confidence limits. These limits would specify bounds on either side of $\hat{p}$ within which the population probability, p, would fall with a certain level of confidence, say 80, 90 or 95 per cent.

To obtain such confidence limits, it is necessary to recognize that samples of size N, drawn from the population, and given a binary choice, satisfy the binomial distribution. In other words, if several samples of size N were drawn, their $\hat{p}$ values would be binomially distributed (Spiegel, 1975). There is a very fortunate aspect of the binomial distribution which can be used to advantage to compute confidence limits, namely, that the mean *and variance* are both a function of a single parameter, p. The mean is p and the variance is

$$\sigma^2 = \frac{p(1-p)}{N}$$

Thus, specification of p determines the variance, and the variance can be used to determine confidence limits.

To develop the confidence limits, the sample value $\hat{p}$ is first converted to standardized units as follows:

$$\frac{\hat{p}-p}{\sigma} =: \frac{\hat{p}-p}{\sqrt{\dfrac{p(1-p)}{N}}}$$

The corresponding values for confidence limits are then specified in terms of number of standard deviations, z_c, and set equal to the standardized units above:

$$z_c = \pm \frac{\hat{p}-p}{\sqrt{\dfrac{p(1-p)}{N}}}$$

Solving this equation for p then yields the confidence limits, given $\hat{p}$, N, and z_c.

$$p = \frac{\hat{p} + \frac{z_c^2}{2N} \pm z_c \sqrt{\frac{\hat{p}(1-\hat{p})}{N} + \frac{z_c^2}{4N^2}}}{1 + \frac{z_c^2}{N}}$$

In this equation z_c is specified as follows for the level of confidence desired:

Confidence level	z_c
50%	0.6745
80%	1.28
90%	1.645
95%	1.96
99%	2.58

These levels are taken from the normal distribution, which provides an excellent approximation to the binomial. If there is some question about the accuracy of the approximation, one can compute values of z_c from the cumulative values of the corresponding binomial distribution itself. However, this computation generally should not be necessary. Furthermore, the binomial distribution is discrete, which would require interpolation of the confidence limits in any case.

As an example, assume once again that 41 of 50 subjects prefer the clockwise direction of motion of a control, and that the 90 per cent confidence limits are desired. These limits are then as follows:

$$p \bigg|_{90\%} = \frac{0.82 + \frac{1.645^2}{100} \pm 1.645 \sqrt{\frac{0.82(0.18)}{50} + \frac{1.645^2}{4(50)^2}}}{1 + \frac{1.645^2}{50}}$$

$$= 0.892, 0.715$$

Note that the confidence interval is not symmetrical about the mean. Nonsymmetry should be expected because the binomial distribution is unsymmetrical for all $\hat{p}$ values except 0.50. The example using a sample size of 50 results in 90 per cent upper and lower confidence *intervals* of 0.072 and 0.105, which appear to be about the maximum acceptable for stereotype work. Thus, sample sizes of 50 or greater appear necessary.

It should be noted that the overall confidence interval shrinks approximately as the inverse square root of N. Thus, to halve the interval requires N^2 subjects instead of N. The inverse square root rule can be used to estimate the sample size needed. The previous equation can then be used to calculate the confidence limits, once the new $\hat{p}$ has been obtained.

It should also be recognized that research subjects are sometimes requested to select from more than two alternatives. For example, they might be asked to choose from among four alternative motions of say, up, down, right or left. This situation can still be handled by converting to dichotomous data. For example, one can treat the data as those preferring movement to the right and those *not* preferring movement to the right. Confidence limits can then be determined for those who prefer movement to the right.

An important footnote in regard to sample sizes is to choose a sample that provides round numbers when used as a divisor, such as 20, 25, 50, 100 and 200. If, for example, experimenters used sample sizes such as 24, 36, 48 or 64, which are often used for reasons involving counterbalancing, they would usually get decimal values with several digits to the right of the decimal point, for example, $43/64 = 0.0671875$. This situation is exacerbated if a sample value is eliminated for any reason and is not replaced. For example, $43/63 = 0.682359$ ... Rounding in either case would decrease the accuracy of the resulting sample probability slightly.

An additional important consideration is obtaining a representative sample. It has already been pointed out that stereotype strength may vary with culture, training and experience. If this is so, then an experimenter must carefully select the sample to be representative of the target population. Assume for example that the target population is the US car driving public. Approximately two-thirds of this population drive domestic brands of vehicles (as their primary vehicle) and the other third drive foreign brand vehicles. In the past, foreign and domestic vehicles have differed greatly in terms of control usage within the vehicles (although more recently, such differences have diminished somewhat).

Typical of such difference is the use of down for on in many European models, while up for on is used in most domestic vehicles. Because of these differences, the sample population must include domestic and foreign car drivers in nearly the same proportions as the car-driving public. Similar statements could be made about gender, age and handedness as examples. While such characteristics may not produce statistically significant differences in results, they can bias the stereotype strengths obtained if they are not taken into account in the sample.

While many of the statistical considerations amount to no more than standard experimental design issues, stereotype data gathering is probably more sensitive to such issues than other types of data gathering. To illustrate, note that in a sample of 50 subjects, a change of response of as few as five subjects produces a difference of 0.10 on a scale that goes from 0 to 1.00. This value is approximately the same as the difference between the mean and one of the 90 per cent confidence limits. The example shows that small differences do indeed matter in stereotype data gathering.

Review of a recent study of automotive stereotypes

An experimental investigation was undertaken recently with the objective of providing as much information as could be obtained from one project. The specific purpose was to determine the nature and strength of automobile direction-of-motion stereotypes for as many present-day controls as possible.

The selection of controls, their locations, and their orientations was based on current usage, prevalence and feasibility of incorporation in the study. To determine prevalence and location of various types of controls, a survey of recent model automobiles was conducted. After reviewing the results of the survey and after discussions with the interested human factors specialists in industry and universities, it was decided that six studies would be performed, one involving each of the following: power mirrors, power windows, manual windows, stalk controls, generic controls and power door locks.

In each of the six studies, various types of controls and control configurations were tested to determine those factors that affect the strength of directional stereotypes. In addition, analyses were conducted to determine if subjects responded differently due to age, gender, handedness or the type of vehicle they drove.

Apparatus

Test equipment

During data collection, subjects were seated in a stationary apparatus similar to a fixed base simulator (Figures 13.1 and 13.2). This apparatus, or buck, contained a contemporary automotive bucket seat mounted on a 0.9 × 1.4 meter wooden platform. A 0.75 meter high base was positioned at the front of the buck. The base, which simulated the frontal portion of a vehicle, contained a steering column, steering wheel, and foot pedals for the accelerator, brake and clutch.

To accommodate the different control fixtures used during the experiment, many hardware additions were made to the buck. Two multi-function stalks were mounted on the steering column, one on the left side, the other on the right side (Figure 13.3). To simulate the location of an instrument panel, angle brackets were mounted on the left- and right-side of the steering wheel (Figure 13.1). The surfaces of the angle brackets were covered with hook and loop self-gripping fastener tape. All of the mounting surfaces and each of the interchangeable controls used in the experiment were affixed with fastener tape. This tape, popularly known as Velcro, was extremely effective in securing the

Figure 13.1. View of buck from the right side

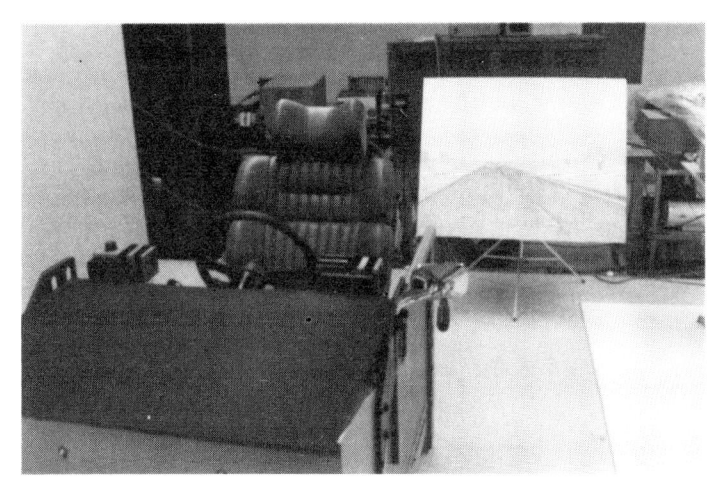

Figure 13.2. Front view of buck and roadscene

controls for actuation. However, since the hold was not permanent, changes in control configuration could be made quickly and easily.

To simulate the driver's door, a panel was attached to the front base and platform on the left side of the buck. A square mounting area was created on the door where controls could be fastened. Figure 13.1 shows the mounting area on the door panel, and one of the power window controls used in the experiment. On the right side of the buck a bracket was configured to accommodate the passenger side manual window control. (This device was held in place by the experimenter for use only in the manual window study.)

Next to the bucket seat, on the lower right side of the buck, a narrow base was used as a center console. The console was designed to allow attachment of

Figure 13.3. View of stalks mounted on steering column

controls at two different angles: horizontal and inclined 45 degrees away from the driver. Figure 13.1 shows the console being used in the horizontal position, and Figure 13.4 shows a wedge being removed so that the console could be used with the control in the −45 degree position.

The power mirror portion of the study used as a side view mirror moved by the experimenter, and a rear view scene as shown in Figure 13.2. These items were included to provide an accurate visual representation of power mirror movements, for which subjects selected their preferred control actuations.

Controls

The controls used in the experiment were configured and mounted on different types of bracketry. Controls mounted on flat carrier plates constituted 14 of the 26 controls. In addition, there were 10 controls mounted in control cubes and there were two permanently mounted multi-function stalks. An angle bracket mounted on a flat carrier plate (used in power door lock study) and a panel section for the passenger side door completed the control hardware.

Painstaking efforts were made to prepare the controls for the research. Many of the power mirror and power window controls were modified and mounted on adjustable brackets so that changes in presentation angle could be made quickly and easily. Labels and any tactile coding (such as bumps and dimples) were also removed to prevent feedback that might indicate *correct* movement direction.

Method

To control for transfer effects, four different groups of subjects were used, each composed of 50 subjects, resulting in a total of 200 for the experiment. Similar conditions were then placed in different subject groups. Also, by using four groups, time length for data gathering could be kept to about one hour, which aided in ensuring that subjects (as well as experimenter) remained fresh.

Figure 13.4. Center console shown with wedge being removed to create −45° condition

There were 107 conditions for the six studies in this experiment, with many of the conditions calling for multiple commands. For example, for power door lock conditions subjects locked and unlocked the doors for each control configuration. Similarly, for each power mirror condition, subjects adjusted the mirror in four ways (up, down, right and left). Hence, 180 commands were needed for the 107 conditions. The 180 commands were divided into four groups, resulting in approximately 45 commands per group (and therefore per subject).

Additional precautions were taken to control for learning and fatigue. Command orders were first randomized. If, however, commands for the same control (in a group) fell near one another, they were manually separated by removal and reinsertion at another point in the sequence. Also, the beginning point in the command sequence was varied from subject to subject in a systematic manner.

Subjects

During recruitment, efforts were made to ensure that subjects met specific demographic requirements. Three factors given consideration were age, type of primary vehicle driven (domestic/foreign) and gender. Each of the four subject groups consisted of 25 individuals in a younger age category (18 to 45 years) and 25 in an older age category (46 to 86 years). Additionally, each age category had 18 members who drove domestic vehicles and seven who drove foreign vehicles. The number of males and females was almost the same for each age category and vehicle type. One of the main goals during recruitment was to obtain a subject sample representative of a new car buying/driving population in the US. While the data were collected in a university laboratory, the great majority of the participants were not students. Care was taken to obtain subjects other than students so that the population would not be unusually skewed.

Procedure

When participants first entered the laboratory they were required to fill out a short questionnaire. They then read and signed an informed consent form. The questionnaire was used to provide general information on the subject's driving background. To assure that all potential participants were licensed, each subject was required to present a valid driver's license. If subjects passed screening requirements they were scheduled for the experiment. Most participants began the session immediately upon completion of the above prerequisites.

When the session began, the subject was first seated in a comfortable position in the buck (using the power seat controls, which were operational) and instructed on the experimental protocol. During the instructions the experimenter stressed that the *direction* of control activation was the important issue in the study. The subject was told to respond in a natural manner, and to move the control in the direction that felt right. The experimenter showed the subject each control to be used during the session. Explanation was provided on how the control operated, and what its function was. However, no feedback was given on *correct* direction for control operation.

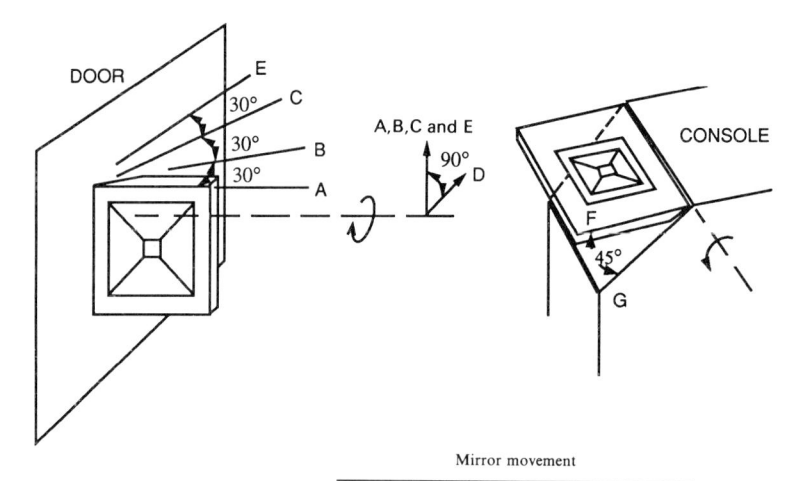

Condition	Orientation of control	Mirror movement				Overall
		Down	Up	Left, away from subj.	Right, toward subj.	
A	Vertical plane 90° to door plane	1.00B	0.98T	0.98L	0.98R	197/200 = 0.99
B	Vertical plane 60° to door plane	1.00B	0.94T	0.98L	0.96R	194/200 = 0.97
C	Vertical plane 30° to door plane	0.96B	0.94T	0.98L	0.98R	193/200 = 0.97
D	Horizontal plane extending from door (e.g., arm rest)	0.80A	0.82F	0.88L	0.94R	172/200 = 0.86
E	Vertical plane flush with door	0.82B	0.86T	0.70L	0.72R	155/200 = 0.78
F	Horizontal plane on console	0.94A	0.96F	0.98L	1.00R	194/200 = 0.97
G	Console plane sloping 45° down and away from driver	0.52F*	0.58F*	0.94L	0.92R	148/200 = 0.74

Chosen direction key
B: Bottom quadrant of pad
T: Top quadrant of pad
L: Left quadrant of pad as viewed by driver
R: Right quadrant of pad as viewed by driver
A: Aft quadrant of pad
F: Fore quadrant of pad
*: Indicates conflict in chosen directions

Figure 13.5. Results summary for power mirrors using a pad control

Once the subject understood all of the instructions, data collection began. The experimenter issued the commands and observed and recorded the subject's selection (where appropriate) and initial direction of control movement. Once the data collection was completed, participants were paid, debriefed and dismissed.

Results

Because of the large number of results obtained, they are presented in a combined pictorial and tabular form for each of the studies. The pictorial portions show the conditions tested and the tabular portions contain the data. The tabulated data show the results for the direction or portion of control with the highest stereotype strength, with 1.00 representing total agreement among subjects. In the following, proportions greater than 0.85 were graded as *strong*, those greater than 0.70 but less than or equal to 0.85 were graded as *moderate*, and those less than or equal to 0.70 were graded as *weak*. This grading scheme is arbitrary, but does allow characterization of the data on a relative basis.

Power mirrors

The conditions tested are shown in Figures 13.5 and 13.6. The power mirror controls tested had moderate to strong stereotypes, with two exceptions. One exception occurred for the pad switch in Condition G. For this configuration the pad switch was mounted on the center console angled 45 degrees below the horizontal (away from the driver). The other exception occurred in Condition E where the pad switch was mounted flush with the driver's door.

Power windows

The conditions tested are shown in Figures 13.7, 13.8 and 13.9. In Conditions J and K a push-pull switch was mounted on the driver's door panel (Figure 13.9). For both of these configurations moderate to weak directional stereotypes were obtained even though there was a strong geometric relationship between the controls and the window locations and directions. It is believed that these results occurred because the push-pull switch is a comparatively new design and subjects were unsure in their choice of activation direction.

As in Condition G for the power mirror controls, the two power window controls that were angled away from the driver proved to have weaker directional stereotypes. Condition D for the toggle switch array (Figure 13.7) and

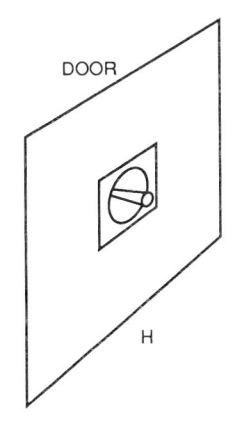

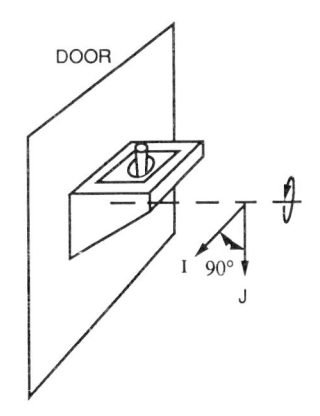

		Mirror movement				
Condition	Orientation of control	Down	Up	Left, away from subj.	Right, toward subj.	Overall
H	Vertical plane on door	0.90D	0.92U	0.84A	0.88F	177/200 = 0.89
I	Horizontal door arm rest	0.92A	0.94F	0.96L	0.96R	189/200 = 0.95
J	Vertical plane 90° to door plane	0.96D	0.98U	0.96L	0.98R	194/200 = 0.97

Chosen direction key
 D: Bat movement down
 U: Bat movement up
 L: Bat movement left as viewed by driver
 R: Bat movement right as viewed by driver
 A: Bat movement aft
 F: Bat movement fore

Figure 13.6. Results summary for power mirrors using a joystick control

![Figure showing door configurations with 2x2 toggle switches at various angles A, B, C, D and Condition E flush with door]

Figure 13.7. Results summary for power windows using a two by two toggle switch array

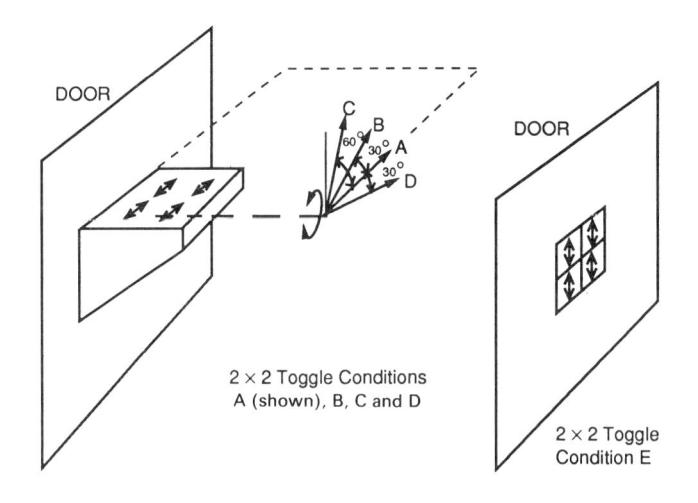

Condition	Orientation of control	Direction selection			Switch selection				
		Raise	Lower	Overall	Driver front	Driver rear	Pass. front	Pass. rear	Overall
A	Horizontal door arm rest	96/100 F	95/100 A	191/200 = 0.96	50/50 LF	49/50 LA	48/50 RF	48/50 RA	195/200 = 0.98
B	30° up from horizontal arm rest	94/100 F	89/100 A	183/200 = 0.92	50/50 LA	50/50 LA	50/50 RF	50/50 RA	200/200 = 1.00
C	60° up from horizontal arm rest	100/100 U	95/100 D	195/200 = 0.98	49/50 LT	49/50 LB	49/50 RT	50/50 RB	197/200 = 0.99
D	30° down from horizontal arm rest	60/100 F	67/100 A	127/200 = 0.64	44/50 LF	42/50 LA	38/50 RF	43/50 RA	167/200 = 0.84
E	Vertical plane flush with door	100/100 U	100/100 D	200/200 = 1.00	29/50 TF*	25/50 TA	29/50 TF*	25/50 BF, BA (equal)	108/200 = 0.54

Chosen direction key
 F: Fore movement
 A: Aft movement
 U: Up movement
 D: Down movement

Chosen switch key
 L: Left side as viewed by driver
 R: Right side as viewed by driver
 F: Fore
 A: Aft
 T: Top
 B: Bottom
 *: Indicates conflict in chosen switches

Condition H for the rocker switch array (Figure 13.8) were both mounted on the driver's door angled 30 degrees away from the driver. On the other hand, all those conditions, except for the push-pull switch, configured on the driver's door in a horizontal plane or angled toward the driver had strong directional stereotypes.

Configurations for Condition E (Figure 13.7) and Condition I (Figure 13.9) were the only positions for which the power window control was mounted flush with the driver's door. For those two cases, a high degree of consistency existed between subjects for actuation direction, however, large discrepancies were noted for control selection (i.e., which switch corresponds to which window).

Control operation: expectancy/supplementary cues

Figure 13.8. Results summary for power windows using a two by two rocker switch array

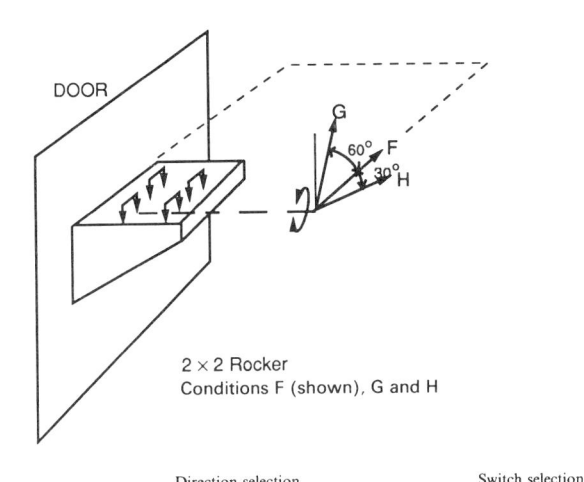

		Direction selection			Switch selection				
Condition	Orientation of control	Raise	Lower	Overall	Driver front	Driver rear	Pass. front	Pass. rear	Overall
F	Horizontal door arm rest	96/100 F	90/100 A	186/200 = 0.93	50/50 LF	48/50 LA	50/50 RF	50/50 RA	198/200 = 0.99
G	60° up from horizontal arm rest	99/100 T	99/100 B	198/200 = 0.99	47/50 LT	49/50 LB	49/50 RT	47/50 RB	192/200 = 0.96
H	30° down from horizontal arm rest	73/100 F	77/100 A	150/200 = 0.75	48/50 LF	48/50 LA	49/50 RF	50/50 RA	195/200 = 0.98

Chosen direction key
F: Fore end activation
A: Aft end activation
T: Upper end activation
B: Lower end activation

Chosen switch key
L: Left side as viewed by driver
R: Right side as viewed by driver
F: Fore
A: Aft
T: Top
B: Bottom

Manual windows

Of the four manual window conditions tested (Figure 13.10), only Condition D displayed an overall strong directional stereotype, while the other three were weak. Even though Condition D had a strong overall stereotype, it had only moderate strength for the raise command. The weak stereotypes obtained for Conditions A, B and C probably occurred because the direction of crank rotation for raising and lowering windows varies among automobiles.

Stalk controls

Multipurpose stalks were mounted on both sides of the steering column (Figures 13.3 and 13.11). Three types of commands were given to the subjects: nonspecific, specific unconstrained and specific constrained (Figure 13.11). For the nonspecific commands, in which subjects turned something on, most subjects pulled on the stalk as compared with pushing it, most raised the stalk as

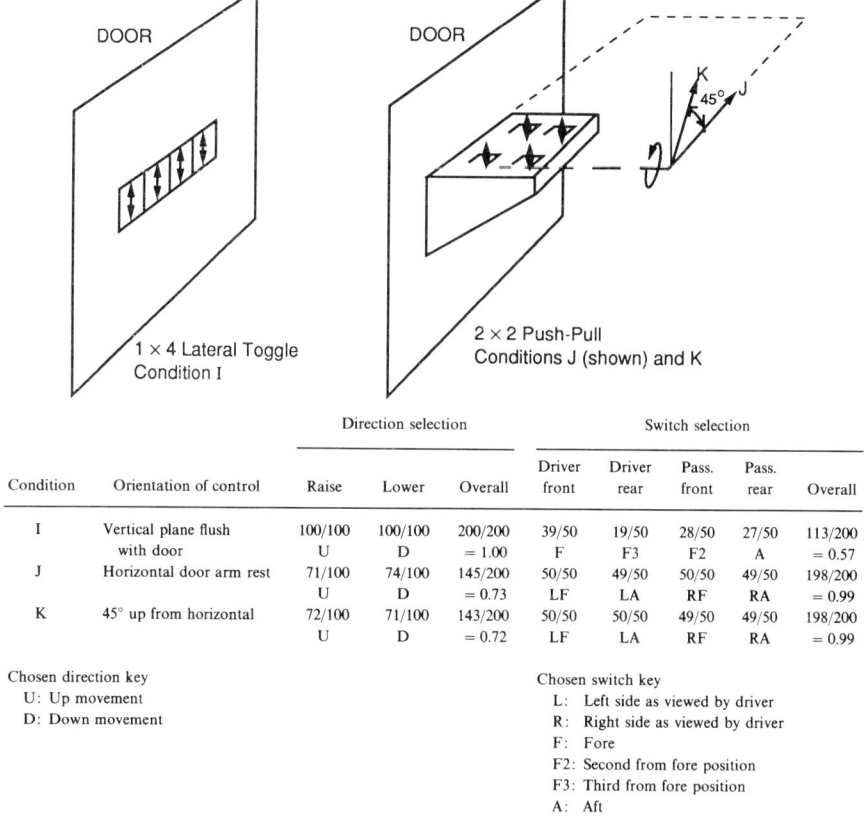

Figure 13.9. Results summary for power windows using a one by four toggle switch array and a two by two push-pull array

Condition	Orientation of control	Direction selection			Switch selection				
		Raise	Lower	Overall	Driver front	Driver rear	Pass. front	Pass. rear	Overall
I	Vertical plane flush with door	100/100 U	100/100 D	200/200 = 1.00	39/50 F	19/50 F3	28/50 F2	27/50 A	113/200 = 0.57
J	Horizontal door arm rest	71/100 U	74/100 D	145/200 = 0.73	50/50 LF	49/50 LA	50/50 RF	49/50 RA	198/200 = 0.99
K	45° up from horizontal	72/100 U	71/100 D	143/200 = 0.72	50/50 LF	50/50 LA	49/50 RF	49/50 RA	198/200 = 0.99

Chosen direction key
U: Up movement
D: Down movement

Chosen switch key
L: Left side as viewed by driver
R: Right side as viewed by driver
F: Fore
F2: Second from fore position
F3: Third from fore position
A: Aft

compared with lowering it, and most rotated the stalk over the top as compared with rotating it under.

When subjects could choose from six directions instead of two (the specific unconstrained conditions), the stereotypes were weakened. However, subjects usually raised, rotated over the top, or pulled on the stalk, as opposed to using the other three directions. This result indicates that there is no clear-cut agreement on which of the three preferred directions (raise, rotate over the top or pull) is dominant. However, because subjects were selecting from among six possible directions, it is not surprising that stereotype strengths are somewhat lower. Condition I, however, for turning high beams on did exhibit moderate stereotype strength for pulling on the left stalk.

The specific, constrained conditions all resulted in moderate to strong stereotypes. Subjects were given a dichotomous choice to turn on a specific function, such as for the wipers. In all cases the subjects preferred pull, rotate over the top, or raise, as in the nonspecific conditions.

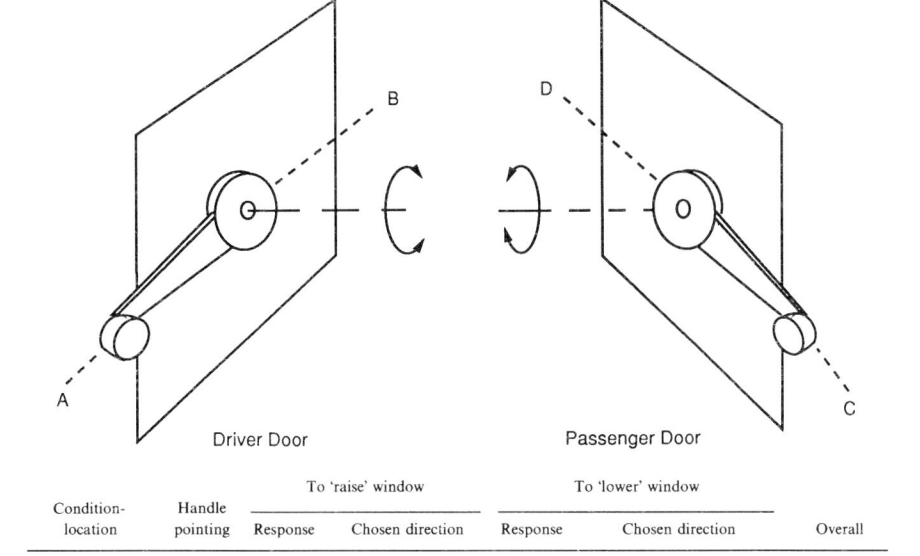

Control operation: expectancy/supplementary cues 291

Figure 13.10. Results summary for manual windows

Condition-location	Handle pointing	To 'raise' window		To 'lower' window		Overall
		Response	Chosen direction	Response	Chosen direction	
A-Driver door	Rear	27/50	Up (or clockwise)	32/50	Down (or counterclockwise)	59/100 = 0.59
B-Driver door	Forward	25/50	(Up and down equal)	27/50	Down (or clockwise)	52/100 = 0.52
C-Passenger door	Rear	34/50	Down (or clockwise)	26/50	Up (or counterclockwise)	60/100 = 0.60
D-Passenger door	Forward	42/50	Up (or clockwise)	45/50	Down (or counterclockwise)	87/100 = 0.87

Generic controls

The equipment used in the generic controls study is shown in Figure 13.12 along with the results. Sixty-one different conditions were tested. With few exceptions, choice of actuation direction observed for generic controls corresponded closely with the recommended motions outlined in the human factors literature: on or increase corresponded to movement to the right, up, forward or clockwise. When conflicts existed between clockwise and over-the-top rotation, subjects chose over the top. The stereotypes obtained varied between moderate and strong.

Only five of the 61 conditions resulted in weak stereotypes (see Figure 13.12). These weak stereotypes were believed due to conflicts among factors such as movement to the right and movement toward center to increase, as examples.

Power door locks

All six power door lock conditions used rocker switches mounted flush with the driver's door or on a simulated arm rest mounted on the driver's door (Figure 13.13). All six conditions resulted in weak stereotypes, which may be due to the mental image of traditional 'post' locks which are locked by manually pushing them downward, and the more generic concept in which movements of up, right, forward or clockwise are ordinarily used to turn something on or increase something.

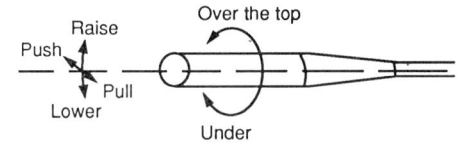

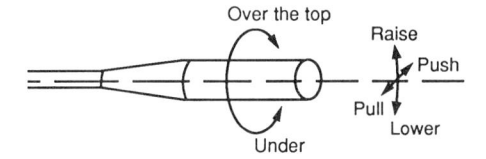

```
                    Over the top
         Raise
   Push      ↑        ⤴
   ←——————→|⊙=========================⟩
         ↓        ⤵
        Pull
         Lower      Under

                Left Stalk


                Over the top
                    ⤴                  Raise
                                        ↑  Push
   ⟨=========================⊙|—————————→
                                        ↓
                    ⤵                  Pull
                    Under               Lower

                Right Stalk
```

Control location	Condition	Instruction action	Direction	Chosen direction	Overall
Nonspecific					
Left	A	Turn on	Raise or lower	Raise	0.96
Right	B	Turn on	Raise or lower	Raise	0.76
Left	C	Turn on	Push or pull	Pull	0.76
Right	D	Turn on	Push or pull	Pull	0.72
Left	E	Turn on	Rotate barrel	Over the top	0.82
Right	F	Turn on	Rotate barrel	Over the top	0.94
Specific unconstrained					
Left	G	Turn on wipers	Unspecified	Over the top	0.50
Right	H	Turn on wipers	Unspecified	Over the top	0.50
Left	I	Turn on hi beam	Unspecified	Pull	0.76
Specific constrained					
Left	J	Turn on wipers	Raise or lower	Raise	0.94
Right	K	Turn on wipers	Raise or lower	Raise	0.94
Left	L	Turn on wipers	Rotate barrel	Over the top	0.72
Right	M	Turn on wipers	Rotate barrel	Over the top	0.72
Left	N	Turn on hi beam	Push or pull	Pull	0.80
Left	O	Turn on headlight	Rotate barrel	Over the top	0.88

Figure 13.11. Results summary for stalk controls

Stereotype strength summary

The stereotype strength results are summarized in Table 13.1 by study. These results demonstrate that there are domains of strong stereotypes for four of the six studies, namely power mirrors, power windows, stalk controls and generic controls. Only the manual window and power door lock studies did not result in usable strong stereotype domains.

Additional statistical results

Statistical tests were performed on the data to determine if there were reliable effects as a function of age (older/younger), primary vehicle driven (foreign/domestic), gender and handedness. Because the data are occurrence data they must be analysed by a nonparametric test, namely the chi-square test (Siegel

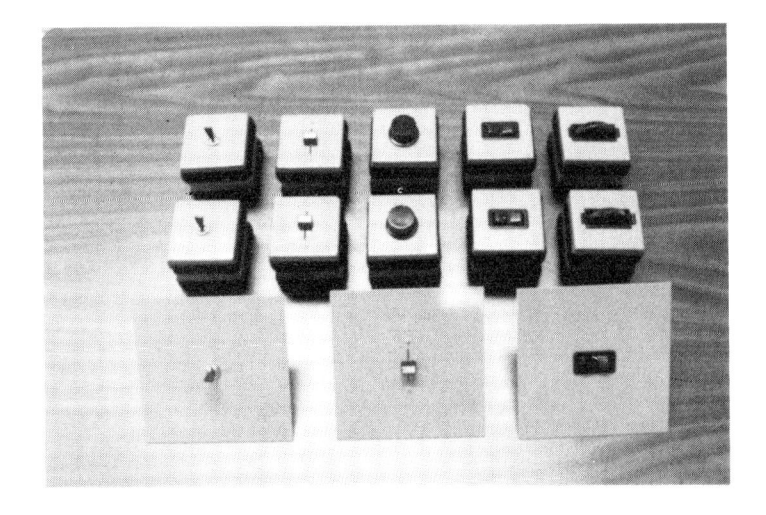

Control operation: expectancy/supplementary cues

	Condition							Overall results		
Control type	Left IP	Right IP	Console	Instruction action	Mounting plane	Control motion	Action taken	Left IP	Right IP	Console
Thumb wheel	A	B		Increase something	Top (X-Y)	Left-Right	Rotate right	0.86	0.90	
	C	D		Increase something	Top	Fore-Aft	Rotate forward	0.78	0.98	
	E	F		Increase something	Front (Y-Z)	Left-Right	Rotate right	0.78	0.92	
	G	H		Increase something	Front	Up-Down	Rotate up	0.96	0.94	
	I	J		Increase something	Outside (X-Z)	Up-Down	Rotate up	0.92	0.80	
Toggle switch	A	B	P	Turn something on	Top	Left-Right	Right	0.72	0.80	0.72
	C	D	O	Turn something on	Top	Fore-Aft	Fore	0.80	0.80	0.88
	E	F		Turn something on	Front	Left-Right	Right	0.78	0.68	
	G	H		Turn something on	Front	Up-Down	Up	0.96	0.98	
	I	M		Turn something on	Inside (X-Z)	Up-Down	Up	0.98	0.94	
	L	J		Turn something on	Outside	Up-Down	Up	0.96	0.92	
	K	N		Turn something on	Outside	Fore-Aft	Fore	0.60	0.66	
Linear slide	A	B		Increase something	Top	Left-Right	Right	0.82	0.80	
	C	D	M	Increase something	Top	Fore-Aft	Fore	0.98	0.94	0.96
	E	F		Increase something	Front	Left-Right	Right	0.76	0.96	
	G	H		Increase something	Front	Up-Down	Up	0.98	0.92	
	K	J		Increase something	Inside	Up-Down	Up	0.92	0.94	
	I	L		Increase something	Outside	Up-Down	Up	0.96	0.96	
Rotary	A	B		Increase something	Top	Rotate	Clockwise	0.84	0.92	
	C	D		Increase something	Front	Rotate	Clockwise	0.90	0.98	
	E	F		Increase something	Outside	Rotate	Over the top	0.90	0.92	
			Horiz. 45°↓							Horiz. 45°↓
Rocker switch	A	B	M O	Turn something on	Top	Left-Right	Right	0.92	0.76	0.96 0.74
	C	D	N P	Turn something on	Top	Fore-Aft	Fore	0.82	0.76	0.84 0.52
	E	F		Turn something on	Front	Left-Right	Right	0.70	0.76	
	G	H		Turn something on	Front	Up-Down	Up	0.92	0.98	
	K	J		Turn something on	Inside	Up-Down	Up	0.82	0.98	
	I	L		Turn something on	Outside	Up-Down	Up	0.92	0.94	

Figure 13.12. Results summary for generic controls, with photograph of controls used

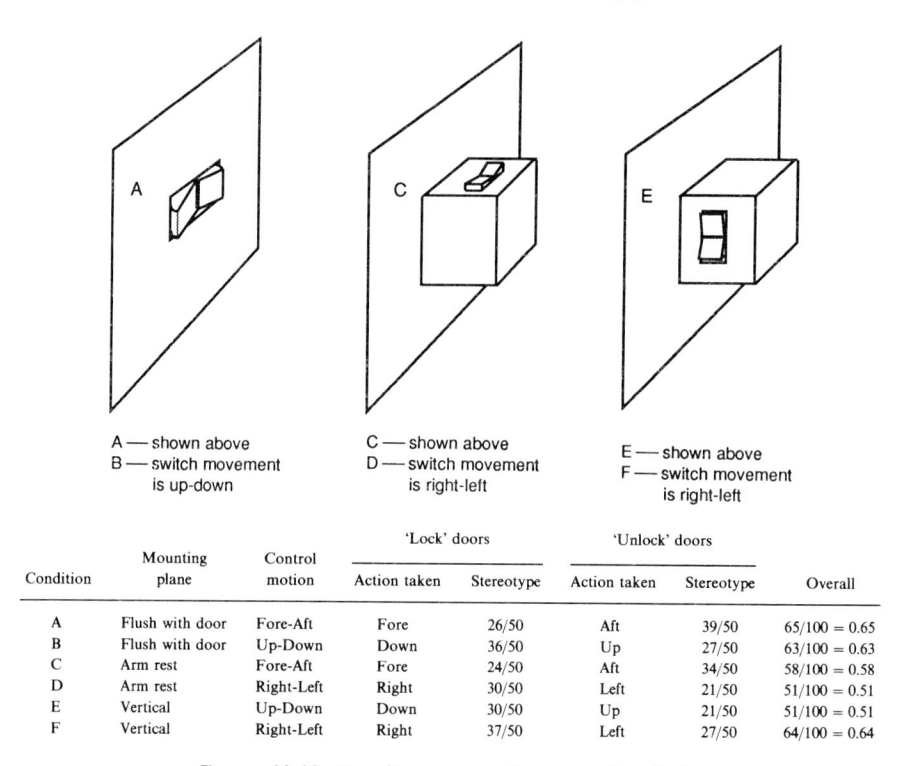

Figure 13.13. Results summary for power door locks

			'Lock' doors		'Unlock' doors		
Condition	Mounting plane	Control motion	Action taken	Stereotype	Action taken	Stereotype	Overall
A	Flush with door	Fore-Aft	Fore	26/50	Aft	39/50	65/100 = 0.65
B	Flush with door	Up-Down	Down	36/50	Up	27/50	63/100 = 0.63
C	Arm rest	Fore-Aft	Fore	24/50	Aft	34/50	58/100 = 0.58
D	Arm rest	Right-Left	Right	30/50	Left	21/50	51/100 = 0.51
E	Vertical	Up-Down	Down	30/50	Up	21/50	51/100 = 0.51
F	Vertical	Right-Left	Right	37/50	Left	27/50	64/100 = 0.64

A — shown above
B — switch movement is up-down

C — shown above
D — switch movement is right-left

E — shown above
F — switch movement is right-left

and Castellan, 1988). In all, 680 tests were performed on the data set. Using a criterion of $\alpha = 0.05$, only 11 significant results were found. It should be assumed that with the large number of tests that were run, approximately 30 would yield significance by chance. Thus, care must be taken not to assume results are reliable unless they occur repetitively and at a much smaller α level.

Of the 11 significant results, six occurred as a function of vehicle, three as a function of age, and one each as a function of gender and handedness. The six significant conditions occurring as a function of vehicle type were examined further and demonstrated that drivers whose primary car was domestic preferred to pull or raise a stalk to turn something on, whereas those whose primary vehicle was foreign had a different pattern. Approximately half had the same preferences as domestic car drivers, while the other half preferred to push or lower to turn something on. This difference is believed to be reliable and is a result of stalk usage on many makes of foreign cars that is different from usage on domestic cars. There appeared to be no reliable pattern in the three significant age effects, nor in either of the significant gender and handedness results. (In all three cases, other similar conditions did not produce significant results.)

The results of the statistical tests generally confirm that there is no appreciable effect of age, gender or handedness on direction of motion stereotypes for automotive controls. Furthermore, whether the driver's primary vehicle is

Table 13.1. *Stereotype results summary for the six studies*

Study	Strong	Moderate	Weak
Power mirror	8	2	0
Power windows			
Direction	7	3	1
Selection	8	1	2
Both	5	—	—
*Manual windows	1	0	3
Stalk controls	5	8	2
Generic controls			
Thumbwheel	7	3	0
Toggle	6	7	3
Linear slide	10	3	0
Rotary	5	1	0
Rocker	7	7	2
**Power door locks	0	0	6

Notes: * Although one of the manual window conditions (D, passenger window) resulted in a strong stereotype, the raise stereotype was only moderate.
** There was no strong stereotype for the power door locks.

foreign or domestic has an effect only on stalk usage and not in other areas of stereotype strength determination.

In spite of the small domain of statistical significance of the independent variables, it should not be assumed that such variables should be neglected in selecting a sample for future population stereotype work. The more representative the sample is, the more accurately the stereotype strength value is likely to be assessed.

Recommendations and conclusions

Recommendations based on study results

The main recommendation coming from this study is that vehicle designers should attempt to use the strong (and moderately strong) stereotypes found in this study, unless they have good reason for doing otherwise. There is a domain of strong stereotypes for four of the six studies. Only the manual window and power door lock studies did not yield strong or moderately strong stereotypes for important conditions.

Designers should particularly avoid inclining controls away from the driver, because it causes stereotype conflicts and resulting weak stereotypes. Such inclined controls resulted in weak stereotypes in all cases in the present experiment.

The fact that the push-pull power window controls resulted in moderate to weak stereotypes suggests that lack of previous experience can have an effect on stereotype strength. The push-pull controls can be considered as well designed, because of their strong geometric correspondence with the windows. However, few of the subjects had experienced them prior to entering the test situation. It is anticipated that further experience with the switch would significantly impact the stereotype strength. Therefore, any future attempts at testing new controls for stereotype strength should include training or learning as a second independent variable.

Because the power door lock study did not result in any strong stereotypes, additional work needs to be done. Other types of controls should be considered and the effects of labels that may aid in overcoming conflicts should be tested. Studies could also be concurrently conducted with regard to key rotation for locking and unlocking various types of controls (i.e. doors, trunk and glove compartment).

It is further recommended, based on results of the manual window study that an attempt should be made to set a guideline on a direction of movement for manual windows for future cars. Consistency across models and manufacturers should result in an eventual increase in stereotype strength, because of consistent experience. This recommendation for setting a guideline can be enhanced by taking maximum advantage of display control compatibility. For example, the window control should have a gear ratio that produces an integral number of turns from the full-up window position to the full-down position. Then by placing the starting point of the crank horizontal, the initial knob movement can be made to correspond to window movement (if the window is full-up or full-down).

Additional perspectives

The experimental study of direction-of-motion stereotypes described in this chapter has taught the authors a great deal about how such stereotypes should be gathered. Most of the material in this chapter resulted from experience in performing the experimental study. It is believed that the methodology was sound and can be recommended to others who wish to perform stereotype work. They are cautioned, however, to note that such a study has a certain critical mass. If too few subjects are tested, transfer and statistical stability problems are likely to result. If too few conditions are studied, results will not generalize easily and the need for a large group of subjects remains, in order that transfer effects might be controlled. Thus, difficulties arise when one attempts to run a study that is smaller than the one described here.

Finally, it is important to give some additional advice on the occurrence of weak stereotypes. What should a designer do when all control concepts for a given function turn out to have weak direction-of-motion stereotypes? It has already been indicated that appropriate coding might be tried. This may increase the strength of the stereotype. Another, more fundamental approach is to examine the relationship between a conceptual control and its corresponding display very carefully, whether real or visualized by the driver. If a control can be developed that has geometric correspondence, that control should result in strong stereotype once the relationship is discovered by the driver. Even though an initial stereotype is weak, within a few attempts it should become ingrained and accepted. Of course the hope is that the driver will in fact discover the relationship quickly. Verbal protocols might be helpful in determining whether or not learning has occurred.

Acknowledgments

This chapter is based on work sponsored by the SAE. Financial contributions to the SAE to support this research were supplied by Ford Motor Company, General Motors Corporation and Toyota Motor Corporation. Portions of the section of this chapter on the experimental project were reported previously in an SAE technical paper (Wierwille and McFarlane, 1991). Additional details on the experimental project are available in a final report (McFarlane and Wierwille, 1990) submitted to the SAE and soon to be published by the SAE.

In regard to the project, the authors wish particularly to thank Dave Benedict (Toyota Motor Corporation), Dr Brian Peacock (General Motors Corporation), Dr Gary Rupp (Ford Motor Company) and Tomoaki Shida (Toyota Motor Corporation), who served as the task force responsible for overseeing this research. These gentlemen helped solve the logistics problems that developed during the course of the study, and they made numerous constructive technical suggestions.

The authors also wish to thank other individuals who contributed their technical and managerial expertise during the project: John Buglioni, Larry Giddell, Tim Kuechenmeister and Ron Roe of the General Motors Corporation; Gene Farber, Lyman Forbes, David Hoffmeister, and Dan Jack of the Ford Motor Company; Seiichi Sugiura of the Toyota Motor Corporation; and Dr Paul Green of the University of Michigan Transportation Research Institute.

Here at Virginia Tech, the authors wish to acknowledge the contributions of Jana Moore. She spent many hours recruiting research subjects and aiding with the literature review. She also carried out a survey of controls used in current automobiles.

References

Anacapa Sciences, Inc., 1974, *SAE Study of Vehicle Controls Location*, Report No. 182-11, March, Santa Barbara, CA: Author.

Anacapa Sciences, Inc., 1976, *Driver Expectancy and Performance in Locating Automotive Controls*, (SAE Special Publication SP-407), February, Warrendale, PA: Author.

Black, T. L., Woodson, W. E. and Selby, P. H., 1977, *Development of Recommendations to Improve Controls Operability*, (Report No. DOT-HS-6-01445), Washington, DC: US Department of Transportation.

Boles, D. and Dewar, R., 1986, Nationality and handedness differences in stereotypes for control movements, in *Proceedings of the 19th Annual Conference of the Human Factors Association of Canada*, pp. 87–90, Vancouver, British Columbia: Human Factors Association.

Chapanis, A. and Gropper, B. A., 1968, The effect of operator's handedness on some directional stereotypes in control-display relationships, *Human Factors*, **10**, 303–20.

Gibbs, C. B., 1951, Transfer of training and skill assumptions in tracking tasks, *Quarterly Journal of Experimental Psychology*, **3**, 99–110.

Green, P., Kerst, J., Otters, D., Goldstein, S. and Adams, S., 1987, Driver preferences for secondary controls, (Technical Report UMTRI-87-47), October, Ann Arbor, MI: University of Michigan Transportation Research Institute.

Humphries, M., 1958, Performance as a function of control-display relations, positions of the operator and locations of the control, *Journal of Applied Psychology*, **42**, 311–16.
Jack, D., 1985, Rocker switch tactile coding and direction of motion stereotypes, in *Proceedings of the 29th Annual Meeting of the Human Factors Society*, pp. 437–41, Santa Monica, CA: Human Factors Society.
Loveless, N. E., 1962, Direction of motion stereotypes: A review, *Ergonomics*, **5**, 357–83.
McCormick, E. J. and Sanders, M. S., 1982, *Human Factors in Engineering and Design*, New York: McGraw-Hill.
McFarlane, J. and Wierwille, W. W., 1990, *Study of Direction-of-Motion Stereotypes for Automobile Controls*, Department of Industrial and Systems Engineering Report No. 90-02, Final Report, August, Blacksburg, VA: Virginia Polytechnic Institute and State University, Vehicle Analysis and Simulation Laboratory.
McGrath, J. J., 1974, *Analysis of the Expectancies of European Drivers and the Commonality of Automotive Control Locations in European Cars*, (Report TM 247-1), September, Santa Barbara, CA: Anacapa Sciences, Inc.
Perel, M., 1976, *Analyzing the Role of Driver/Vehicle Incompatibilities in Accident Causation Using Police Reports*, USDOT, Report No. DOT-HS-801-858, March, Washington, DC: National Highway Traffic Safety Administration.
Salvendy, G., 1987, *Handbook of Human Factors*, New York: Wiley.
Siegal, S. and Castellan, N. J., 1988, *Nonparametric Statistics for the Behavioral Sciences*, 2nd Edn, New York: McGraw-Hill.
Society of Automotive Engineers (SAE), 1970, *Supplemental information-driver hand controls location for passenger cars, multi-purpose passenger vehicles, and trucks*, J-1139, Warrendale, PA: SAE.
Spiegel, M. R., 1975, *Theory and Problems of Probability and Statistics*, New York: McGraw-Hill Book Company.
Turner, C. H. and Green, P., 1987, *Human Factors Research on Automobile Secondary Controls: A Literature Review*, Technical Report UMTRI-87-20, October, Ann Arbor, MI: University of Michigan Transportation Research Institute.
Wierwille, W. W. and McFarlane, J., 1991, *Overview of a Study of Direction-of-Motion Stereotype Strengths for Automotive Controls*, Warrendale, PA: Society of Automotive Engineers, Paper No. 910115, presented at the 1991 International Congress and Exposition, Detroit, MI, January.
Woodson, W. E. and Conover, D. W., 1971, *Human Engineering Guide for Equipment Designers*, Berkeley, CA: University of California Press.

14

Visual and manual demands of in-car controls and displays

Walter W. Wierwille

Introduction

The task of driving a modern automobile is a complex one. The driver provides loop closure within the driver/vehicle/roadway system, while also performing information detection, analysis and implementation for travel. As the driver performs these functions, the vehicle itself provides comfort and convenience to the driver, *en route* to the destination. These interactions result in a tightly coupled operator/machine system, with many types of communication and control links.

Visual and manual loading

In driving the primary information gathering sense modality is visual. The driver views the forward scene, glances to either side when necessary and also watches the rear view mirrors and instruments (mainly the speedometer). While the driver does use auditory and other sensory cues, such as touch in many situations, these senses play a minor role in information gathering. In terms of human operator outputs to control the vehicle, the major outputs are manual and pedal. (Voice and other forms of output are not presently used in automobiles.) The hands are used for steering (lateral-directional) control, for gear changes, and for outputting commands to in-car controls, such as those on the instrument panel (IP). The feet provide outputs for longitudinal control (acceleration/deceleration).

As obvious as it may seem after it is stated, it is important to recognize that the driver has a single visual resource. The eyes do not operate independently of one another. Rather, the eyes operate synergistically as a single sensory system, resulting in much better depth perception and slightly better peripheral vision and visual acuity than either eye can produce individually. This visual resource has foveal and peripheral capabilities, but it does not allow simultaneous foveal vision of two objects at different angular locations. (Olson, this volume, provides a detailed treatment of visual perception and focus of attention). The driver cannot uncouple the two eyes and fixate on, say, a road sign with one eye and the speedometer with the other. The main implication of this single visual resource is that the driver *must* share vision temporally. There is but one foveal visual resource and it must be moved about to gather detailed information.

In contrast, the driver has two hands that can perform different functions simultaneously. For example, one hand can be kept on the steering wheel while the other is used to actuate a control on the IP. This redundancy in hand motion makes it possible for the driver to provide commands for vehicle control while at the same time providing commands for other in-car tasks. Nevertheless, there are occasions when hand use must be time-shared. Whenever the driver makes a sharp turn, for example, both hands are required on the steering wheel. If the driver is performing an IP task with one hand, that task should be interrupted temporarily while the turn is made if good control is to be maintained. These simple statements demonstrate that drivers are required to time-share vision input continuously and manual output occasionally.

As a driver drives, the forward scene becomes the primary information gathering source. The driver may occasionally look elsewhere, but must still pay close attention to the forward scene because both the input for guidance and the input for hazard detection are obtained largely from the forward scene. In heavy traffic, it is hazardous to glance away from the forward scene for any appreciable length of time. On a hilly, winding road without traffic, it is similarly hazardous to glance away from the forward scene. More docile conditions allow longer looks away from the forward scene, but even under such conditions the driver must return to the forward scene frequently. Failure to do so may result in running out of lane, off the road, or into another object. In other words, attending to the forward view (and associated vehicle control cues) is the primary task of the driver. Because of the need for time sharing, IP and other in-car tasks that require vision must be considered as secondary.

Comparison with other vehicles

It is instructive to compare automobile driving with aircraft piloting. There are two main regimes in flying: IFR (instrument flight rules) and VFR (visual flight rules). The main difference between IFR and VFR is the weather in which they can be used. VFR can only be used when visibility is good and likely to be unobstructed by clouds and precipitation in the region through which the flight is to pass. In IFR the pilot observes the instruments and rarely looks through the windscreen. In fact training for IFR is usually carried out in simulators that have no visual scene generation capability, or in which the scene generation equipment is blanked temporarily. Pilots must be able to obtain all of their flight and control information from the instruments, visually and auditorily. In VFR, the pilot may time-share between the windscreen scene and the instruments. However, there remains heavy reliance on the instruments throughout most of the flight. The altimeter, compass, artificial horizon and radios, for example, must be closely monitored.

This situation differs greatly from automobile driving. IFR flight relies almost completely on the use of the cockpit instruments, VFR flight relies on a combination of instruments and outside scene, and automobile driving relies very heavily on the outside scene. Indeed, an automobile can be driven for long periods of time without glancing to the in-vehicle displays.

Another contrast in these situations is collision danger. Once an aircraft has become airborne and as long as the pilot carries out instructions from ground control, there is little chance of a collision with other objects or aircraft. The pilot's job is to carry out the instructions given, instructions that are already devised to avoid collision. Difficulties are likely to arise only when the pilot (or some other pilot) fails to carry out instructions. The automobile driver on the other hand is constantly faced with potential collision situations. The lane in which the vehicle must be kept is relatively narrow and there usually are other vehicles in the lane and in adjacent lanes. The distance between vehicles traveling in opposite directions can be as little as one meter, and in some cases even less. Because of the proximity of stationary and moving objects near the path and in the path of the driver's vehicle, the driver has no choice but to monitor the driving scene very carefully. In fact, under normal driving conditions, the driver looks into the vehicle only occasionally, say, to observe the speedometer or fuel gauge, or to adjust another system, such as the radio.

The reason for discussing the sensory inputs and control outputs in such detail is that the driving task is often misunderstood in terms of its priorities and uniqueness. There is a very well-defined primary task, whose inputs are almost exclusively via *head up* information gathering and whose outputs are via the lateral-directional (steering) and longitudinal (pedals) controls. All other in-car displays and controls can be considered as subsidiary or secondary. When they are used by the driver, they have the potential of competing for driver resources, and may therefore interfere with the primary task.

Conceptual models of in-car tasks

Task classification

In-car tasks that the driver performs can be separated into five categories, based largely on the driver resources needed. These tasks make different levels of visual and manual demands and can be classified accordingly.

Certain tasks can be performed by one of the driver's hands without visual reference, after sufficient practice. Good examples are sounding the horn, setting the directional signal lever, pressing the set or resume on the cruise control, shifting gears (manual transmission), adjusting the power seat position (in some cases), and adjusting the high/low beam control. The corresponding controls might require an initial glance until the driver becomes familiar with them. Once their locations and directions of movement are memorized, they can usually be activated without visual reference. These tasks can be classified as *manual only*.

A closely related category is that in which the driver uses vision to find a control and possibly determine its present setting. Thereafter the driver moves one hand to the control, and as the hand approaches the control, vision is no longer required. The driver then executes the remainder of the task manually, regardless of task length or number of steps. Often these tasks have auditory or other sensory feedback associated with them. Examples include turning on the radio and adjusting the volume, changing fan speed on the heater or air conditioner, adjusting seat position, and adjusting the direction of air flow from one of the air conditioning vents. These tasks are termed *manual primarily*.

Analogously, there are tasks that are completely or largely visual in terms of resources used. Tasks that require no manual input are always information gathering tasks. In some cases they are used for checking, while in other cases they are used for gathering more complex information. Examples of *visual only* tasks are reading the speedometer or chronometer, determining whether or not the high beams are on (using the IP idicator), and determining the current mode of the HVAC (heating, ventilating and air conditioning) system. Certain futuristic tasks may be visual only, such as determining from a map display whether the vehicle is headed in the correct direction to reach a specified destination.

Tasks that rely heavily on vision, but require a degree of manual input are termed *visual primarily*. Vision in such tasks is used both to locate and to gather information. However, to be able to gather information, some degree of manual input is required. Examples in this category include: determining the station frequency on the radio when the display initially provides the time and has to be switched to frequency, accessing the correct format of the trip monitor display to determine remaining miles, and accessing a compass display in a navigation system to determine direction of travel.

The final classification of in-car tasks is *visual-manual*. These tasks are distinguished by their interactive visual and manual demands. The driver gathers information and uses it for making additional manual inputs, or the driver makes manual inputs sequentially to access desired information. Examples include manually tuning a radio to a specified frequency, adjusting the commanded temperature in an automatic HVAC system, operating a cellular telephone, making mirror adjustments, and finding the correct page in a menu-driven display and then adjusting a specific quantity appearing on the display. For navigation with a map display, the driver might make several zoom level changes to bring up the right combination of area coverage and detail needed for the situation at hand. Visual-manual tasks are often complex and possibly demanding of driver resources because the driver performs loop closure between the displays and controls on either a sequential or continuous-adjustment basis.

Relationship to cognitive load

The five categories just described are based upon measurable inputs and outputs, namely, visual and manual demands. Little has been said about cognitive load levels. These loads can be quite taxing, particularly as automotive instrumentation becomes more complex. Cognitive load is more difficult to measure than visual and manual load. Generally one must resort to opinion or possibly other methods of mental workload estimation, including dual task methods.

This chapter includes discussion of cognitive load only insofar as it impinges upon visual and manual demands. The reason for this is that drivers can shift attention quickly, provided their vision is in the appropriate direction. To illustrate, consider the situation in which the driver is observing the forward scene when an emergency in the scene occurs. Chances are excellent that the driver would detect the emergency and take action to counteract or ameliorate it. There would be an abrupt shift of attention to the emergency, regardless of the cognitive load or its content just prior to the emergency. On the other hand, if the driver is attending to an IP task and is fixated on the IP, the likelihood is relatively low that an emergency in the forward scene would be detected. Detection would probably only occur after the driver returns glance direction to the forward scene.

This discussion demonstrates that diversion of the visual resource to secondary tasks within the vehicle must be given particular emphasis in design and devaluation. While cognitive load and manual load may influence visual load, it is the visual load of in-car tasks that must be considered to be of utmost importance.

Elementary sampling models of the driver

When a driver performs an in-car task (a secondary task), that task must be time-shared with the driving task (the primary task). As previously discussed, the driver has but one foveal visual resource which can only gather detailed information from a single source at any given time.

The process of time-sharing can be modeled simply (Wierwille, 1987). Consider that a driver is to perform a specific in-car task. Then assuming the task requires vision, the driver samples the task until the necessary visual aspect is completed, as shown in Figure 14.1. The driver samples the task, returns to the forward view, samples the task, returns to the forward view, etc., until the visual aspect is completed. In some cases a single sample of the in-car task is sufficient. In other cases, more than one sample is required.

Drivers normally realize that they must not glance away from the forward view for too long a period of time. They also realize that they cannot accomplish the in-car task without vision. They therefore develop a time-sharing strategy, which meets a competing set of requirements. The strategy consists of employing short glances away from the driving task, to gather visual information associated with completion of the in-car task.

Generally, drivers do not time-share *manual* demands. They simply remove a hand from the steering wheel and move it toward the in-car control or controls. They then activate the control(s) until the manual portion of the task is completed, whereupon they may return their hand to the wheel. Only in rare circumstances do drivers return the hand to the wheel before the manual demands of a task are met. Such circumstances include performing an emergency driving maneuver, turning a sharp corner, realigning the car in the lane assuming lane deviation or yaw deviation are unacceptably large, or possibly postponing or deciding to discontinue an in-car task. Also, on rare occasions

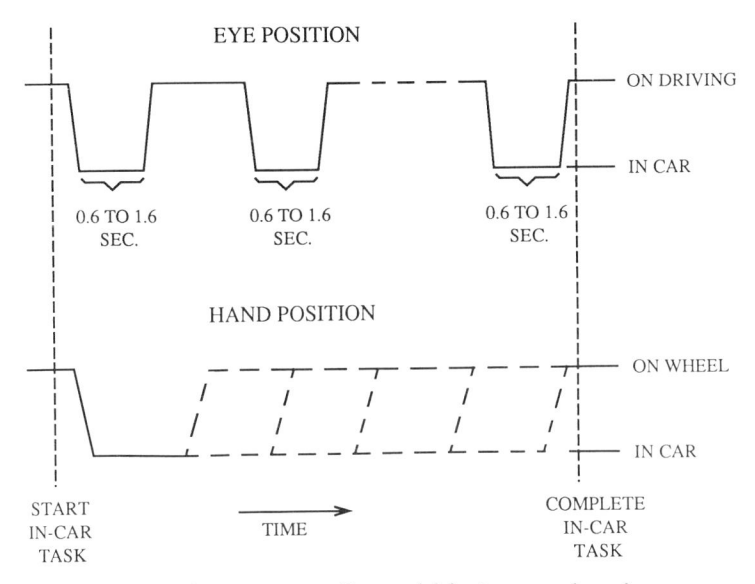

Figure 14.1. Elementary sampling model for in-car task performance

the driver may arbitrarily return the hand to the wheel while interpreting visual information, even though the task is not complete.

The model for visual sampling can be further specified as shown in Figure 14.2. This is a logic model and is based on a variety of experimental data to be described in the next section of this chapter. The model is a normative, deterministic model. Referring to Figure 14.2, when a driver begins to perform an in-car task he or she does so by glancing to the appropriate location. Information extraction begins as elapsed time passes. If the information can be chunked at about one second or less, the driver will do so and will then return glance to the forward scene. On the other hand, if chunking takes longer, the driver will continue to glance at the location for a bit longer. However, in doing so, the driver senses time pressure to return to the forward scene. If the glance to the in-car location continues up to about 1.5 seconds and the information cannot be obtained (or chunked), the driver will return glance to the forward scene anyway, and will try again later. On the other hand, if chunked information can be obtained within about 1.5 seconds, the driver will extract the information and return to the forward scene. Additional samples would be handled in exactly the same way, until all required visual information is obtained.

Given a free choice, drivers would prefer not to have to glance to the appropriate location for more than about one second. They will do so, however, under many circumstances if they must. They will not under most circumstances glance away for more than about 1.6 seconds (on the average). While the upper strategic bound does vary somewhat with the individual driver and the driving conditions, there is nevertheless an upper bound for each driver. Time pressure and forward scene uncertainty build to the point that the driver is compelled to return glance to the forward scene.

The model depicted in Figure 14.2 is admittedly too deterministic. Drivers do not measure elapsed time accurately and they do not consciously decide

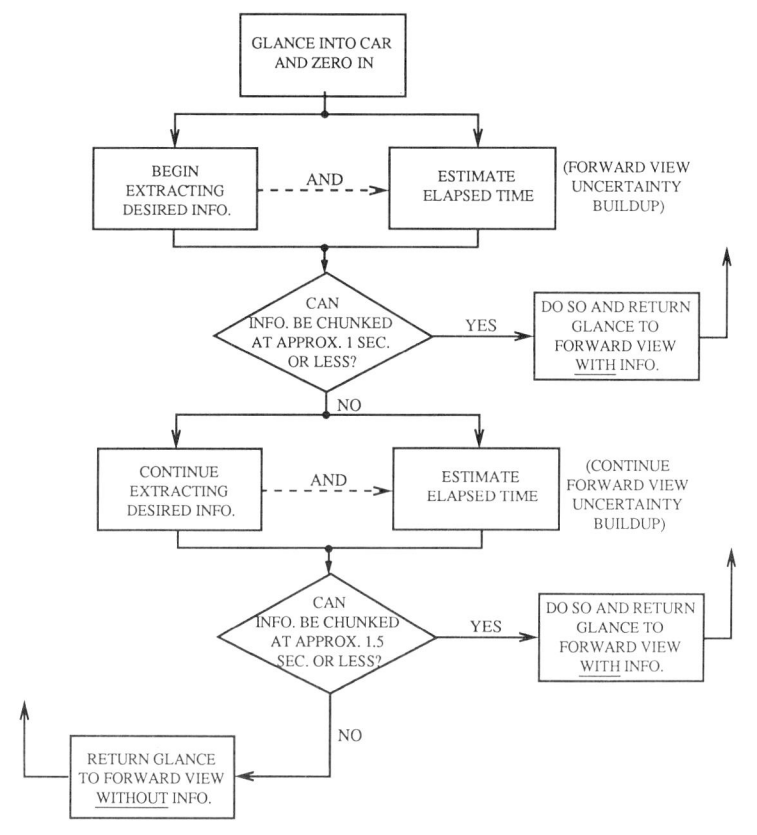

Figure 14.2. A more detailed model of visual sampling for in-car task performance.

whether the available information can be chunked. Furthermore, they do not use the logic flow as specifically as it is shown. Nevertheless, the model does explain the processes that must be taking place that limit the lengths of glances into the car.

It would seem that fuzzy logic would be ideally suited for describing glance length associated with in-car tasks (Yager, 1987). Fuzzy logic has the advantage that it allows *blurring* and the use of imprecise statements about processes. If a fuzzy logic model could be developed, it could potentially overcome much of the criticism that can be leveled against the deterministic model shown in Figure 14.2. In the meantime, however, the existing model can be used to explain the essential processes associated with in-car visual glances.

There are sufficient data available to show how parameters in the sampling model vary as a function of several independent variables, including type of task, age, clutter and forward view (primary task) demand. In addition to variation as a function of these independent variables, there is the usual inter- and intra-subject variability that always occurs when human beings perform tasks. Thus, mean values and deterministic models must be interpreted with caution, because there is a great deal of variation about the means.

In this chapter, the model of visual sampling has been provided first, so that it could be explained easily. In so doing, however, the reader has had to follow the discussion while having been provided with little or no data to support the model. The next section of the chapter provides an overview of some of the data, and in general shows support for the model.

Literature supporting sampling models

Early work

Early work supporting sampling models of driving was directed toward understanding the attentional demand of driving. Senders *et al.* (1967) developed a theoretical model based on information theoretic concepts. Fundamentally, the idea was that drivers sample the forward view and that between samples there is an uncertainty buildup with time. Eventually, an uncertainty threshold is reached and the driver must then glance to the forward view. The Senders *et al.* work was demonstrated by using a vision occlusion device. This helmet-mounted apparatus blocked the driver's vision on a periodic basis. The view and occlusion times could be adjusted. They found when using the device that there was indeed a minimum tolerable information sampling frequency for each given type of road and vehicle speed. If the sampling frequency was set below this minimum, drivers would immediately decelerate to compensate for lack of forward view information.

The Senders *et al.* work was directed at determining and modeling the attentional demand of driving, that is, the primary task of driving. However, the research demonstrates that drivers must have samples of the forward view at approximately equally-spaced intervals. It also therefore helps to explain the visual time-sharing that takes place when drivers attempt to perform in-car tasks while driving.

Before leaving the Senders *et al.* work it is important to note that the experimental portions of this research were performed on closed roadways in which there was no traffic. It is clear that the addition of traffic would have made higher sampling rates necessary for safe vehicle operation. In fact, one could argue that glances away from the forward view should be short and infrequent so that drivers can detect hazards promptly and reliably.

Two additional studies using occlusion have also been performed. Both of these were directed at assessing driving workload and spare capacity. Farber and Gallagher (1972) used a variation of the technique in which the roadway samples were viewed through filters of various neutral densities, while the between-sample portions totally occluded vision. They found that increasing filter densities and higher speeds both required more frequent samples. Hicks and Wierwille (1979) used the occlusion technique in a moving-base driving simulator. They blanked the scene generation system as a means of achieving occlusion. They found that the technique was not as sensitive as other workload measures; however, they used ordinary drivers who probably had difficulty adapting to the apparatus. Nevertheless, both of these latter studies on occlusion demonstrate that drivers can drive while sampling the forward view.

It appears that the first study to examine eye-fixations on in-car displays and controls was that of Mourant, Moussa-Hamouda and Howard (1977; see also Mourant, Herman and Moussa-Hamouda, 1980). Their study compared five different types of stalk controls and three different types of panel controls. The results of the study indicated that the frequency of direct looks increased as the reach distance increased, and that direct looks were slightly longer for stalk controls than for panel controls (owing to hand orientation requirements and eye accommodation).

Heinz, Bouis and Haller (1982) measured eye movements while drivers in a simulator performed three- and four-digit data entry. One data entry device was a telephone keypad and the other was a sequential device in which the driver pressed a single button a number of times to enter a corresponding digit. Drivers took longer to complete the sequential entry task, but had shorter in-car display glance times.

In pioneering studies, Rockwell (1987) examined the in-car glance durations and number of glances for radio and mirror tasks. He used a cross-section of drivers in traffic. He found that individual glance times into the car clustered around 1.25 seconds, as shown in Table 14.1, and that for radio tasks such as tuning, four or five glances were required. Based on the results of the experiments, Rockwell concluded that there was a fairly consistent time-sharing strategy for performing IP tasks while driving, that individual glance lengths were relatively consistent, and that for complex tasks more glances were required.

Bhise, Forbes and Farber (1986) extended the work of Rockwell to a greater variety of tasks. In summarizing their results, they indicated that single glance times to the IP do in fact vary somewhat with the type of task and number of glances varies greatly with the type of task. Table 14.2 shows some of the results they reported.

Recent work

Recently, Wierwille, Antin, Dingus and Hulse (1988), performed a study in which in-car conventional and navigational tasks were compared in terms of visual glance times and number of glances (see also Dingus, Antin, Hulse and Wierwille, 1989 and Antin, this volume). This research was performed in an instrumented vehicle on public roads with an operational, computerized moving-map navigation display. Some of the results of the study are summarized in Tables 14.3 and 14.4.

Table 14.1. Summary of glance duration data for conventional tasks
(Source Rockwell, 1987)

	Study	No. Runs	$\bar{x}$	Median	s	5%	95%
Radio	A	35	1.27	1.20	0.48	0.82	2.16
	B	100	1.28	1.29	0.50	0.89	1.83
	C	72	1.42	1.30	0.42	0.80	2.50
Left mirror	A	35	1.06	0.96	0.40	0.80	0.20
	*B	100	1.22	1.15	0.28	0.94	1.80
	C	72	1.10	1.10	0.33	0.70	1.70

Note: * Commanded mirror looks of discrimination.

Table 14.2. Summary of results presented by Bhise, Forbes and Farber (1986)

Tasks requiring a single glance	
Task	Mean glance duration (seconds)
Read Analog Speedometer	
• Normal	0.4 to 0.7
• Check	0.8
• Exact value	1.2
Read Analog Fuel Gauge	1.3
Read Digital Clock	1.0 to 1.2

Tasks requiring several glances		
Task	Number of glances	Mean glance duration (seconds)
Turn on radio, find station, adjust volume	2 to 7	1.1
Read all labels on a 12-button panel	7 to 15	1.0

Table 14.3 shows the means and standard deviations of total in-car glance times for a variety of tasks. Total glance time is computed by summing the individual glance lengths into the car until the task is completed. The table shows very clearly that total visual demand varies markedly with the task; it ranged from 0.78 seconds for reading speed up to 10.63 seconds for determining the name of the roadway at which the next turn was to be made in reaching a programmed destination. Of course, in most cases, drivers did not glance continuously. Rather, they sampled, as shown in Table 14.4. This table shows the average length of glances and number of glances into the car, along with corresponding standard deviations. The table further shows that single glance lengths varied somewhat more than previous studies had indicated, with a range of 0.62 to 1.66 seconds. The table also shows that average number of glances ranged from 1.26 to 6.64. This latter result is in agreement with earlier work, but is more comprehensive.

The results of these studies demonstrate several important relationships. First, they show the relatively narrow range of single glance times, and the relatively broad range of number of glances. The results also show that gathering information for several navigation tasks and certain other conventional tasks (radio tuning for example) results in large visual demands. These tasks are in the visual primarily and visual-manual categories, as described previously. Most importantly, however, the results clearly support the sampling model previously presented. Drivers do not, on the average, allow their single glance times to exceed about 1.6 seconds, even for complex information gathering tasks. Instead, they return to the forward scene, attend to the driving task, and then return to gather additional in-car information. This process continues until the task is completed.

Follow-up work has been done to determine the effects of driving task (*per se*) demand on the visual sampling process (Wierwille, Hulse, Fischer and

Table 14.3. *Total in-car glance times for a variety of conventional and navigation tasks* (Source: Dingus *et al.*, 1989)

Task	Mean	Standard deviation
Speed	0.78	0.65
Following traffic	0.98	0.60
Time	1.04	0.56
Vent	1.13	0.99
Destination direction	1.57	0.94
Remaining fuel	1.58	0.95
Tone controls	1.59	1.03
Info. lights	1.75	0.93
Destination distance	1.83	1.09
Fan	1.95	1.29
Balance	2.23	1.50
Sentinel	2.38	1.71
Defrost	2.86	1.59
Fuel economy	2.87	1.09
Correct direction	2.96	1.86
Fuel range	3.00	1.43
Cassette tape	1.59 + 1.64*	0.96 (0.59)*
Temperature	3.50	1.73
Heading	3.58	2.23
Zoom level	4.00	2.17
Cruise control	4.82	3.80
Power mirror	5.71	2.78
Tune radio	7.60	3.41
Cross street	8.63	4.86
Roadway distance	8.84	5.20
Roadway name	10.63	5.80

Note: * Time required to search for and orient cassette tape.

Dingus, 1988; Kurokawa and Wierwille, 1990). The results of these studies show reliable adaptive trends by the driver as driving task demands increase. The following are general statements supported by the studies:

1 For drivers using an in-car navigation system and as roadways become more difficult to drive, the probability that the driver's eyes will be on the roadway increases (as shown in Figure 14.3). Under the same conditions, the probability that the driver's eyes will be on the navigation display decreases by approximately the same amount (as shown in Figure 14.4).
2 For drivers using an in-car navigation system and for varying traffic conditions, glance length to the forward view increases with traffic density and the possibility of an impending conflict (Figure 14.5). Under the same conditions, the probability that the driver's eyes will be on the navigation display decreases (Figure 14.6).
3 As crosswind disturbance level increases while the driver performs various conventional in-car tasks, single glance length to the roadway increases and single glance length into the car decreases (Figure 14.7).

Table 14.4. *Average length and number of in-car glances for a variety of conventional and navigation tasks*

Task	In-car single glance length		Number of glances	
	Mean	Standard deviation	Mean	Standard deviation
Speed	0.62	0.48	1.26	0.40
Following traffic	0.75	0.36	1.31	0.57
Time	0.83	0.38	1.26	0.46
Vent	0.62	0.40	1.83	1.03
Destination direction	1.20	0.73	1.31	0.62
Remaining fuel	1.04	0.50	1.52	0.71
Tone controls	0.92	0.41	1.73	0.82
Info. lights	0.83	0.35	2.12	1.16
Destination distance	1.06	0.56	1.73	0.93
Fan	1.10	0.48	1.78	1.00
Balance	0.86	0.35	2.59	1.18
Sentinel	1.01	0.47	2.51	1.81
Defrost	1.14	0.61	2.51	1.49
Fuel economy	1.14	0.58	2.48	0.94
Correct direction	1.45	0.67	2.04	1.25
Fuel range	1.19	1.02	2.54	0.60
Cassette tape	0.80	0.29	2.06	1.29
Temperature	1.10	0.52	3.18	1.66
Heading	1.30	0.56	2.76	1.81
Zoom level	1.40	0.65	2.91	1.65
Cruise control	0.82	0.36	5.88	2.81
Power mirror	0.86	0.34	6.64	2.56
Tune radio	1.10	0.47	6.91	2.39
Cross street	1.66	0.82	5.21	3.20
Roadway distance	1.53	0.65	5.78	2.85
Roadway name	1.63	0.80	6.52	3.15

(Source: Dingus *et al.* 1989)

These results suggest that drivers undergoing increased visual loading due to (the primary task of) driving adapt their visual sampling strategy; they are under greater pressure to return their glance to the forward view sooner and maintain it on the forward view a greater proportion of total time. The models shown in Figures 14.1 and 14.2 do not account for this adaptation process. However, they certainly could be modified to include this as well as other adaptation processes. In Figure 14.2 for example, the pressure to return to the forward view could be included by shortening or lengthening the chunking times, or by replacing the decision model with a more complex one.

Age effects

A number of driving research studies involving in-car tasks have included age as an independent variable (Rackoff, 1974; Poynter, 1988; Hayes, Kurokawa and Wierwille, 1989; Imbeau, Wierwille, Wolf and Chun, 1989). The results

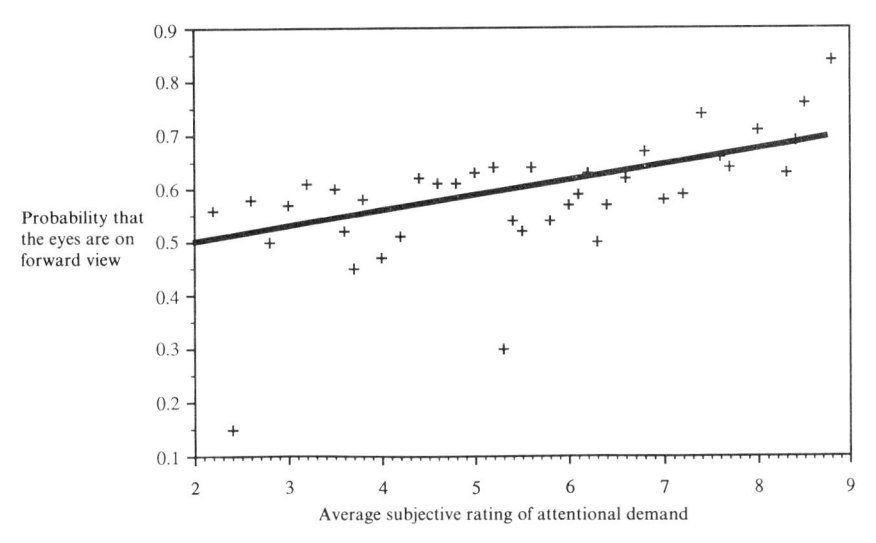

Figure 14.3. Probability that the eyes are on the forward view as a function of rated roadway demand (regression line and averaged data shown)
(Source: Wierwille, Hulse, Fischer and Dingus, 1988).

generally show that deterioration of vision and slowing of cognitive processes have an effect on in-car task completion times. Furthermore, if inadequate character size or contrast ratios are used in displays, even greater decrements will occur. In regard to glance times, there is strong evidence that in-car single glance times and number of glances increase with age. In addition, transition

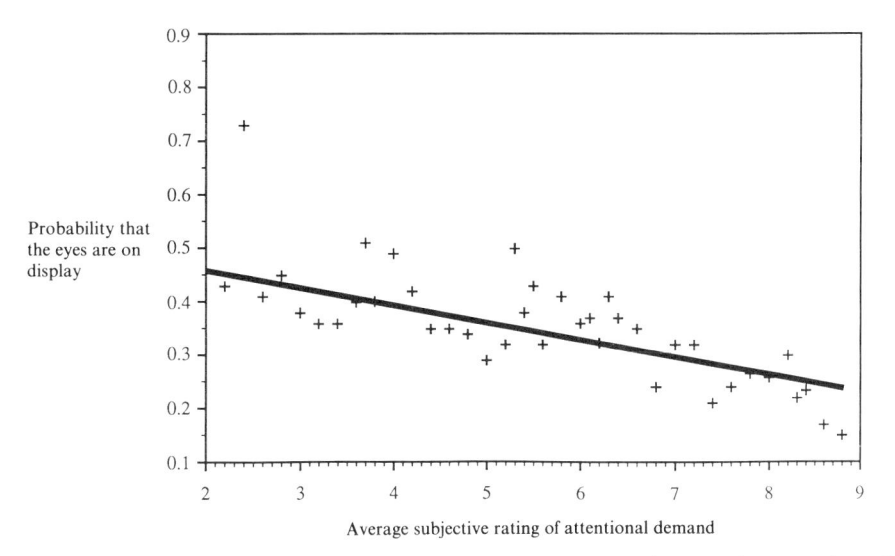

Figure 14.4. Probability that the eyes are on the navigation display as a function of rated roadway demand (regression line and averaged data shown)
(Source: Wierwille, Hulse, Fischer and Dingus, 1988).

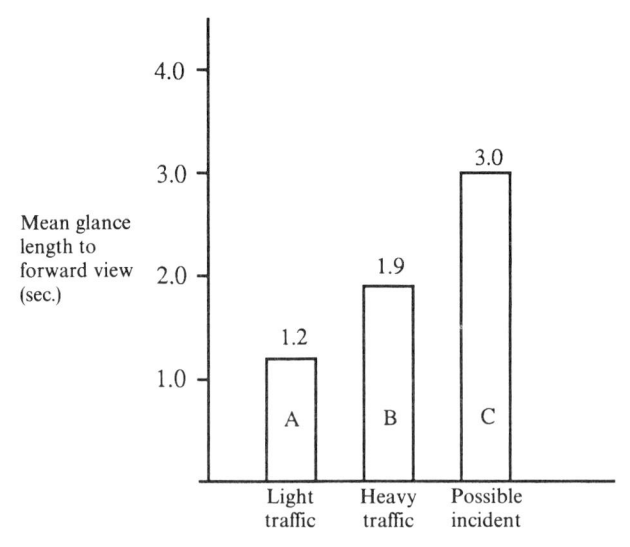

Figure 14.5. Glance length to forward view as a function of traffic type
(Source: Wierwille, Hulse, Fischer and Dingus, 1988).

time between the in-car task and forward view increases with age. Figures 14.8 through 14.10, taken from Hayes *et al.* (1989), illustrate some of these effects for drivers performing conventional in-car tasks.

Clutter

Additional work recently completed (Kurokawa and Wierwille, 1991) shows that various types of instrument panel clutter cause an increase in the number

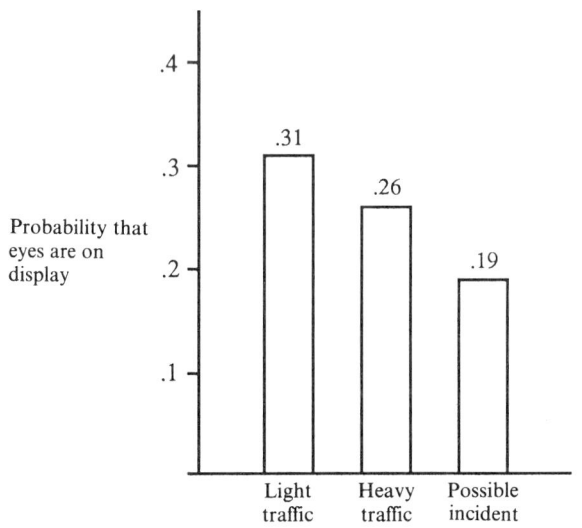

Figure 14.6. Probability that the eyes are on the navigation display as a function of traffic type
(Source: Wierwille, Hulse, Fischer and Dingus, 1988).

In-car controls and displays 313

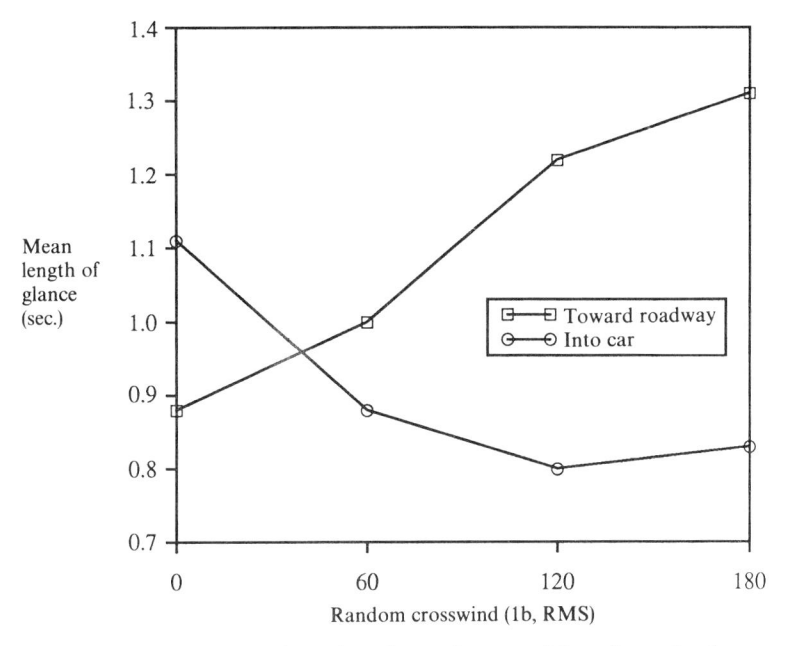

Figure 14.7. Mean glance length as a function of disturbance level
(Source: Kurokawa and Wierwille, 1990).

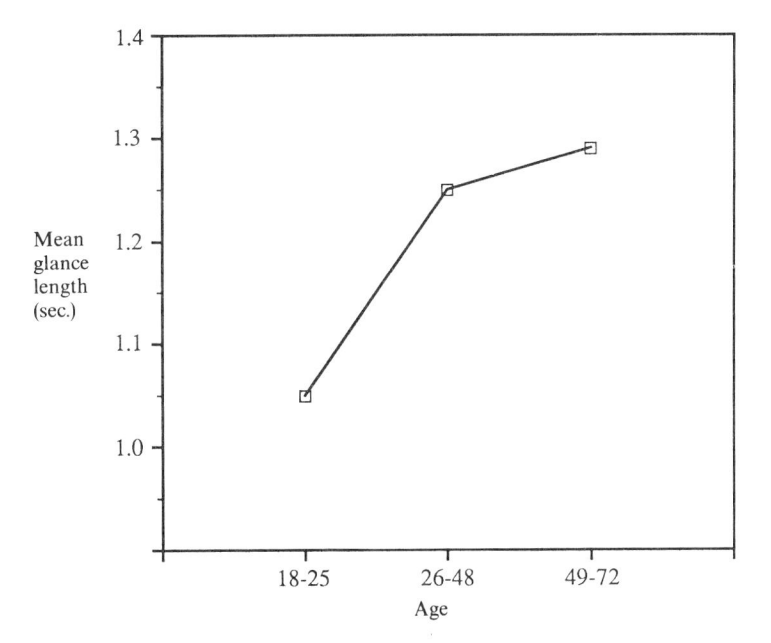

Figure 14.8. Mean in-car glance length as a function of age
(Source: Hayes *et al.*, 1989).

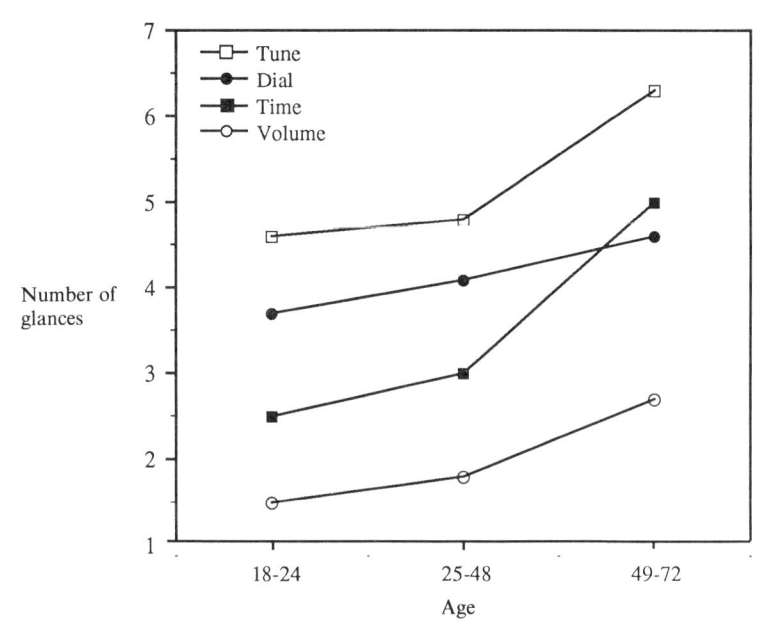

Figure 14.9. Mean number of in-car glances as a function of age for four typical tasks
(Source: Hayes *et al.*, 1989).

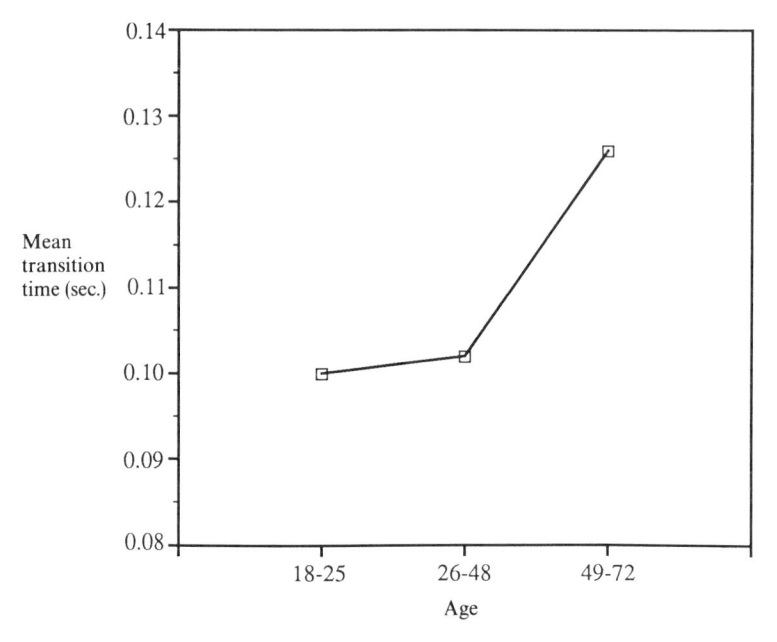

Figure 14.10. Mean transition time between forward view and in-car task as a function of age

(Source: Hayes *et al.*, 1989).

of in-car glances. (Figure 14.11 shows results for 'macroclutter', when similar-appearing panels of controls are present. Figure 14.12 shows results for 'microclutter', where similar-appearing individual buttons within a control panel are present. In the latter figure, both the number of glances and the single glance times increase with the number of buttons present. Additional results of Kurokawa and Wierwille's (1991) work show that well-chosen abbreviations for labels provide shorter single glance times than do their fully-spelled counterparts.

The literature thus demonstrates the effects of several independent variables on the sampling process that occur for in-car tasks. Not only do these results substantiate the model, they also show how an appropriate model could be enhanced to account for a greater number of independent variables.

One shortcoming of the present literature is the lack of theoretical explanation of the processes involved in sampling. The literature shows *what* is occurring but not to any great extent *why* it is occurring. Referring to Table 14.4, for example, it would be important to determine more precisely *why* single glance times are longer for some tasks than others, and how one designs an in-car task to induce shorter chunking. Lack of such concepts makes it difficult to develop a design philosophy. Nevertheless, the literature does point the way toward *evaluating* existing tasks experimentally.

Remedies and recommendations

Remedies

The literature review demonstrates that some in-car tasks have relatively high visual demands, and, as mentioned, this trend is likely to continue as in-car displays (and possibly controls) become more complex. Are there any remedies that can be applied to reduce visual load? A few have been suggested and are discussed briefly in this section.

Specialized displays

It is generally recognized that as the angle between the forward view and the in-car task increases, transition time for the eyes also increases. Furthermore, accommodation time (time to refocus from the forward scene to the in-car task) increases with change of distance from the viewer and also with age. Finally, the further down into the car the task is, the less likely it is that peripheral vision could be used to detect a hazard in the forward scene. These well-known results suggest that it is best to locate the in-car task display as high on the IP as possible (or even above the IP) and the focus distance to the in-car display should be further away rather than nearer. There are ways of accomplishing such objectives, namely through the use of head-up displays, other virtual image displays, and variable message information centers mounted at or near the top of the IP.

The head-up display puts the information at a distance in front of the driver and includes it in the visual scene. This arrangement means that transition time and accommodation time can be minimized, while at the same time peripheral vision can be used to detect emergencies and hazards. The advantage of the head-up display must be weighted against the possible disadvantages,

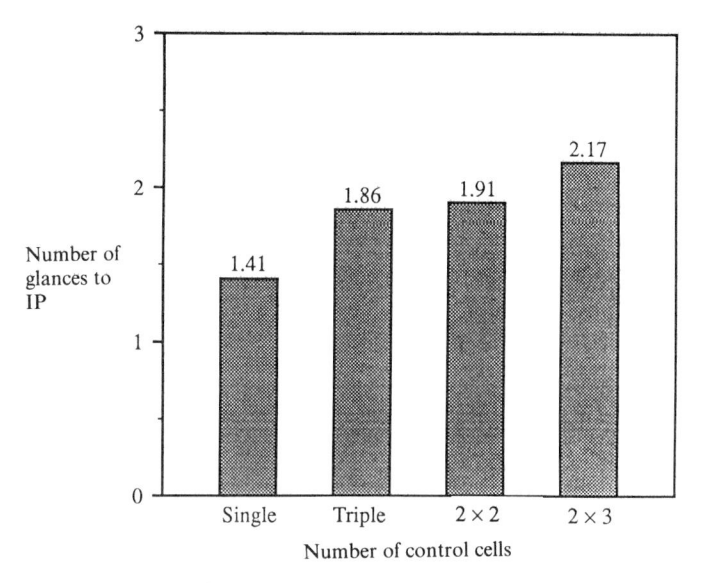

Figure 14.11. Effect of macroclutter on number of in-car glances
(Source: Kurokawa and Wierwille, 1991).

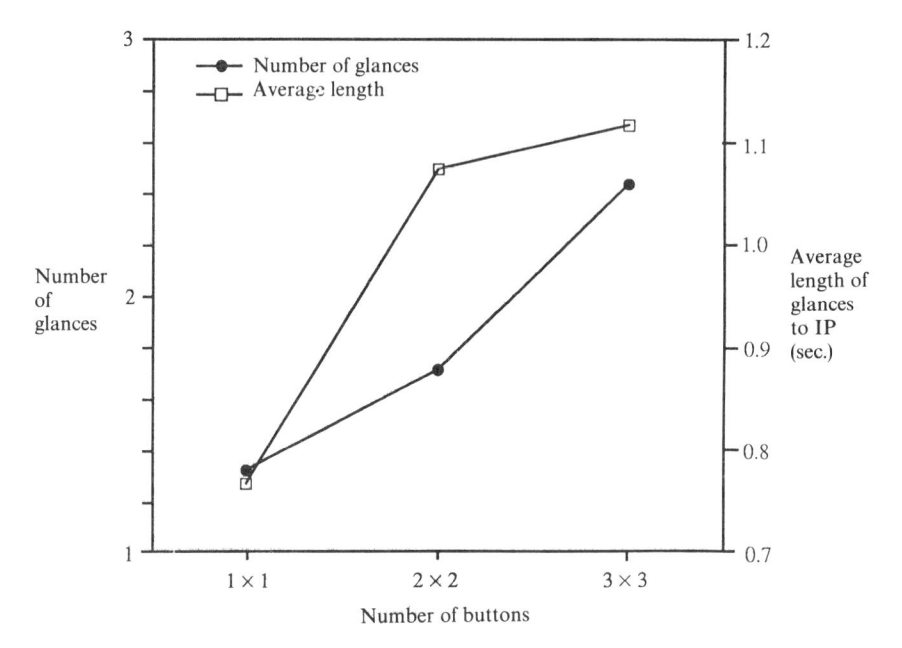

Figure 14.12. Effect of microclutter on number of in-car glances and single glance lengths
(Source: Kurokawa and Wierwille, 1991).

namely, possible distraction caused by its presence in or near the driving scene and possible obscuration of scene information 'behind' the virtual image (Weintraub, 1987). Research is needed on finding the configuration that maximizes the advantages and minimizes the disadvantages of the head-up display, that is, that optimizes the tradeoffs.

A virtual image display is similar to a head-up display in that it places the image at a distance. However, such displays are usually in the upper portions of the dash (Swift and Freeman, 1985). While they produce an image at an apparent distance, they create an illusion for the driver that the display is deep in the engine compartment or near the front bumper. The ramifications of this on driver acceptance and performance are not well understood. Nevertheless, a virtual image display, if properly designed, should reduce accommodation time because the image distance is closer to the primary task object distance.

Information centers placed near the top of the IP have the advantage of allowing a variety of messages to be presented from one location, a location that rapidly becomes familiar to the driver. Reconfigurable displays, if properly designed, should provide the same type of advantage.

Auditory displays

One way to reduce the visual load of driving and performing in-car tasks is to rely more heavily on auditory input. The technology is now at hand to provide high-quality synthesized or digitally recorded (and accessed) human speech. In addition, certain distinct sounds can be used as driver inputs (Wolf, 1987). Unfortunately, speech synthessis has a poor record of consumer acceptance. The way in which it was previously introduced has caused it to be viewed as gadgetry, and even an annoyance, because the messages conveyed were too common and contained little unknown information. Unloading the visual resource by use of auditory inputs thus represents a formidable challenge to the designer and human factors engineer.

Auditory input to the driver is not a panacea. Only certain visual displays can be replaced by auditory displays, and the driver's ability to comprehend auditory messages is limited. Because simultaneous messages can be difficult to understand, auditory inputs should be presented serially, and they should be adjustable in terms of volume, equalization and possibly speaker's voice. This means that all in-car auditory displays must be connected together in such a way that messages are coordinated on the basis of priority.

In some cases, auditory displays may require a different design concept. For example, in navigation, an in-car computer generated map and command display could be supplemented by an auditory command display, which could make it possible for the driver to navigate in congested areas without visual diversion. Later, when conditions permit, the driver could update perceived location and direction using the visual display. Such ideas are already under investigation (Streeter, Vitello and Wonsiewicz, 1985).

Training

There are other ways that visual diversion to in-car devices can be minimized. One of these methods is training. We have observed in our experiments that some drivers are much poorer at visual sampling than others. In fact, they may even fixate on a control or display for a short period of time after the task is completed. They tend to be less concerned about the forward view than they should be, and their glance times into the car tend to be relatively long (perhaps two seconds or more). Such individuals are prime candidates for training or retraining.

The training program need not be a major effort. In fact, such training could conceivably be handled in perhaps an hour, with printed material provided to enhance the experience. The training program might include the following elements:

1 Encouraging and demonstrating set-up procedures for any in-car tasks that can be performed before beginning actual driving. Typical examples would be setting the radio station memory buttons and other radio controls, as well as memorizing the on/off, volume, memory button control positions and directions of operation.
2 Demonstrating visual sampling at higher frequency and explaining the importance and need for faster sampling. Thereafter have the trainee practice faster sampling until proficient.
3 Introducing the concept of comprehension *after* in-car sampling. This concept is one of glancing briefly into the car, and then on returning to the forward view, processing what was observed. Such an approach can reduce the sample length appreciably even for simple tasks such as reading the speedometer.
4 Demonstrating how tasks can often be postponed until there is plenty of headway, or the car has stopped at a stoplight that has just turned red, or until other low-hazard conditions exist.

In regard to training, it may be time to recognize that driving is becoming a complex skill and that only so much can be accomplished by hardware design. Because of complexity, training or retraining may be necessary, particularly for those individuals who exhibit deficits in needed skills. Of course, legal issues may have to be resolved before remedial training may be implemented on a wide scale.

Recommendations

Research results demonstrate conclusively that drivers do use a visual sampling strategy to perform in-car tasks, that their behavior is relatively consistent, and that models can be developed that show the nature of the sampling process. In addition, a good deal is already known about how specific independent variables affect the visual sampling process. Quite clearly, a primary goal of automotive human factors research and design must be to minimize the visual demands of in-car tasks. Additional goals should be to minimize cognitive and manual demands. However, this is not as important as the goal of minimizing visual demands, because, as stated, visual diversion into the car competes *directly* with the visual aspect of driving.

To minimize visual demands of in-car displays and controls there are several rules that can be developed. Many amount simply to good human factors. For example, any labels or legends should possess adequate size and contrast ratio, as determined from already available research data. Appropriate, easily interpreted, nonambiguous words or abbreviations should be used for labels. In addition, care should be taken to avoid clutter, which has been demonstrated to increase glance time because of the need to search. Also, whenever possible, in-car displays and controls should be designed so that visual information gathering can be chunked into segments of not more than one second.

It should be recognized that future cars are likely to make higher in-car demands on driver resources, because such cars will have greater capabilities and more exotic features. The IVHS (Intelligent Vehicle/Highway System) now under development through various projects and agencies will rely on informing the driver what actions to take. Outputs to the system by the driver are also likely to be necessary. Such systems may require moving-map displays with a variety of menu options, which may create large visual and cognitive (attention) loads.

IVHS is but one system likely to increase in-car demands. Night vision enhancement, drowsiness detection, and proximity warning systems are also being developed. In addition, improved and more versatile forms of current systems may add to the in-car task load. It is therefore very important to develop and refine methods for reducing this load, particularly visual load. Very little work has been done on integrating in-car tasks and on minimizing the combined load of all such tasks.

There are, then, two areas where future research is needed. The first is in refinement and expansion of models of in-car task performance by the driver. This chapter has provided a simple model and a brief overview of the available data base. However, it is only the beginning. The models need to be improved in terms of accuracy, degree of generality and determination of effects of independent variables. Also, a better *understanding* of what is causing parameter changes must be obtained. The second area of further research is development of better guidelines for in-car task communications with the driver and associated developments that allow minimization of demands on driver resources. Both of these areas of research represent great challenges, but promise substantial rewards in terms of driver-vehicle efficiency and accident prevention.

Acknowledgments

This chapter is based in part on research sponsored by various divisions of the General Motors Corporation including the Research Laboratories, C-P-C Advanced Vehicle Engineering and GM Systems Engineering.

The opinions expressed are those of the author.

References

Bhise, V. D., Forbes, L. M. and Farber, E. I., 1986, Driver behavioral data and considerations in evaluating in-vehicle controls and displays, paper presented at the Transportation Research Board, National Academy of Sciences, 65th Annual Meeting, Washington, DC, January.

Dingus, T. A., Antin, J. F., Hulse, M. C. and Wierwille, W. W., 1989, Attentional demand requirements of an automobile moving-map navigation system, *Transportation Research*, **A23**(4), 301–15.

Farber, E. I. and Gallagher, V., 1972, Attentional demand as a measure of the influence of visibility conditions on driving task difficulty, *Highway Research Record*, **414**, 1–5.

Hayes, B. C., Kurokawa, K. and Wierwille, W. W., 1989, Age related decrements in automobile instrument panel task performance, *Proceedings of the Human Factors Society 33rd Annual Meeting, Vol. 1*, Santa Monica, CA: Human Factors Society, October, pp. 159–63.

Heinz, F., Bouis, D. and Haller, R., 1982, *Safer Trip Computers by Human Factors Design*, SAE Paper 820105, Warrendale, PA: Society of Automotive Engineers.

Hicks, T. G. and Wierwille, W. W., 1979, Comparison of five mental workload assessment procedures in a moving base driving simulator, *Human Factors*, 21(2), 129–43.

Imbeau, D., Wierwille, W. W., Wolf, L. D. and Chun, G., 1989, Effects of instrument panel luminance and chromaticity on reading performance and preference in simulated driving, *Human Factors*, 31(2), 147–60.

Kurokawa, K. and Wierwille, W. W., 1990, Validation of a driving simulation facility for instrument panel task performance, *Proceedings of the Human Factors Society 34th Annual Meeting*, pp. 1299–303, October, Santa Monica, CA: Human Factors Society.

Kurokawa, K. and Wierwille, W. W., 1991, Effects of instrument panel clutter and control labeling on visual demand and task performance, paper submitted for presentation at the *Society for Information Display International Symposium*, May.

Mourant, R. R., Herman, M. and Moussa-Hamouda, E., 1980, Direct looks and control location in automobiles, *Human Factors*, 22(4), 417–25.

Mourant, R. R., Moussa-Hamouda, E. and Howard, J. M., 1977, Human factors requirements for fingertip reach controls, Technical Report DOT-HS-803-267, Washington, DC: US Department of Transportation.

Poynter, D, 1988, *The effects of aging on perception of visual displays*, SAE Paper 881754, Warrendale, PA: Society of Automative Engineers.

Rackoff, M. J., 1974, An investigation of age-related changes in drivers' visual search patterns and driving performance, and the relation to tests of basic functional capabilities, *Proceedings of the Human Factors Society 19th Annual Meeting*, pp. 285–8. Santa Monica, CA: Human Factors Society.

Rockwell, T. H., 1987, Spare visual capacity in driving—revisited: New empirical results for an old idea, in Gale, A. G. (Ed.) *Vision in Vehicles II*, pp. 317–24, Amsterdam: North Holland Press.

Senders, J. W., Kristofferson, A. B., Levison, W. H., Dietrich, C. W. and Ward, J. L., 1967, The attentional demand of automobile driving, *Highway Research Record*, 195, 15–32.

Streeter, L. A., Vitello, D. and Wonsiewicz, S. A., 1985, How to tell people where to go: Comparing navigational aids, *International Journal of Man-Machine Studies*, 22, 549–62.

Swift, D. W. and Freeman, M. H., 1985, The application of head-up displays in cars, in Gale, A. G. (Ed.) *Vision in Vehicles*, pp. 249–55, Amsterdam: North-Holland Press.

Weintraub, D. J., 1987, HUDS, HMDS, and common sense: Polishing virtual images, *Human Factors Society Bulletin*, 30, October, (10), pp. 1–3.

Wierwille, W. W., 1987, Can dash instrumentation visual attentional demand be predicted using the design driver concept? Paper presented at the Transportation Research Board Annual Meeting, Washington, DC, January.

Wierwille, W. W., Antin, J. F., Dingus, T. and Hulse, M. C., 1988, Visual attentional demand of an in-car navigation display system, in Gale, A. G. (Ed.) *Vision in Vehicles II*, pp. 307–16, Amsterdam: North-Holland Press.

Wierwille, W. W., Hulse, M. C., Fischer, T. J. and Dingus, T. A., 1988, Strategic use of visual resources by the driver while navigating with an in-car navigation display system, *XXII FISITA Congress Technical Papers; Automotive Systems Technology: The Future*, Vol. 11, SAE P-211, Paper No. 885180, September, Warrendale, PA: Society of Automotive Engineers, pp. 2.661–2.675.

Wolf, L. D., 1987, The investigation of auditory displays for automotive applications. *Proceedings of the 1987 Society for Information Display International Symposium*, New York: New Orleans, LA, May, pp. 49–51.

Yager, R. R. (Ed.), *Fuzzy Sets and Applications: Selected Papers by L. A. Zadeh*, New York: John Wiley and Sons.

15

Informational aspects of car design: Navigation

Jonathan F. Antin

Introduction

In the vast majority of cases the primary purpose in driving is to get from a particular starting point (A) to a particular destination (B) in the context of large-scale space (i.e., space of which only a relatively small portion can be observed from any one vantage point, Kuipers, 1978). Since the destination will almost always not be visible at the starting point, this definition implies that at each node or intersection in the roadway network a path selection or navigation decision must be made with the currently available information which might include prior knowledge, maps, landmarks, street names, etc.[1]

In recent years advances in microprocessor technology in terms of size and price reduction, power efficiency, durability, enhanced storage and memory capacity and portability, as well as similar advances in the associated display technologies have made new forms of onboard navigation aids feasible. In addition to private sector efforts, governments around the world have also been involved in the development of these onboard navigation aids as well as centrally located traffic management systems. This development is due to the global recognition that the current traffic problems in many metropolitan areas are intractable, demand will continue to outstrip the ability to build new roads, and so more efficient use must be made of the existing roadway infrastructure (Koltnow, 1989; Shibata, 1989; Willis, 1990).

These technological advances and cooperative efforts have spawned discussion of the fundamental issues related to automobile navigation such as what and how much navigation-oriented information is needed by the driver, how and when should it be presented, what are society's needs regarding navigation and traffic management, and what is the most cost-effective way to meet those needs? The goal of this chapter is to consolidate this discussion in the context of the human factors issues that must be considered if automobile navigation is to become safer and more efficient for the driving public.

Driver characteristics related to navigation

Spatial ability

Spatial ability characterizes the facility with which one stores, organizes, recalls and utilizes information pertaining to the relationships among objects in a given space. For example, when considering a city as a space, the objects might be buildings, roadways, intersections, etc. There are essentially two reference frames from which the navigator can view objects in space: the ego reference frame (ERF) and the world reference frame (WRF; Stokes, Wickens and Kite, 1990). The ERF represents the viewpoint of the navigator, and is completely dependent on the location and orientation of that individual in space. The WRF is a global perspective of a given space which remains essentially unchanged as the navigator moves within the space. Aretz (1990) has identified four cognitive operations, based on the relationships between these two frames of reference, that are required for successful navigation: *triangulation* establishes the geometries of the ERF and WRF; *mental rotation* aligns the WRF and the ERF; *image comparison* confirms alignment of the WRF and the ERF; and *translation* which monitors motion through the WRF. It is the manner in which the driver performs these sorts of spatial-cognitive activities which determines to a large extent how well he or she can navigate within that space.

With regard to map learning, Thorndyke and Stasz (1980) found that successful performance depended upon the use of high quality, organized approaches. Furthermore, they found that those with greater visual memory ability benefited more from instruction in these quality approaches to map learning than those with less ability. Antin, Dingus, Hulse and Wierwille (1988) reported on an in-vehicle navigation study whose results suggested that those with higher spatial ability (determined by a combined score on three factor-referenced tests) spent a smaller proportion of driving time looking at available map information relative to a group with lower spacial ability.

Selective attention ability

Avolio, Kroeck and Panek (1985) found that those with relatively less ability to attend to relevant sources of information selectively, while ignoring sources defined to be irrelevant, have more prolific accident histories. The importance of this type of finding is magnified in the use of any onboard navigation aid, especially any visual aid. If one has less ability to attend selectively to multiple, simultaneous sources of information, then introducing a source which competes with the primary driving tasks for the driver's attention may increase the likelihood of accident involvement beyond the already increased levels which might be expected for the typical driver when an additional visual task is added to the driving milieu.

Cognitive mapping

When an individual develops a mental representation of the relationships among objects within a given space, a cognitive map has been formed. It is an individual's inherent spatial ability as well as experiences with a given space (or its representation) which determine the content and extent of the cognitive map and how it may be used to solve real navigation problems associated with that space. For instance, if a landmark were missed by a low ability navigator, he or she might become lost, whereas the high ability individual would have a greater likelihood of knowing his or her general location in the space, despite having missed any particular landmark (Streeter, Vitello and Wonsiewicz, 1985).

Levels of cognitive mapping

Thorndyke and Hayes-Roth (1982) have proposed that cognitive maps are developed to different levels depending on the amount of time spent in or studying the space and the manner in which it is experienced (i.e., via travel, simulated travel, map study, etc.). They identified two categories of cognitive mapping knowledge: *procedural descriptions* and *survey knowledge.*

Procedural descriptions convey knowledge of how to get from A to B, including how to respond to important landmarks along the route. This type of knowledge is typically acquired through actual navigation, and so may also include information regarding time and distance of the overall route as well as individual legs; it also may include other route features such as additional landmarks, signal lights, terrain features, etc. Procedural descriptions are essentially based on the ERF.

Survey knowledge is based on the WRF in which a general picture or concept of an entire space, including Euclidean relationships among objects in that space, is developed. This type of knowledge is acquired most efficiently through map study, but usually would become much more well established by extensive navigation experience within the space in question. However, Thorndyke and Hayes-Roth do make the distinction that survey knowledge acquired via navigation is still effected from the perspective of the traveller, as opposed to an overall or bird's-eye view which would be acquired through map study.

Kuipers (1978) has claimed that the cognitive map is developed primarily via direct navigational experience, but that maps can contribute to the content of the cognitive map. He later divided the spatial awareness contained in a cognitive map into three categories: *sensorimotor procedures*: a sequence of actions required to get from A to B; *topological relations*: knowledge of properties such as connectivity, containment and order; and *metrical relations*: knowledge of and ability to mentally manipulate quantities such as distance, direction, angles and absolute and relative location of objects in the space (Kuipers, 1983). A similar conceptualization has been brought forward by Byrne (1979). Using his terminology, a *network-map* is one in which the topological relations as defined by Kuipers are preserved in the mental representation, but not necessarily the metrical relations, which Byrne has termed the *vector-map.*

The difference between such conceptualizations is illustrated by the process of rectilinear normalization (Chase and Chi, 1980, as cited in Wickens, 1984) in which the we tend to perceive and encode the roadway network as an interconnected grid of straight paths and right angles between intersections. To make a navigation decision requires mainly topological relations, which are essentially preserved in the context of rectilinear normalization, but not necessarily metrical relations which are distorted by the same process. A similar construct is that of *alignment* in which figures in an array tend to be perceived as being aligned (Tversky, 1981). Also, Wickens (1984) has noted that we tend to orient this array with respect to the four main points of the compass, a tendency termed *rotation* by Tversky.

Driver/navigator information requirements

From the above discussion it can be argued that the basic type of information needed for navigation is what has been variously referred to as procedural knowledge (Thorndyke and Hayes-Roth, 1982), sensorimotor procedures and topological relations (Kuipers, 1978), or the network-map by Byrne (1979). If one is travelling for the first time in an unfamiliar area without the benefit of explicit directions, then this level of cognitive map will not have had a chance to develop. In these circumstances, maps and other navigation aids may be required to avoid a trial and error search for the desired destination.

Paper maps primarily show the complete network of roadways and the type of roadway, and sometimes provide additional information regarding landmarks such as schools, railroad lines, bodies of water, tunnels and the like. Since the complete network of roadways is continuously available with the local paper map, the entire route to the destination can be planned at the outset; also, it can be referred to at any time. If a navigation error is committed, a new route from the current location to the destination can also be planned, once the new location has been ascertained. These factors may in some respects represent a significant advantage for the standard paper map over some competing technological solutions which will be discussed below (see Antin, Dingus, Hulse and Wierwille, 1990).

Problems with paper maps

There may be several difficulties encountered when using paper maps. Because the paper map must be a static presentation of information, the entire network of roads in a local area is usually included, and so there are considerable clutter and scale problems. Labels must be relatively small to fit them all within the given space; conversely, the map itself must be large, resulting in handling and manipulation problems. Size can be especially troublesome if the driver is attempting to use the map while driving.

Yet the paper map may present even more formidable challenges to the spatial ability of the navigator. In a survey of mobile telephone users, 15 activities commonly performed in the automobile were ranked on their perceived danger (Smith, 1978). Reading a map while driving was perceived as being the most dangerous (7.7 rating on a scale from 1, 'not at all dangerous', to 10 'extremely dangerous'). Use of a paper map requires the user to determine vehicle location and orientation mentally, and select a suitable route. Research performed by Aretz (1988, 1989) in the context of aviation has led him to conclude that since a moving-map display in a heading-up orientation (i.e., where the map information rotates around and moves past the vehicle location marker in such a way that the ERF is preserved on the display) performs the location and orientation tasks for the navigator, the navigation task is made much simpler. Figure 15.1a illustrates the heading-up orientation for a vehicle travelling south; Figure 15.1b is the same information shown in north-up orientation. In this way we can begin to see how limitations in traditional paper maps may possibly be overcome with new technological solutions.

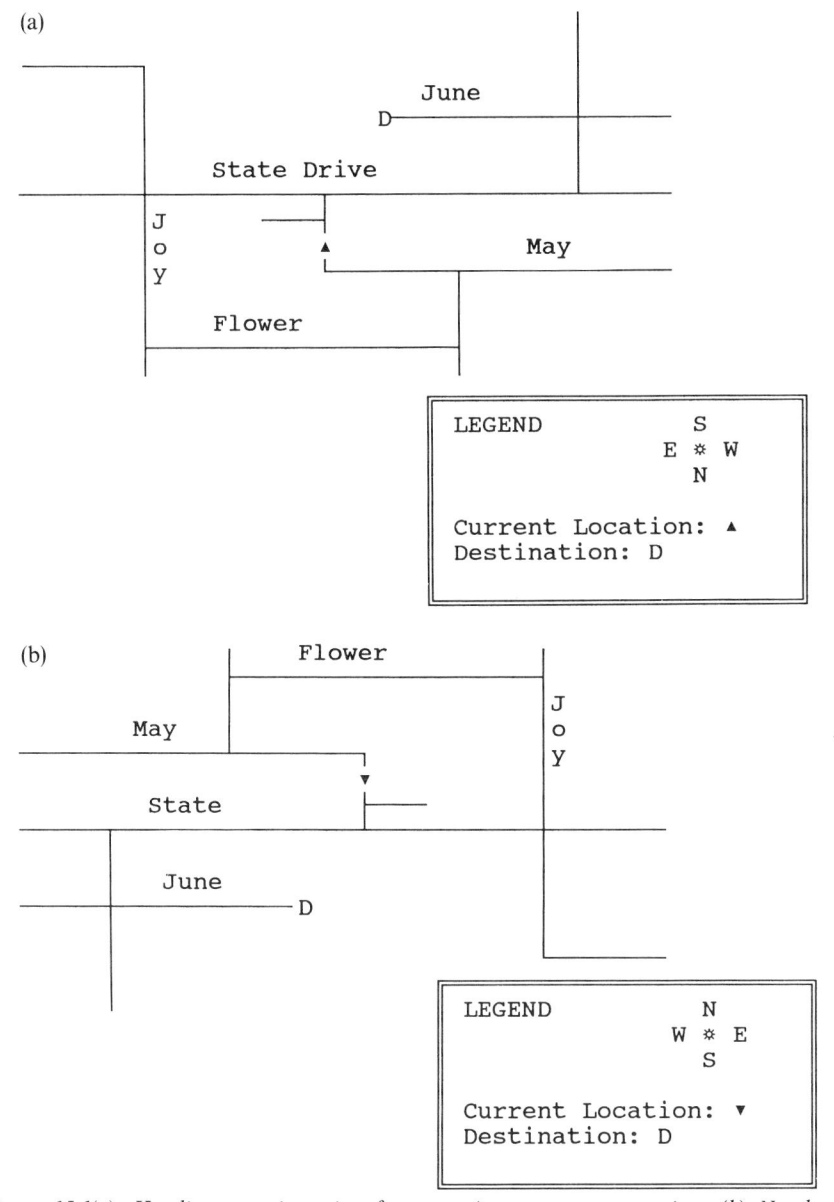

Figure 15.1(a). Heading-up orientation for a moving-map representation; (b) North-up orientation for a moving-map representation

Vehicle navigation aids: Information presentation approaches

King (1986a) gave subjects a chance to rank several items such as: information availability, skills improvement and navigation aids in terms of importance for successful navigation. Only information availability received ratings of important or better. King said the low ratings on navigation aids may be due to their novelty. He also found that, in general, subjects' willingness to pay for such aids is much less than some current cost estimates. These data may be based on his finding that drivers have a relatively high opinion of their route planning and following skills; although the greatest number felt that they had average skills, 12 times as many said that they had good or very good skills compared with the number of those who claimed to have bad or very bad skills. Although some people may be satisfied with their navigation abilities, others are clearly less comfortable with this activity. Perhaps a well designed navigation aid can make the process easier and more effective for a wide range of drivers. Some of the information presentation approaches for vehicular navigation aids are described below.

Basic information

An electronic navigation aid will tend to have, at a minimum, two basic pieces of information: distance and direction from the current location to the programmed destination. Compass information may also be included. An example is the VDO City Pilot by VDO Adolf Schindling AG; it uses earth magnetic field sensors and an odometer distance sensor to calculate and display direction and as the crow flies distance to the destination (French, 1987).

Moving-map display

The Etak Navigator® is an example of the next level of information that can be provided by an onboard navigation aid. This system includes a moving-map representation showing the specific location and orientation of the vehicle on the map, and eight pushbutton-selectable scale levels (from about one-eighth mile to a 40 mile radius around the vehicle) on which the map information can be displayed. To reduce display clutter, each scale level only shows a restricted set of roadways (i.e., the greater the area shown, the less detail shown). Detail in this context refers to the extent to which connecting, residential and otherwise minor streets are shown on the display. This system also includes a direction north indicator, and it presents information on euclidean distance and direction from the current location to a programmed destination which is highlighted as a star on the display. The Etak system is represented in Figure 15.2. The display is shown on the one-half mile scale, and the vehicle is about 0.3 miles northwest of the destination. The system can be operated in either heading-up or north-up modes.

Heading-up vs. north-up

It has already been argued that one potential benefit of moving-map displays is the fact that they can be operated in heading-up mode, in which case the

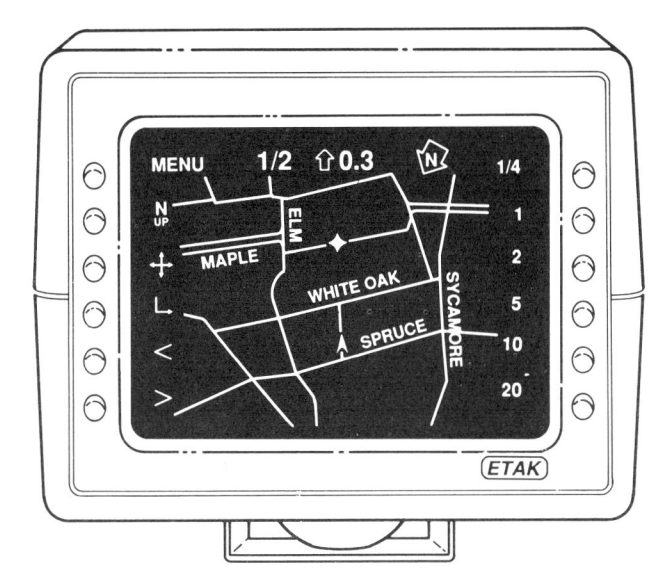

Figure 15.2. Representation of Etak Navigator®

ERF always coincides with the WRF, requiring no online mental rotations. Harwood (1989) has recommended using a north-up orientation for helicopter applications of the moving-map display, but points out that navigation is only best with north-up maps when travelling north, that is, when no mental rotation is required. For example, Figure 15.1a shows a southbound vehicle on a map display in heading-up orientation. In this case the vehicle would have to turn right on State Drive to reach the destination, which is consistent with how it appears on the map display. However, Figure 15.1b shows the same information in a north-up orientation which may cause it to falsely appear to some that a left turn on State Drive is required to reach the destination (i.e., it is a left turn *on the display*). This example demonstrates the fundamental advantage of heading-up displays; straight or on the display is always straight ahead of the vehicle, regardless of the direction of travel. Also, right or left relative to the display always corresponds to the same direction with respect to the vehicle itself.

On the other hand, the north-up display has an obvious advantage over the heading-up orientation with regard to the consistency between a user's sense of compass directions (i.e., the WRF), and the display of these directions. In the heading-up orientation these are decoupled, and can easily be reversed. Refer again to Figure 15.1a; here the requisite *right* turn corresponds with a turn to the *west* which may be confusing to some. In the north-up representation shown in Figure 15.1b, a right turn always corresponds with a turn to the east which should be compatible with driver expectations. This is especially salient when travelling on expressways which are often labeled in terms of compass directions (e.g. Interstate 40, West).

Aretz (1990) has claimed that research is equivocal in choosing between the heading-up and north-up electronic map configurations. His approach was to introduce heading wedge vectors onto a north-up map in an effort to compare the two basic orientations to one which combined the best aspects of both. His results showed that navigation tasks involving a coupling of ERF and WRF are better with heading-up than north-up orientations. The north-up view produced better results for tasks not involving the ERF, such as map reconstruction (which is not navigation *per se*).

Route guidance

Even when a moving-map of the area is provided, this still does not imply that explicit route guidance is provided. If a route selection algorithm is implemented, then a complete set of instructions is generated and presented to the driver. Such algorithms can be designed to select the shortest path or can be augmented to incorporate data related to route simplicity or shortest time to destination. These data might also be based on the following: major vs. minor roadways, the number of traffic lights, expected traffic problems at specific times, and if it is part of a larger system, real-time traffic information. Route information can be presented in several ways: the correct path on a map could be indicated by color changes, highlighting, or simply by directional arrows which appear at the appropriate time. Directions could also be verbal, either presented aurally or in written form.

Streeter *et al.* (1985) have demonstrated that people select routes by integrating information on such things as perceived route length, estimated travel time, traffic density estimates based on time of day, etc. In this regard they found that novices tend to choose more major routes, whereas experienced travellers in a particular area tend to choose the shortest route irrespective of road type. The question then arises, do we design route selection algorithms to coincide with the way we would expect experts to select routes? This would accrue the benefits of route selection expertise to all, even though these routes may be the most difficult for local strangers to follow. Another option is to use novice-oriented routes for all, even though these may not be the shortest routes in terms of time or distance.

Spatial vs. verbal navigation aids

A study by Streeter and Vitello (1986) determined that self-rated poor navigators prefer verbal directions employing landmark information. They concluded that map reading is a difficult task, especially while driving, and that, for the general population, verbal directions should be employed. Streeter *et al.* (1985) tested this hypothesis in a subsequent study in which performance of drivers using several modes of navigation was compared. These modes included: 1) taped verbal instructions, 2) customized route maps, 3) a combination of 1 and 2 above, and 4) standard road maps. Their results generally showed that the taped verbal instructions provided the shortest routes in terms of time and distance and produced the fewest navigation errors. The conclusion that was drawn from this study was that due to the potentially problematic visual distraction which might result from use of a visual display, verbal directions should be presented aurally to the driver. For example, Streeter *et al.* (1985) actually found that when the customized route map was combined with verbal instructions, navigation performance actually suffered relative to that demonstrated with verbal instructions alone. Stokes, Wickens and Kite (1990) have pointed out that the benefits associated with either spatial or verbal navigation aids may depend strongly on whether the navigator is in the planning or route-following phase of the navigation process.

Navigation technology: safety issues

In addition to the work of Streeter et al. (1985), other researchers have also looked at the safety-related aspects of different navigation technologies. A series of in-vehicle studies was performed at the Vehicle Analysis and Simulation Laboratory at Virginia Tech to evaluate the relative safety associated with use of a moving-map navigational display (see Antin et al., 1990; Dingus et al., 1989; Wierwille et al., 1990). Dingus et al. (1989) performed a study to compare the visual attentional demand associated with tasks typically performed in the automobile environment (e.g., tuning the radio, activating the turn signal indicator, etc.) with specific tasks which were associated with navigating with the Etak Navigator®. The Etak system was chosen because it was felt that it possessed the basic features which could be expected to be included in a moving-map display (see McGranaghan, Mark and Gould, 1987). While 32 driver/subjects were guided along a specified route, they were instructed to perform a particular task involving the moving-map display or one of the more conventional dashboard tasks. Results showed that three of the most complex but important tasks associated with the moving-map display required greater visual attention than any other task as measured by mean single glance duration and total glance duration for each task. When the data were separated based on whether or not the information was currently available when the task was presented, then these three tasks became more comparable in demand to some of the conventional tasks. Based on this reasoning, Dingus et al. (1989) concluded that the scale level should be automatically updated in such a way that the destination star remains displayed at all times.

Antin et al. (1990) performed a study (using the same subjects and apparatus as were used in Dingus et al., 1989) which was designed to evaluate the relative effectiveness and efficiency (in terms of time taken in reaching the destination, route selection quality and level of intrusion on driving behaviors) associated with the use of three navigation methods: moving-map display, conventional paper map and memorized route. One of the important findings in this study was that, on average, driver/subjects spent a significantly greater proportion of driving time looking at the moving map, compared to that spent looking at the paper map (Table 15.1), implying that the selective attention abilities of subjects would be strained much more in use of the moving-map display.

Although the two individuals who evidenced a proportion of driving time glancing to the moving map greater than 0.5 were both women over the age of

Table 15.1. Proportion of driving time spent glancing to map information (Antin et al., 1990)

Condition	N	Mean	S.D.	Maximum	Minimum
Paper map	32	0.068	0.045	0.183	0.0
Moving map	32	0.331	0.095	0.586	0.148

50, there were virtually no overall differences attributable to age or gender. It is plausible that any age-related decrements in perception or central processing were made up for by navigation experience and a reduced willingness to take risks by those older individuals.

Also, those participating in this study must be considered practiced novices in use of the moving-map display, therefore it is possible that with greater experience with such a system, the documented 'eyes-off-the-road' time may decrease. However, with the system tested, new and more detailed information is continually provided on the display as the destination is approached (i.e., as the user zooms into larger, more detailed scale levels); this information must be continually attended to if navigation is to be successful. So, some portion of the disparity between the moving and paper maps in terms of the visual attention demanded may still exist after practice effects are factored out of the evaluation. It is also possible that the driver/subjects felt that, due to its convenient location on the dashboard, more visual attention could be allocated to the moving map while still maintaining a comfortable safety margin.

Hughes and Cole (1986) have suggested that from 30 to 50 percent of visual attention of the driver may be allocated to things unrelated to the driving task. They have claimed that this amount represents spare capacity on the part of the driver (i.e., indicating that the driver must only allocate 50 to 70 percent of visual attention to the driving task). Such findings indicate that the time-sharing demands of the moving-map display found in this study would be acceptable for most individuals. This is to say that the demands could theoretically be met with spare resources, leaving the driving task largely undisturbed. Yet it remains to be determined how the additional attentional demands might affect driving performance in areas such as the detection of salient visual cues representing potentially hazardous situations. An example is the sudden appearance of a child's ball a short distance in front of the vehicle, possibly to be followed by a child who may be oblivious to any potential danger.

Wierwille *et al.* (1990) set out to investigate this problem; they monitored driver's reactions to salient cues or *incidents* while navigating with a moving-map display. The forward roadway was videotaped, and incidents were defined by the experimenters as any situation imposing high and immediate demand if a hazardous situation was to be avoided. They found that even when using a moving-map display to navigate, if high attentional demand can be anticipated (e.g., when approaching a known, large place of business where entering traffic may be expected), then visual sampling rate increases, whereas if there is a sudden, unanticipated situation requiring immediate attention (e.g., a vehicle running a stop sign ahead), then sampling rate decreases with longer glances being devoted to the forward roadway. In both situations drivers increased the proportion of time devoted to the forward roadway, and decreased the time spent on the moving-map display. This reaction suggested to Wierwille *et al.* (1990) that drivers were adopting a rational, adaptive behavioral pattern to the changing task demands that are imposed while driving and navigating with a moving-map display.

The use of guidance messages (written or auditory) associated with different levels of map completeness was studied by Labiale (1989). He found that the highest percentage of drivers were able to recall the correct route when the maps were associated with written guidance messages. However, more drivers

Table 15.2. *Average time per glance to available map information (Antin et al., 1990)*

Condition	N	Mean (s)	S.D.	Maximum	Minimum
Paper map	32	1.481	0.683	3.942	0.0
Moving map	32	1.367	0.221	1.993	1.026

preferred using maps with auditory guidance messages than maps with written guidance messages, or maps alone. Reasons given by those preferring the auditory messages indicated that they felt that this was a safer alternative. Drivers who preferred maps alone felt more comfortable developing their own set of instructions, possibly indicating that this group had higher spatial ability than the other groups. In this study the mean visual time per glance to any of the map displays was 1.28 seconds, with 92.3 percent of all glances being less than or equal to two seconds. This compares to the mean single glance time to the moving-map display found by Dingus et al. (1989) of 1.4 seconds, (range 1.06 to 1.66 seconds). Antin et al. (1990) found similar results (Table 15.2). All of these sets of results support the adaptive time-sharing behavior noted by Wierwille et al. (1990) in general, since they show how drivers tend to respond to complex visual tasks with increased glance frequency, not relying as heavily on increased glance duration.

Zwahlen and DeBald (1986) performed a study to evaluate the effects of visual distraction on lateral lane deviations. Their subjects were asked to read continuously from a CRT while driving on a test track. Results showed that the percent chance of laterally deviating from a 10 foot lane for reading times of 2, 4, and 6 seconds, were 1.25, 6.3 and 18.14 percent, respectively. Zwahlen and DeBald recommended based on these possibly alarming findings that the development and implementation of complex in-vehicle displays should be halted pending further serious review. However, it is crucial to note that these data do not attempt to account for the natural way in which drivers tend to time-share visual attention when driving, rationally adapting their gaze patterns according to the situation (Wierwille et al., 1990).

Labiale (1990) validated his earlier findings in a subsequent study; in addition he found that when maps were associated with verbal directions, that the visual and auditory modalities produced approximately equal percentages of drivers correctly recalling the route (87.5 and 90.6 percent, respectively) for relatively simple routes. However, when the routes became somewhat more complex (including three turns as opposed to one), then the percentage of those using the visual modality who correctly recalled the route dropped to 75 percent compared to 40.6 percent for those exposed to the auditory messages.

Navigation and traffic management technology

Recoverable navigation waste estimates

It has been estimated that 6.4 percent of all travel distance and 10 to 12 percent of time in non-commercial vehicular travel is recoverable navigation waste. Costs associated with this waste were estimated to be 45 billion dollars, considering vehicle operating and accident costs and time (French, 1987; Van Aerde and Case, 1988). Reed (1989) has estimated that rush hour delays may increase by as much as 400 percent in the 1990s!

King (1986b) has pointed out that navigation waste can come from: choice of route selection criteria, route planning, route following and destination sequencing (i.e., with multiple destinations and no obvious or given order). The routes selected by local strangers were compared by King to those selected by local experts. He employed different levels of trip planning in order to determine more precisely the source of navigation waste. King determined that considerable excess driving occurs as a result of approximately equal contributions of inefficiencies in trip planning and route following. The following is a description of some government, industry and often combined efforts in Europe, Japan and the United States to produce navigation systems not only designed to make navigating more efficient and safe for the individual driver, but to improve the efficiency in our use of the roadway system.

European efforts

Walzer and Zimdahl (1988) have discussed some recent efforts in traffic management and navigation in Europe. One of the major ongoing efforts is the PROgraM for a European Traffic with Highest Efficiency and Unprecedented Safety or PROMETHEUS. The program has several modules, some of which emphasize the development of computerized in-vehicle systems to support navigation, as well as the development of communication systems between vehicular and road-based computers for advanced traffic management. Philips' CARIN represents an attempt to use voice synthesized route instructions in addition to a color CRT map. The Radio Data System (RDS) uses a subcarrier to convey traffic information on regular FM broadcasts; these data could be decoded by an addition to the standard radio receiver (French, 1987).

The United Kingdom's AUTOGUIDE is an attempt to use proximity beacons which transmit route information to the vehicle receiving information on vehicle type, destination and route preference (e.g., shortest distance, time, or no expressway). This system uses modern technology to update the FHA's 1960s electronic route-guidance system (ERGS), which utilized radio-emitting proximity beacons that provided visual route guidance in the form of written instructions and directional arrows at key points along the route (Jeffery, Russam and Robertson, 1987; Rosen, Mammano and Favout, 1970).

EVA by Bosch-Blaupunkt is an electronic 'traffic pilot' which utilizes an LCD and voice synthesized directions to tell the driver, in a self contained unit, where she or he is, direction and distance to the destination, and what is the shortest route (Pilsak, 1986). Autodriver Leading and Information System (ALI-SCOUT) is a German effort that is dependent upon proximity beacons and dead reckoning with map matching, which allows navigation between far-spaced beacons; route guidance is presented with simple graphics (French, 1987).

Japanese efforts

Japan has special problems related to navigation. There are currently not enough roads, so construction activity is high, but updating the map database will be difficult, since few roads are named, and street names are not used in the address system in Japan (Shibata, 1989).

The Road/Automobile Communication System (RACS) is aimed toward improving roadway and traffic information for drivers, as well as communication between drivers and those outside the roadway system in Japan. The location system uses magnetic sensors and odometers for dead reckoning (with map matching) between beacons which provide exact location when the vehicle is nearby. Communication to and among vehicles is also to be implemented regarding traffic status (Shibata, 1989). Another advanced approach being taken in Japan, the Advanced Mobile Traffic Information and Communications System (AMTICS) combines real-time traffic information, routing advice and traffic overlays (Willis, 1990).

United States efforts

The Etak Navigator®, which has already been described in some detail, represents a private effort which utilizes dead reckoning to locate the vehicle. With this type of system, the current location of the vehicle is programmed into the computer by the driver, and the system keeps track of location from that point forward based on the distance and direction travelled using a magnetic compass and differential odometry (French, 1986). The major advantage of the dead reckoning paradigm is that no central external online source of information is required, and so such systems can be produced and sold as a self-contained navigation unit to individual consumers.

Dead reckoning systems tend to accumulate slight errors over time. As a result, artificial intelligence in the form of map matching is usually employed. The map-matching or augmentation comes into play when, for example, the system indicates that the current location is consistently parallel to but not directly on one of the roadways in the mapped area. After a short time, the augmentation system 'assumes' that the vehicle is on the nearby road and correctly updates its position; this type of augmentation is illustrated in Figure 15.3.

Willis (1990) has discussed intelligent vehicle/highway systems or IVHS which includes a variety of advanced traffic management systems (ATMS) and advanced driver information systems (ADIS). They may reduce urban traffic, travel time, pollution, fuel use and traffic related accidents, thus increasing highway transportation productivity.

The Navstar Global Positioning System (GPS) was developed by the Department of Defense and considered for concept cars by several US automobile manufacturers (for example, it was the basis for Chrysler's Laser Atlas and Satellite System, CLASS). The GPS system has 18 satellites which are placed such that at least four will always be in range from any point on earth to identify the location of the vehicle and synchronize the time accurately. Some form of dead reckoning will still probably be required due to signal distortions caused by buildings and other large structures (French, 1987). PATHFINDER is an effort put forth by the FHA as well as the US automobile industry to investigate the feasibility of combining in-vehicle navigation with motorist information to allow the most efficient use of the least crowded arteries (Koltnow, 1989).

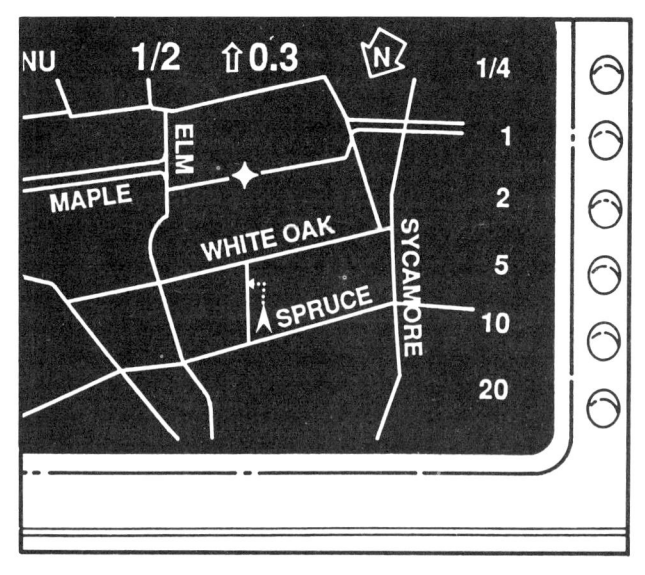

Figure 15.3. Intelligent map-matching augmentation

Conclusion

Although paper maps have been successfully used by drivers for a long time, they have certain limitations which can be especially daunting for some. These weaknesses and driver apprehensions may be overcome via the implementation of electronic vehicular navigation aids. Another reason for the push in recent years for new navigation technology is overflowing traffic which has been a serious problem throughout the world's metropolitan centers; without radical technological or regulatory change, this problem can only worsen.

Future directions for research

Vehicle location technology

Navigation technologies can employ any of several methods of locating the vehicle within the defined area. Some require massive infrastructure investment, whereas others rely on self-contained dead reckoning units which may accumulate error over time, requiring intelligent augmentation and occasional recalibration. Still others make use of satellite location, which can be highly accurate, but the signal can be distorted due to vehicle proximity to large structures such as bridges and tall buildings. Because of these technical advantages and limitations, successful systems of the future will likely make use of two or more of these location technologies in a complementary fashion (French, 1987).

Whatever methods are implemented, the overall savings in improved traffic efficiency would have to justify the per car expense (probably paid for directly by the auto consumer) as well as any infrastructure investment (e.g., beacons, central computers, transmitters, etc.). However, it is important to remember that traffic flow is smoothed even if only some of the drivers in an area are making use of efficiency enhancing navigation techniques.

Human factors issues

Research must also continue to search for the best way to present navigation information to the driver. Navigation aids are packaged in a variety of ways; some include only basic information, whereas others include a moving-map display, while others actually determine the best route to take to reach the destination.

Some of the following issues have been addressed, but all want definitive answers. One issue concerns weighing the relative merits of visual-spatial vs. auditory-verbal information presentation methods. It could be argued that spatial information is more compatible with the navigation task, yet if explicit directions are provided, then a purely auditory system could guide the driver along the route obviating the potential for visual distraction posed by a visual display.

Also, if a map is used, then what properties should it have (i.e., how many scale levels, what is shown at each level, is there automatic scale level selection software, etc.)? Should the orientation be heading-up, north-up, or some combination (see Aretz, 1990), and should these options be user selectable? Is the head-up display (HUD) format suitable for the presentation of complex map information, for written verbal directions? If it is suitable, then any visual distraction may be substantially reduced, but this idea needs to be proven before it is implemented.

Whatever direction is taken, if a visual display is used and especially if it is any type of map display, then the impact on safety must be considered. Several in-vehicle studies have been conducted, but results so far are inconclusive. They generally show that the gaze patterns of drivers are somewhat changed when using map displays. Still, the ultimate effects are unclear, since drivers seem to be incorporating these additional visual attentional demands into the driving task in a reasonably safe fashion.

Acknowledgments

I would like to thank Dr Walter W. Wierwille for his invaluable direction in this area of research, Dr Ruth J. Arnegard for proofreading the manuscript, and Mark Ransom for his help with the figures.

Note

1   The psychomotor act of steering, considered a separate behavior from the spatial-cognitive act of navigation, will not be discussed in this chapter.

References

Antin, J. F., Dingus, T. A., Hulse, M. C. and Wierwille, W. W., 1988, The effects of spatial ability on automobile navigation, In Agazadeh, F. (Ed.) *Trends in Ergonomics/Human Factors V*, pp. 241–8, North-Holland: Elsevier.

Antin, J. F., Dingus, T. A., Hulse, M. C. and Wierwille, W. W., 1990, An empirical evaluation of the effectiveness and efficiency of an automobile moving-map navigational display, *International Journal of Man/Machine Studies*, **33**, 581–94.

Aretz, A. J., 1988, A model of electronic map interpretation, in *Proceedings of the Human Factors Society 32nd Annual Meeting*, pp. 130–4, Santa Monica, CA: Human Factors Society.

Aretz, A. J., 1989, Spatial cognition and navigation, In *Proceedings of the Human Factors Society 33rd Annual Meeting*, pp. 8–12, Santa Monica, CA: Human Factors Society.

Aretz, A. J., 1990, Map display design, in *Proceeedings of the Human Factors Society 34th Annual Meeting*, pp. 89–93, Santa Monica, CA: Human Factors Society.

Avolio, B. J., Kroeck, K. G. and Panek, P. E. 1985, Individual differences in information-processing ability as a predictor of motor vehicle accidents, *Human Factors*, **27**, 577–87.

Byrne, R. W., 1979, Memory for urban geography, *Quarterly Journal of Experimental Psychology*, **31**, 147–54.

Chase, W. and Chi, M., 1980, Cognitive skill: Implications for spatial skill in large-scale environments, (Technical Report No. 1), Pittsburgh, PA: University of Pittsburgh Learning and Development Center.

Dingus, T. A., Antin, J. F., Hulse, M. C. and Wierwille, W. W., 1989, Attentional demand of an automobile moving-map navigation system, *Transportation Research-A*, **23**, 301–15.

French, R. L., 1986, In-vehicle route guidance in the United States: 1910–1985, *IEE Second International Conference on Road Traffic Control*, pp. 6–9, London: IEE.

French, R. L., 1987, The evolving roles of vehicular navigation, *Navigation*, **34**, 212–28.

Harwood, K., 1989, Cognitive perspectives on map displays for helicopter flight, in *Proceedings of the Human Factors Society 33rd Annual Meeting*, pp. 13–17, Santa Monica, CA: Human Factors Society.

Hughes, P. K. and Cole, B. L., 1986, What attracts attention when driving? *Ergonomics*, **29**, 377–91.

Jeffery, D. J., Russam, K. and Robertson, D. I., 1987, Electronic route guidance by AUTOGUIDE: The research background, *Traffic Engineering & Control*, **28**, 525–9.

King, G. F., 1986a, Driver attitudes concerning aspects of highway navigation, *Transportation Research Record 1093 Visibility and Highway Navigation*, pp. 11–21, Washington, DC: Transportation Research Board National Research Council.

King, G. F., 1986b, Driver performance in highway navigation tasks, *Transportation Research Record 1093 Visibility and Highway Navigation*, pp. 1–11, Washington, DC: Transportation Research Board National Research Council.

Koltnow, P. J., 1989, Advanced vehicle and highway technology and research leadership opportunities, *Transportation Quarterly*, **43**, 495–509.

Kuipers, B., 1978, Modeling spatial knowledge, *Cognitive Science*, **6**, 129–53.

Kuipers, B., 1983, Modeling human knowledge of routes: Partial knowledge and individual variation, in *Proceedings, AAAI 1983 Conference*, pp. 216–9, Menlo Park, CA: The American Association of Artificial Intelligence.

Labiale, G., 1989, *Influence of In-car Navigation Map Displays on Drivers' Performances* (SAE Tech Paper 891683), Warrendale, PA: Society of Automotive Engineers.

Labiale, G., 1990, In-car road information: Comparisons of auditory and visual presentations, in *Proceedings of the Human Factors Society 34th Annual Meeting*, pp. 623–7, Santa Monica, CA: Human Factors Society.

McGranaghan, M., Mark, D. M. and Gould, M. D., 1987, Automated provision of navigation assistance to drivers, *The American Cartographer*, **14**, 121–38.

Pilsak, O., 1986, EVA—an electronic traffic pilot for motorists, *Electronic Displays and Information Systems, SP-654*, No. 860346, pp. 99–106, Warrendale, PA: Society of Automotive Engineers.

Reed, D., 1989, Washington report intelligent vehicles, *Automotive Engineering*, **97**, 19.
Rosen, D. A., Mammano, F. J. and Favout, R., 1970, An electronic route-guidance system for highway vehicles, *IEEE Transactions on Vehicluar Technology*, **VT-19**, 143–52.
Shibata, M., 1989, Development of a road/automobile communication system, *Transportation Research-A*, **23**, 63–71.
Smith, V. J., 1978, What about customers? A survey of mobile telephone users, in *Proceedings of the 28th IEEE Vehicular Technology Conference*, pp. 190–3, New York: Institute of Electrical and Electronic Engineers.
Stokes, A., Wickens, C. and Kite, K., 1990, *Display Technology Human Factors Concepts*, Warrendale, PA: Society of Automotive Engineers, Inc.
Streeter, L. A. and Vitello, D., 1986, A profile of driver's map-reading abilities, *Human Factors*, **28**, 223–39.
Streeter, L. A., Vitello, D. and Wonsiewicz, S. A., 1985, How to tell people where to go: Comparing navigational aids, *International Journal of Man-Machine Studies*, **22**, 549–62.
Thorndyke, P. W. and Hayes-Roth, B., 1982, Differences in spatial knowledge acquired from maps and navigation, *Cognitive Psychology*, **14**, 560–89.
Thorndyke, P. W. and Stasz, C., 1980, Individual differences in procedures for knowledge acquisition from maps, *Cognitive Psychology*, **12**, 137–75.
Tversky, B., 1981, Distortions in memory for maps, *Cognitive Psychology*, **13**, 407–32.
Van Aerde, M. and Case, E. R., 1988, Individual driver route guidance systems and their implications for traffic control, *Canadian Journal of Civil Engineering*, **15**, 152–6.
Walzer, P. and Zimdahl, W., 1988, European concepts for vehicle safety, communication and guidance, *1988 Convergence International Congress on Transportation Electronics Proceedings*, pp. 91–95, VTS-IEEE and SAE, Inc.
Wickens, C. D., 1984, *Engineering, Psychology and Human Performance*, Columbus, OH: Merrill.
Wierwille, W. W., Hulse, M. C., Fischer, T. J. and Dingus, T. A., 1990, Visual adaptation of the driver to high-demand driving situations while navigating with an in-car navigation system, In Gale, A. G., *et al.* (Eds) *Vision in Vehicles—III*, Amsterdam: North Holland Press.
Willis, D. K., 1990, IVHS Technologies: Promising palliatives or popular poppycock? *Transportation Quarterly*, **44**, 73–84.
Zwahlen, H. T. and DeBald, D. P., 1986, Safety aspects of sophisticated in-vehicle information displays and controls, in *Proceedings of the Human Factors Society 30th Annual Meeting*, pp. 256–60, Santa Monica, CA: Human Factors Society.

16

Age, display design and driving performance

Daniel Imbeau, Walter W. Wierwille and Yves Beauchamp

Introduction

Following advances in microprocessors, electronics and communications, new technologies in the form of driver information systems are being integrated into the automobile at an increasing rate (Wierwille and Peacock, 1989). These new systems may offer important opportunities to enhance safety, improve driver convenience and reduce travel time and fuel consumption. However, the success of these new technologies is highly dependent on the ergonomics of the driver interface.

The temptation to display even simple information with the new systems (e.g., navigation and route guidance systems, head-up displays, cellular phones, multiple function CRTs, etc.) appears to be hard to resist since such an approach increases significantly the quantity and type of information related to driving that can be made available to the driver (Stokes, Wickens and Kite, 1990). This approach, virtually inevitable, also permits the reduction of the number of isolated displays and controls by integrating these devices into the instrument panel. However, instrument panel display reading requires visual attention resources, and hence is likely to interfere with the primary driving task (Stokes *et al.*, 1990; Zwahlen and DeBald, 1986). Zwahlen and Balasubramanian (1974) have shown that vehicle lateral deviation from normal lane position increases with time spent looking away from the external visual scene. Instrument panel displays of word or text information must therefore be designed to minimize visual attentional demand as much as possible.

Extensive data on how to design visible and legible alphanumeric information to increase acquisition performance and minimize fatigue is currently available (Human Factors Society, 1988; Society of Automotive Engineers, 1984; Mourant and Langolf, 1976; Helander, 1986; Easterby and Zwaga, 1984). For instance, character size, luminance and contrast are generally considered to be key parameters in visual display design (Mourant and Langolf, 1976; Fowkes, 1984). Some quantitative information relating the isolated effect of each of the various display design parameters (e.g., character size, contrast, color, etc.) on performance can also be found in this literature. However, quantitative information about the combined effect of several of these parameters as well as information on the interactions between these parameters is scant. For instance, what is the increase in reading time associated with simultaneous and fixed decreases in both character size and display contrast (chromatic and luminance)? Moreover, little of the data available relate display design parameters and information acquisition performance in a driving context specifically. A variation of several hundredths of a second in glance time at an instrument panel while driving could make a difference in terms of accident rates for the vehicle fleet.

The proportion of older people in North America has been increasing steadily since the turn of the century and this trend is expected to continue. More specifically, within the next 10 years, the forecast increase for the population segment aged 55 and over is 11.6 percent, whereas for the segment aged 75 and over, the increase will be 26.2 percent (US Department of Commerce, Bureau of the Census, 1988). Since driving is the most common form of transportation over age 65, the proportion of older persons on the road will increase greatly (Czaja and Guion, 1990).

There is much literature documenting the commonly known 'slowing-with-age phenomenon' that is, the slowing of behavior commonly associated with decrements observed with increasing age in motor skills, information processing, vigilance, the ability to focus attention and the visual function (Salthouse, 1985; Sekuler, Kline and Dismukes, 1982; Kline and Schieber, 1985; NRC, 1988; Czaja and Guion, 1990). However, as recently suggested, human factors information on aging currently available in the literature '... is often prescriptive and seldom in the quantitative form most needed to answer application questions' (Smith, 1990, p. 512). Most of the research on aging has been aimed primarily at understanding age differences and the aging process rather than at human factors applications. The result is an extensive body of literature on aging that provides relatively little quantitative human factors data that can be used by equipment and systems designers (Smith, 1990). This situation is particularly true in the automobile driving context (Smith, 1990). The literature available on the design of vehicles for the older drivers describes most of the pertinent problems and prescribes 'threshold' design guidelines, whenever possible, that should ensure acceptable performance for this segment of the population (Yanik, 1989; Mortimer, 1989; Rockwell et al., 1988). However, this literature does not relate these design guidelines to actual performance estimates which are necessary to truly evaluate and compare design alternatives. Also, this literature does not adequately address the effect of color on performance, despite the fact that a wide variety of colors are used in automobile instrument panels (Stokes et al., 1990).

Recently, Imbeau et al. (1989) investigated the effect of several display design parameters including color, brightness, character size and word complexity on various performance measures with drivers in three age groups (young, middle and older) during a simulated nighttime driving task. Previously, the parameters investigated (with the exception of color) had been shown to affect information acquisition performance greatly (Olzak and Thomas, 1986; Carr, 1986; Mourant and Langolf, 1976; Klare, 1963). For instance, Welford (1980) reported research results that demonstrate an age related increase of about 27 percent both in reaction time (choice and initiation of a response) and in digit identification time between age groups of 18–28 years and of 65–75 years.

The experimental paradigm in the simulator study was a word recognition task that simulated the acquisition of a simple legend or label information from the instrument panel while driving. Since the glance time at an automobile instrument panel determines the amount of visual attention left for the driving task, it was speculated that the display design parameters studied would directly affect driving performance (Zwahlen and Balasubramanian, 1974; Zwahlen and DeBald, 1986). The goal of this study was to quantify the variations in both information acquisition performance and in driving performance associated with the manipulation of these design parameters for different age groups.

The Imbeau et al. (1989) paper did not stress quantitative performance data. Instead, it described and commented on the various effects that were significant in the ANOVAs. Based on this discussion, several 'threshold' design recommendations were drawn. One of the salient results of the study was that relatively important age-related decrements occurred in the various performance measures. Also, as expected, the design parameters investigated affected performance, each to a different extent and also in the form of interactions. Table 16.1 summarizes the effects that were significant in the study.

This chapter is based on the work of Imbeau et al. (1989). It is a further analysis of the simulator study data. It describes quantitatively the effects on performance of the variables studied. The main goal here is to provide designers with integrated quantitative performance data that will help them answer design questions and evaluate design alternatives. These data are presented in the form of regression models that can be used in human factors design applications including computer simulation.

Overview of the study

Forty subjects (20 males and 20 females) participated in the study, 16 (eight males and eight females) in a pilot study and 24 in the main experiment. Those who participated in the pilot study did not take part in the main experiment. All had valid Virginia driver's licenses. Subjects in the main experiment were

Table 16.1. *Significant effects from the analyses of variance in the Imbeau et al. (1989) study*

Effect	Dependent measures		
	Glance[1]	Vocal[2]	Lanedev.[3]
Age	✓	✓	✓
Character size	✓	✓	✓
Color	✓	✓	
Brightness	✓	✓	✓
Word complexity	✓	✓	
Age × character size		✓	
Age × brightness	✓	✓	
Character size × color	✓	✓	
Character size × brightness	✓	✓	✓
Color × brightness		✓	
Character size × word complexity	✓	✓	
Age × character size × brightness	✓	✓	✓
Character size × brightness × color		✓	

Notes: [1] Glance time at the display.
[2] Vocal response time.
[3] Standard deviation of lane position.

distributed within three age groups: younger (20–30), middle (31–50), and older (51–73). The corresponding mean ages were 23.6, 41.9, 58.4 years, respectively with standard deviations of 2.7, 6.4 and 7.0 years. All had a minimum of 20/40 static visual acuity (corrected where necessary) and normal color vision, as verified with a Titmus II vision tester. These are the minimum requirements to obtain a driver's licence in most of the United States. A short hearing test was given to the subjects to ensure they could hear the click sound of the projectors' signaling a new stimulus in the experimental conditions.

In the main experiment, the task of the subjects was to read words aloud that were presented at eight-second intervals on two displays while driving a simulated vehicle in nighttime conditions. (The words were displayed for the entire eight-second interval). Random lateral wind gusts with constant spectral characteristics and probability density function were introduced to provide light task loading. The displays on which the words were presented had been positioned on each side of the driver, emulating written legends found on automobile instrument panels (i.e., about 20 degrees horizontal and 15 degrees vertical (downward) from the 'straight ahead' line of sight). The words were presented in different chromaticities (eight levels), brightnesses (two levels), character sizes (four levels) and word complexity (two levels). These independent variables were the within-subject factors while age group (three levels) and gender were the between-subject variables.

The brightness levels used on the displays had been determined in a pilot study by subjects representative of the older and younger segments of the driver population (Imbeau and Wierwille, 1989). In this pilot study, each driver selected a brightness level that was subjectively comfortable for nighttime driving and that was equivalent across all chromaticities. The reader is referred to Imbeau et al. (1989) for the chromaticity coordinates of the eight colors used. Luminance of the stimuli ranged from 1.65 cd/m^2 to 49 cd/m^2. The luminance contrast ratios (L_{max}/L_{min}) on the displays were very high. (The high luminance contrast ratios occurred because subjects were instructed to set the brightness at levels that were appropriate for the nighttime driving scene, which had a maximum luminance in the roadway of 4.5 cd/m^2. Since the background of the slide was black, with luminance of $L_{min} = 0.005$ cd/m^2, very high contrast ratios resulted). The four character sizes used subtended 7, 11, 17, and 25 minutes of arc (arcmin) at the driver's eye, respectively. The 7 arcmin size is smaller than currently used in automobiles and was included in the experiment to provide additional information on performance under extreme conditions. The font used corresponded closely to the MIL-STD-18012B (a capitalized type) with a width to height ratio of 78 percent and a stroke width to height ratio of 11 percent. Word complexity was defined and used as a combination of frequency of occurrence in the English language and number of syllables. In the experiment, low complexity consisted on one syllable words occurring more than 20 times per million words read, while high complexity consisted of two-syllable words occurring between six and 20 times per million words read.

Upon presentation of each word, the following performance measures were obtained:

- time to vocalize the response (*vocal*, in seconds);
- glance time at the display (*glance*, in seconds);
- Standard deviation of lane position (*lanedev*, in m).

In total, 3072 data points were generated in this experiment. The reader is referred to the Imbeau et al. (1989) study for more detail on the experimental setting and procedure.

Results

Since several regression analyses were performed on the same data, that is, one regression analysis for each dependent variable, a conservative Bonferroni criterion (Keppel, 1982; Finkelman, Wolf and Friend, 1977) of three (for the three dependent variables) is used throughout the results section to adjust the alpha level in the statistical tests. The alpha levels reported are thus corrected.

Variable recoding

For the purpose of the regression analyses, the categorical variables (except word complexity and gender) were recoded into original interval type variables. Age group was replaced by age of the subject while the character size was replaced by the corresponding angle subtended at the eye in minutes of arc (arcmin). The inverse of the subtended angle was actually used in the regression analyses since it correlated better with the dependent variables.

Brightness (which had been determined subjectively) and color were replaced by a corresponding ΔE distance, that is, a measure of the perceptual difference between the display background (black) and the color of the characters appearing on the display and represented by the distance between two points in a three dimensional color space (Wyszecki and Stiles, 1982). Such a distance integrates the three dimensions of the visual stimulus, that is, luminance, dominant wavelength and excitation purity, and therefore constitutes a comprehensive metric of the difference between two visual stimuli.

Based on earlier work (Lippert, 1986; Lippert and Snyder, 1986), the Human Factors Society (1988) suggested the use of a ΔE (CIE Yu'v') metric derived for the 1976 CIE Uniform Color Diagram to assess the legibility of characters on a CRT display. However, ΔE distances as computed in the 1976 CIE Uniform Color Space (CIE UCS L*u*v*) were found to be better overall predictors of the dependent measures in the simulator study and therefore were retained for the regression analyses. The CIE UCS L*u*v* metric has been used extensively as a basic tool for the design of self luminous color displays (Merrifield and Silverstein, 1986). The inverse of ΔE was actually used in the regression analyses since it correlated better with the dependent measures.

Color discrimination is influenced by field size of the colored image (Wyszecki and Stiles, 1982; Carter, 1989). Given the fact that the character sizes used in the experiment are much smaller than the 2- and 10-degree standard observer data that form the basis of current predictive color models, important errors in estimated color difference can result. Therefore, a small-field correction assuming three weighting factors ku*, kv*, and kL* that represent the relationship between field size angular subtense and the sensitivity of the red/green, violet/green-yellow, and light/dark visual channels, respectively was used (Merrifield and Silverstein, 1986). The reader is referred to Merrifield and Silverstein (1986) or to Carter (1989) for the exact values of these weighting factors.

The small field correction described above was originally developed for fairly symmetrical fields such as symbols, not for characters. However, since no such correction for characters was found in the literature on color, the correction reported by Merrifield and Silverstein (1986) was used directly. The corrected ΔE distances correlated better with the performance measures than the uncorrected distances. For instance, the uncorrected ΔE distances yielded a Pearson correlation of 0.09 (statistically different from 0, $p < 0.005$) whereas the corrected distances yielded a correlation of 0.35 (statistically different from 0, $p < 0.005$). The corrected (CIE UCS L*u*v*) ΔE metric equation for small fields that was used is as follows:

$$\Delta E = ((kL^* \Delta L^*)^2 + (ku^* \Delta u^*)^2 + (kv^* \Delta v^*)^2)^{0.5}$$

where:

ΔL^* = difference between L* target and L* background
Δu^* = difference between u* target and u* background
Δv^* = difference between v* target and v* background
$L^* = 116 (Y/Ym)^{0.33} - 16$
$u^* = 13 L^* (u' - u'_n)$
$v^* = 13 L^* (v' - v'_n)$
Y is the luminance
Ym is the maximum luminance ($= 50$ cd/m^2 in our experiment)
u' and v' are the target chromaticity coordinates from the CIE 1976 Uniform Color Diagram
$u'_n = 0.1978$ (CIE 1976 UCS, u' of a D65 illuminant)
$v'_n = 0.4684$ (CIE 1976 UCS, v' of a D65 illuminant)

The resulting ΔE distances ranged from 5 to 28, for the 7 arcmin character size, from 9 to 45 for the 12 arcmin character size, from 13 to 68 for the 17 arcmin character size, and from 17 to 90 for the 25 arcmin character size. It is interesting to note that the Human Factors Society (1988) does not provide any small-field (or target size) correction for its ΔE metrics for both legibility and color discrimination.

Behavioral variable

During the regression analyses, it became necessary to include a categorical variable to account for a behavior effect that resulted in a model shift at some of the data points (Myers, 1986). These data points corresponded to unusually large responses for both the *glance* and *vocal* dependent measures. These responses were associated with the subject bending towards the visual display in order to read the low luminance and/or small character sizes presented. The time required to bend represented a fairly constant delay that was systematically added on top of both the *glance* and the *vocal* measures. These data points accounted for less than 4 percent of the database and were spread across several of the treatment conditions.

Regression models

Least squares multiple regression analyses were performed using a backward elimination procedure for each dependent variable (Wilkinson, 1989). For a given measure, the analysis began by fitting the most complete model, that is, a model including all candidate predictors including the two- and three-way interaction effects. Then, working backwards the effect that was least useful was eliminated and a smaller model was fitted. For a given size model, the effects previously removed were replaced one at a time to see if they then contributed significantly. This iterative procedure was pursued until a satisfactory model was obtained. The final models include only predictors for which coefficients (βs) are significantly different from zero at the $\alpha = 0.05$ level. The final models are presented in Tables 16.2, 16.3 and 16.4. ANOVAs performed on each of the four models showed that they were all highly significant ($p < 0.0001$).

It is worth noting that the total number of degrees of freedom for the regression model of *vocal* is smaller than for *glance* and *lanedev*. This situation occurred because a number of data points for which a glance time had been recorded had no corresponding vocal response, that is, the subject did not

Table 16.2. Regression results for glance (*glance time at the display*)

Source	df	SS	MS	F
Model	7	442.7	63.2	999.9***
Error	2962	161.9	0.05	
Total	2969	604.6		

Note: *** $p < 0.0001$

$R^2 = 0.732$

Term[1]	Estimate	df	SS	F
Intercept	0.6661			
Gender	−0.0247	1	1.8	32.4***
Inverse of character size	−2.7306	1	5.9	108.1***
Inverse of ΔE	−2.3667	1	1.8	33.0***
Age × inverse of character size	0.1066	1	59.1	999.9***
Inverse of character size × inverse of ΔE	44.4700	1	9.6	175.8***
Body movement	1.0632	1	115.3	999.9***
Word complexity	−0.0362	1	1.0	17.8***

Notes: $p < 0.0001$
[1] Gender (multiply by: 0 = female, 1 = male), A = age (years), CS = character size (arcmin), WC = word complexity (multiply by: 1 = low, 0 = high), body movement (multiply by: 0 = normal, 1 = long), ΔE = distance (see text for reference).

Table 16.3. Regression results for vocal (vocal response time)

Source	df	SS	MS	F
Model	7	2413.4	344.76	873.9***
Error	2949	1163.3	0.39	
Total	2956	3576.6		

Note: *** $p < 0.0001$

$R^2 = 0.675$

Term[1]	Estimate	df	SS	F
Intercept	2.7869			
Gender	0.0671	1	13.1	33.1***
Age	−0.0105	1	13.2	33.4***
Inverse of character size	−8.1698	1	24.0	60.9***
Inverse of ΔE	−3.9937	1	5.1	12.9**
Age × inverse of character size	0.3125	1	92.2	233.7***
Inverse of character size × inverse of ΔE	77.7869	1	28.6	72.5***
Body movement	−1.2356	1	820.1	999.9***

Notes: ** $p < 0.001$
*** $p < 0.0001$
[1] Gender (multiply by: 0 = female, 1 = male), A = age (years), CS = character size (arcmin), body movement (multiply by: 0 = long, 1 = normal), ΔE = distance (see text for reference).

respond within the eight-second time interval during which the word was displayed. These data points were not included in the regression analyses for *vocal*.

Discussion

Glance time

The regression model for *glance* explained 73 percent of the variance in the data as shown by the multiple R^2 (Table 16.2). Both the *body movement* effect and the *age × inverse of character size* interaction effect were major contributors as shown by the large sum of squares. The root mean square error for glance time was 0.23 second for a mean response of 0.94 seconds. Glance time at the display ranged from 0.27 second to 4.00 seconds.

Body movement

The bending movement towards the display added on average a little over one second to the total glance time at the display. It can be expected that subjects above the age of 50 are likely to bend towards the display to read character

Table 16.4. Regression results for lanedev (standard deviation of lane position)

Source	df	SS	MS	F
Model	8	33.3	4.16	92.5***
Error	2961	133.2	0.04	
Total	2969	166.5		

Note: *** $p < 0.0001$

$R^2 = 0.199$

Term[1]	Estimate	df	SS	F
Intercept	0.2568			
Gender	0.0346	1	3.4	75.6***
Age	0.0226	1	8.7	193.2***
Age × Age	−0.0003	1	9.1	201.6***
Inverse of ΔE	−2.2598	1	0.9	19.0***
Inverse of character size × inverse of ΔE	11.5535	1	0.7	16.5***
Age × inverse of character size	0.0091	1	0.2	4.2*
Age × inverse of ΔE	0.0362	1	1.1	24.2***
Body movement	0.0888	1	0.8	17.9***

Notes: * $p < 0.05$
*** $p < 0.0001$
[1] Gender (multiply by: 0 = female, 1 = male), A = Age (years), CS = character size (arcmin), body movement (multiply by: 0 = normal, 1 = long), ΔE = distance (see text for reference).

sizes smaller than 11 arcmin at low luminance levels (i.e., less than 7 to 8 cd/m² on a dark background). For adequate design, an engineer would normally consider that the subject does not have to bend towards the display to read it accurately and therefore, would set this variable to zero in the regression model.

Both the *age × inverse of character size* and the *inverse of character size × inverse of ΔE* interactions each accounted for a portion of the variance that was more important than the proportion explained by the simple effects (Table 16.2). Figure 16.1 shows the *age × inverse character size* interaction effect on glance time as predicted by the model. As shown, even with the largest character sizes the predicted age difference between a 20-year-old and a 70-year-old is about 0.2 second for this effect alone. This difference increases markedly for sizes smaller than 16 arcmin. Figure 16.2 shows the *inverse of character size × inverse of ΔE* interaction effect on glance time predicted by the model. This figure shows that for character sizes above 16 arcmin and ΔE values above 30, performance averaged over age generally reaches an asymptote. This ΔE CIE L*u*v* value is somewhat lower than that specified by the Human Factors Society (1988) for color coding—ΔE CIE L*u*v* = 40—but is larger than the more recent value of 20 reported by Carter (1989). The larger ΔE value found here may have occurred because letters do not constitute complete shapes like symbols and therefore require a larger color difference for

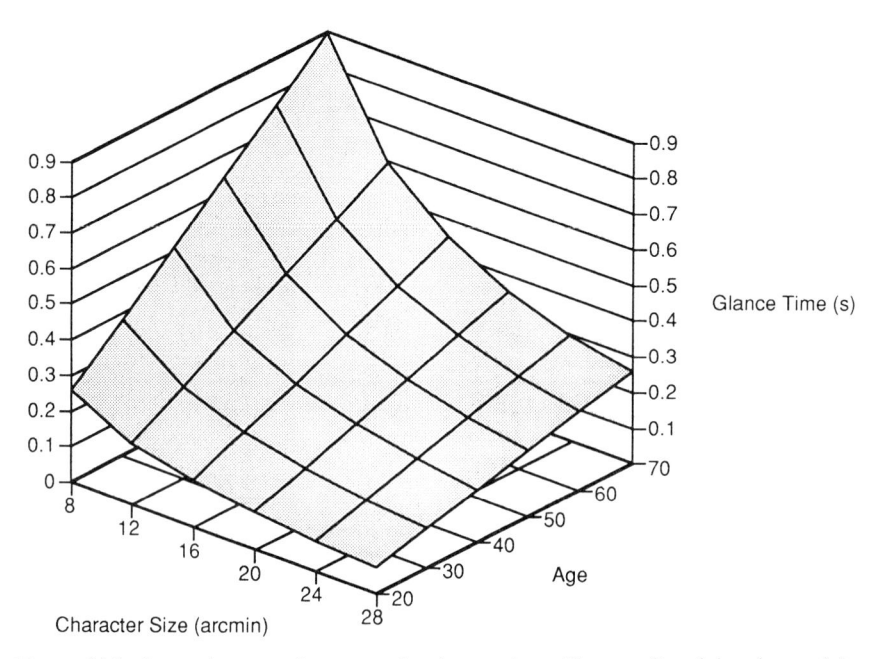

Figure 16.1. Age × inverse character size *interaction effect predicted by the model on glance time* (glance)

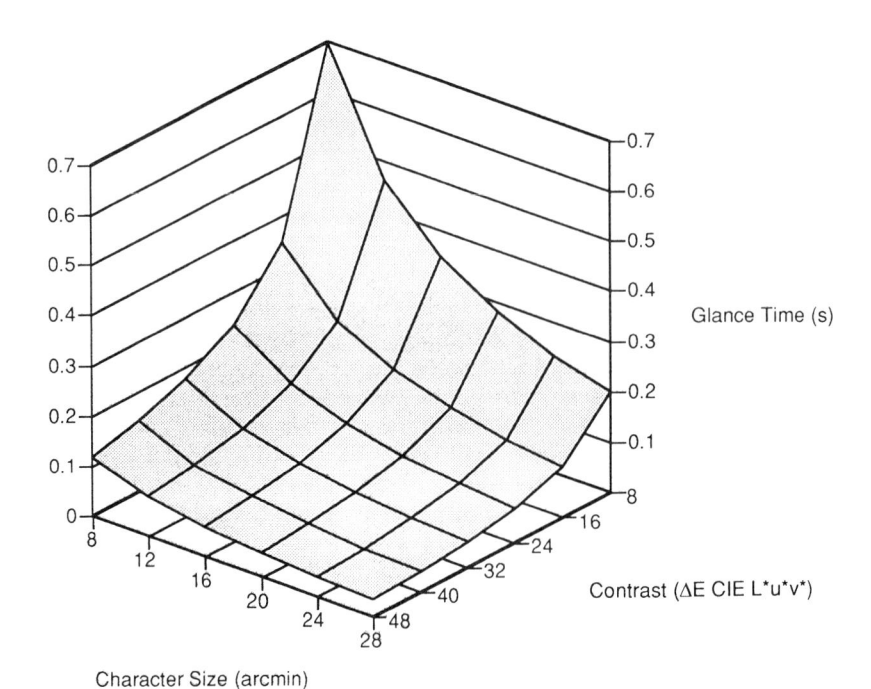

Figure 16.2. Inverse of character size × inverse of ΔE *interaction effect predicted by the model on glance time* (glance)

equivalent symbol discrimination. More research is needed to clarify this situation. This CIE L*u*v* value of 30 is difficult to compare with the value specified by the Human Factors Society (1988) for adequate legibility since the latter was derived for a different color space (ΔE Yu'v' $= 100$) and also does not consider any small field correction.

Simple effects

Both the *gender* and the *word complexity* effects each had a small influence on glance time (less than 0.07 second combined). The same conclusion does not apply to the *inverse of character size* effect, however.

Given display parameters typical of good reading conditions comparable to those found in laboratory experiments on word recognition (e.g., 25 arcmin character size, one two-syllable word displayed, and amber 21 cd/m^2 characters displayed on a dark background: $\Delta E \approx 60$), the predicted glance time at the display assuming no bending of the torso is involved, for accurate acquisition of the word is 0.63 second for a 25-year-old male and 0.82 second for a 70-year-old male. This represents a 30 percent performance difference. Such a result is in general agreement with laboratory studies on word recognition and age that report an average 27 percent difference between groups of age 18–28 and 65–75 (Welford, 1980). This difference would be primarily associated with the slowing of the perceptual and information processing functions.

For a 16 arcmin character size, the corresponding values are 0.64 second for the 20-year-old and 0.94 second for the 70-year-old, a 47 percent difference. This time, the increase in the difference can certainly be linked to age related differences in the visual function (e.g., speed of accommodation); which is more taxed as visual conditions degrade (e.g., decrease in character size; Kline and Scheiber, 1985). This result underscores the importance to provide large character sizes to aging subjects in order to minimize performance degradation associated with reduced visual capabilities. As use of the model demonstrates, increased contrast (chromatic and achromatic) may compensate to a certain extent for a smaller character size.

The young subject glance time value for the 25 arcmin character size (0.63 second) coincides fairly well with the upper bound of the visual dwell time range generally reported in the literature for word recognition tasks performed in laboratory conditions, that is, 260–620 msec depending on reading material and experimental conditions (Carr, 1986; Card, Moran and Newell, 1986; Cavanaugh, 1972). For instance, the Model Human Processor (Card, Moran and Newell, 1986) would predict an average glance time of 570 msec for this display with lower and upper bounds of 410 and 720 msec, respectively. It appears that the literature provides enough information to estimate the glance time required for a subject to recognize a short message displayed in good reading conditions. However, the information provided does not allow one to estimate precisely the global performance degradation that will result from precise decreases in both instrument panel display contrast (i.e., ΔE) and character size for an older driver. The model presented for glance time helps to fill this gap for the driving context.

Kurokawa and Wierwille (1990) conducted a study in which they measured the average glance time associated with instrument panel tasks (e.g., activation of a pushbutton, high accuracy adjustment) for varying levels of random crosswinds in a simulator environment. The glance times they measured in their study were slightly longer than those predicted by the model presented here, probably because their instrument panel tasks involved eye-hand movement coordination requiring more visual attention (i.e., hand movement guiding) than visual acquisition of information alone. Their results showed a significantly shorter glance time at the instrument panel when crosswinds were present during the simulated driving task than when none was present. Also, the average glance times associated with the varying crosswind levels they tested were not significantly different from one another. The average difference was about 33 percent with respect to the conditions where crosswinds were present.

These results would indicate that a driver will spend more time glancing at the instrument panel when the driving task is easier (no crosswinds) whereas with a more difficult driving task, glance time will be shortened by a relatively fixed amount that appears independent of the level of driving difficulty past a certain level. In the Imbeau *et al.* (1989) study, constant level random crosswinds were present during the driving task to introduce a light task loading, thereby ensuring that the subject would maintain attention to the primary driving task. If one assumes that the results of the Kurokawa and Wierwille (1990) study can be transferred to the Imbeau *et al.* (1989) study, then the glance time predictions from the model described above could be generalized to driving tasks of varying difficulty levels as described by varying levels of random crosswinds. This result would also indicate that for 'easy' driving conditions (no crosswinds), the glance times predicted by the model would likely be increased by about 33 percent.

If one assumes that acquisition of information from a distant road sign at night is similar to acquisition of information from a low luminance self luminous automobile instrument panel display in terms of visual attentional demand, then the model developed for *glance* can be used to estimate the letter height on road signs that will ensure enough time for adequate information acquisition. For instance, assume that the ΔE value resulting from a road sign being illuminated by automobile headlamps is on the order of 15 between 12 arcmin characters and their background (i.e., 24 cd/m^2 white letters on a standardized green 6 cd/m^2 background, with the 12 arcmin size being chosen because it would not require any bending in difficult reading conditions were it presented on an instrument panel). Then assuming that atmospheric visibility is good, the average time required for the acquisition of only one familiar two-syllable word by a 60-year-old driver is about 1.04 seconds. If the sign is designed for a 100 km/h maximum speed highway, the minimum distance this sign must be from the vehicle when the driver starts glancing at it is at least 44 m (144 feet). This assumes that the vehicle will be at 15 m (50 feet) from the sign when the driver finishes reading it. At this distance, letters must be about 15 cm high (5.9 inches) to subtend 12 arcmin at the eye, that is, a legibility standard of 5.1 cm per 15 m of sight distance (2 inches per 50 feet). If 95 percent of the population of that age were to be accommodated, then the letters would have to be 22 cm (8.6 inches) high, that is, a legibility standard of 5.3 cm per 15 m of sight distance (2.1 inches per 50 feet). The current road sign letter height standard is based on the assumption that 2.5 cm is legible at 15 m (1 inch at 50 feet; NRC, 1988). The value proposed by the model for the acquisition of a single word is twice the standard in current use for road signs of any length. The current standard defines a distance at which the characters are legible but does not consider whether or not the driver will have time to acquire the information. Figure 16.3 shows the predicted glance time for a single two-syllable word displayed as a function of contrast and character size for a 60-year-old driver. The performance degradation resulting from a combination of low contrast and small characters can be appreciated from this figure.

If the length of the message increases, then the time to acquire it will be longer if the driver is not familiar with it. This time may be estimated by assuming that one fixation is necessary for each word (Carr, 1986), the duration of each fixation being estimated with the model for *glance*. When the message to be acquired from the sign is long, visual attention must be shared between the road and the display to ensure proper control of the vehicle. Hence, fixations back to the road must be considered. The transition time between the display and the road can be estimated to be about 0.11 second (Hayes, Kurokawa and Wierwille, 1989). The average glance time to the road in normal driving conditions can be estimated to be about 1.0 seconds (Kurokawa and Wierwille, 1990). As the length of the information to be acquired increases, the distance at which the driver must initiate a visual glance also increases, which translates to even larger letters on the sign. Clearly, from these results the current standard is inappropriate for older drivers. This conclusion is supported by various other studies and reports

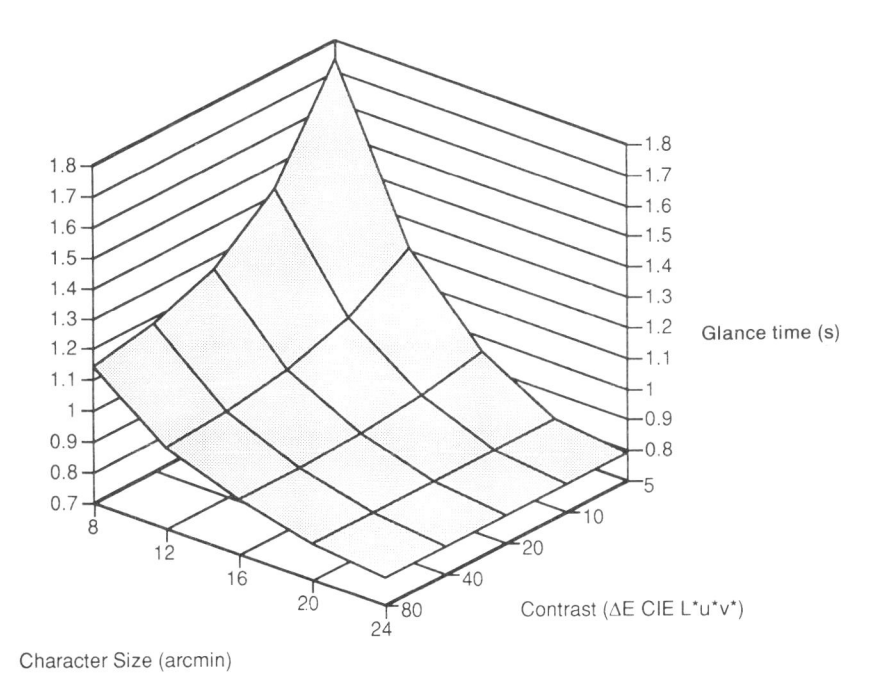

Figure 16.3. Predicted glance time (glance) *for the acquisition of a single two-syllable word as a function of character size and contrast for a 60-year-old driver*

(Sivak, Olson and Pastalan, 1981; NRC, 1988; Czaja and Guion, 1990). It appears that designing an adequately legible road sign with a complex message that is to be read by older drivers generally results in a sign with unfeasibly large dimensions.

Vocal response time

The regression model for *vocal* explained 67.5 percent of the variance in the data as shown by the multiple R^2 (Table 16.3). As for the *glance* measure, the *body movement* effect accounted for a substantial portion of the variance. Three other effects, two interactions (*age* × *inverse of character size* and *inverse of character size* × *inverse of* ΔE) and one simple effect (*inverse of character size*) also contributed significantly to the proportion of variance explained. The root mean square error for vocal response time was 0.62 second for a mean response of 1.83 seconds. Vocal response time ranged from 0.55 to 3.79 seconds, excluding the no responses which would have given *vocal* values of at least 8 seconds.

The general interpretation of the model for *vocal* is very similar to that of *glance*. The main difference is that *vocal* responses are longer than *glance* responses since they take into account the following times: time required by the subject to perceive the click sound of the projector, time to initiate eye movements to read the display, time to go back to the road and then back to the display when more than one fixation was required (e.g., small character size at low luminance), and finally, time to initiate a vocal response.

For the 'good' display parameters and conditions specified earlier, the predicted vocal response time for the younger drivers is 1.33 seconds and 1.43 seconds for the older drivers (a difference of about 8 percent). This difference, much smaller than the general 27 percent that could have been expected, might be due to the fact that younger subjects 'took their time' before vocalizing a response, as compared with older subjects. In these conditions, the difference between vocal response time and glance time (i.e., time to perceive the click + time to move the eyes towards the display + time to initiate a response) is about 600 ms for both groups, indicating that older subjects were just as quick as younger subjects to respond; in these good conditions, the age difference appears to be due only to glance time. In degraded reading conditions, however (small characters and low contrast), the difference between vocal response time and glance time was much larger for old subjects because they needed more than one eye fixation to read the display.

Standard deviation of lane position

The regression model for *lanedev* explained only 20 percent of the variance in the data as shown by the multiple R^2 (Table 16.4). The *age* variable explained most of the variance as shown by the sums of squares of the first and second order *age* effects. The second order effect indicates that the older drivers had slightly better driving performance than middle age drivers. The root mean square error for *lanedev* was 0.21 m for a mean response of 0.74 m. *lanedev* ranged from 0.26 m to 1.58 m.

Although the display parameters and age contribute significantly to the regression, the relatively low R^2 value for the *lanedev* regression would indicate that other variables could potentially contribute to describe this driving performance measure. Wierwille and Gutmann (1978) found that difficulty level of a secondary digit shadowing task–a task similar to word recognition in terms of attentional resource utilization—affected driving performance significantly for low levels of primary task loading, that is, normal driving conditions. Primary task loading levels were provided by lateral crosswinds in their study. This result is compatible with those of Zwahlen and Balasubramanian (1974) and of Zwahlen and DeBald (1986) who observed that vehicle lane deviation increased with visual time spent away from the road. In their experiments, the subjects read text while driving a car on a deserted aircraft landing strip (i.e., easy driving conditions) or they drove the car on the landing strip with their eyes closed. Such results would suggest a strong correlation between secondary task difficulty and driving performance.

On the other hand, Wierwille and Gutmann (1978) found that secondary task difficulty did not affect driving performance significantly for high levels of primary task loading. This result is compatible with those of Noy (1990) who found that difficulty level for various cognitive secondary tasks did not affect driving performance as measured by standard deviation of lane position, in simulated driving. Such results (probably caused by a change in attention allocation strategy) would suggest a very weak correlation between secondary task difficulty and driving performance.

Based on these results, it can be speculated that if the crosswind level had been lower in the Imbeau *et al.* (1989) study, the display parameters would have contributed more importantly to the *lanedev* regression (i.e., explained a larger portion of the variance). On the other hand, if the crosswinds had been higher, then the opposite effect would have been observed. The model provided for *lanedev* should therefore allow one to estimate the standard deviation of lane position attributable to display parameters and to age for normal driving conditions.

If more than one word must be read from a display, it may be more appropriate for a designer to estimate the total glance time with the model presented above, and then use this value to estimate the standard deviation of the vehicle lane position (Zwahlen and Balasubramanian, 1974; Zwahlen and DeBald, 1986). The following equation ($R^2 = 0.13$, $F_{1, 2968} = 433.3$, $p < 0.0001$) derived from the results of the Imbeau *et al.* (1989) experiment allows this computation for a driving speed of 55 mph:

$$\text{Standard deviation of lane position} = 0.57 + \text{Total glance time} \times 0.19$$

The *lanedev* values predicted by this equation and the model presented in Table 16.4 are generally somewhat larger than those reported by Wierwille and Gutmann (1978), by Hicks and Wierwille (1979), and by Zwahlen and DeBald (1986). This difference may certainly be the result of the wider variety in the subjects who participated in the Imbeau *et al.* (1989) study. The use of different experimental conditions may also be part of the explanation. In the case of the Zwahlen and DeBald (1986) study, the difference may also be attributable to radically different driving conditions. The designer should keep in mind that other variables not accounted for in the regression can potentially influence the predicted value. For instance, variables such as driving experience, distance driven per year or experience with particular displays (primary and secondary), would probably have contributed significantly to the *lanedev* regression. However, such variables are difficult to measure objectively. Since there is psychological evidence to show that individuals recall selectively, embellish facts and exhibit other anomalies of human recall, conclusions drawn from such information are tentative.

Conclusions

The results from previous studies clearly demonstrate that the introduction of a secondary task requiring visual attention degrades driving performance (Wierwille and Gutmann, 1978; Noy, 1990; Zwahlen and Balasubramanian, 1974; Zwahlen and DeBald, 1986). Considering that subjects had been formally instructed about the importance of maintaining the vehicle properly under control in the Imbeau *et al.* (1989) study, it must be concluded that in normal driving conditions drivers are tempted to perform the word recognition task even though degradation of driving occurs, or they do not perceive the degradation in their driving performance. Hence, the introduction of information systems in automobiles should be considered carefully. The models presented in this chapter should contribute to help the designer identify the design options that minimize visual attention.

Body movement, age, contrast (chromatic and luminance) and character size were all found to be major determinants of word recognition performance and of visual attention as measured by glance time at the instrument panel. These parameters were also found to affect driving performance significantly as measured by standard deviation of vehicle lane position in normal driving conditions. The effects of character size and body movement are already fairly well documented in the literature. Those of age and of contrast are less well documented.

The models presented provide variable weighting factors that can be applied to the performance measures studied which are of typical interest in the context of automobile driving. The models also provide quantitative information concerning interaction effects between various display design parameters and age on information acquisition performance during driving. The models should therefore prove useful to designers of automobile instrument panels as well as to designers of road signs and road displays. These equations can be used in computer simulation models of man-machine systems involving recognition of displayed labels and words. It must be noted however, that the models should be used for display conditions similar to those on which they were based.

The need for design data and models relating the effects of the various design parameters and driver characteristics to performance in the driving context is pressing and it will become more so as new technologies are introduced in the automobile. This chapter constitutes one further step in filling the data gap.

References

Card, S. K., Moran, T. P. and Newell, A., 1986, The model human processor, in Boff, K. R., Kaufman and Thomas, J. P., (Eds) *Handbook of Perception and Human Performance—Volume II Cognitive Processes and Performance*, ch. 45, New York: Wiley.

Carr, T. H., 1986, Perceiving visual language, in Boff, K. R., Kaufman, L. and Thomas, J. P. (Eds) *Handbook of Perception and Human Performance—Volume II Cognitive Processes and Performance*, ch. 29, New York: Wiley.

Carter, R., 1989, Calculate (don't guess) the effect of symbol size on usefulness of color, in *Proceedings of the Human Factors Society Meeting, 33rd Annual Meeting*, pp. 1368–72, Santa Monica, CA: The Human Factors Society.

Cavanaugh, J. P., 1972, Relation between the immediate memory span and the memory search rate, *Psychological Review*, **79**, 525–30.

Czaja, S. J. and Guion, R. M., 1990, *Human Factors Research Needs for an Aging Population*, Washington, DC: National Academy Press.

Easterby, R. and Zwaga, H., 1984, *Information Design—The Design and Evaluation of Signs and Printed Material*, New York: Wiley.

Finkelman, J. M., Wolf, E. H. and Friend, M. A., 1977, Modified discriminant analysis as a multivariate post-comparison extension of MANOVA for interpretation of simultaneous multimodality measures, *Human Factors*, **19**(3), 253–61.

Fowkes, M., 1984, Presenting information to the driver, *Display Technology*, **5**, 215–23.

Hayes, B. C., Kurokawa, K. and Wierwille, W. W., 1989, Age-related decrements in automobile instrument panel task performance, in *Proceedings of the Human Factors Society Meeting, 33rd Annual Meeting*, pp. 159–63, Santa Monica, CA: The Human Factors Society.

Helander, M. G., 1986, Design of visual displays, in Salvendy, G. (Ed.) *Handbook of Human Factors*, pp. 507–48, New York: Wiley.

Hicks, T. G. and Wierwille, W. W., 1979, Comparison of five mental workload assessment procedures in a moving-base driving simulator, *Human Factors*, **21**(2), 129–43.

Human Factors Society, 1988, *American National Standard for Human Factors Engineering of Visual Display Terminal Workstations*, ANSI/HFS Standard No. 100-1988, Santa Monica, CA: The Human Factors Society.

Imbeau, D. and Wierwille, W. W., 1989, Effects of automobile instrument panel luminance and color on word recognition, in *1989 SID International Symposium Digest of Papers*, pp. 368–71, Playa del Rey, CA: Society for Information Display.

Imbeau, D., Wierwille, W. W., Wolf, L. D. and Chun, G. A., 1989, Effects of instrument panel luminance and chromaticity on reading performance and preference in simulated driving, *Human Factors*, **31**, 147–60.

Keppel, G., 1982, *Design and analysis—A researcher's handbook*, Englewood Cliffs, NJ: Prentice Hall.

Klare, G. R., 1963, *The Measurement of Readability*, Ames, IA: Iowa State University.

Kline, D. W. and Schieber, F., 1985, Vision and aging, in Birren, J. E. (Ed.) *Handbook of the Psychology of Aging*, pp. 296–331, New York: Van Nostrand Reinhold.

Kurokawa, K. and Wierwille, W. W., 1990, Validation of a driving simulation facility for instrument panel task performance, in *Proceedings of the Human Factors Society Meeting, 34th Annual Meeting*, pp. 1299–303, Santa Monica, CA: The Human Factors Society.

Lippert, T. M., 1986, Color-difference prediction of legibility performance for CRT raster imagery, *Digest of the Society for Information Display*, pp. 86–89. Playa del Rey, CA: Society for Information Display.

Lippert, T. M. and Snyder, H. L., 1986, *Unitary suprathreshold color-difference metrics of legibility for CRT raster imagery*, Virginia Polytechnic Institute and State University Technical Report HFL/ONR 86-3.

Merrifield, R. M. and Silverstein, L. D., 1986, *The Development and Evaluation of Color Systems for Airborne Applications: Fundamental Visual, Perceptual, and Display Systems Considerations*, Technical Report no. NADC-86011-60, Seattle: Boeing Commercial Airplane Company, Crew Systems Technology.

Mortimer, R., 1989, Older drivers' visibility and comfort in night driving: Vehicle design factors, in *Proceedings of the Human Factors Society Meeting, 33rd Annual Meeting*, pp. 154–8, Santa Monica, CA: The Human Factors Society.

Mourant, R. R. and Langolf, G. D., 1976, Luminance specifications for automobile instrument panels, *Human Factors*, **18**, 71–84.

Myers, R. H., 1986, *Classical and Modern Regression with Applications*. Boston, MA: Duxbury Press.

National Research Council (NRC), 1988, *Transportation in an Aging Society— Improving Mobility and Safety for Older Persons*, Special Report 218, Washington, DC: Transportation Research Board, NRC.

Noy, Y. I., 1990, Selective attention with auxiliary automobile displays, in *Proceedings of the Human Factors Society Meeting, 34th Annual Meeting*, pp. 1533–7, Santa Monica, CA: The Human Factors Society.

Olzak, L. A. and Thomas, J. P., 1986, Seeing spatial patterns, in Boff, K. R., Kaufman, L. and Thomas, J. P. (Eds) *Handbook of Perception and Human Performance— Volume II Cognitive Processes and Performance*, ch. 7, New York: Wiley.

Rockwell, T. H., Augsburger, A., Smith, S. W. and Freeman, S., 1988, The older driver—A challenge to the design of automotive electronic displays, in *Proceedings of the Human Factors Society Meeting, 32nd Annual Meeting*, pp. 583–7, Santa Monica, CA: The Human Factors Society.

Salthouse, T. A., 1985, Speed of behavior and its implications for cognition, in Birren, J. E. (Ed.) *Handbook of the Psychology of Aging*, pp. 296–331, New York: Van Nostrand Reinhold.

Sekuler, R., Kline, D. and Dismukes, K., 1982, *Aging and Human Visual Function*, New York: Alan R. Liss Inc.

Sivak, M., Olson, P. L. and Pastalan, L. A., 1981, Effect of driver's age on nighttime legibility of highway signs, *Human Factors*, **23**(1), 59–64.

Smith, D. B. D., 1990, Human factors and aging: An overview of research needs and application opportunities, *Human Factors*, **32**(5), 509–26.

Society of Automotive Engineers, 1984, *Ergonomic Aspects of Electronic Instrumentation: A Guide for Designers*, Warrendale, PA: Author.

Stokes, A., Wickens, C. D. and Kite, K., 1990, *Display Technology—Human Factors Concepts*, Warrendale, PA: Society of Automotive Engineers.

US Department of Commerce, Bureau of the Census, 1988, *Statistical Abstract of the United States*, Washington, DC: US Government Printing Office.

Welford, A. T., 1980, *Reaction Times*, New York: Academic Press.

Wierwille, W. W. and Gutmann, J. C., 1978, Comparison of primary and secondary task measures as a function of simulated vehicle dynamics and driving conditions, *Human Factors*, **20**(2), 233–44.

Wierwille, W. W. and Peacock, B., 1989, Human factors and the automobile of the near future, *Human Factors Bulletin*, **32**(11), 1–5.

Wilkinson, L., 1989, *SYSTAT: The System for Statistics*. Evanston, IL: SYSTAT Inc.

Wyszecki, G. and Stiles, W. S., 1982, *Color science—Concepts and methods, quantitative data and formulae*, New York: Wiley.

Yanik, A. J., 1989, Factors to consider when designing vehicles for older drivers, in *Proceedings of the Human Factors Society Meeting, 33rd Annual Meeting*, pp. 164–8, Santa Monica, CA: The Human Factors Society.

Zwahlen, H. T. and Balasubramanian, K. N., 1974, A theoretical and experimental investigation of automobile path deviation when driver steers with no visual input, *Transportation Research Record*, **520**, 25–37.

Zwahlen, H. T. and DeBald, D. P., 1986, Safety aspects of sophisticated vehicle information displays and controls, in *Proceedings of the Human Factors Society Meeting, 30th Annual Meeting*, pp. 256–60, Santa Monica, CA: The Human Factors Society.

17

Driver mental workload

Robert E. Schlegel

Introduction

Picture yourself driving along a quiet stretch of highway in the middle of the afternoon. There is no one else in the car, no one else on the road and everything is going smoothly. You're feeling great, the sun is shining, and you haven't a care in the world. You are experiencing that wonderful state of *optimal driver workload*. Enjoy it—it won't last. You arrive in your home town and pick up your 10-year-old son—on blasts the sound system. You pick up the 4-year-old daughter at day-care—85-decibel screams fill the vehicle. All you want to do is get home. To make matters worse, it is now rush hour, the beautiful sunshine has metamorphosed into a hail storm, and you realize why you should have changed your wiper blades last week. Don't look back but you have just crossed into the zone of *high driver workload*.

The above scenario points out two important features of driver mental workload. First, it can be affected by a vast array of factors related to the driver, the vehicle, the task and the environment. Second, driver workload is a constantly changing state of affairs. The amount of physical effort needed to drive a late model automobile is a mere fraction of what was required by previous generations. On the other hand, the modern automobile is a complex system with accessories and features which may distract the driver rather than complement safe vehicle operation. Sound systems that rival home stereos, mobile telephones and advanced navigation systems place demands on the same information processing resources that are needed for the primary task of vehicle guidance, and thereby increase the level of mental workload. As stated by Kantowitz and Sorkin (1983), human factors research has repeatedly demonstrated that workload is one important factor in the occurrence of human error. Workload that is excessive, too low, or rapidly and unexpectedly changing encourages driver error and, depending on the situation, may encourage an accident.

This chapter is intended to provide a background in the concepts and measurement of mental workload. By examining various elements of the driving task and the cognitive skills and abilities required to perform those elements, it is hoped that the reader will appreciate the complexity of the mental workload concept and the difficulty of its measurement. Individual driver characteristics (age, driving experience, physical and mental state) and environmental factors (temperature, noise, vibration, social interaction) will also be examined. Finally, a review of mental workload research specific to automobiles will be provided along with implications for vehicle design. A genuine understanding of this dynamic concept can potentially result in improved driver performance and fewer accidents through the design of vehicles that encourage efficient workload allocation.

Mental workload

So, what is mental workload? A variety of definitions have been proposed (Hart, 1985; Kantowitz, 1985). When it comes to mental workload, the one thing upon which researchers agree is the lack of a single definition (Moray, 1979). Wiener (1985) has commented that there have been so many reviews of workload that a review of the reviews may be needed. Within human factors the concept has become so common that it is included in the standard texts (Sanders and McCormick, 1987; Kantowitz and Sorkin, 1983) and handbooks (Salvendy, 1987; O'Donnell and Eggemeier, 1986). An excellent treatise on the subject is provided by Hancock and Meshkati (1988) who also provide over 500 reference citations, a number of which have been summarized in an annotated bibliography by Wierwille and Williges (1980).

In general most researchers view mental workload as a multidimensional interaction of task and system demands, operator capabilities and effort, subjective performance criteria, and operator training and experience (Eggemeier and O'Donnell, 1982). In one sense, mental workload is an expression of the demand that a task places on an individual and the individual's capacity to meet this demand and produce an acceptable level of performance. The concepts of *stress* and *strain* as used to represent the properties of materials or human physiological responses to physical workload provide a good analogy

CONCEPTUAL FRAMEWORK FOR WORKLOAD

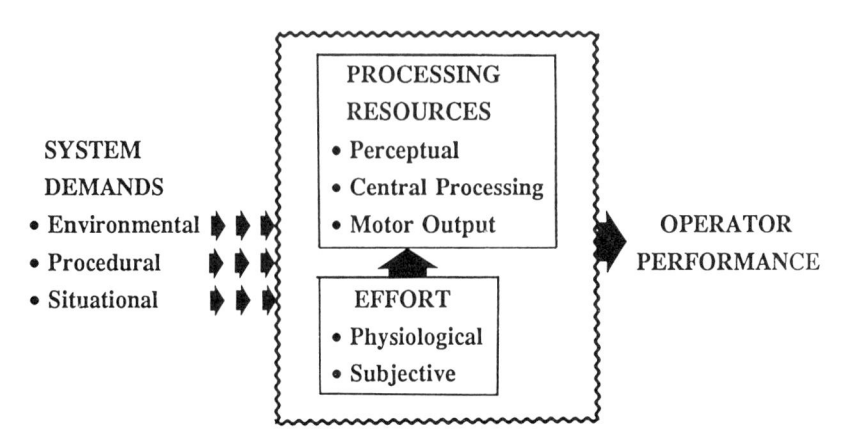

Figure 17.1. Simplified model of mental workload

for defining mental workload. Task demands and the operating environment determine the level of *stress*. The impact on the particular individual represents the *strain* and is reflected in task performance and other measures. Thus, the same level of stress does not result in the same amount of strain for all drivers.

The level of mental workload impacts the amount and strategic allocation of resources invested by the operator to achieve an acceptable level of performance (Stokes, Wickens and Kite, 1990). Thus, mental workload reflects how 'busy' the driver is and how much attention the driving task requires.

The multiple dimensions of mental workload include subfactors of task demands and procedures, operator variables including mental and emotional capabilities, and environmental variables. A simplified model of mental workload is presented in Figure 17.1. A major implication of the numerous definitions is that no single measurement technique will provide a comprehensive means for the assessment of workload in every situation. What is needed is a battery of techniques and a methodology for selecting the appropriate subset for a given situation.

Workload assessment

Research on the nature of human information processing capacity has generated a variety of proposed workload measures in the following categories: performance and behavioral measures, subjective measures and physiological or biocybernetic measures. Performance measures inlude primary task measures and secondary task measures. Primary task measures quantify overall system performance and describe performance characteristics of the task whose workload is measured. Increases in the difficulty of the task are expected to increase the level of workload and thereby decrease performance. For example, changes in traffic conditions, driver fatigue or lane width might be reflected in driver performance measures such as side-to-side weaving (lateral standard deviation), the root mean square (RMS) distance from the lane center or the rate of steering wheel reversals (Hicks and Wierwille, 1979). Problems with these measures include the difficulty of making comparisons across tasks and of identifying what task difficulty really does to workload (Wickens, 1990). More importantly, human beings are able to adapt to increased task demands by investing more resources, and thus performance does not always worsen. By the time mental workload is at a level where a performance decline is evident, catastrophic results may occur.

Secondary task measures estimate mental workload by measuring the performance change on an additional well-defined and controlled task performed simultaneously with the primary task. A large performance decrease for the secondary task indicates high resource requirements for the primary task (Brown, 1978; Ogden, Levine and Eisner, 1979; Knowles, 1963). For example, a digit detection task has been used to measure the spare mental capacity of car drivers (Brown and Poulton, 1961) and to evaluate changes in vehicle handling characteristics (Hoffman and Joubert, 1966) and the performance of trainee drivers (Brown, 1966).

In general, the secondary task is irrelevant or unrelated to the primary task. However, to provide a more realistic situation, researchers have also used *embedded secondary tasks* which are an integral (though less important) part of the total task (Shingledecker et al., 1980; Hart and Wickens, 1990). In essence, the use of instrument panel controls and displays always constitutes an embedded secondary task in comparison with the primary task of stabilization, control and navigation of the car (Heintz, Haller and Bouis, 1982).

Ideally, secondary task performance is inversely proportional to the primary task resource demands. Secondary tasks may reflect differences in task resource demands, automation or practice not reflected in primary task performance. They typically provide more information regarding the specific resources demanded by a task. Secondary task techniques possess a high degree of face validity and can be used with different tasks. The costs associated with secondary tasks include interference with primary task performance, and the burden of finding a secondary task that taps the same resource pool that is used for the primary task (Wickens, 1990).

As the name implies, behavioral measures reflect the behavior of the operator. An example is the description of eye and head movement behavior while driving (Antin et al., 1988). Behavioral measures are often easily quantifiable and may overlap with primary task measures. As an example, the rate of steering wheel reversals may be considered a behavioral measure (MacDonald and Hoffman, 1980), but may correlate highly with a performance measure related to lane-keeping ability.

Subjective or self-report measures allow the individual to rate feelings of effort. These measures include the Cooper-Harper scale, Sheridan's dimensional scale, the NASA Task Load Index (TLX; Hart and Staveland, 1988), the Subjective Workload Assessment Technique (SWAT; Reid and Nygren, 1988) and other multidimensional scaling techniques. Subjective measures are sensitive to total demand without the level of intrusiveness that physiological measures possess. However, they rely on an individual's *perception* of workload that might be biased by unrelated variables. Overall, subjective ratings of task difficulty represent perhaps the most acceptable measure of workload from the operator standpoint. Some researchers have argued that these measures come nearest to tapping the essence of mental workload (Sheridan, 1980; Johansen, 1979). The measures usually do not disrupt primary task performance and are relatively easy to formulate. The only drawback is the uncertainty with which an operator's subjective rating truly reflects that operator's mental workload (Wickens, 1990).

Physiological measures attempt to quantify the physical, electrical or chemical influence that mental workload has on the body, much the same as one might use heart rate or oxygen consumption as measures of physical workload. These measures include electrodermal (GSR), electrocardiograph (ECG), electroencephalograph (EEG), eye movement (EOG) and analysis of body fluids. Physiological measures are generally less precise than secondary task measures. However, they can provide a relatively continuous record of data over time, and are generally non-disruptive of primary task performance. On the other hand, they impose physical equipment constraints and may be obtrusive in a physical sense. These constraints may influence user acceptance of the measuring devices (Wickens, 1990).

It is essential to repeat that while each of the above methods measures one or more aspects of mental workload, there is no single methodology that indicates the absolute level of mental workload for every given task. For example, it has been shown that for some laboratory tasks (e.g., Sternberg memory search and unstable tracking) and for some operational tasks, a SWAT rating of 45 to 50 coincides with rapid performance deterioration (Schlegel, Schlegel and Gilliland, 1988). However, that does not mean that the SWAT is appropriate for other tasks or settings (Nygren, 1991). The same is true of other measures.

A number of criteria have been proposed for evaluating and selecting an appropriate workload assessment technique for a given situation. According to Wickens (1990), the following criteria should be satisfied by any workload index:

Sensitivity. The index should be sensitive to changes in task difficulty or resource demand. Changes in workload should be reflected by changes in the numeric index. Related to sensitivity is the concept of precision or resolution, that is, the smallest increment on the workload scale (i.e., the scale unit).
Diagnosticity. The index should not only identify when workload varies, but also indicate the cause of such variation in terms of the specific resource demands on the operator.
Selectivity/Focus. The index should be selectively sensitive only to differences in workload and should not reflect changes in factors unrelated to mental workload.
Reliability. A workload index should provide stable values across repeated administrations. It is also important that the reliable index be provided rapidly enough to allow estimation of the transient changes, thus providing a valid measure as outlined below.

Additional criteria that often go unstated have been listed by Meshkati (1988) as follows:

Validity. The index should embody face, construct and content validity.
Intrusion/Obtrusiveness. To the extent possible, the assessment technique should not interfere with or degrade the performance of the task being evaluated.
Ease of field utilization. The technique should be easy to administer in the task environment. Factors affecting this criterion include instrumentation, analyst and operator training, data recording and analysis.
Operator and user acceptance. The technique must appear valid to both the operator and the user of the technique, and should not interfere to the extent that the operator will not actively comply with the assessment requirements.

Processing resources

A discussion of the fundamental concept of mental workload is not complete unless one addresses the concepts of processing resources and the time-sharing of tasks. If humans possessed unlimited information processing capabilities, the question of excessive workload would be a moot point, but this is not the case. Humans are limited in terms of the speed with which they can process information. Early theories of information processing proposed that humans had a single information processing channel with a limited capacity on the order of 10 bits/second. More recent theories (Wickens, 1990) are based on the concept that humans possess different pools or types of resources (spatial abilities, verbal skills, visual scanning ability, etc.). To the extent that two tasks require different resources, they can be time-shared effectively. As the required resources of the tasks overlap, time-sharing becomes less efficient. If the demand for resources common to both tasks is increased for one of the tasks, there will be increasing interference, and performance on one or both tasks will suffer. Due to the potentially interactive nature of secondary task techniques, experimental designs and data analyses should allow one to examine the impact of the secondary task on the primary task. The experimenter should not rely on measuring secondary task performance changes alone.

Thus, two major features of a complex task will affect the level of workload and the expected level of performance. One feature is task difficulty. As an individual task or subtask becomes more difficult, additional resources must be invested to maintain a constant level of performance. If the available amount of resources is insufficient, performance will decline. The second feature is the structure of the complex task in terms of potential overlap in the type of resources required by individual subtasks as discussed above.

A primary example of these two features with respect to the driving task involves the visual attentional resources. Visual scanning of the forward road scene is much easier on an uncrowded highway than on a busy residential street with numerous parked cars and heavy traffic. Maintaining the same level of accident-free driving requires more visual attention in the second case because the task is more difficult. Adding an additional task of visually monitoring the speedometer on the instrument panel or looking at a map on the passenger's seat diverts visual attention from the road scene and increases the likelihood of an accident. Both tasks demand visual resources. In other words, they have overlapping task structures.

A commonly accepted scheme of identifying processing structures is the multiple resources model of Wickens (1990). The scheme isolates three resource dimensions: processing modalities, processing codes and processing stages. Examples of different processing modalities include auditory vs. visual perception, and manual vs. vocal control. This dimension explains why it may be easier to time-share an auditory task and a visual task (listening to directions while driving the car) than two visual tasks (reading a road map while driving the car). Processing codes are typically either verbal or spatial. Driving a car is a predominantly spatial task (essentially a tracking task), allowing tasks with a verbal orientation (e.g., conversing on a mobile phone) to take place simultaneously.

Processing stages contrast perceptual and cognitive activities with response activities. This dimension is based on the earlier stage classification of input, central processing, and output (Sternberg, 1969). Two tasks that involve perception are less easily time-shared than a perceptual task in combination with a response task (Peacock, 1972).

The previous discussion related to the psychological concepts of attention and the allocation of attentional resources provides a good starting point for the description and evaluation of driver mental workload. The next section attempts to apply these concepts to the driving task.

Driving task elements

To develop a better appreciation of driver workload assessment, one should start with a basic understanding of the nature of the driving task and its elements. It is important to develop a macroview of the overall driving system in order to comprehend the numerous interactions of vehicle, driver, road and environment. An appropriate framework may be borrowed from the aviation sector, in particular, since a vast amount of mental workload research has been conducted in this setting.

Federal Aviation Regulation 25 deals with aircraft crew workload and lists basic workload functions of 1) flight path control, 2) collision avoidance, 3) navigation, 4) communications, 5) operation and monitoring of aircraft engines and systems, and 6) command decisions. In addition, FAR 25 outlines the following workload factors:

1   The accessibility, ease and simplicity of operation of all necessary flight, power and equipment controls . . .
2   The accessibility and conspicuity of all necessary instruments and failure warning devices . . . The extent to which such instruments or devices direct the proper corrective action is also considered.
3   The number, urgency and complexity of operating procedures with particular consideration given to the specific fuel management schedule . . . (Kantowitz and Casper, 1988).

A similar approach may be taken for establishing a standard for ground vehicles. Keep in mind that this approach addresses the vehicle and task elements without considering in detail the equally important driver traits and states and environmental characteristics (at least for the present).

Vehicular guidance

Analogous to flight path control and collision avoidance is the primary automobile task of vehicular guidance. Kramer and Rohr (1982) list the three main tasks of vehicular guidance as: 1) keeping to a lane, 2) adapting speed, and 3) reacting to obstacles, traffic signs and other road users. Turn negotiation requiring an assessment of vehicle size, turning radius and steering ratio in relation to spatial turning restrictions may be added as a variation of Task 1. Tasks 1 and 2 are essentially tracking functions involving visual-motor coordination. Task 3 involves visual scanning (and auditory monitoring), decision making, and response selection and execution. According to Kramer and Rohr (1981), the perceptual system must process the information relating to all three tasks in a parallel manner since performance on Tasks 1 and 2 must be maintained while sufficient capacity is reserved for responding to discrete events (Task 3). The motor system must supply the control inputs to the steering, acceleration and braking systems while sharing time with the activation of secondary controls for the lighting, HVAC, sound system and other components.

Navigation

As with flight navigation, ground vehicle navigation is a higher order process involving spatial abilities. It requires the recall of an internal map of the desired course or extraction of this information from a printed map, human navigator, information signs, written instructions or an electronic navigation display system. There are several automotive navigation systems in various stages of development. Some are self-contained in terms of car location data while others rely on input from various satellites (Stokes, Wickens and Kite, 1990). Reviews of the technology are provided by French (1986) and McGranaghan, Mark and Gould (1987).

The research of Zwahlen and his colleagues provides some insight to the potentially adverse impact of these navigation systems on driver workload. A major difference between aircraft and automobile control is the high level of visual attention to the outside world that is required in driving. The driver can not point the vehicle in a particular direction and assume that it will remain on the roadway for any length of time. Instead, the driver needs almost constant eye contact with the road in order to maintain proper lane position (Zwahlen, 1985). Zwahlen and Balasubramanian (1974) have shown that lateral deviation from the lane center increases as a function of the attentional time spent away from the external visual scene.

Related to this result is the time-to-line-crossing (TLC) concept, in which driving strategy is modeled as a mixture of active and passive steering strategies (Godthelp, Milgram and Blaauw, 1984). The length of the passive steering interval is determined by the driver's internal prediction of when the vehicle would reach either edge of the driving lane or a personally established actualization of this limit. This period of time enables the driver to divert attention to other visually related aspects of the driving task (or other activity), such as monitoring displays and aiming for controls. Thus the lane width determines steering precision and consequently the TLC value. If the attentional demands of another task, such as viewing the navigational display, exceed the actual available TLC, the potential for an accident increases.

Zwahlen and DeBald (1986) compared driving performance under conditions of normal visual scanning, eyes closed, and reading an item of text or a section of a road map. The average lateral deviation for the text reading condition was 10 feet for 700 feet of travel. This amount greatly exceeded the deviation likely to cause an accident on the average highway. Antin *et al.* (1988) found that drivers spent a significantly longer proportion of time glancing to a moving-map display (0.331) than to a conventional paper map (0.068), and that this proportion was affected by the driver's spatial ability. These results point to the need for developing alternative means of presenting navigation information. However, Wierwille *et al.* (1988) also determined that as the driving task demands increase, drivers are able to adapt and shift their visual scanning strategy appropriately. More on the topic of navigation displays is presented in the chapters by Antin and Wierwille in this volume.

Communications and the social environment

Communication within the automobile does not share the same criticality with communication aboard aircraft but nonetheless provides as much or more distraction. An older study of accidents in the US (Treat *et al.*, 1977) indicated that three of the most common human error causes in traffic accidents were improper lookout (23 percent), inattention (15 percent), and internal distraction (9 percent). Conversations with passengers, listening to the radio and the use of mobile phones diverts at least a portion of the attentional resources away from the driving task. Add to that the unwanted communications of arguing children and adults, 'noise' from the sound system, and the inevitable comments from 'backseat drivers', and one has the potential for rapidly increasing workload. Visual 'communication' from road signs and vehicle displays adds to the loading.

The non-immediate social environment also has a tremendous impact on performance and accident potential. As pictured in the introductory scenario, being in a hurry or worrying about personal or family problems may lead to inattention to the task at hand.

Operation and monitoring of systems

As with the airplane, there are a number of automobile systems for which the driver is responsible. Some of these systems (e.g., engine, braking, tire pressure, fuel) are more critical than others. Monitoring may be accomplished with the aid of visual displays, the use of other senses (auditory, vestibular, tactile/vibration), or the help of others (e.g., traffic citation for a non-working taillight, passing driver pointing to a low tire). Dewar (1988) provides an excellent summary paper describing the perceptual and information processing skills needed to operate modern vehicles safely and efficiently. With basic examples, he presents some of the potential difficulties associated with the presentation of information in vehicles, especially with high-technology displays and controls.

The comments in the section on navigation displays apply equally well to the installation of sophisticated information displays and controls. As Zwahlen (1985) points out, a driver has some spare visual capacity on long straight road sections, under low traffic conditions, no adverse weather and low navigational demands. Unfortunately, a significant portion of driving occurs in a multi-vehicle environment, under adverse weather conditions, under traffic signal control, and on curves. In these situations, the information acquisition and processing rates and the eye scanning activities are high, and the spare visual capacity is low. Displays that are confusing and controls that are difficult to use increase a driver's time demands and mental effort, two key elements of workload.

System operation is typically accomplished through activation of the numerous controls. Important parameters of control design such as location, size, shape, method of actuation, expectancy of location and operation (stereotypes), labeling, and clustering (functional grouping and crowding) are discussed in the chapters by Green, Wierwille, and Graesser and Marks of this book. It is sufficient to say that the attention and information processing abilities of the driver can be strained by poorly designed controls as well as displays. Difficulties in locating, identifying or operating controls without looking at them can add to the time a driver must divert attention from the roadway. On the other hand, this activity utilizes an independent tactile resource and in that sense eases the visual attentional demands. In either case, the necessity of the numerous small buttons on instrument groups of modern automobiles (especially the sound system) is questionable for the average driver (Dewar, 1988), who is likely to ignore many of the available functions.

Graesser (1989) and Schlegel (1989) reported the development of a model to predict hand movement times to automobile secondary controls as a function of the distance from the steering wheel, the size of the control and a number of other factors. Recent work by Chukwu and Schlegel (1990) confirms that the

Fitts' Law (Fitts, 1954) model of hand movement time is applicable to automobile control research and demonstrates that predicted times are accurate under conditions of restricted and occluded vision. These studies are helpful in furthering the prediction of workload and the corresponding performance level for a particular instrument panel configuration.

In recent years, a combined reconfigurable display/control device has become available. This device uses a flat panel or CRT display with a programmable, overlaid touch entry device which permits the driver to touch the display information as a means of control input (Snyder and Monty, 1985). As with the navigation display, the driver must look at the display to determine the current information state, to guide finger movement, and to verify that specific changes have been made. Here again the visual verification of touch location and display change of state require visual time-sharing with the driving task. Evaluation of this technology is essential to determine its impact on driver workload. A methodology for this evaluation has been provided by Snyder and Monty (1985). Research has shown that practice is very important in reducing part of the negative impact of novel displays and controls in vehicles (Kiefer, 1991).

Command decisions

Command decisions within the driving system may be categorized as tactical planning and execution vs. strategic planning. Examples of tactical decisions include planning and executing a passing maneuver on a two-lane road and establishing the approach speed to an intersection based on the predicted future state of the signal light. Traffic assessment and recall of expected traffic patterns in order to select a specific or alternate routing is an instance of strategic planning. For example, a driver may select different routes from work to home depending on the departure time and may update those routes based on current traffic conditions. Performance on higher level activities such as these is more difficult to quantify. The relationships to driver workload are even more elusive but perhaps even more critical in terms of leading to catastrophic human error (e.g., improper passing or running a red light).

Workload factors

The workload factors for aircraft as mentioned in FAR 25 transfer with minimal modification to the automobile driving task. As outlined previously, the factors address the vehicle controls, instruments/displays, and operating procedures. Some of these points have been addressed in the previous sections. As in aviation, the major issue is how the workload components related to the design of these three major areas within the vehicle subsystem can be assessed in conjunction with the other major subsystems of driver and environment.

Driver traits and states

The analogy with the aviation setting is greatly weakened when one looks at driver characteristics. People who operate aircraft are selected, highly trained individuals who tend to be well-educated, highly motivated and physically fit. The general US driving population, however, spans a wide range of ages and abilities, is culturally diverse, possesses a high degree of variation in behavior and attitude, and is not subject to regular proficiency or medical checks. Each individual feels as if driving is a constitutionally guaranteed right rather than a privilege. As a result, design engineers have a difficult time agreeing on the range of physical and information processing characteristics for which to design.

There are no regulated limits on the number of hours behind the wheel. Limitations in the number of law enforcement personnel and in the effectiveness of educational programs prevent the total elimination of driving under the influence of alcohol or drugs. Permit testing typically involves an elementary vision check, a multiple-choice test of traffic laws, and a cursory examination of basic driving skills. However, there is little evidence that basic sensory ability is related to accident rates (Henderson and Burg, 1974) in comparison with appropriate scanning behavior, attention to relevant stimuli, correct decision making and taking proper action. Advanced training is essentially non-existent or self-conducted without supervision for the general driving public. As such, the results of much of the research in aviation are not applicable or may even conflict with the automotive setting (Saunby, Farber and Jack, 1986).

In their model of driver behavior, Kramer and Rohr (1982) distinguish between *global states* (e.g., vehicle design, weather, driver experience) which show little dependence on time and *actual states* (e.g., momentary speed, course of the road, driver steering movements) which characterize instantaneous processes. This classification is perhaps most relevant with respect to driver states.

Driver experience

The amount of driver experience affects the level of automaticity of the driving task at skill-based, rule-based, and knowledge-based levels of functioning. According to Fitts, learning a perceptual-motor task involves three stages: cognition, fixation and automation. Cognition is the element of learning which involves familiarization with the task. Once familiarity is established, basic responses can be integrated. The second stage, fixation, involves the kinesthetic (joint and muscle) and visual feedback provided when a task is performed. The appropriate physical response is associated with the stimulus in this stage as the learning of the task progresses. In the automation stage of learning, information sampling strategies and behavior may change, and less and less conscious effort is required to perform the task while speed and accuracy continue to improve. Although one may continue to practice the task, the focus has changed to improving fine details and organizing responses into larger units. An example of this is the fact that novice drivers tend to fix their gaze close to the vehicle and toward the right-hand side of the roadway (presumably to perform precise lane tracking), and they perform relatively narrow sampling of the environment. With experience, drivers look further up the road, scan the environment more broadly and efficiently, and potentially obtain more information (Dewar, 1988).

Hale, Quist and Stoop (1988) evaluated the use of a behavior model incorporating the three levels of human functioning to explain errors in routine driving tasks. Kobayashi (1988) recommended the following forms of education and training for each of the three levels. For skill-based elements, people should learn through hands-on experience using simulators and mockups and the training should include how to handle emergency situations. With respect to rule-based procedures, checklists, manuals and other materials should be used to minimize inattention to rules. These are not always successful as evidenced by the crash of a Northwest Airlines jet in which the pilot had failed to configure the plane for takeoff properly while going through his checklist. For knowledge-based elements, a thorough knowledge of the details of the driving system provides a firm basis for accident prevention.

A companion issue is that of vehicle familiarity. A disproportionately high number of accidents involve drivers who are unfamiliar with their vehicles because they are rented, borrowed or newly acquired. Unfamiliar drivers are two to three times more likely to be involved in an accident than are familiar drivers (Perel, 1983). One reason for this is the lack of standardization of display and control locations and operation (McGrath, 1975), a concept often referred to as 'control commonality'.

Age

A growing concern in vehicle design is how to respond to the expanding segment of the driving population consisting of elderly drivers. It is well documented that elderly individuals have significant limitations in sensory, cognitive and psychomotor skills (Welford, 1958; Welford, Norris and Shock, 1969). Mourant and Langolf (1976) evaluated the minimum level of illumination for reading automobile instrument panels. They determined that older people deserve special consideration in the design of control panel displays and illumination levels and concluded that the lighting practices existing at that time were adequate for younger drivers but caused problems for older people.

The elderly are typically slower at initiating eye movements and because of poor visual acuity take longer to obtain relevant information (Dewar, 1988). Other problems, such as higher sensitivity to glare and slower accommodation of the lens result in elderly drivers taking longer to process information. Gramberg-Danielsen (1967) reported that the total time required for the eye to move from an outside fixation point to the speedometer on the instrument panel (a distance of 20 m) and for the accommodation of the lenses to read the scale was 0.4 to 0.8 second for drivers below age 40 and was 2.0 to 2.7 seconds for drivers above age 60. Drivers wearing bifocals have even more trouble and must often adjust their heads to view vehicle displays. Smith, Meshkati and Robertson's chapter in this volume discusses the topic of the aging driver population further.

Alcohol and other chemical substances

Drivers under the influence of alcohol tend to have longer eye fixations with less active eye movement. They are less efficient at scanning the environment and thus take in less information. In addition, their ability to shift attention from one input to another is impaired, and their reaction time is slowed (Dewar, 1988). An excellent review of the effects of various drugs on performance of the driving task is provided by O'Hanlon and deGier (1986), with the chapter by Sanders (1986) of particular interest.

Other individual differences

It is important to point out that aside from the factors mentioned in the preceding sections, there are a number of individual differences or personality variables that may affect workload assessment and the actual level of workload itself. Adaptation level, personality factors, motivation, time of day (Hancock, 1988) and other endogenous variables must be considered. The concepts of stress and strain in defining mental workload were introduced earlier. Aspects of the driving task (stress) affect individual drivers in differing ways (strain) based on each person's response capacity and motivation. Leplat (1978) pointed out that characteristics of personality could intervene substantially in the determination of workload. Hopkin (1979) also considered personality variables as potentially relevant to mental workload and stated that individual differences have often precluded general judgments on whether the workload associated with a task was excessive or merely high. As Moray (1984) pointed out, the concept of 'individual differences in workload research is far more important than has hitherto been acknowledged. Without taking this into account, we are seriously delaying the development of a useful measure'. Meshkati, Hancock and Rahimi (1990) provide a good summary of relevant studies examining the impact of individual differences on workload measures.

It has been shown that differences in perceptual style (field independent vs. field dependent) and selective attention ability are related to automobile accident involvement (Mihal and Barrett, 1976; Kahneman, Ben-Ishai and Lotan, 1973). Extension of this relationship to differing levels of perceived workload is logical.

Individual differences in secondary task performance have been addressed by a number of researchers through association with personality constructs. Gibson and Curran (1974) and Huddleston (1974) have shown that the introduction of an additional task may increase arousal which has been shown to affect differing personality types in contrasting ways. Motivation is a related factor that may affect secondary task performance (Kalsbeek and Sykes, 1967).

Damos (1988) reviewed six studies that examined the relationships between established individual differences constructs and subjective estimates of workload. She pointed out that the major personality trait which has been studied is cognitive complexity, a trait which reflects the abstractness of thought. Using a projectile firing task with four levels of difficulty, Robertson (1984) examined the relation between subjective estimates of difficulty and objective task difficulty. She found that objective and subjective difficulty were correlated for abstract thinkers but not for concrete thinkers. Robertson and Meshkati (1985) confirmed these results and also found that concrete thinkers tended to adopt less complex decision styles.

Damos and Bloem (1985) found that under dual-task conditions, Type A (coronary prone) subjects reported significantly higher frustration on the Frustration Level subscale of the NASA TLX subjective workload measure.

However, under single-task conditions, the Type B subjects reported more frustration. In a separate study by Damos (1985), this result could not be confirmed.

There are several indications of the relationship between personality traits and physiological reaction parameters. Ray and Lamb (1974) and Gatchel (1975) found that internal locus of control subjects were better able to modify their cardiac responses, thus making physiological measures of workload rather suspect for those subjects. Meshkati and Loewenthal (1988) found that an operator's individual information processing behavior affects sinus arrhythmia and the subjective rating of task difficulty.

Finally, general individual differences and differences in the adaptivity of the driver to the system, the driving task, and the resulting impression of the task as viewed by the driver, may cause higher than normal ratings (Williges and Wierwille, 1979). To accommodate differences among subjects in terms of time-sharing abilities, Wickens (1979) suggests that particular workload measurement techniques be calibrated for different operators. To summarize, the level of mental workload is in fact very individualized and this human variability is a major obstacle to workload assessment.

The driving environment

As pointed out at the beginning of this chapter, the concept of workload involves an interaction of the task, the operator and the environment in which the task is performed. For the driving task, one may consider two distinct environments—internal and external. The major elements of the internal environment include temperature, sound and noise, lighting, vibration, acceleration and the social environment. The interaction of these many elements greatly affects driver workload and driving performance. In a properly working modern automobile, a well-controlled climate is provided for the driver and passengers. On some vehicles, a thermostatically controlled HVAC system even reduces the amount of attention demanded of the driver. However, in cars lacking air conditioning or cars with malfunctioning HVAC systems, temperature extremes may have detrimental effects on performance.

The environment external to the vehicle includes such factors as roadway design, traffic signals and speed limits, road surface, time of day (or night), lighting, weather conditions, traffic conditions and the movement patterns of other drivers.

Driver workload research

Based on epidemiology, several factors impact automobile accident statistics and by implication produce a corresponding impact on the level of driver workload. These factors include day vs. night driving, automobile velocity, road type, visibility, driver expectancy, and so on. However, from an empirical workload assessment perspective, it has been more difficult to quantify the relevant factors.

The overwhelming majority of workload assessment research has been conducted in the aviation setting. Advanced cockpit automation and air traffic control technology have in many ways placed additional mental processing demands on the pilot and controller. Workload certification for aircraft cockpits has also driven the research programs. That does not mean, however, that there has been a total lack of workload assessment research on the automobile driving task. As early as the 1960s, Brown (1962a, 1962b) conducted a series of studies to evaluate the attentional demands placed on drivers in various situations. Since then, many researchers have examined driver workload of one type or another using some combination of primary task, secondary task, subjective and physiological measures. A survey of major studies follows.

Primary task studies

As pointed out earlier, specific primary task measures may not accurately reflect the level of workload since people may adjust their level of motivation to match the difficulty of the task and thus maintain the same level of performance. Thus, primary task measures may not be sensitive to changes in workload or task difficulty. One method of controlling for motivation is the forced time-sharing technique of occlusion (Senders *et al.*, 1967). Senders *et al.* used a translucent visor to occlude the driver's view of the roadway for an adjustable frequency and duration. The visor was raised and lowered by the experimenters in some cases and by the subjects in others. When the drivers were allowed less frequent and shorter observations, they tended to drive at slower speeds. When the drivers had control of the visor, they tended to make longer and more frequent observations as the speed at which they were instructed to drive increased. The minimum time that viewing was obstructed was used to determine the percent workload.

Farber and Gallagher (1972) also used the occlusion technique to measure the attentional demands of the driving task, and to determine the effect of visibility conditions on driving task difficulty in a slalom course. The primary task measures of driving performance (maneuver smoothness and cones knocked down) were insensitive to task difficulty, but attentional demand as measured by frequency of looks was sensitive.

Secondary task studies

The early studies of Brown and his colleagues concentrated on the use of secondary task measures to evaluate 'spare mental capacity'. Brown and Poulton (1961) used an auditory subsidiary (secondary) task of determining which digit in an eight-digit number differed from the preceding eight-digit number. In comparing two different traffic density situations (a residential area vs. a shopping area) using average drivers, they found that secondary task performance was somewhat poorer in heavy traffic. Brown and Poulton obtained the same results using advanced drivers and a mental addition secondary task.

Brown (1966) used a similar auditory task to study the performance of successful and unsuccessful trainee drivers of public service vehicles. Success on an independent driving test was related to previous driving experience and to an intermediate progress check. More importantly, secondary task performance was significantly better for the successful group of drivers compared with the unsuccessful group. Objective measures of driver control movements (steering wheel reversals, longitudinal decelerations, etc.) did not differ between groups. Brown (1965) also compared attention and memory tasks as secondary tasks to evaluate driver fatigue.

Hoffman and Joubert (1966) used a secondary task involving the detection of numerical odd-even-odd sequences in an evaluation of the effects of vehicle handling parameters on driver steering performance. Vehicle parameters included vehicle response time, steering gear ratio, and near-sight and far-sight distances. The driving task consisted of tracking through a narrow winding course marked by traffic cones. The secondary task possessed the same level of sensitivity in measuring spare mental capacity as did the objective measure of the number of traffic cones hit during the obstacle avoidance task. However, the changes in secondary task performance as a function of relatively small changes in handling and steering ratio were not statistically significant.

McDonald (1973) and McDonald and Ellis (1975a) used a hood-mounted digit display to present a secondary task in an evaluation of driver attention related to lane curvature and vehicle speed. They demonstrated that the level of workload in tracking curves increased with increasing degree of curvature and increasing vehicle speed in passing through the curves.

Wierwille et al. (1977) used a secondary task of reading random digits aloud from a single-digit dashboard display in a simulator study to evaluate vehicle handling parameters. They found that workload increased significantly with degraded vehicle handling but that steering ratio did not produce a significant change in workload. Increased disturbance levels in the form of simulated wind gusts and roadway curvature produced a significant decrease in secondary task performance. The results of this study, combined with the results of Hoffman and Joubert (1966), support the use of secondary tasks to estimate driver workload but illustrate that the changes in secondary task performance are not as large as might be desired, further suggesting that more sensitive measures are needed.

An important question about secondary task measures concerns the degree to which the secondary task intrudes on the primary task. Brown (1965) found that secondary tasks significantly increased the time needed to complete a test driving circuit. Brown (1966) and Brown, Simmonds, and Tickner (1967) found that in prolonged driving tasks, the presence of a secondary task increased steering wheel activity and the mean driving time to complete a circuit.

Wierwille and Gutmann (1978) analyzed additional performance data (steering reversals, high-pass steering deviation, yaw deviation and lateral deviation) from their previous study. They found that the introduction of the digit-reading secondary task produced poorer performance on the standardized performance measures at lower primary task workload levels. However, the measures were not affected by the secondary task at high workload levels. The authors concluded that there was some non-redundant information in the secondary task that was not directly obtainable from the primary task measures, even though the secondary task measure was somewhat insensitive. These conclusions support the concept of employing multiple measures of driver workload (i.e., primary, secondary, subjective and physiological).

To address the issue of multiple measures, Hicks and Wierwille (1979) compared five mental workload assessment procedures in a moving-base simulator. The methods included primary task performance (lateral deviation, yaw deviation, steering reversals), secondary task performance using the previously described digit-reading task, visual occlusion, subjective opinion scales, and cardiac sinus arrhythmia. Different levels of workload were generated by the placement of wind gusts along the vehicle. Gusts at the front of the vehicle produced high workload, those toward the center represented lower workload. Significant differences among workload levels were detected with the subjective opinion scales, and with the primary performance measures of lateral deviation, yaw deviation and steering reversals. From highest to lowest sensitivity, the methods would be ranked as steering reversals and yaw deviation, subjective ratings and lateral deviation. The occlusion, cardiac arrhythmia, and secondary task measures were *not* sensitive to the workload manipulation.

Behavioral studies

Unema, Rotting and Luczak (1988) examined the relationship between eye movements and mental workload for novice and experienced city bus drivers. They hypothesized that novice drivers would show a larger number of very short fixations and a shorter mean fixation duration than experienced drivers, and that the same would be true for all drivers when confronted with difficult or complex situations. They found that experience and situational difficulty did affect the number of short fixations. In general, fixation duration decreased with increasing situational complexity, and there was a tendency toward longer fixations with more experience.

Subjective ratings and verbal protocols

Aside from the rating scales mentioned previously, very little research has been done in this area with respect to driver workload assessment. Hicks and Wierwille (1979) used two 11-point Likert scales to rate the effect of crosswind disturbance (from 'extremely harsh and troublesome' to 'extremely small and imperceptible') and the attention required to keep in the right lane (from 'extremely high attention needed' to 'extremely low attention needed'). Both rating scales demonstrated highly significant differences across the three levels of workload, which supports the contention that a driver's or pilot's direct perception of feelings, exertion or condition may provide the sensitive and reliable indicator of workload. Alternatively, the use of a scale similar to the Borg Scale for physical workload may have some potential.

Physiological measures

Few studies have examined physiological indices of driver workload. Most of these studies have employed galvanic skin response (GSR), alternatively known as electrodermal response (EDR). Michaels (1960) measured the galvanic skin response of subjects driving on arterial routes and alternate routes. Although he found a significantly higher GSR on the arterial routes, the measure was not sensitive to design factors along the roadway. Other researchers who have used GSR include Hulbert (1957), Taylor (1964) and Preston (1969).

Helander (1975) examined electrodermal response, heart rate and muscle activity (EMG) of two muscles in the driver's right leg, in addition to several vehicle parameters. He found that the difficulty of a traffic event is reflected in both the driver's use of the brake and in the physiological responses. EDR was most sensitive while the average heart rate followed a smoother change and was less responsive to swift changes in the environment. The EMG variables quantify primarily the physical workload aspects associated with exercising longitudinal control of the vehicle.

Designing to reduce workload—driving in the future

From the previous review, the reader may draw the conclusion that substantial research in driver mental workload has been conducted. While this is the case, one might also question the progress of the research. On the one hand, several workload measures have been tested and effectively used to isolate variables that affect driver workload. However, as in other applications of workload assessment, there is still no universal measuring stick for driver workload. One of the major advantages of considering driver mental workload is that it provides a conceptual framework within which the designer and the human factors engineer can develop vehicles that offer reduced perceptual and cognitive requirements. Attention to the concepts of shared multiple resources and the various resource dimensions should enable designers to balance the requirements of the various elements of the driving task better. Unfortunately, the increased complexity of new display and control systems seems to work against the best intentions of the designer and human factors engineer.

Wierwille and Peacock (1989) discuss the automotive advances likely to take place in the 1990s. With respect to ride and handling, they mention a move toward systems that will result in more effective and active control of the vehicle. These systems will permit tailoring the handling of the vehicle to the driver, vehicle speed, type of road and other conditions. When thoroughly tested and implemented, these systems have the potential for reducing driver workload.

Much of the current and near-future advanced technology in automobiles involves the instrument panel displays and controls, navigation systems and virtual image displays. The negative impacts on workload from the increased visual requirements of these devices have been mentioned earlier. Attention to these negative impacts is likely to result in the same advanced features without their current disadvantages.

Automated longitudinal guidance (speed control and braking) and automated lateral guidance (steering) have enormous potential for reducing the visual requirements of driving. Successful implementation of these systems would decrease the dissimilarity of driving tasks and flying tasks where the pilot has an autopilot feature. However, these systems would not reduce the visual attentional demands of responding to events. Systems for automated collision avoidance could help reduce this aspect of driver workload. All of these systems must undergo thorough performance evaluation, particularly in terms of switching between manual and automatic control, and training drivers to perform monitoring functions in place of active control.

In moving to a system where one enters the vehicle, inputs the desired destination, and allows the car to proceed with little or no intervention, it becomes obvious that the role of the driver will change from primarily manual control to supervisory control. The major task will be one of monitoring the system and perhaps providing backup. Driver workload in this setting represents a completely different concept which presents new and difficult challenges to the engineering psychologist and human factors engineer. In this new environment, there may be little concern about entering the zone of high mental workload. It is likely that one will have to be alert to the boredom of insufficient workload!

References

Antin, J. F., Dingus, T. A., Hulse, M. C. and Wierwille, W. W., 1988, The effects of spatial ability on automobile navigation, in Aghazadeh, F. (Ed.) *Trends in Ergonomics/Human Factors V*, pp. 241–8, Amsterdam: North-Holland.

Brown, I. D., 1962a, Measuring the spare 'mental capacity' of car drivers by a subsidiary auditory task, *Ergonomics*, **5**, 247–50.

Brown, I. D., 1962b, Studies of component movements, consistency and spare capacity of car drivers, *Annals of Occupational Hygiene*, **5**, 131–43.

Brown, I. D., 1965, A comparison of two subsidiary tasks used to measure fatigue in car drivers, *Ergonomics*, **8**, 467–73.

Brown, I. D., 1966, Subjective and objective comparisons of successful and unsuccessful trainee drivers, *Ergonomics*, **9**, 49–56.

Brown, I. D., 1978, Dual task methods of assessing work-load, *Ergonomics*, **21**, 221–4.

Brown, I. D. and Poulton, E. C., 1961, Measuring the spare 'mental capacity' of car drivers by a subsidiary auditory task, *Ergonomics*, **4**, 35–40.

Brown, I. D., Simmonds, D. C. V. and Tickner, A. H., 1967, Measurement of control skills, vigilance, and performance on a subsidiary task during 12 hours of car driving, *Ergonomics*, **10**, 665–73.

Brown, I. D., Tickner, A. H. and Simmonds, D. C. V., 1966, Effects of prolonged driving upon driving skill and performance of a subsidiary task, *Industrial Medicine and Surgery*, **35**, 760–5.

Chukwu, N. A. and Schlegel, R. E., 1990, 'Aimed hand movements under direct vision, limited vision and non-visual conditions', presentation at the Human Factors Society 34th Annual Meeting, Orlando, FL.

Damos, D. L., 1985, The relation between the Type A behavior pattern, pacing, and subjective workload under single- and dual-task conditions, *Human Factors*, **27**, 675–80.

Damos, D. L., 1988, Individual differences in subjective estimates for workload, in Hancock, P. A. and Meshkati, N. (Eds) *Human Mental Workload*, pp. 231–7, Amsterdam: North-Holland.

Damos, D. L. and Bloem, K., 1985, Type A behavior pattern, multiple-task performance, and subjective estimation of mental workload, *Bulletin of the Psychonomic Society*, **23**, 53–56.

Dewar, R. E., 1988, In-vehicle information and driver overload, *International Journal of Vehicle Design*, **9**, 557–64.

Eggemeier, F. T. and O'Donnell, R. D., 1982, A conceptual framework for development of a workload assessment methodology, in *Text of the Remarks Made at the 1982 American Psychological Association Annual Meeting*, Washington DC: American Psychological Association.

Farber, E. and Gallagher, V., 1972, Attentional demand as a measure of the influence of visibility conditions on driving task difficulty, *Highway Research Record*, **414**, 1–5.

Fitts, P. M., 1954, The information capacity of the human motor system in controlling the amplitude of movement, *Journal of Experimental Psychology*, **47**, 381–91.

French, R. L., 1986, Historical overview of automobile navigation technology, in *Proceedings of the 36th IEEE Vehicular Technology Conference*, New York: IEEE.

Gatchel, R. J., 1975, Change over training sessions of relationships between locus of control and voluntary heart rate control, *Perceptual and Motor Skills*, **40**, 424–6.

Gibson, H. B. and Curran, J. B., 1974, The effect of distraction on a psychomotor task studied with respect to personality, *Irish Journal of Psychology*, **2**, 148-58.

Godthelp, H., Milgram, P. and Blaauw, G. J., 1984, The development of a time-related measure to describe driving strategy, *Human Factors*, **26**, 257–68.

Graesser, A. C., 1989, 'A human factors tool for the prediction of driver performance with hand controls', Final Report for General Motors Corporation C-P-C Group, Memphis, TN: Memphis State University, Department of Psychology.

Gramberg-Danielsen, B., 1967, *Sehen und Verkehr*, Springer Verlag.

Hale, A. R., Quist, B. W. and Stoop, J., 1988, Errors in routine driving tasks: A model and proposed analysis technique, *Ergonomics*, **31**, 631–41.

Hancock, P. A., 1988, The effect of gender and time of day upon the subjective estimate of mental workload during the performance of a simple task, in Hancock, P. A. and Meshkati, N. (Eds) *Human Mental Workload*, pp. 239–50, Amsterdam: North-Holland.

Hancock, P. A. and Meshkati, N., 1988, *Human Mental Workload*, Amsterdam: North-Holland.

Hart, S. G., 1985, Theory and measurement of human workload, in Zeidner, J. (Ed.) *Human Productivity Enhancement*, New York: Praeger.

Hart, S. G. and Staveland, L. E., 1988, Development of NASA-TLX (Task Load Index): Results of empirical and theoretical research, in Hancock, P. A. and Meshkati, N. (Eds) *Human Mental Workload*, pp. 139–83, Amsterdam: North-Holland.

Hart, S. G. and Wickens, C. D., 1990, Cognitive workload, in Booker, H. (Ed.) *People, Machines, and Organizations: A Manprint Approach to System Integration*.

Heintz, F., Haller, R. and Bouis, D., 1982, Safer trip computers by human factors designs, in *Electronic Displays and Information Systems and On-Board Electronics*, pp. 113–7, Warrendale, PA: Society of Automotive Engineers.

Helander, M. G., 1975, Physiological reactions of drivers as indicators of road traffic demand, *Transportation Research Record*, **530**, 1–17.

Henderson, R. L. and Burg, A., 1974, 'Vision and audition in driving', Report No. TM(L)-5297/000/00, System Development Corporation.

Hicks, T. G. and Wierwille, W. W., 1979, Comparison of five mental workload assessment procedures in a moving-base driving simulator, *Human Factors*, **21**, 129–43.

Hoffmann, E. R. and Joubert, P. N., 1966, The effect of changes in some vehicle handling variables on driver steering performance, *Human Factors*, **8**, 245–63.

Hopkin, V. D., 1979, General discussion based upon interactive group sessions, in Moray, N. (Ed.) *Mental Workload: Its Theory and Measurement*, pp. 484–7, New York: Plenum.

Huddleston, H. F., 1974, Personality and apparent operator capacity, *Perceptual and Motor Skills*, **38**, 1189–90.

Hulbert, S. F., 1957, Drivers' GSRs in traffic, *Perceptual and Motor Skills*, **7**, 304–15.

Johansen, G., 1979, Workload and workload measurement, in Moray, N. (Ed.) *Mental Workload: Its Theory and Measurement*, pp. 3–11, New York: Plenum.

Kahneman, D., Ben-Ishai, R. and Lotan, M., 1973, Relation of a test of attention to road accidents, *Journal of Applied Psychology*, **58**, 113–5.

Kalsbeek, J. W. H. and Sykes, R. N., 1967, Objective measurement of mental load, *Acta Psychologica*, **27**, 253–61.

Kantowitz, B. H., 1985, Channels and stages in human information processing: A limited analysis of theory and methodology, *Journal of Mathematical Psychology*, **29**, 135–74.

Kantowitz, B. H. and Casper, P. A., 1988, Human workload in aviation, in Wiener, E. L. and Nagel, D. C. (Eds) *Human Factors in Aviation* (San Diego, CA: Academic Press), pp. 157–187.

Kantowitz, B. H. and Sorkin, R. D., 1983, *Human Factors: Understanding People–System Relationships*, New York: Wiley.

Kiefer, R. J., 1991, 'Effect of a head-up versus head-down digital speedometer on visual sampling behavior and speed control performance during daytime automobile driving', SAE Report 910111, Warrendale, PA: Society of Automotive Engineers.

Knowles, W. B., 1963, Operator loading tasks, *Human Factors*, **5**, 155–61.

Kobayashi, T., 1988, Human factors in driving, *International Journal of Vehicle Design*, **9**, 586–99.

Kramer, U. and Rohr, G., 1981, Psycho-mathematical model of vehicular guidance based on fuzzy automata theory, in *Proceedings of the European Annual Manual*, Delft, The Netherlands: Delft University of Technology.

Kramer, U. and Rohr, G., 1982, A model of driver behaviour, *Ergonomics*, **25**, 891–907.

Leplat, J., 1978, Factors determining workload, *Ergonomics*, **21**, 143–9.

MacDonald, W. A. and Hoffman, E. R., 1980, Review of relationships between steering wheel reversal rate and driving task demand, *Human Factors*, **22**, 733–9.

McDonald, L. B., 1973, 'A model for predicting driver workload in the freeway environment: A feasibility study', Unpublished doctoral dissertation, Texas A&M University, College Station, TX.

McDonald, L. B. and Ellis, N. C., 1975a, Driver work load for various turn radii and speeds, *Transportation Research Record*, **530**, 18–30.

McDonald, L. B. and Ellis, N. C., 1975b, Stress threshold for drivers under various combinations of discrete and tracking workload, in *Proceedings of the Human Factors Society 19th Annual Meeting*, pp. 488–93, Santa Monica, CA: Human Factors Society.

McGranaghan, M., Mark, D. M. and Gould, M. D., 1987, Automated provision of navigation assistance to drivers, *The American Cartographer*, **14**, 121–38.

McGrath, J. J., 1975, Driver expectancy and performance in locating automotive controls, *Vehicle Research Institute Report*, **14.1**.

Meshkati, N., 1988, Toward development of comprehensive theories of mental workload, in Hancock, P. A. and Meshkati, N. (Eds) *Human Mental Workload*, pp. 305–14, Amsterdam: North-Holland.

Meshkati, N. and Loewenthal, A., 1988, The effects of individual differences in information processing behavior on experiencing mental workload and perceived task difficulty: An experimental approach, in Hancock, P. A. and Meshkati, N. (Eds) *Human Mental Workload*, pp. 269–88, Amsterdam: North-Holland.

Meshkati, N., Hancock, P. A. and Rahimi, M., 1990, Techniques in mental workload assessment, in Wilson, J. R. and Corlett, E. N. (Eds) *Evaluation of Human Work*, pp. 605–27, London: Taylor & Francis.

Michaels, R. M., 1960, Tension response of drivers generated on urban streets, *HRB Bulletin*, **271**, 29–43.

Mihal, W. L. and Barrett, G. V., 1976, Individual differences in perceptual information processing and their relation to automobile accident involvement, *Journal of Applied Psychology*, **61**, 229–33.

Moray, N., 1979, *Mental Workload: Its Theory and Measurement*, New York: Plenum.

Moray, N., 1982, Subjective mental workload, *Human Factors*, **23**, 25–40.

Moray, N., 1984, Mental workload, in *Proceedings of the 1984 International Conference on Occupational Ergonomics*, pp. 41–46.

Mourant, R. R. and Langolf, G. D., 1976, Luminance specifications for automobile instrument panels, *Human Factors*, **18**, 71–84.

Nygren, T. E., 1991, Psychometric properties of subjective workload measurement techniques: Implications for their use in the assessment of perceived mental workload, *Human Factors*, **33**, 17–33.

O'Donnell, R. and Eggemeier, T., 1986, Workload assessment methodology, in Boff, K. R., Kaufman, L. and Thomas, J. P. (Eds) *Handbook of Perception and Human Performance*, New York: Wiley.
Ogden, G. D., Levine, J. M. and Eisner, E. J., 1979, Measurement of workload by secondary tasks, *Human Factors*, **21**, 529–48.
O'Hanlon, J. F. and deGier, J. J., 1986, *Drugs and Driving*, London: Taylor & Francis.
Peacock, J. B., 1972, 'Information retention in manual control', Unpublished doctoral thesis, University of Birmingham, Department of Engineering Production, Birmingham, UK.
Perel, M., 1983, 'Vehicle familiarity and safety', NHTSA Technical Note DOT HS-806-509, Washington, DC: US Department of Transportation.
Preston, B., 1969, Insurance classifications and drivers' galvanic skin response, *Ergonomics*, **12**, 437–46.
Ray, W. J. and Lamb, S. B., 1974, Locus of control and the voluntary control of heart rate, *Psychosomatic Medicine*, **36**, 180–2.
Reid, G. and Nygren, T. E., 1988, The Subjective Workload Assessment Technique (SWAT): A scaling procedure for measuring mental workload, in Hancock, P. A. and Meshkati, N. (Eds) *Human Mental Workload*, pp. 185–218, Amsterdam: North-Holland.
Robertson, M., 1984, Personality differences as a moderator of mental workload behavior: Mental workload performance and strain reactions as a function of cognitive complexity, in *Proceedings of the Human Factors Society 28th Annual Meeting*, pp. 690–4, Santa Monica, CA: Human Factors Society.
Robertson, M. and Meshkati, N., 1985, Analysis of the effects of two individual differences classification models on experiencing mental workload of a computer generated task: A new perspective to job design and task analysis, in *Proceedings of the Human Factors Society 29th Annual Meeting*, pp. 178–82, Santa Monica, CA: Human Factors Society.
Salvendy, G., 1987, *Handbook of Human Factors*, New York: Wiley.
Sanders, A. F., 1986, Drugs, driving and the measurement of human performance, in O'Hanlon, J. F. and deGier, J. J. (Eds) *Drugs and Driving*, pp. 3–16, London: Taylor & Francis.
Sanders, M. S. and McCormick, E. J., 1987, *Human Factors in Engineering and Design*, New York: McGraw-Hill.
Saunby, C. S., Farber, E. T. and Jack, D. T., 1986, A computerized evaluation system for automotive displays and controls, in *Proceedings of the Human Factors Society 30th Annual Meeting* pp. 916–20, Santa Monica, CA: Human Factors Society.
Schlegel, B., Schlegel, R. E. and Gilliland, K., 1988, Determining excessive mental workload with the Subjective Workload Assessment Technique, in *Proceedings of the Tenth Congress of the International Ergonomics Association*, Vol. II, pp. 475–7, Sydney, Australia: Ergonomics Society of Australia.
Schlegel, R. E., 1989, 'A human factors tool for the prediction of driver performance with hand controls', Final Report under contract PB276743 with General Motors Corporation C-P-C Group, Norman, OK: University of Oklahoma, School of Industrial Engineering.
Senders, J. W., Kristofferson, A. B., Levison, W. H., Dietrich, C. W. and Ward, J. L., 1967, The attentional demand of automobile driving, *Highway Research Record*, **195**, 15–33.
Sheridan, T. B., 1980, Mental workload, what is it? Why bother with it?, *Human Factors Society Bulletin*, **23**, 1–2.
Shingledecker, C. A., Crabtree, M. S., Simons, J. C., Courtright, J. F. and O'Donnell, R. D., 1980, 'Subsidiary radio communications tasks for workload assessment in R&D simulations: I. Task development and workload scaling', Technical Report AFAMRL-TR-80-126, Dayton, OH: Wright-Patterson Air Force Base Aerospace Medical Research Laboratory.

Snyder, H. L. and Monty, R. W., 1985, Methodology and results for driver evaluation of electronic automotive displays, in *Proceedings of the Ninth Congress of the International Ergonomics Association*, pp. 514–6, London: Taylor & Francis.

Sternberg, S., 1969, The discovery of processing stages: Extensions of Donders' method, in Koster, W. G. (Ed.) Attention and Performance II, *Acta Psychologica*, **30** (Amsterdam: North-Holland), 276–315.

Stokes, A., Wickens, C. and Kite, K., 1990, *Display Technology—Human Factors Concepts*, Warrendale, PA: Society of Automotive Engineers.

Taylor, D. H., 1964, Drivers' galvanic skin response and the risk of accident, *Ergonomics*, **7**, 439–451.

Treat, J. R., Tumbas, N. S, McDonald, S. T., Shinar, D., Nume, R. D., Mayer, R. E., Stansifer, R. L. and Castellan, N. J., 1977, 'Tri-level study of the cause of traffic accidents', Report No. DOT HS-034-3-535-77 (TAC), Washington, DC: US Department of Transportation.

Unema, P., Rotting, M. and Luczak, H., 1988, Eye movements and mental workload in man–vehicle interaction, in *Proceedings of the Tenth Congress of the International Ergonomics Association*, Vol. II, pp. 463–5. Sydney, Australia: Ergonomics Society of Australia.

Welford, A. T., 1958, *Ageing and Human Skill*, Oxford University Press for the Nuffield Foundation.

Welford, A. T., Norris, A. H. and Shock, N. W., 1969, Speed and accuracy of movement and their changes with age, in Koster, W. G. (Ed.) Attention and Performance II, *Acta Psychologica*, **30** (Amsterdam: North-Holland) 3–15.

Wickens, C., 1979, Measures of workload, stress and secondary tasks, in Moray, N. (Ed.) *Mental Workload; Its Theory and Measurement*, pp. 79–99, New York: Plenum.

Wickens, C., 1990, *Engineering Psychology and Human Performance*, Glenview, IL: Scott Foresman.

Wiener, E. L., 1985, Beyond the sterile cockpit, *Human Factors*, **27**, 75–90.

Wierwille, W. W. and Gutmann, J. C., 1978, Comparison of primary and secondary task measures as a function of simulated vehicle dynamics and driving conditions, *Human Factors*, **20**, 233–44.

Wierwille, W. W. and Peacock, B., 1989, Human factors and the automobile of the future, *Human Factors Society Bulletin*, **32** (11), 1–5.

Wierwille, W. W. and Williges, B. H., 1978, 'Survey and analysis of operator workload assessment techniques', Report No. S-78-101, Blacksburg, VA: Systemetrics, Inc.

Wierwille, W. W. and Williges, B. H., 1980, 'An annotated bibliography on operator mental workload assessment', Report SY-27R-80, Patuxent River, MD: Naval Air Test Center.

Wierwille, W. W., Gutmann, J. C., Hicks, T. G. and Muto, W. H., 1977, Secondary task measurement of workload as a function of simulated vehicle dynamics and driving conditions, *Human Factors*, **19**, 557–65.

Wierwille, W. W., Hulse, M. C., Fischer, T. J. and Dingus, T. A., 1988, Strategic use of visual resources by the driver while navigating with an in-car navigation display system, in *XXII FISITA Congress Technical Papers; Automotive Systems Technology: The Future* (SAE P-211; vol. II, pp. 2.661–2.675), Warrendale, PA: Society of Automotive Engineers.

Wildervanck, C., Mulder, G. and Michon, J. A., 1978, Mapping mental load in car driving, *Ergonomics*, **21**, 225–229.

Williges, R. C. and Wierwille, W. W., 1979, Behavioral measures of air crew mental workload, *Human Factors*, **21**, 549–74.

Zwahlen, H. T., 1985, Driver eye scanning, the information acquisition process and sophisticated in-vehicle information displays and controls, in *Proceedings of the Ninth Congress of the International Ergonomics Association*, pp. 508–10, London: Taylor & Francis.

Zwahlen, H. T. and Balasubramanian, K. N., 1974, A theoretical and experimental investigation of automobile path deviations when driver steers with no visual input, *Transportation Research Record*, **520**, 25–37.

Zwahlen, H. T. and DeBald, D. P., 1986, Safety aspects of sophisticated in-vehicle information displays and controls, in *Proceedings of the Human Factors Society 30th Annual Meeting*, pp. 256–60, Santa Monica, CA: Human Factors Society.

18

Models that simulate driver performance with hand controls

Arthur C. Graesser and William Marks

When engineers design the controls and displays on the panel of a vehicle, they need to know how the drivers would perform on tasks that use the panel components. They need some estimate of the error rates and task completion times for such actions as turning on the windshield wipers, turning off the heater, and adjusting the rear-view mirror. If the vehicle is already designed, human factors engineers can provide these estimates by collecting performance data from a representative sample of drivers. However, this empirical approach is frequently impractical. The collection and analysis of human performance data may be too time-consuming to be effectively coordinated with production schedules. The empirical approach is impossible when the engineer is at the design stage and needs some idea of how performance will be affected by a particular design feature.

Both designers and human factors specialists would benefit enormously from a computer tool that simulated performance data. The engineer would first specify the components on the panel (e.g., each display and control) and the attributes of the components (e.g., the type, size and location of a particular control). The engineer would specify the characteristics of the driver, such as the driver's sex, age and any visual deficits. The engineer would declare a particular action or action sequence to be simulated, such as turning on the windshield wipers and then turning off the defrost. With this information, the computer program would simulate a) the mean and distribution of transaction times and b) error rates in completing the two-step transaction. The simulated times and errors would be based on research in human factors and cognitive science. The computer tool would be useful because it projects performance data during the design stage and thereby minimizes the likelihood of manufacturing panels that are difficult for drivers to use.

This chapter describes two classes of models that provide a computational foundation for simulating the actions that drivers perform on hand controls in vehicles. The hand controls include buttons, knobs, levers, slides, stalks and other physical components that operate the radio, the HVAC, the windshield wiper, the turn signal, the ignition, and so on. Given that our examples involve the driver's hand controls and that the computer tool would no doubt be useful, we refer to these two classes of models as HANDY-1 and HANDY-2. HANDY-1 has quantitative foundations in additive multicomponent models from the field of information processing (Sternberg, 1969; McClelland, 1979; Meyer et al., 1985). HANDY-2 was inspired by qualitative, strategic models in cognitive science that have adopted production system architectures (Anderson, 1983; Card, Moran and Newell, 1983; Jacob, 1983; Kieras and Bovair, 1984; Klahr, Langley and Neches, 1987). This chapter presents the advantages and shortcomings of these two classes of models.

HANDY-1: An additive multicomponent model

This first class of models is the easiest to conceptualize and to implement in the form of a computer program. In additive multicomponent models, there are a set of N components and each component (C_i) increases overall processing time by some amount (T_i). The overall transaction time is an additive combination of a constant (k) and the contributions for all N components, as shown in equation 1.

$$\text{Transaction time} = k + T_1 + T_2 \ldots T_N \quad (1)$$

In essence, the researcher identifies all of the major components that predict transaction time and determines how much time it takes to execute each component.

The magnitude of T_i (perhaps measured in milliseconds) is a product of the value of C_i and its temporal weighting (b_i), as shown in equation 2.

$$T_i = b_i * C_i \quad (2)$$

For example, suppose that the component of interest is the number of functions associated with a particular control. A control on a radio may either have a single function (turning on the radio) or multiple functions (turning on the radio and adjusting the volume). Suppose that the temporal weighting for this component is 100 milliseconds. According to this weighting, overall transaction times would be incremented by 100 milliseconds for a control that had one function, 200 milliseconds for a control with two functions, and 400 milliseconds for a control with four functions. It should be noted that the magnitude of the temporal weighting can be estimated by raw regression coefficients (i.e., b-weights) in multiple regression analyses.

The value of C_i may be a product of a complex mathematical function. In Hick's law, for example, the time to decide among n equally likely alternatives is a logarithmic function (Hick, 1952), as shown in equation 3.

$$\text{Time} = a + b*[\log_2(n + 1)] \quad (3)$$

Therefore, if $b = 150$ milliseconds, then it would take 300 more milliseconds to decide among 15 alternatives than among three alternatives. The values of some components are discrete (0 or 1), e.g., specifying whether a control does or does not have a word label. The values of other theoretical components are on a continuous scale, e.g, the distance between one control and another control.

The assumption that times combine additively has different explanatory foundations. A literal interpretation of HANDY-1 would state that each component takes a total of T milliseconds to process and that the total time is computed by simply adding up contributions from all N components. The time to process a particular component is independent in the sense that it would not be influenced by the set of other components. This literal interpretation is probably not psychologically plausible because components frequently have context-sensitive interactions with other components (McClelland, 1979; Meyer et al., 1985; Welford, 1980).

A more plausible interpretation of HANDY-1 is that the empirical temporal parameters (i.e. b-weights) underestimate the actual processing time needed to execute a particular component (i.e., from the beginning to the completion of the processing component). This is because processing components can be executed in parallel to some extent (although not entirely). In cascade models (McClelland, 1979), for example, stage X provides output that is needed for a subsequent stage Y so stage X must begin its execution before stage Y; however, stage Y can begin after a partial output from X and before stage X is finished being processed. A plausible interpretation of a b-weight in HANDY-1 is that it represents the increment in time to execute a processing component, after considering interactions with other components and after averaging over a set of representative processing contexts.

A concrete application

This section provides a concrete example of how the additive multicomponent model would be applied to vehicle panels. This example simulates the time that an average driver needs to execute a single action (e.g., turn on a heater) or an action sequence. The simulated transaction times would be generated after the engineer specified all of the controls and display elements on the panel.

The components of HANDY-1 are summarized in Figure 18.1. The transaction times are simulated by two subnetworks of components: the learning subnetwork and the asymptotic subnetwork. The learning subnetwork is operative while the driver is learning how to execute a particular transaction, such as turning on the heater in a particular vehicle. The learning subnetwork has three stages:

1 Find the device (i.e., the HVAC unit);
2 Find the control (i.e., the 'heater' button);
3 Operate the control (i.e., press the 'heater' button).

These three stages are ordered sequentially. It is logically impossible to find a control on a device without first finding the device; it is impossible to operate the control on a device without first finding the control. After the driver has hundreds of experiences turning on the heater, the learning subnetwork atrophies in its impact on transaction times and performance is determined entirely by the asymptotic subnetwork. The asymptotic network captures performance when the transaction is overlearned and automatized. According to the 'power law of practice' (de Jong, 1957; Rosenbloom and Newell, 1987a, b), the learning subnetwork prevails during early trials and dampens exponentially as a function of number of trials. The dampening is computed as t^{-r}, where t is the number of trials and r is the learning rate.

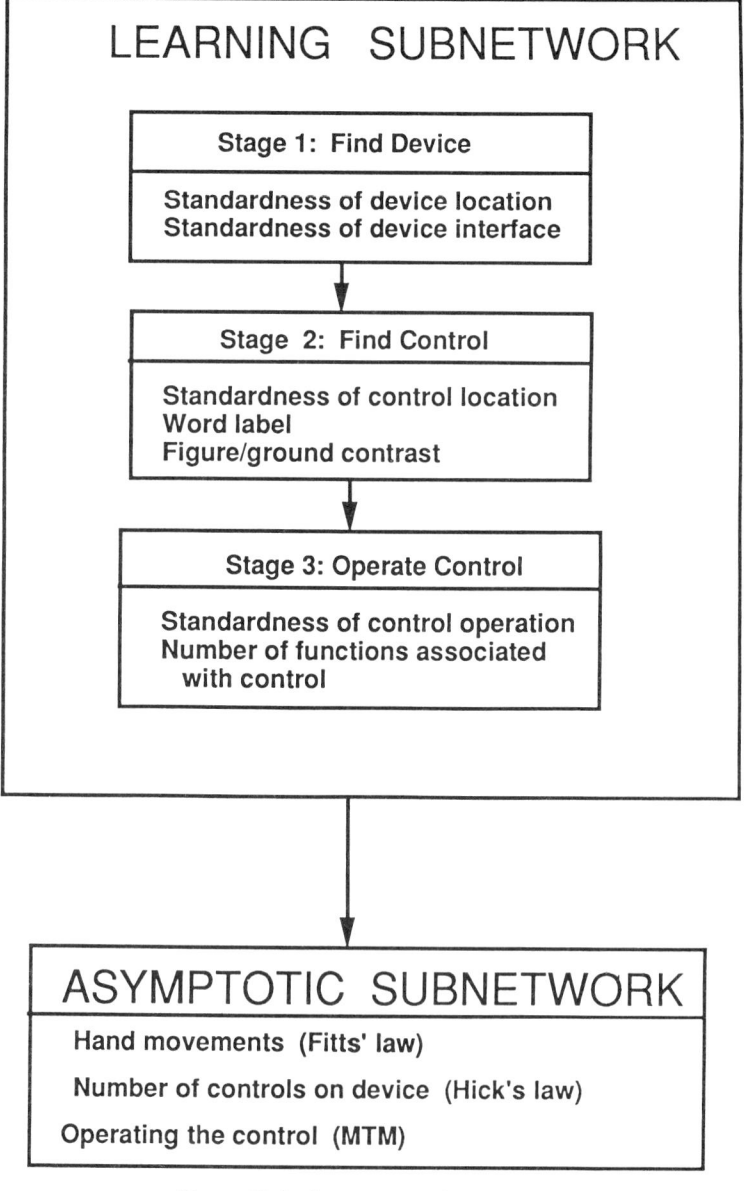

```
┌─────────────────────────────────────────────┐
│          LEARNING SUBNETWORK                │
│                                             │
│   ┌─────────────────────────────────────┐   │
│   │     Stage 1: Find Device            │   │
│   ├─────────────────────────────────────┤   │
│   │ Standardness of device location     │   │
│   │ Standardness of device interface    │   │
│   └─────────────────────────────────────┘   │
│                    │                        │
│                    ▼                        │
│   ┌─────────────────────────────────────┐   │
│   │     Stage 2: Find Control           │   │
│   ├─────────────────────────────────────┤   │
│   │ Standardness of control location    │   │
│   │ Word label                          │   │
│   │ Figure/ground contrast              │   │
│   └─────────────────────────────────────┘   │
│                    │                        │
│                    ▼                        │
│   ┌─────────────────────────────────────┐   │
│   │     Stage 3: Operate Control        │   │
│   ├─────────────────────────────────────┤   │
│   │ Standardness of control operation   │   │
│   │ Number of functions associated      │   │
│   │    with control                     │   │
│   └─────────────────────────────────────┘   │
└─────────────────────────────────────────────┘
                     │
                     ▼
┌─────────────────────────────────────────────┐
│         ASYMPTOTIC SUBNETWORK               │
├─────────────────────────────────────────────┤
│  Hand movements (Fitts' law)                │
│  Number of controls on device (Hick's law)  │
│  Operating the control (MTM)                │
└─────────────────────────────────────────────┘
```

Figure 18.1. Components of HANDY-1

Asymptotic subnetwork

The three components in the asymptotic subnetwork were recommended in a previous report that attempted to quantify transaction times for hand controls in automobiles analytically (Green, 1979). The time that it takes to move the hand from a source location (e.g., the steering wheel) to the target location (e.g., the heater button) can be estimated by Fitts' law (Fitts, 1954). Fitts' law is shown in equation 4.

$$T = a + b*[\log^2(D/S + 0.5)] \quad (4)$$

T denotes movement time wheras D is the distance the hand moves and S is the size of the target button. The parameter a is an intercept constant, a value that is not of immediate concern. The important parameter is the temporal weight b; the magnitude of this parameter is surprisingly constant across tasks, subject populations and contexts (approximately 100 milliseconds according to Card et al., 1983).

The second component of the asymptotic network selects the appropriate control and is influenced by the number of controls on the device. The time to select the control is approximated by Hick's law (1952), which was provided in equation 3 under conditions in which each control is equally likely to be selected. Alternative formulas are available which take into consideration those situations in which controls are not equally likely, the discriminability of the controls, and the signal strength (Card et al., 1983; Hick, 1952; Welford, 1980). A reasonable estimate of the temporal parameter b is 150 milliseconds.

The third component of the asymptotic network computes the time to operate the control. This time has been estimated by Methods-Time-Measurement (MTM), as discussed by Maynard (1971). MTM decomposes the operation into parameters (e.g., the intensity of required finger pressure) and subactions (e.g, grasp control, push control). There are tables of values which estimate the mean time to execute each subaction and to incorporate each parameter.

Once again, the overall transaction time for the asymptotic subnetwork is an additive combination of the above three components. This subnetwork prevails when the driver has extensive experience performing a particular transaction.

Finding the device

The driver must figure out where the device is located during this stage of the learning subnetwork. If the transaction involves turning on the heater, the driver would scan the panel until the HVAC unit is recognized. Finding the device should take less time if the HVAC unit is at a standard location on the panel (to the right of the steering wheel) than if it is at a nonstandard location (e.g, near the floor, to the left of the steering wheel). Similarly, a very typical HVAC interface should take less time to recognize than a very unusual HVAC interface.

It is possible to scale devices on a vehicle panel on standardness, ranging from 0 (very nonstandard) to 1 (very standard). Consider the standardness of device location for a HVAC unit. The researcher could select a sample of cars and record the location of the HVAC units in each car. The locations could be mapped on a rectangular grid. A frequency score could be computed for each zone in the grid, corresponding to the number of vehicles with the HVAC unit in that zone's location (see Anacapa, 1976). Based on these data, a standardness score could be computed for the location of a particular device on a particular vehicle. The standardness score for an HVAC in a particular car would simply be the proportion of vehicles in the sample that have the HVAC unit in the same zone as that car. In a similar manner, the 'standardness of the device interface' would measure the extent to which vehicles in the vehicle sample have the same interface attributes as the particular device in target vehicle.

Finding the control

Once the device is found, the driver needs to find the appropriate control associated with the transaction to be executed. The time needed to find the control depends on the standardness of the control location. Controls at nonstandard locations should require more time to find than controls at standard locations. The time to find a control should be faster when there is a word, letter string, or graphic symbol that distinctively conveys the function of the control. The time should be faster when there is a contrast in color, texture and brightness between the controls and the background. Temporal parameters (b-weights) need to be estimated for these components that influence the course of finding a control.

Operating a control

Given that the proper control is located, it takes time for the driver to operate the control in order to achieve the desired function. There are at least two components that influence this time during the learning phase. First, the time should be longer if there is a nonstandard method of operating the control. For example, pulling the heater button is nonstandard whereas pushing it is standard. Second, the time should increase as a function of the number of functions associated with the control. Each function would involve manipulating the control in a different manner, so the different methods of manipulating the control might get confused during the learning phase. Once again, temporal parameters need to be estimated for these components associated with operating the control during the training phase.

Simulating transaction times on computer

Transaction times can be simulated by the components in the HANDY-1 network (see Figure 18.1). One temporal parameter (b-weight) is associated with each of the seven components in the learning subnetwork and each of the three components in the asymptotic subnetwork. The parameter values in the learning subnetwork are attenuated to the extent that the driver has more experience with the transaction. The decrease is computed according to the power law of practice discussed earlier (one parameter per component). In addition to these parameter values that estimate means for the 10 components, there are parameters that determine the extent to which the parameters are dispersed (i.e., variance parameters). We assume that the parameter values are distributed according to a gamma function, rather than a normal curve; it is impossible to have negative times so that the distributions for temporal durations are normally skewed to the right. Given that there are 10 mean parameter values, 10 variance parameters, and 10 learning rate parameters, HANDY-1 has a total of 30 parameter values. These same 30 parameters could be used to simulate hundreds of transactions and thousands of transaction sequences after the driver has had different amounts of experience with particular transactions.

One way of simulating the transaction times adopts Monte Carlo methods and discrete event simulation techniques. During each trial, an execution time is produced for each of the 10 components. This time is generated randomly from the theoretical distribution, that is, the gamma distribution defined according to the mean, variance and learning parameters associated with the component. Once the 10 times are generated by this Monte Carlo method (i.e., one time for each of the 10 components), an overall transaction time is generated by adding these times to an overall constant (i.e., the base time value). Therefore, a single transaction time is produced as a result of the simulation of a single trial. The simulation is repeated for hundreds of trials, yielding a distribution of transaction times. The computer program ultimately prints out the mean, standard deviation and frequency distribution for these transaction times. In addition to these values, the computer prints out a statistical upperbound in the transaction time. That is, 5 percent of the simulated times are above the upperbound and 95 percent are below.

The program could also segregate the overall transaction time into the 10 theoretical components. This would allow the engineer to observe those components that are particularly time-consuming and to perform a sensitivity analysis on the simulation. The engineer could then modify the design in order to decrease the times of the problematic components.

Problems with HANDY-1

Additive multicomponent models enjoy a number of advantages over alternative models and computational architectures. They are easy to implement on a computer. It is easy to coordinate these models with standard statistical procedures. In the above example, the researcher could perform a multiple regression analysis that assesses the extent to which transaction times are predicted by the 10 components of HANDY-1. The b-weights from the regression analysis are assumed to be quantitatively close to the theoretical temporal parameters. Additive multicomponent models are tractable in the sense that it is easy to derive quantities analytically and easy to troubleshoot mathematical errors. These models also have a rich tradition of research in human factors and cognitive psychology. For these reasons, additive multicomponent models are pursued more often than alternative classes of models.

Nevertheless, there are potential problems inherent in additive multicomponent models. Perhaps the most serious problem is that systematic and theoretically interesting interactions may occur between processing components (McClelland, 1979; Meyer *et al.*, 1985). In other words, the processing components are not always independent, autonomous processing modules that can be combined in a simple additive fashion. Indeed, researchers frequently are fascinated with failures in independence and additivity among processing components. For example, Meyer *et al.* (1985) reported that an additive multicomponent model may be adequate for transactions that involve a limited number of response alternatives, such as a binary choice, but there is a failure in additivity when there are mutliple response alternatives.

One type of interaction exists when two or more processing components can be executed in parallel. Suppose that the driver is learning how to use a compact disk player. The driver may proceed in finding the CD device (see stage 1 in Figure 18.1) simultaneously with finding the CD player control (see stage 2). That is, locating the controls may be inextricably bound to the process of recognizing the interface in the first place. Whenever there is a 'parsing paradox', it is difficult to determine whether comprehension of the parts is needed to identify the whole, or whether comprehension of the whole is needed to identify a part (Norman and Rumelhart, 1975). The issue of global versus local precedence in processing visual displays and in executing transactions presents a substantial challenge in specifying component processes (Navon, 1977). Consider another example in which components may be partially executed in parallel. In the asymptotic network, the process of selecting a control may be accomplished while the driver is moving the hand to the control. If these two processes are executed simultaneously, then Fitts' law would not combine additively with Hick's law (Beggs *et al.*, 1972). Whenever there is an overlap in the times to execute two different components, the researcher faces a decision: How should the overlapping time be partitioned between the two components?

Another type of interaction occurs when components are executed strategically, under context-sensitive conditions. For example, the driver of a new car might need to execute the 'find device' stage in HANDY-1 under some conditions, but not in others. In some conditions, there is a distinctive function that can be applied to only one device in the car, such as turning on the heater; the HVAC unit is the only device in the car that would have a control involving a heater. Under these conditions, there is no need to find the HVAC device (stage 1) because the driver can directly scan the panel for the 'heater' control label (stage 2). In other conditions, however, the desired function is not unique to a particular device, e.g., turning on the device. A control is needed to turn on many devices in a vehicle; the HVAC, the radio, the CD player, and so on. Under these conditions, the driver must find the appropriate device before finding the appropriate control on the device.

These context-sensitive, strategic processes are easy to miss in the additive multicomponent models. An *ad hoc* approach to handling context-sensitive processing is to add on an additional discrete stage whenever it is needed. However, if this approach is pursued, it would be necessary to have some theoretical foundation for explaining when a particular stage is included versus deleted from the processing stream. Unfortunately, the existing discrete additive models provide no clear basis for context-sensitive processing.

Although interactions may conceivably occur among processing components, the robustness of the interactions is an empirical question. For example, Haberlandt and Graesser (1985) investigated the extent to which word reading times for words in passages are influenced by processing components at the word, sentence and passage levels. Using multiple regression techniques, they found that approximately a dozen components accounted for over 50 percent of the reading time variance. They assessed whether this percentage increased when pairwise interaction terms were added to the regression equation. For example, if components A and B predicted times significantly, would an A × B interaction term predict even more variance? Haberlandt and Graesser (1985) found that all of the interaction terms together improved the amount of explained variance by only 1 percent. In this case, interactions among components did not impose a serious threat to an additive multicomponent model.

The components of HANDY-1 discussed so far hardly exhaust the total set of processing components needed to account for transaction times. For example, additional components are needed to account for problem solving mechanisms that are needed when a driver tries to figure out a complicated control or a difficult action sequence. As researchers accumulate additional processing components, they would simply add them onto HANDY-1 in a modular fashion and thereby account for more variance in transaction times.

HANDY-2: A production system architecture

Production system architectures have been adopted in models that account for strategic cognitive activities, such as problem solving and reasoning (Klahr, Langley and Neches, 1987; Newell and Simon, 1972; Laird, Newell and Rosenbloom, 1987). Production system models have also been developed to account for automatized activities, such as reading (Thibadeau, Just and Carpenter, 1982), editing texts on computer (Card et al., 1983), memory retrieval (Anderson, 1983), and motor skills (Rosenbloom and Newell, 1987a, b). HANDY-2 embraces the production system architecture.

A production system contains a working memory and a set of production rules in its knowledge base. The working memory is a temporary, dynamically changing workspace which receives input from the environment and from production rules that get activated. The working memory contains an agenda of goals that the driver wants to achieve. Each production rule is an IF-THEN expression which captures the notion that a specific action is activated (i.e., triggered) by a particular configuration of states and events in working memory. A production has an IF(condition)-THEN(action) form, as expressed below.

IF [state S_1] THEN [execute action A_1]

Multiple states and actions can be combined in a single production, as shown below.

IF [state S_1 and state S_2 and ... state S_n]
THEN [action A_1, then action A_2, then ... action A_m]

Whenever the condition elements are satisfied by the contents of working memory, the actions in the production are executed and the results update working memory. Whenever the condition elements are not satisfied by the contents of working memory, the actions in the production are not executed.

The production rules in a production system are evaluated and executed in parallel. All of the production expressions are evaluated during a given 'cycle', with the states and events being updated in working memory at the end of each cycle. If several productions are activated within a cycle, the productions can be executed in parallel as long as the actions do not compete for resources or conflict. When conflicts occur, conflict resolution strategies are needed to referee the conflict. Behavior in the system unfolds over time as a function of these cycles and the dynamically changing world. A particular state/event can trigger several alternative actions; a particular action can be triggered by several states/events.

There are some restrictions on what types of information may be included in the condition expression (i.e., with the IF) and the action expression (with the THEN). The states and events in the condition expression may have three different kinds of referents.

1 Ongoing states and events in the physical world that the driver perceives,
2 States of knowledge in the mind of the driver,
3 Goal states in the mind of the driver.

The contents of the action expression include four different referents.

1 A goal that becomes activated in the mind of the driver.
2 A single basic action (e.g, movement of the eye, head, hand or foot; searching memory, making a decision).
3 An overlearned sequence of basic actions which may be executed in a smooth, ballistic fashion (e.g., turn head to the left and then to the right).
4 An embedded plan which may or may not be achieved by physical actions (e.g., figuring out how to operate a heater control involves the embedded plan of manipulating the control and observing whether heat is produced).

When the action expression is a plan, the plan can normally be broken down into subplans (or subactions). For example, manipulating a control has several alternative subactions, such as pull, push, turn and slide. These subactions may be executed in a particular order or may be sensitive to feedback during the course of manipulating a control. Indeed, it is convenient to view an embedded plan as an embedded production system that is locally constrained. Similarly, the act of observing whether heat is coming from a vent could be represented as an embedded production system that involves hand movements, directing attention to skin sensations, head movements and eye movements.

Basic actions, whether physical or cognitive, take time to execute. This time can be measured empirically or estimated theoretically (Card et al., 1983). It is important to acknowledge that timing parameters are governed by strategic components, namely the production rules, in addition to indices of the physical world. For example, Fitts' law (equation 4) specified that there is a logarithmic relationship between hand movement time and the distance the hand is moved. A strategy-based account of the hand movement would not merely assign a logarithmic function (Meyer et al., 1988; Schmidt, 1975). Instead, the movement time is explained by a sequence of discrete acts which end up being executed ballistically when they are overlearned:

1 Move the hand near the target control;
2 Observe discrepancy between hand location (H1) and target (T);
3 Move hand from H1 to a location near T;
4 Observe discrepancy between the new hand location H2 and T;
5 Repeat steps 3 and 4 until the hand reaches the target T.

Each hand movement is a linear function of distance whereas each 'observation' act takes a constant amount of time (240 milliseconds, Card et al., 1983). The two-step correction mechanism (i.e., steps 3 and 4) is repeated more often as distance increases and the size of the target decreases. According to the production system architecture, the number of two-step correction cycles would explain the logarithmic function observed by Fitts (see also Keele and Posner, 1968; Meyer et al., 1988).

The overall time to complete a transaction is determined by a) the evaluation of production rules and b) the execution of actions in those productions that are activated. When a production rule is evaluated, the condition elements are matched to the contents in working memory. If all elements in the condition expression are satisfied by a matching code in working memory, then the production rule is activated. The production rules in the production system were evaluated in a sequential order (from the top of the list to the bottom) in early production system models (Newell and Simon, 1972) whereas the production rules are evaluated in parallel in the most recent models (Anderson, 1983; Thibadeau, Just and Carpenter, 1982; Klahr et al., 1987; Laird et al. 1987). During a single 'cycle', the amount of time it takes to evaluate all productions in the production system is assumed to be very rapid (70–200 milliseconds). This evaluation phase is not costly in terms of processing time because these matching processes are pattern recognition mechanisms, which are known to be rapid. Therefore, it takes only 200 milliseconds to evaluate the condition elements of hundreds of productions in a production system during a single cycle; 10 cycles would take 2000 milliseconds to complete. Whenever a production rule is activated (i.e., its condition elements match the contents of working memory), time is needed to execute the actions in the production. The time to execute these actions is long compared to the time to evaluate the condition elements. The movements of the hand, finger, eyes and head range from several hundred milliseconds to several seconds. Moreover, additional time is needed to resolve conflicts whenever there are conflicts in the actions of two or more activated productions.

It is important to emphasize that the contents of working memory get updated as the productions are activated, actions are executed, new input from the world is perceived, and elements are removed from working memory. The overall time to complete a transaction can be estimated by computing the number of evaluation cycles and the duration of all basic actions that are executed in activated productions, until the transaction is completed. Needless to say, a sophisticated computer model is needed to generate the predicted transaction times because of the dynamic interactions among working memory, perceptual input and the production system.

Cognitive scientists have proposed mechanisms of learning in the production system architecture (Anderson, 1983; Klahr et al., 1987; Laird et al., 1987; Rosenbloom and Newell, 1987a, b). The production system is modified systematically as a person gains more experience and receives feedback on the outcomes of actions. Some ways that productions are modified are specified below.

1 *Chunking*. When sequences of actions from different productions end up leading to successful outcomes, the groups of actions are chunked into a single production.
2 *Tuning*. There is a tuning of the exact condition elements that activate a production and the tuning of the appropriate actions to execute. Generalization and discrimination are forms of tuning.
3 *Accretion*. Additional productions are added as experience accumulates and finer distinctions are acquired.

4 *Module formation.* A small set of productions form a natural, semi-autonomous module that is locally constrained.

In a production system architecture, learning is accomplished on a production-by-production basis (Anderson, Conrad and Corbett, 1989; van Lehn, 1989) rather than there being a diffuse improvement on the system as a whole.

A concrete example

Table 18.1 presents an example production system for operating a heater control. This would be a locally constrained production system rather than a global production system for operating a vehicle. The five productions achieve the goal of turning on the heater. This goal is triggered by the driver's state of being cold (production 1). Productions 2–5 handle different states of knowledge in the driver. That is, the driver may or may not know where the heater control button is. The driver may or may not know how to operate the heater control button. Production 2 handles the situation when the driver has already learned the location and operation of the heater button. Productions 3–5 handle situations in which the driver is in the process of learning.

In order to provide some idea of how transaction times are generated theoretically, we will assign some hypothetical times to components and walk through some examples. The following time estimates will be assumed.

Table 18.1. Production system for operating heater control

1 IF [driver is cold] THEN [Goal: turn on heater]
2 IF [Goal: turn on heater
      and driver know where heater control button is
      and driver know how to manipulate heater control button]
   THEN [Move hand to heater control button
         and then operate heater control button]
3 IF [Goal: turn on heater
      and driver not know where heater control button is]
   THEN [Scan panel for location of heater control]
4 IF [Goal: turn on heater
      and driver know where heater control button is
      and driver not know how to manipulate heater control]
   THEN [move hand to heater control
         and goal: fugure out how to operate heater control]
5 IF [Goal: turn on heater
      and goal: figure out how to operate heater control]
   THEN [Manipulate control
         and observe whether heat is coming from vent]

Simultaneous evaluation of the condition elements of all production rules
= 200 milliseconds per cycle.
Execution of action in production 1 = 500 milliseconds.
Execution of action in production 2 = 1500 milliseconds.
Execution of action in production 3 = 2000 milliseconds.
Execution of action in production 4 = 1500 milliseconds.
Execution of action in production 5 = 4000 milliseconds.

Consider first the situation in which the driver already knows the location of the heater control and knows how to operate the control. The transaction would be completed in only two cycles: production 1 would be activated during cycle 1 whereas production 2 would be activated during cycle 2. Given there are two cycles, the production evaluation time would be assessed at 400 milliseconds. The total time is computed below.

Total time (full knowledge) = 400 (production evaluation)

+ 500 (action in P − 1) + 1500 (action in P − 2)

= 2400 milliseconds

Consider next the situation in which the driver does not know the location of the heater control, but does know how to operate it. The sequence of productions being activated would be 1, 3 and 2. The processing during these three cycles yield a total time of 4600 milliseconds. Suppose that the driver knew where the heater control was located, but did not know how to operate the control. The sequence of productions would be 1, 4, 5 and 2. These four cycles generate a total time of 8300 milliseconds. Finally, there is the situation in which the driver knows neither the location nor the operation of the heater control. The resulting sequence of five productions would be 1, 3, 4, 5 and 2, yielding a total transaction time of 10 500 milliseconds. It should be noted that the driver's knowledge of control location and control operation have additive effects on transaction time in this example. However, there are substantial deviations from additivity in more complex production systems.

In a serious production system model, each basic action in the production system could be further decomposed. For example, the basic action of moving the hand would be computed according to the strategic explanation of Fitts' law that we discussed earlier. The action of scanning the panel for a heater control would involve several cycles of eye movements; attention would initially and primarily focus on panel zones that correspond to expected locations of HVAC units. There would be an embedded production system module that handles scanning operations in search of target patterns. In production 5, the actions involve a) manipulating the control and b) observing whether heat is produced. The action involves complex, perhaps coordinated sequences of eye movements, hand movements, shifts of attention and time sharing strategies. Moreover, the time to complete action 5 depends on the delays imposed by mechanical characteristics of the automobile, such as the temperature setting on the HVAC unit and the amount of time it takes for hot air to be released from the vents.

The production system architecture can accommodate a large class of motor skill theories which make specific claims about the cognitive structures and representations that appear to be needed to explain organized behavior. Several models of motor skill would endorse the notion that there are modules of locally constrained production rules. For example, 'schema' models (Schmidt, 1975) assume that schema units are natural modules of perception and action. A schema is a self-contained cognitive unit that consists of a number of perceptual and action elements, whose values mutually constrain one another; that is, the value of a perceptual element has an impact on acceptable values of other elements. Any schema theory can be translated into a set of production rules. Several models of motor skill have assumed that recursive embedding exists among theoretical units which organize actions, i.e., schema units embedded within other schema units, plan units embedded within other plan units, or subgoals embedded within goals (MacKay, 1987; Pew, 1974; Rosenbaum, 1987). This recursive embedding is necessary for any model which claims that a hierarchical organization of constituents underlies organized action sequences.

Once again, the production system in Table 18.1 is a locally constrained production module that is presented merely for purposes of illustration. A sophisticated production system for driving would have thousands of production rules that handle dozens of driving tasks. Such a model would accommodate diverse states of the world and the execution of multiple tasks simultaneously (e.g., steering, pushing the accelerator and turning on the heater).

Problems with HANDY-2

The production system architecture of HANDY-2 has three important advantages over HANDY-1. Production systems provide a general architecture of human cognition that is psychologically plausible and is grounded theoretically. In contrast, additive multicomponent models are constrained by the required additivity and independence of processing components; these two features cannot be ascribed to some psychological mechanisms without sacrificing the integrity of the mechanisms. HANDY-2 furnishes a dynamic model of driving that is sensitive to the states of the world and the knowledge of the driver. In contrast, HANDY-1 is a static model that changes only by virtue of the nonexplanatory power law of practice. HANDY-2 can account for situations in which drivers perform multiple tasks simultaneously. In contrast, HANDY-1 can explain performance on only a single task at a time. HANDY-2 naturally captures complex interactions among world states, the knowledge of the driver and tasks. In contrast, HANDY-1 is not suited to handle such complex interactions. The best that HANDY-1 can do is permit the researcher to assess statistical interactions among variables (i.e., deviations from additivity). Unfortunately these statistical interactions do not provide an adequate foundation for investigating interactions among components in complex dynamical systems and even less sophisticated systems (McClelland, 1979). Complex interactions among theoretical components may generate additive effects among measured variables! Conversely, additive combinations of theoretical components may generate interactions among measured variables.

Nevertheless, a production system model would encounter a number of problems when applied to driving and hand controls. Some of these problems are inherent in any production system model. Production system models are very difficult to develop. We suspect that a team of programmers, engineers and human factors specialists would need to concentrate on a project for two or more years in order to develop a sophisticated version of HANDY-2. Developing HANDY-2 would be on a par with the development of an expert system. HANDY-2 would have thousands of production rules. Because of the volume of information and the dynamic properties of the architecture, HANDY-2 would encounter problems of intractability. That is, it would be difficult (but not impossible) for designers and programmers to trace the processes that lead to a generated transaction time and to diagnose problems in the program when computer errors occur.

One serious problem that the designers of HANDY-2 would face is incomplete knowledge about the durations of particular basic actions and of plans that involve interactions with the environment. The duration of executing some basic actions has been reviewed in Card et al. (1983), such as a single eye movement, a gaze duration, a hand movement, a cognitive act of making a decision, search of working memory, and so on. However, there are no accurate estimates for some basic actions, such as head movements and manipulation of controls (e.g., turning a knob, sliding a lever, pulling a button out). It is extremely difficult to estimate transaction times on embedded plans that depend on states of the environment. Consider, for example, the time it takes to change a station on the radio with an analog control. The time would depend on a) the distance between the current station and a station that the driver likes and b) how long it takes the driver to evaluate each candidate station. Arriving at accurate estimates of the b-weights for such components is probably beyond the scope of cognitive science and human factors. The designer would need to resort to empirical estimates of these parameters.

It is important to acknowledge that there is some flexibility in the design of production system models (Klahr et al., 1987). Some of these design decisions are enumerated below.

1 Should the productions be evaluated sequentially or in parallel?
2 Should there be one large list of productions that are evaluated during each cycle, or should there be sets of smaller production systems that are under local control?
3 Should goals be distributed throughout the knowledge base of production rules, or should there be a planning module (with an agenda of goals) that is separate from the production system?
4 Should all condition elements be satisfied before a production rule is activated, or can activation result from partial matches between elements in working memory and elements in the condition expression?
5 Should there be extensive use of conflict resolution strategies, or should the production rules be sufficiently complex (i.e., detailed, precisely tuned) so that conflicts virtually never arise?

The fact that there are alternative design decisions makes it difficult to evaluate whether a production system architecture provides a valid and reliable account of transaction times. Although it is difficult to test the very general claim that HANDY-2 accounts for driving performance, it is possible to compare HANDY-2 with HANDY-1 and with other models of driving performance.

In spite of the problems discussed in this section, we believe that production system models should be pursued vigorously in future research and development because there is a solid psychological foundation for production systems. It is unfortunate that it takes so long to develop and test these models.

However, the long-range payoffs should be enormous, even if particular aspects of the model are intractible or challenging to pin down.

References

Anacapa Sciences, 1976, *Driver Expectancy and Performance in Locating Automotive Controls*, (SAE Special publication SP-407) February, Warrendale, PA: Society of Automotive Engineers.
Anderson, J. R., 1983, *The Architecture of Cognition*, Cambridge, MA: Harvard University Press.
Anderson, J. R., Conrad, F. G. and Corbett, A. T., 1989, Skill acquisition and the LISP Tutor, *Cognitive Science*, **13**(4), 467–505.
Beggs, W. D. A., Graham, J. C., Monk, T. H., Shaw, M. R. W. and Howarth, C. I., 1972, Can Hick's law and Fitts' law be combined? *Acta Psychologica*, **36**, 348–57.
Card, S. K., Moran, T. P. and Newell, A., 1983, *The Psychology of Human-Computer Interaction*, Hillsdale, NJ: Erlbaum.
DeJong, J. R., 1957, The effects of increasing skill on cycle time and its consequences for time standards, *Ergonomics*, **1**, 51–60.
Fitts, P. M., 1954, The information capacity of the human motor system in controlling the amplitude of movement, *Journal of Experimental Psychology*, **47**(6), 381–91.
Graesser, A. C., Lang, K. and Elofson, C. S., 1986, Some tools for redesigning system-operator interfaces, in Berger, D., Pezdek, K. and Banks, W. (Eds) *Applications of Cognitive Psychology: Computing and Education*, Hillsdale, NJ: Erlbaum.
Green, P., 1979, Automobile multifunction stalk controls, *Literature, Hardware and Human Factors Review* (UM-HSRI-79-78), December, Ann Arbor, MI: University of Michigan Highway Safety Research Institute.
Haberlandt, K. and Graesser, A. C., 1985. Component processes in text comprehension and some of their interactions, *Journal of Experimental Psychology: General*, **114**, 357–74.
Hick, W. E., 1952, On the rate of gain of information, *Quarterly Journal of Experimental Psychology*, **4**, 11–26.
Jacob, R. J. K., 1983, Using formal specifications in the design of human-computer interface, *Communications of the ACM*, **26**, 259–64.
Keele, S. W., 1968, Movement control in skilled motor performance, *Psychological Bulletin*, **70**(6), 387–403.
Kieras, D. E. and Bovar, S., 1984, The role of a mental model in learning to operate a device, *Cognitive Science*, **8**(3), 255–73.
Klahr, D., Langley, P. and Neches, R. (Eds), 1987, *Production System Models of Learning and Development*, Cambridge, MA: Bradford Books/MIT Press.
Laird, J. E., Newell, A. and Rosenbloom, P. S., 1987, SOAR: An architecture for general intelligence, *Artificial Intelligence*, **33**, 1–64.
MacKay, D. G., 1987, *The Organization of Perception and Action: A Theory for Language and Other Cognitive Skills*, New York: Springer-Verlag.
Maynard, H. (Ed.), 1971, *Industrial Engineering Handbook*, 3rd edn, New York: McGraw Hill.
McClelland, J. L., 1979, On the relations of mental processes: An examination of systems of processes in cascade, *Psychological Review*, **86**, 287–330.
Meyer, D. E., Abrams, R. A., Kornblum, S., Wright, C. E. and Smith, J. E. K., 1988, Optimality in human motor performance: Ideal control of rapid aimed movements, *Psychological Review*, **95**(3), 340–70.
Meyer, D. E., Yantis, S., Osman, A. M. and Smith, J. E. K., 1985, Temporal properties of human information processing: Tests of discrete versus continuous models, *Cognitive Psychology*, **17**, 445–518.

Navon, D., 1977, Forest before trees: The precedence of global features in visual perception, *Cognitive Psychology*, **9**, 353–83.
Newell, A. and Simon, H. A., 1972, *Human Problem Solving*, Englewood Cliffs, NJ: Prentice-Hall.
Norman, D. A. and Rumelhart, D. E., 1975, *Explorations in Cognition*, San Francisco: Freeman.
Pew, R. W., 1974, Human perceptual-motor performance, in Kantowitz, B. (Ed.) *Human Information Processing: Tutorials in Performance and Cognition*, Hillsdale, NJ: Erlbaum.
Rosenbaum, D. A., 1987, Successive approximations to a model of human motor programming, in Bower, G. (Ed.) *The Psychology of Learning and Motivation*, Orlando, FL: Academic Press.
Rosenbloom, P. S. and Newell, A., 1987a, An integrated computational model of stimulus response compatibility and practice, in Bower, G. (Ed.) *The Psychology of Learning and Motivation*, Orlando, FL: Academic Press.
Rosenbloom, P. S. and Newell, A., 1987b, Learning by chunking: A production-system model of practice, in Klahr, D., Langley, P. and Neches, R. (Eds) *Production System Models of Learning and Development*, Cambridge, MA: Bradford Books/MIT Press.
Schmidt, R. A., 1975, A schema theory of discrete motor skill learning, *Psychological Review*, **82**(4), 225–60.
Sternberg, S., 1969, The discovery of processing stages: Extensions of Donder's method, *Acta Psychologica*, **30**, 276–315.
Thibadeau, R., Just, M. A. and Carpenter, P. A., 1982, A model of the time course and content reading, *Cognitive Science*, **6**(2), 101–55.
Welford, A. T. (Ed.), 1980, *Reaction Times*, New York: Academic Press.
van Lehn, K., 1989, Learning events in the acquisition of three skills, *Proceedings of the 11th Annual Conference of the Cognitive Science Society*, Hillsdale, NJ: Erlbaum.

19

Informational aspects of vehicle design: A systems approach to developing facilitators

Lila F. Laux and David L. Mayer

Introduction

What is a facilitator?

Materials which assist users in their interactions with complex devices or systems are called *facilitators*. In this chapter we will discuss the development and evaluation of instructions and warnings to support vehicle users in non-driving functions. An *instruction* communicates procedures along with relevant critical context information; it helps the user determine when to act and it tells what to do and why. A *warning* communicates that a hazard exists, what the hazard is, how to avoid the hazard and the consequences of failing to take the proper action. Instructions and warnings make up an interdependent facilitator set, each of which has the common function of communicating critical information to users.

Facilitators are provided to inform and instruct users about nondriving vehicle interactions so that they do not injure or kill themselves or others, do not damage the vehicle and do interact with their vehicle in ways that are most efficient with regard to the user-vehicle system. To be effective a facilitator must communicate *critical information*: information the user must know to act in the safest and most efficient way. Of course, giving a user the critical information does not guarantee that he or she will act appropriately. But when users act without having that critical information, the probability of an accident to the user and/or the vehicle is increased.

Why are facilitators important?

In 1972 the National Transportation Safety Board (NTSB) estimated that approximately 1.5 million people were injured annually in maintenance-related, nonoperating motor vehicle accidents. This figure did not include accidents which resulted in damage to vehicles, which far exceeds the number of accidents which produce injuries.

One way the incidence of accidents which occur during nondriving interactions can be reduced is to increase the effectiveness of facilitators such as instructions and warnings. Users have shown in a variety of ways, including purchase decisions and product liability actions (particularly failure-to-warn cases) that they expect manufacturers to provide adequate warnings and instructions. The best way to do that is to take a user-oriented approach when developing facilitators. The purpose of this chapter is to describe the process of user-centered facilitator development and the variables which have an impact on the effectiveness of any communication.

Who should develop facilitators?

Developing an effective facilitator system requires the efforts of a team. A technical writer and a graphics designer are essential parts of the team. Before they know what to write or draw, however, engineers, designers, safety professionals and others must determine what the user needs to know to behave appropriately, exactly what procedure the user must follow, what the potential hazards in the system are, and how they can be avoided. Other team members, perhaps marketing and human factors specialists, must determine who the users will be, what they understand and how they are likely to behave. Technical personnel responsible for the design and implementation aspects of a system should *not* be the ones to write the instructions, warnings and labels although they must be a part of the development team.

The systems approach

The vehicle is a complex device. We conceptualize the user, vehicle and the associated facilitators as a dynamic, interdependent *system*. Just as changes in any one part of the system can have an effect on the other parts of the system, weak aspects of the user-vehicle system can make interaction difficult or dangerous.

Almost all user-vehicle nondriving accidents result from a lack of understanding on the user's part which leads to inappropriate action. The reason for providing facilitators such as warnings and instructions to inform the user, but a facilitator is only effective if the information gets through to users so it can affect their actions.

The communications model

We approach the development and evaluation of the facilitator itself in terms of a communication system. The facilitator consists of a message and a medium for communicating it. The facilitator system includes the sender and receiver of the message as well. In the user-vehicle system the vehicle owner or user is the intended receiver.

Before a message can motivate users, it must first be communicated. This means that the message not only has to be sent, it has to be received (sensed and attended to) and understood (the user must assign the same meaning to the message as the facilitator developer). A communication system includes a *sender* (in this case the vehicle manufacturer/facilitator developer), a *message* (an instruction or warning), a *medium* (booklet, cassette tape, sticker, light, buzzer), and a *receiver* (the user of the vehicle).

This chapter describes procedures for deciding what is to be communicated to facilitate the user, how best to communicate specific information, and the specifics of the communication design. None of these decisions can be made, however, until the person who will use the facilitator has been identified and characterized.

The user

One critical step in the development of facilitators is identifying the user population and determining what user characteristics might affect their ability to take in and understand information. Then the facilitator can be designed to help those who need it most.

Almost every person aged 15 or older is a potential user of a vehicle facilitator. This diversity means that facilitators must communicate critical information to a group of users whose knowledge, experience, language skills, intellectual ability and dexterity differ enormously.

In addition, drivers as a group are changing. The United States is experiencing a 'graying' of vehicle users. The fastest growing group of new drivers is women over 55. By about 2020, half of all licensed drivers will be 50 or older and one in 12 will be over 80 (Transportation Research Board, 1988). Older drivers' changing sensory, physical and psychomotor functioning are associated with changing practices and abilities which may affect how they use facilitators (Laux and Brelsford, 1990).

So who is *the* user that facilitators are intended to serve? 'The user' for whom facilitators are developed is anyone who needs information and instructions to perform an unfamiliar task.

When this user must carry out a nonoperating vehicle function like jump-starting a battery, an effective facilitator may be the only thing that allows the task to be completed safely and efficiently. A facilitator which works for this user will also facilitate users who have more understanding and experience.

Engineers and designers, like many people who have good mental models of how vehicles and other machines work and have a good understanding of the physical laws which affect many nondriving functions, may assume that much of the knowledge users need is 'common sense'. But many users simply do not know much of what they need to know to perform nondriving functions safely and efficiently. Many users are deficient in knowledge about the forces and processes involved in nondriving functions. One of the most common reasons why facilitators fail is that they assume too much knowledge or understanding on the part of users.

Beliefs, attitudes, and expectations

Users are more likely to comply with safety instructions when they believe that what they are doing may be hazardous. Vehicle users often have highly inaccurate beliefs about how hazardous it is to perform any particular nondriving function. They may also have inappropriate beliefs about whether following safety recommendations will 'pay off'. Other factors which affect the rate at which they comply with safety instructions include whether or not it is effortful to comply, and whether or not they see others comply.

Beliefs, attitudes and expectations can be more powerful in determining the effectiveness of facilitators than knowledge. A facilitator may tell users that batteries can explode, but if a user does not *expect* it to happen, or does not *believe* it can cause injury or damage they will be unlikely to comply. If users do not think that by complying with safety recommendations they can avoid injury or damage, or if they do not think the risk is high enough to make it worth taking the effort to follow directions, the facilitator simply will not be effective because it will not influence behavior.

Assessing a representative goup of users for attitudes, beliefs and expectations about the hazards and risks involved in any nondriving function and what they believe and expect with regard to these hazards and risks is important. In a number of personal injury lawsuits, the courts have found that simply stating safety information is not adequate, particularly when inappropriate user beliefs and expectations are widespread or foreseeable. More will be said about assessing the user.

User centered design

Because the function of the facilitator is to communicate critical safety and maintenance information to the person interacting with the vehicle, the processs of developing effective facilitators must take into account the users *as they are* and not as they 'ought' to be. User centered design is a design approach which allows the cognitive capabilities and informational needs of the users to guide the developers of facilitators (or for any system for that matter). Figure 19.1 shows an overview of a human factors approach to the development of facilitators.

Since a facilitator cannot contain all of the information needed to understand every hazard completely, the first priority for the facilitator developer is to determine which gaps in knowledge are most likely to result in serious consequences for the user. User centered design will guide this decision process.

Information which vehicle users need includes:

- what the different pieces of equipment are and how to operate them (e.g., radio, air-conditioner, cruise control, windshield wipers, seatbelts, head restraints),
- what routine vehicle maintenance must be followed to maintain the safe working condition of the vehicle (checking tire tread, lubricating the chassis, etc.),
- what safety behaviors should be routinely practiced (seatbelt procedures, head restraint use, locking doors, etc.), and
- what to do in nondriving emergency situations (changing a tire, jumpstarting a battery).

This chapter will focus more on the last type of information because the risk of injury and damage associated with these situations is typically more immediate and more severe, and because communication is made more difficult by the fact that users rarely need this information and when they do it is often under stressful or unusual circumstances. This means that users may have no experience with or understanding of the function, no memories or mental

Informational aspects of vehicle design 405

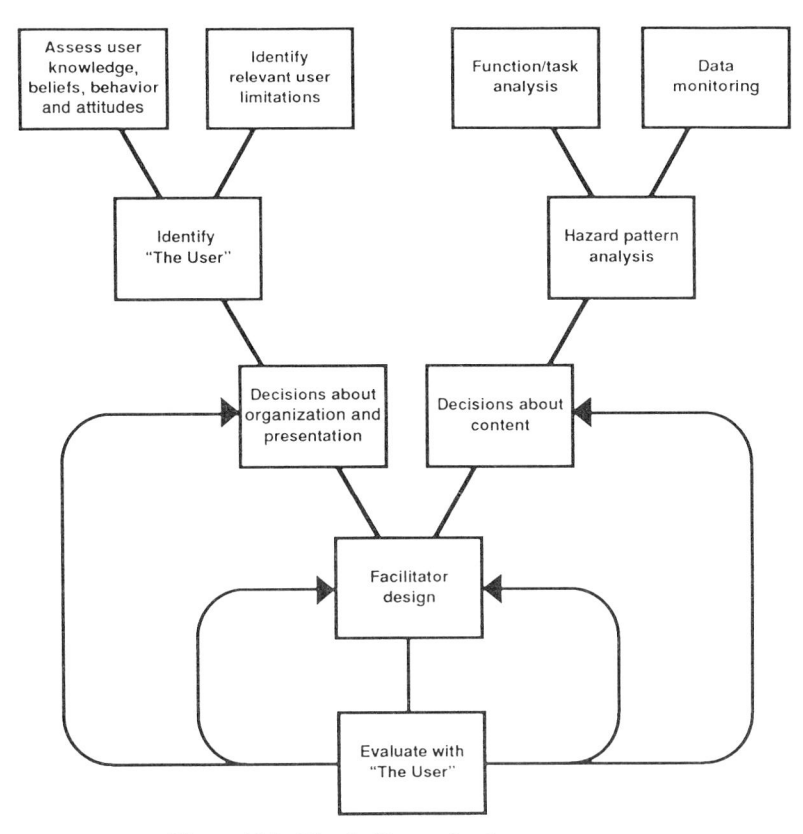

Figure 19.1. The facilitator development process

models to help them understand the instructions or warnings. It may also mean that the user is frightened or angry—conditions which hinder effective communication.

Many emergency situations occur because users do not follow routine maintenance procedures like checking battery fluid levels. Thus, it is important to safety to develop facilitators which communicate about safety practices, routine maintenance and routine operations. The vast majority of vehicle users today use self-service gas stations, which has enormous implications for safety. Many routine maintenance functions are simply not being performed at all, increasing the risk of injury and damage to vehicles. Other users perform maintenance procedures which involve hazards of which they are unaware, also raising the risk of injury or damage (Mayer and Laux, 1990). It is important to know the foreseeable behaviors of users when they interact with their vehicles in order for facilitators to address these problems.

What hazards are associated with nondriving functions in the user-vehicle system? What kinds of accidents are occurring or can be expected to occur given vehicle characteristics, the laws of science, and human limitations? For any given function, facilitator developers need to determine:

- what accidents *have* occurred,
- *how* they happened,

- their relative frequency,
- their relative severity, and
- how human characteristics and behavior contributed to them.

Hazard pattern and failure modes analyses

Answering the first four questions comprises what is called a *hazard pattern analysis*. The last question is addressed through a modified *failure modes analysis* to include failures due to human action and inaction (see Hammer, 1972, p. 148). It is often argued in the context of litigation that systems are defective or unsafe because designers failed to anticipate and make allowances for foreseeable user behavior. Hazard pattern analysis and failure modes analysis are the tools used to anticipate user behavior. Their goal is to gain a better understanding of what may be expected from human interaction with the system.

Hazard pattern analysis

Data about the types, causes, frequency and severity of injuries associated with non-driving user vehicle interactions can be accumulated from a variety of sources.

Owners' manuals. Owners' manuals from a variety of vehicle manufacturers can be used to identify many hazardous situations. These common hazards are associated with 1) emergency situations such as changing a flat tire and jump-starting a battery, 2) situations arising when users fail to follow recommended maintenance procedures (e.g., a worn brake hose ruptures), 3) the maintenance procedures themselves (getting caught in a fan belt while working under the hood), and 4) failure to practice routine safety recommendations (unrestrained child, person with improperly adjusted seatbelt or head restraint).

Accident databases. Manufacturers' and users' associations and private and governmental agencies collect and organize various kinds of accident and injury data according to their individual needs and interests. The following databases record this type of information:

- *Consumer Product Safety Commission (CPSC).* CPSC collects data through the National Electronic Injury Surveillance System (NEISS) about injuries associated with the use of consumer products.
- *Departments of Public Safety (PDSs).* DPSs or state police departments for each state collect data about vehicle accidents and may have data needed to determine the pattern of hazards associated with maintenance failures and with failure to practice routine safety behaviors. This data is often difficult to extract from their data bases, however.
- *Dealershp service departments.* Dealership service departments observe system failures which have occurred due to user actions or inactions. Their records contain the evidence of which failures occur with the most frequency and what the most serious failures are. They can evaluate which maintenance functions are being performed and whether this is leading to hazardous situations.

Regulations, guidelines and standards. Another kind of evidence that a hazard exists is found in voluntary standards and guidelines and in manufacturing regulations. The existence of such standards and regulations shows that certain hazards occur with enough frequency to be recognized as something about which users need to be warned.

Technical experts. Technical experts employed by vehicle manufacturers are another source of information about hazards available to facilitator developers. Marketing experts can provide information on user characteristics which may relate to hazard patterns for particular vehicles such as luxury sedans, sports cars, vans or pickups. The legal department can tell you what kinds of non-driving accidents have resulted in litigation against car manufacturers. Engineers who test products can tell you where users have difficulty.

Users. Users themselves are an often untapped source of information about common human actions or inactions that increase the likelihood of failure. They can tell you what kinds of system failures and 'near misses' they have experienced while performing non-driving functions and what precipitated those failures.

Failure modes analysis

Vehicle manufacturers have access to a wealth of information about hazard patterns and user behavior. How will you assimilate and use the information you have gathered? Once the data have been collected, a modified failure modes analysis should be carried out. We have modified the traditional procedure (see Rowland and Moriarty, 1983, Ch. 22), developed to evaluate system failure due to mechanical failure, to focus on the interface between the human and the vehicle and to analyze failure which results from action or inaction on the part of the user.

Function/task analysis

The first step is to perform a task analysis for each nondriving function. Analyze what the user must do to perform each function correctly and compile a sequential list of task statements which describe each action.

Writing task statements. Task statements should begin with an action verb which describes the required user behavior (e.g., set, lift, open, fill, attach, pull, remove, turn). Part two of the statement will be the object which is acted upon (handbrake, nozzle, nut, jack, cap, handle) and may include some qualifier of the action (how high, how much, how long, how fast). The last part of the task statement should explain why the action needs to be done.

The failure modes associated with each specifc task can now be identified. Consider a task analysis for jacking an automobile to change a tire: If a task statement is 'Set handbrake firmly to keep car from rolling off jack', the human actions or inactions which could lead to failure would include not setting the handbrake or not setting it firmly enough. These are the user behaviors (failure modes) which can lead to system failure. For each task, identify all of the human actions or inactions which could lead to failure, and describe how the facilitator may contribute to the likelihood of these human failures. Then describe how an effective facilitator could reduce the failure rate. The example in Figure 19.2 shows how this process works.

The hazard pattern analysis may reveal other failures which result from human action but which are not directly related to any specific task. For instance, a user may lean on the vehicle or let someone else lean on it while it

From the function/task analysis: Jacking a Car (automatic transmission)

Task Statement 1: Park on level ground a safe distance from the roadway.

Task Statement 2: Put transmission in 'Park'.

Task Statement 3: Set parking brake firmly to prevent vehicle from rolling off jack.

User action that can cause failure	How facilitator design may be related to human actions	How to facilitate
1 Does not set brake.	Did not read instruction (could not find it).	Provide user-specified index and table of contents.
2 Does not set brake firmly enough.	Did not understand (what do 'set' and 'firmly' mean?); Did not understand consequences.	Specify what is meant by terms; Spell out cause-and-effect relationship or use 'if ... then ...' wording. Develop adequate warning to accompany instructions.

Figure 19.2. Example of a failure modes analysis for jacking a car

is jacked, causing it to fall. These failure modes can be listed in a final separate general section.

Although a failure modes analysis may seem tedious it is critical to carry out the analysis thoroughly. Decisions about how to facilitate and what to include in facilitators will be based on this analysis. In addition, it will serve as a guide to determining what users must know in order to interact with their vehicles safely and effectively.

The new design or device

When a facilitator is being developed to support interaction with a new version of an old device or for a new device, you cannot depend on data about past use to predict future failure patterns. There are three sources of information about potential user problems with regard to new or redesigned devices:

1. Find data that indicate how users have interacted with similar types of devices including earlier versions of the device;
2. Get the designers and developers of the devices to do a function/task analysis and develop task statements, do your own function/task analysis; and,
3. Observe naive users interacting with the device, to see where and how they make mistakes, what they do and do not do, and where they are puzzled or do not understand.

Monitoring hazard patterns

Just as it is never safe to assume that engineers, designers or other experts have correctly identified or anticipated all of the hazards associated with a new product or have identified the appropriate hazard pattern, it is never safe to assume that once a thorough hazard pattern analysis and a failure modes analysis have been done the job is finished. People are infinitely creative and continually find new ways to interact with devices in ways which designers and manufacturers never intended.

In other situations, the environment in which the product is used changes, or a change in one device or aspect of the vehicle alters the hazard pattern associated with some other function. The expertise or other characteristics of the user population may also change. Laws and regulations governing users or manufacturers may cause changes in behavior. Any change in user behavior, the environment, the product, user beliefs, etc., can lead to new patterns of user-vehicle interaction and may lead to new or different failure modes and hazard patterns. Thus, the process of evaluating failure modes and hazard patterns should be ongoing and facilitators need to be updated to fit the appropriate hazard pattern.

Assessing user knowledge

The function/task analysis and the failure modes analysis have provided the information you need to determine what specific (critical) information the user must have to carry out a function. These analyses also suggest what general or background knowledge users must have to understand the critical knowledge. No facilitator can provide all of the background and specific information that users need in order to understand hazards and the implications of hazards for their behavior. To decide what must be included, you need to know what users already know about the system and its associated hazards. In order to evaluate whether your facilitator is effective in communicating critical information, you also need to be able to assess what users know *after* they are exposed to the facilitator (Mayer and Laux, 1990).

Facilitator developers should never take it for granted that users know something, especially if it is important to safety. Many users are not well informed and, more importantly, do not even understand that they need to be informed. They do not have all the facts and do not understand many hazards, nor do they understand which of their own behaviors puts them at risk.

Methods of assessment

There are two typical ways of assessing what people know: asking them if they recognize information (for example, responding to true or false statements such as 'A car battery produces an explosive gas' or selecting the correct response when asked 'Which of the following is a safety device? Headrest, turn signal, radio') or asking them to recall information ('How can you get hurt when jumpstarting a battery?' or 'Name the safety devices in your vehicle'). Recall tets are often called 'open-ended' or discussion questions. Both of these methods of assessing knowledge will be useful to the facilitator developer in some situations.

Recognition and recall tests can be administered either in paper-and-pencil form or by interview; interviewing may be the only way to get accurate information from users who have reading disabilities. In addition to assessing what people know by asking them to recognize or recall information, we can collect performance information. One of the best ways to assess procedural knowledge is to observe users carrying out the procedure.

The 'best' method. None of these methods is always the 'best' method of assessing what users know. The 'best' method depends on what you need to find out. There are some important considerations in deciding which method should be used. Responses to recognition tests are often based on the response choices—a user may not be sure of an answer but can eliminate all other choices or can do reasonably well by simply guessing (people should get half the items on a true/false test correct just by guessing). Furthermore, people can recognize much more information than they can recall. Care must be taken, especially with recognition-type questions, to avoid biasing users' responses by 'suggesting' correct answers. Users may respond by supplying answers which they think are expected of them rather than by giving more natural responses.

One of the most important jobs of facilitators, especially warnings, is to get people to recall what they know about the situation and to use the relevant information they have. It is useful to employ open-ended questions to find out what people can recall because that is typically what users are required to do when interacting with a vehicle. To determine how many people know the proper adjustment tension for the shoulder strap of the seatbelt, it is appropriate to ask the user to tell you what the proper criterion is or to ask him or her to show you. If you use a recognition test (either a true/false or a multiple choice) you will conclude that more users know than really do.

Who should be assessed? When determining what users know, you need to test users who are likely to need the information in a facilitator. For most vehicles, this means testing 'naive' users: newly licensed teenagers, senior citizens (particularly women), naturalized Americans, etc. There are some vehicles whose user population may not include such a wide variety of foreseeable users. The marketing and sales departments can tell you if the user population of interest to you is predominantly young or old, male or female, experienced or inexperienced, likely or unlikely to perform its own maintenance, etc.

Developing facilitators

Good facilitation requires that complex technical information be translated into plain English. It also requires that facilitator developers know exactly what needs to be communicated and communicate it in ways that users will understand, believe and use. A facilitator should communicate to users so effectively that hazardous situations do not arise because of user behavior and users can avoid an accident in situations which are inherently hazardous (changing a tire on a busy highway, jumpstarting a battery).

Most of the guidelines in this chapter are just that—they are rules of thumb, not laws. In many cases, the answer to any question about what the content of a message should be or how it should be communicated will be the frustrating, 'It depends'. But, in fact, it *does* depend—on what the user knows, his or her experiences and expectations about the facilitator and the task, how hard it will be to comply, how hazardous it is, what people normally do, what kinds of system failures occur and with what frequency.

Facilitator failure

An understanding of how facilitators often fail to influence the behavior of users is crucial for the design of effective facilitators. Facilitators can fail for the following reasons:

1 Users do not comprehend the information,
2 The 'information' is actually not information,
3 Users are unable to follow the instructions, or the cost of following the instructions is too high,
4 The information is presented in such a way that it is not believable and credible, and
5 The information is ambiguous.

We will examine each of these areas.

Comprehensibility. One of the most important considerations for facilitator developers is the level of language to be used when communicating to a diverse group of users. There is a sizeable group of Americans who do not read even at the fourth grade level. It is estimated that at the present time, 10 percent or more of the adult population of the US falls into this group (Miller, 1988). For users who do read, but whose reading ability is limited, it is important to keep the readability level of the text as low as possible. The Army recommends writing procedural information at the fourth grade level. Technical writing experts suggest that the seventh grade level is appropriate for trained users. Both of these estimates suggest that facilitator developers will miss significant numbers of their intended target audience if they do not keep reading levels low. Clear, simple language is not entirely a function of the grade level at which information is written, but certainly information written at a lower readability level should be understandable to a larger number of users.

There are a number of algorithms for predicting the readability of a text passage. Readability is generally determined by considering such factors as the average length of the sentences, the number of infrequent or uncommon words and the average number of syllables in a word. These three factors have been found to affect the comprehensibility of text passages. Klare (1974) presents a good review of the techniques available for predicting readability.

The following list of suggestions helps assure that the material is written at the lowest reasonable grade level:

1 Use short sentences.
2 Do not use technical language and jargon.
3 Use common (to the user population) short words.

Following these rules means that more people, whether good or poor readers, will be able to comprehend and therefore remember and act on, the information in the facilitator. If you follow these rules, for instance, you will not create a sentence like, 'The protrusion on top of the transaxle generally makes a good point for this final ground attachment, see illustration', which was found in a 1987 owners' manual.

Another substantial group of users either do not speak English or do not have an adequate command of English to read facilitators. This group represents a serious challenge to facilitator developers who must depend on non-verbal visual communication or auditory messages to communicate critical safety information. One solution for the illiterate or non-English speaking user is to use pictorials and graphics. This solution sounds easier than it is. Choosing good graphic symbols which are universally understood is tricky. More is said about selecting or developing pictorials and graphics for facilitators in Chapter 12.

To increase comprehensibility, keep the message straightforward: Use concrete rather than abstract language (say 'flame' or 'spark' instead of 'source of ignition'), use the active voice (say 'turn the dial', not 'the dial should be turned'); present information in the correct temporal order (say 'turn the dial left, then pull lever B', not 'before pulling lever B, turn the dial left'), tell what to do instead of what not to do (say 'place the device in location B' not 'do not put the device in location A'). One invaluable source of information about effective and concise writing which every writer should have is *The Elements of Style*, by Strunk and White (1979). Finally, do not refer the reader to other parts of the manual to get information unless it absolutely cannot be helped—it is better to have the same information in two places than to put any additional load on user's memory when the situation is unfamiliar, and it makes reading the information more effortful for the user.

Non-information. Reduce non-information by avoiding using terms like 'now and then', 'adequate', 'snug', and 'serious'—there is no common understanding of the magnitude of these terms; what is snug to a facilitator developer may not be snug to a user. Furthermore, avoid technical jargon which the user is not likely to recognize (e.g., torque wrench, 'ram' air, or assist starting). It is critical to determine and use the terms that users themselves use when describing these processes.

Instructions which cannot be followed. To avoid giving impossible instructions or instructions which are simply too difficult to follow, you must know your users. Many users do not have a flashlight to look under the hood at night; users almost never have an air pressure gauge with them to check tire pressure after changing a tire; most will not have goggles to allow them to work safely when jumpstarting a battery, and due to memory limitations it is not likely that they will remember to do things like 'check periodically to see that latches on folding rear seats work properly'. These instructions may be well-intended but are impossible to follow.

Some instructions simply require more effort, time or expense than most people are willing to expend without more justification. One example might be, 'As soon as possible after installing any wheel, have a technician tighten wheel nuts with a torque wrench to the torque shown in "Specifications", Section 6'. Referring the user to another section (not even a page number) to find out what the specifications are increases the cost of complying and does not indicate that there is really any strong reason for doing it.

Credibility. To increase believability and credibility avoid statements like this one which we found in a 1987 automobile owners' manual: 'To reduce risk ... an occupied reclining seat should not be reclined anymore than *needed for comfort*' (italics ours). This statement conveys a mixed message about whether or not it actually is dangerous to ride with the seat reclined. When information about carbon monoxide poisoning is presented in exactly the same style and format as this message about reclining seats, and both are called to the user's attention by the word **CAUTION**, it is not too surprising if users do not take that information seriously either.

There are three cardinal rules to remember when designing information for users:

1 State what you want to say briefly, but explicitly;
2 Use words familiar to the user; and
3 State everything the user needs to know, *do not expect the user to infer anything.*

Ambiguity. English is an incredibly ambiguous language. Differences in people are one source of the ambiguity of words. Another related source of ambiguity comes from the fact that words have different meanings in different contexts.

To minimize ambiguity, facilitator developers should solicit feedback from potential users early in the design phase. As soon as a symbol is selected, a pictorial designed, or a warning or procedural passage written, four or five uninvolved users should be asked to explain what the warning label or instruction means to them. Labels and warnings are particularly susceptible to ambiguity because of the need for brevity. Remember that meanings will also be inferred from context. A sign reading 'watch your head' would certainly be perceived differently if it were posted in a barber shop than when posted near a low passageway.

Types of information

Facilitators provide three basic types of information: general information, procedural information, and critical information.

General information. This includes naming parts of the vehicle and telling what they are for. It also includes such information as what the caution triangle means, what the red light by the oil pressure symbol means, what kind of gasoline to use, where to find the Vehicle Identification Number, what everything is on the instrument panel, what kinds of tires to use, whether or not a trailer can be towed. One primary criterion for facilitators is that this type of information be readily accessible, and for that reason it is often placed 'on site' (warnings, instructional stickers) as well as in the owners' manual.

Devices and processes should be called by terms which are meaningful to the user ('user preferred terms'). Calling a battery which no longer produces electrical current a 'discharged' battery is technically correct, but the vast majority of users think of a discharged battery as a 'dead' battery. They will never think to look for 'discharged battery' in an index. Preferred user terms can be determined by surveying users. Terms commonly employed by users must be incorporated into the facilitator system. By including both the user preferred term and the technical term, users can be educated about correct terminology.

Procedural information. This information contains instructions on how to perform non-driving functions. This requires that the user know the names of parts and devices involved and how they work. Referring users to other sections of a manual or from a diagram on-site to the manual for general information when giving procedural information puts a heavy memory load on users and causes confusion and frustration. Sometimes critical information must simply be placed in more than one location.

The function/task analysis spells out the steps which must be described in giving procedural information. The failure modes analysis will reveal where users might act or not act to cause system failure. Hazard pattern analysis reveals what kinds of failures have been happening. All of this information must be considered when deciding what information to include in describing a procedure.

Critical information. In order to interact safely with a vehicle in carrying out non-driving functions, users must also possess 'critical knowledge' about each of the functions. This is specific information critical to understanding how to perform that function safely. The facilitator should use appropriate language and provide enough general information so that the critical information can be understood. If the user does not understand the phrase 'source of ignition' (a phrase commonly used in vehicle facilitators) he or she will not know how to behave to avoid a 'source of ignition'.

Facilitator production

Content, organization and presentation are the three major concerns of facilitator developers. Decisions on the content will include what topics to include and how much to say about each topic. These decisions should be based on the hazard and function/task analyses. Obviously, facilitators must supply all of the general prerequisite declarative and contextual information the user needs as well as all procedural information necessary to guide appropriate behavior.

The following guidelines for the actual generation of facilitators are based on a distillation of the recommendations and research findings of safety experts, psychologists and communications experts as well as guidelines published by ANSI and other organizations. No guidelines can provide a 'cookbook' which spells out the answer to every decision which must be made about facilitator design. No guidelines can guarantee that your facilitator will be 100 percent effective. How effective the facilitators you develop are will depend upon the extent to which you are guided by the cognitive capabilities and informational needs of the user.

Ask these questions about every proposed facilitator:

- Will the user see it when he or she needs to see it?
- Will the user get the same message I want to send?
- Will it motivate the user?

You will be more likely to be able to answer those questions affirmatively if you have developed your facilitator based on information the user needs and limitations, task demands and hazards. If you also adopt the following general rules of thumb when generating a facilitator, you will successfully communicate with the largest number of users:

- Design for the user who has no expertise (never changed a tire, does not know what the coolant temperature symbol means);
- Design for the user who finds him or herself in a hostile environment (dark, isolated, rainy, etc.);
- Design for the user who is experiencing psychological stress (anxiety, frustration, anger, fear).

What we have said so far about developing facilitators applies to facilitators of all types. We will now specifically discuss issues relevant to the design of owners' manuals and warnings.

Developing warnings

Warnings are necessary because most hazards are not obvious to users. Even when the product designers and engineers make every effort to design and manufacture a product which cannot be misused, or they try to design safeguards to protect the user from interacting with the product in a hazardous

Informational aspects of vehicle design 415

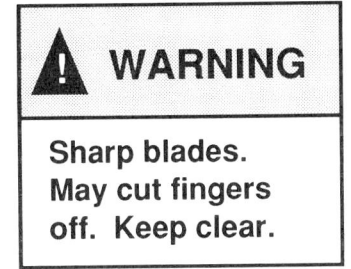

```
┌─────────────────────────┐  ┌─────────────────────────┐
│  ⚠ WARNING              │  │  ⚠ WARNING              │
├─────────────────────────┤  ├─────────────────────────┤
│  Toxic fumes. Can       │  │  Sharp blades.          │
│  damage lungs.          │  │  May cut fingers        │
│  Wear respirator.       │  │  off. Keep clear.       │
└─────────────────────────┘  └─────────────────────────┘
```

Figure 19.3. Two examples of warnings created using FMC guidelines

way, there are often situations which simply cannot be eliminated. These are latent hazards—people do not know about them and do not expect them. For this reason, simply instructing users in the proper way to interact with a product is not sufficient to ensure their safety. A warning is an instruction which is designed primarily to call attention to a latent hazard, identify it, and indicate the urgency of following the instruction. FMC Corporation publishes an execellent guide to development of warning labels and signs (FMC Corporation, 1990) and Easterby and Zwaga (1984, Ch. 21) also discusses their approach to warnings. Figure 19.3 depicts two warnings which have been developed using the FMC guidelines. Finally, Miller, Lehto and Frantz (1990) have published a useful annotated bibliography of research on warnings and instructions.

The communication function of a warning is to alert users to the presence of a latent hazard, let them know how hazardous it is, and tell them what to do to avoid the hazard and what will happen if they do not act appropriately. Safety experts agree that all warnings should include a *signal word* conveying the intensity level of the hazard, a *hazard statement* which tells what the hazard is, a *consequence statement*, and a *safety instruction message*. The signal word and the hazard statement *must* be part of a warning; depending upon the type of hazard and where the warning is located, the extent of the last two components may vary.

Signal words

The first question which developers must answer is: What is the hazard intensity? That is, how serious are the potential consequences of the hazard? The more serious the possible consequences, the greater the safety responsibility of the manufacturer.

The principal function of the signal word is to attract the users' attention to the hazard and indicate to him or her at a glance just how severe the hazard is. ANSI standards and those written by other hazard communication specialists have made the following signal words standard for communicating hazard intensities:

- **DANGER**—immediate hazard which will result in severe injury or death;
- **WARNING**—hazard or unsafe practice which *could* result in severe injury or property damage;
- **CAUTION**—hazard or unsafe practice which could result in minor injury or property damage.

Other common signal words required by some regulatory agencies include **POISON** and **FLAMMABLE** which communicate specific hazards. These words are typically used in conjunction with the hazard intensity signal words. ANSI recommends that signal words be prefaced by the hazard warning triangle.

Content

Once the correct signal word has been chosen to represent the hazard intensity level, the content of the warning must be developed. The content must include a statement of the hazard, the potential consequences of noncompliance, and an instruction message telling how to avoid the hazard. However, increasing the length of any part of a facilitator increases the likelihood that poor readers will not read it or that users will simply decide that the investment in reading time is too great. More information can be included to supplement the warning in a label or manual, but even in these situations the warning itself should be brief and to the point and clearly set apart from the rest of the text.

Statement of the hazard. The statement of the hazard can be in text format or in pictorial/symbolic form. For on-site warnings, symbols or pictorials are highly recommended because pictorials and symbols have high attention attracting characteristics and can overcome language problems of users. Symbols differ from pictorials in that they are abstractions, not normally depicting any recognizable event. Examples are the skull-and-crossbones and the radiation hazard symbol. In order for symbols to be effective, users must have learned the meaning of the symbol. A number of symbols already exist which reprsent hazards; however, our research shows that many of them are not universally recognized (Mayer and Laux, 1989; Mayer, 1990). When a symbol is to be used in a vehicle facilitator, designers must be sure that users know its meaning, which can only be determined by field testing on users. This certainty is particularly important for any symbol which must stand alone, without accompanying text.

Pictorials, on the other hand, can be developed to depict hazards in ways that most users recognize because they represent specific situations or events. A warning label on a vehicle at the site of the hazard often has severe space constraints. For this reason, symbols and pictorials may be particularly useful as a hazard statement. Before any symbol or pictorial is used as a hazard statement, however, it should be tested (usually with a recognition test) to determine if users recognize the hazard implied. The hazard statement must state the hazard in terms the user can understand, which, again, is best determined by user testing.

The consequences. A warning should also inform users of what can happen if they do not follow the safety instructions. In many instances, users are simply not able to infer from the hazard statement what can happen if they do not follow instructions. It is important to spell out the specific consequences for most hazards, and it is important to communicate these consequences without trying to minimize the risk.

A phrase frequently overused in facilitators is *personal injury*. This is an informationless phrase—it is used for hazards where the consequences range from a bruise to death. 'Personal injury' is abstract, impersonal. It has no impact, fails to suggest any particular consequences or any particular need to take preventive action. Personal injury is a 'fudge phrase'. It replaces those explicit, concrete, personally salient words like death, disfigurement, blindness, deafness, paralysis. The safety responsibilities of manufacturers do not allow facilitator developers to 'fudge'—if users can get killed or maimed doing something, facilitator developers must be sure this is brought to their attention, particularly in situations where they are not expecting a hazard and not looking for a warning.

Pictorials can sometimes present the hazard statement and the consequence statement combined, as in the 'bloody hand' pictorial which demonstrates a cutting hazard and shows the consequences. Unfortunately, few hazards and consequences can be portrayed as concretely and with as much recognizability with one pictorial.

The safety instruction message. The fourth critical part of a warning message is the safety instruction message. This part of the communication tells the user what to do and what not to do to avoid the consequences of the hazard or to avoid causing a hazard. If users are to act appropriately, the facilitator must tell them what appropriate behavior is, and how to determine whether or not they are in compliance.

Why would users not know if they are doing what the facilitator says to do? Sometimes it is simply a matter of misunderstanding. But at other times, it is because the facilitator does not tell the user what 'yardstick' to use to evaluate his behavior. If the facilitator says to make something snug, how tight is that? We asked a group of users to specify how they would know if their shoulder belt is snug. There were many answers from 'can just slip your open hand between belt and chest' to 'can lean across seat to operate something on dashboard'. It was clear to us that users did not have a common understanding of what snug meant and this means that many of them think they are following safety instructions when in fact they are not.

When telling users about appropriate behavior to avoid hazard consequences you may need to communicate both 'do's and don'ts'. Normally, warnings should focus on what the user *should* do, not on what they should not do. However, when data indicate that certain inappropriate behaviors are foreseeable, either because of the design of the system or because of the human tendencies, users must be specifically warned against doing them.

In sum, the content of a warning tells users the nature of the hazard and explicitly informs them of its potential consequences. Warnings also tell users how to behave safely to avoid the consequences of the hazard.

Legibility and presentation

For on-site warnings, white lettering on a black background is recommended for content information because it produces the highest contrast and readability. Size of print must be determined as a function of the distance from which the warning will be read. Choose plain fonts or letter styles such as Helvetica bold which reduce the likelihood that letters or numbers will be confusable. A ratio of letters to blank space of around one-half will also enhance legibility. (For further reading about typographic issues see Sanders and McCormick, 1987, Ch. 4), as well as Easterby and Zwaga, 1984, Chs 9 and 10). The FMC manual (FMC Corporation, 1990) may also be consulted for some specific recommendations.

The appearance of an on-site warning must be such that the message will be seen and attended to. It must be clearly visible to the user and placed close to the hazard, or at least in the users' line of sight when approaching the hazard. The warning should stand out against the background and be large enough to be readable from the distance from which it will be viewed. Lighting under which the warning will be read as well as other environmental factors (e.g., dirt, grease, corrosive substances, water) must be taken into account. Finally, typographic features such as letter size, font and case, and contrast and spacing, must be chosen to enhance readability. Warnings that will appear as part of labels or other facilitators must be prominently located on them. They should be set off by surrounding it with a border or other attention getting device. Each of these factors can affect the readability and attention attracting properties of a warning.

Warning effectiveness and criteria

There are no hard and fast rules when it comes to designing and developing good warnings. There are no design principles which will ensure that a warning will be effective. There are no principles which apply to the design of every warning. The guidelines in this chapter must be applied in conjunction with an analysis of the particular warning situation to formulate hypotheses about how to warn most effectively given the circumstance. You must also keep the user always in mind.

The following general questions will help you evaluate whether you have developed a good warning message.

- Have you used the right signal word?
- Have symbols or pictorials and color been employed where possible to enhance recognizability and comprehensibility?
- Does the warning clearly identify the hazard?
- Does the warning describe appropriate behaviors to avoid hazard exposure?
- Does the warning clearly represent the severity of the hazard and consequences (no 'fudge phrases')?
- Is the language/terminology/symbology used understood by the user group?
- Is the warning brief and concise?

Once you produce a warning which gets high marks when evaluated against these criteria, you still have no basis to assume that your warning will be more effective than a warning already in existence. To determine if in fact the warning is effective, you must test it to determine what it means to real users and what kinds of behaviors it suggests to them.

Evaluating warning effectiveness

The effectiveness of any facilitator has to be evaluated in terms of system outcomes. We have discussed many characteristics of warnings and instructions which determine their potential to be effective. Based on this information, we can construct checklists which represent what we think are *minimum criteria* for adequacy. Do not confuse checklists with evaluating effectiveness, however. To evaluate effectiveness you must test the facilitator with the appropriate users. This kind of evaluation normally overestimates the effectiveness of facilitators because users are likely to try harder and they know their performance is being assessed in some way. Nevertheless, field testing provides a good estimate of potential effectiveness.

There are several procedures for evaluating effectiveness, each of which contributes only part of the evidence needed to determine if a warning is effective. As a first step, it is useful to check for recognition and comprehension: Can the user explain the meaning of the warning? Can he or she identify the hazard, tell you the appropriate hazard intensity, tell you under what circumstances the hazard is likely to be manifested? Can he or she tell you what the consequence of inappropriate behavior could be and what should be done to avoid the consequences? Can the user tell you what should be done if exposed to the hazard?

Paper and pencil effectiveness evaluations. Paper and pencil tests and structured interviews can be effectively used to collect information about the degree to which the warning was effective as a communication. Both have their advantages and shortcomings. Paper and pencil tests can be administered in groups and cost less to administer and analyze. Their shortcoming is that user motivation may be low and users may not make the effort to answer the questions in depth. Using so-called 'objective' questions (for example multiple choice, matching, or true/false quetions) is not generally appropriate for the reasons discussed earlier. Furthermore, this type of question provides no insight into the way the user is thinking about the system and why he or she does not understand.

It is often very informative to look at the kinds of errors users make. Since a warning should provide all of the information needed to answer the questions, why would any user not get all the answers correct? Sometimes the failure can be directly traced to wording or language. Sometimes the pictorial may not be understandable. Sometimes we assume that users have knowledge that they do not have. Error patterns allow facilitator developers to identify what information is not getting communicated and help them to identify why.

Open-ended questions, or 'discussion questions' provide a great deal more information and allow analysis of user errors, but they are time-consuming to administer and analyze and require high levels of user motivation. If written protocols are used, the user will not always understand the question and may answer the wrong question or only answer part of the question. In addition, many users have poor written communication skills. Without an interviewer to probe for additional information, you will not be sure what and how much is understood. Some answers will be incomprehensible because of language or terminology—you will not be able to tell if the user understands the warning or not. This is always a problem with open ended questions—by what criteria will you decide that the user understands? Remember, this is a test where the person answering is not being graded, it is the warning that is being evaluated.

Interview methods of evaluating effectiveness. A structured interview will generally provide more complete and better information than paper and pencil tests, but it is even more time-consuming to collect and analyze information this way. The interview must be carefully planned in advance. The questions must be ordered and phrased in such a way that they do not give away answers, 'lead' the user, or indicate what response the interviewer wants. They must also be planned for content—what do you want to find out? What question can you ask that will indicate to the user what you want to know without giving away the answer?

Interviewers must be trained not to show disapproval or frustrations with users but to encourage them to give a complete response. It should be decided in advance how and when it will be acceptable for the interviewer to offer suggestions or probe for additional information. Interviewers also need to know when to stop probing and how to go on to another question without making the user feel like a failure.

Advance plans also need to be made for analyzing the responses: What will constitute a correct answer? Is an answer either correct or not correct or is there some level of partial correctness? What specific response will convince you that the user understands the warning in the same way that you do? These questions are not trivial, and they are not always easy to answer. Before conducting a large number of interviews, collect data from two or three people and see whether or not you are getting the information you need in a form that you can use. A good way to check your criteria is to tape an interview and have someone else listen to the tape and rate the correctness and completeness of the responses, then compare those ratings with ratings from the original interviewer.

Performance tests of warning effectiveness. Performance evaluations of warnings must be carried out by observing users interacting with the vehicle. Ask the user to perform the function for which the warning was generated and observe performance. Observe whether the user reads the warning, if possible, and record whether he or she performs behaviors which indicate that the warning was indeed read and understood.

At the end of the testing, check for memorability by asking the user if there was a warning present and if so what it looked like and what it said. You can also solicit user opinions and comments at this point. Did the user like the appearance of the warning? The content? Were the instructions easy to follow? Was it easy to see? If your data indicate that the user did not see or did not read the warning, ask why. Users often have very good insight into what would be effective in attracting their attention and motivating them to read and heed warnings.

Developing Owners' Manuals

Owners' manuals provide various types of information which users need to interact with their vehicles safely and efficiently. The factors we have mentioned previously which affect readability and comprehensibility are important in the generation of instructions. We will not reiterate those issues and the human factors approach to dealing with them as we have discussed this extensively in other sections.

What we will focus on in this section is generating good procedural information—writing good instructions for carrying out functions. This is one of the hardest tasks the facilitator developer has. Instructional manuals that accompany products are not generally held in high esteem by users because they are so often a stumbling block rather than a facilitator. Users actually express surprise and joy when they discover a users' manual which they can understand and follow. As technological advances make vehicles more sophisticated and complex, writing good instructions is becoming both more difficult and more important.

How can good instructions which people will use be developed? Once again, we advocate approaching the design of a facilitator based on users' needs and task requirements. What does the user need to know and what does he or she need to be able to do to get the job done? Engineers and technicians might seem at first glance to be the appropriate resource for this information. In fact, they are probably not the right people to write instructions. They know too much, and they have their own vocabulary; too much is intuitively obvious to them. In short, they are not at all like the user who will have to read and understand the instructions. Technical writers and graphic artists, while not technically expert in the design and functioning of vehicles, are expert at communicating. They are trained to evaluate the intended audience for any communication and tailor the material for that audience. For further reading about writing manuals, consult Schoff and Robinson (1984), Leonard and McGuire (1983) or Easterby and Zwaga (1984, Chs 24–27).

Content

An instruction works best when it breaks the procedure down into meaningful behavioral units, or action steps. The action steps which a user must accomplish in performing any task can be determined from a function/task analysis. Technical experts can assist in performing this analysis, but the person who will be writing the instructions should conduct a function/task analysis by actually performing the task and writing down each identifiable action. For complex tasks, several people should do this and then the action steps should be compared for agreement. Once the necessary steps have been determined, write instructions for each action step.

Writing good instructions for each action step. Identifying what actions must be performed is one step, writing the instructions which tell the user what to do and how to do it is another. The following guidelines will help you in writing action step instructions:

- Start each sentence with a verb (e.g., 'Turn off the ignition');
- Use short, simple, positive sentences in active voice (not 'To avoid personal injury the seat should not be adjusted while driving');
- Do not use abbreviations or jargon; use user preferred terms;
- Use simple, unambiguous wording; be consistent in naming objects.

Organization of the action steps. To be effective, instructions must follow the appropriate order in which the tasks need to be carried out. The appropriate temporal and logical order can be determined by having several people who are *skilled* at performing the task perform it and explain what they are doing and why they are doing each step at that particular point in the sequence.

Once you have determined what steps must be performed and in what order, and have written down instruction statements, the instructions should be tested on several 'naive' users—that is, uninvolved users who are no more knowledgeable about the function than a user would be expected to be. During this kind of try-out, any undetected ambiguities in the instructions should surface. Users will be able to tell you and show you what they do not understand, or you will be able to observe where they appear to be having particular difficulty or doing something improperly. At this point the writers will know from their own efforts which action steps are particularly difficult to write instructions for. This is the time to consider including diagrams, pictorials, flow-charts or other graphic materials to support the written instructions.

Graphics. Diagrams, pictorials and other graphic materials have been found to increase user understanding of unfamiliar objects and procedures in instructions. However, they take up space and it is not always easy to be unambiguous in instructing actions with pictorials. For that reason, we recommend that diagrams and pictorials be chosen only after you have identified information that is difficult to communicate verbally. Then they must be tested.

Choosing or designing graphics to support instruction. One of the most effective formats for communicating procedural information is highly pictorial with related text. Instructional materials which show the focus of each action step accompanied by short verbal statements have been found to support both fast and accurate performance.

Instructions include both action information (e.g., 'Put air in the tire') and condition information (e.g., 'the correct pressure is 34 PSI'). Diagrams and pictorials can speed recognition of objects and clarify meanings of complex instructions. It may be almost impossible to tell users how to find an object without a diagram. Diagrams make it easier to present condition information (e.g., to show readings on gauges and dials). It is harder to present action information with diagrams.

What constitutes a good graphic? A number of studies have been conducted to determine what kind of diagram or pictorial is best for procedural instructions (e.g., Kolers, Wrolstab and Bouma, 1979, 1980; Szlichcinski, 1979). The consensus is generally that photographs and detailed line drawings do not provide enough focus to make it easy for the user to locate the object of the action step. Recognition and comprehension are best when diagrams *suggest* the larger system context (by 'ghosting', dotted lines, or inclusions of 'landmark' information only) and provide more details of the focus area. System drawings which show the focus areas 'blown-up' have also been found to improve recognition and comprehension. In general, moderate levels of detail and the use of locational aids are the most effective in aiding unskilled and untrained users to identify objects and perform complex procedures. Too much pictorial detail can easily overload users.

Obviously, a good graphic should be shown in the perspective from which the user will see it unless that perspective makes it ambiguous— for instance, if you are looking down on something cylindrical, it will appear round. As there may be any number of things which could appear round, a better perspective, although not the perspective from which the user will work, would depict it at an angle from which the user can view it to check that he or she is looking at the right object. There should be enough detail in a diagram or pictorial to allow the user to orient him or herself but not so much that it obscures the focus of the action step.

When choosing or designing graphics to accompany instructions it is imperative that the diagrams or pictorials be user-evaluated *early and often*. Many users do not have strong spatial visualization skills; drawings which require users to visualize objects in three dimensions may be incomprehensible to these users. Check to see if naive users can locate and identify objects as needed to perform the procedure correctly. Ask these users to point out and identify the objects in the diagram. Have the graphics artists interact with the users and revise the diagrams or pictorials as needed for clarification; graphics artists are trained to abstract graphic information and suggest ways of enhancing comprehensibility of graphic material. Chapter 23 of Easterby and Zwaga's *Information Design* (1984) discusses graphical instructions.

The index. Information is not always easy to find. One way to improve this situation would be for all vehicle facilitator developers to adopt some consistent rules. If all owners' manuals for similar types of vehicles used the same terms and locations this would facilitate users in using facilitators! The most important criterion for selecting a term for the index is that index terms should include user preferred terms.

Facilitator developers tend to use the terms that they or their technical experts would use and often do not include terms most frequently looked for by the user. For example, when we asked 66 users what index term they would look for to find information about what to do if their battery was discharged, 60 percent said 'battery', 33 percent 'jumpstarting', and 6 percent said 'emergency'. In one owners' manual, the only entry related to this situation was 'Assist starting'. Under 'battery' the user could only find out what kind of battery the vehicle was provided with. Assist starting is not a user preferred term.

In order for users to read the information in a facilitator they have to see it, and they cannot see it if they cannot find it. To identify user preferred terms, the user should be asked, 'If you wanted to find information about what to do in (any) situation, what would you look for in the index? Do *not* say, 'would you look for assist starting, jumpstarting, etc.' To find out what users think up when on their own, the question must be phrased so that no hints or clues are given about 'right' answers or what is expected.

Other considerations

Typographic and style considerations should enhance legibility. Paper should be thick enough that drawings and text do not show through on the other side. Paper should be of a quality that will last as long as the car. Bindings and covers should be sturdy. Print size and contrast should be large enough to be easily read under moderate lighting conditions. Print quality should be good, providing high contrast. All of these guidelines may seem obvious, but in reviewing a number of 1987 vehicle owners' manuals, we found that these guidelines were often violated.

Presentation of instructions. Make sure that appropriate text directly accompanies each diagram or pictorial. The material should be arranged on the page so that it is easy to reenter the sequence after the user has executed an action step; there should be sufficient white space so that the text does not look dense, and there should be few distracting headings and highlighting devices. Instructions in flow-chart format or diagrams which show a series of steps should proceed from left to right and from top to bottom. Steps should be numbered to help users to remember where they are in the sequence.

Do not overuse italic type because it is harder to decipher. Do not use all uppercase letters, this slows reading speed and makes it harder to locate oneself in the sequence. Use right and left justified text for text passages with proportional spacing of letters. Use double columns in manuals that are larger than pocket-book size, particularly where the pages are wider than they are long. Use no more than two highlighting techniques—and be consistent. Bold print and underlining are attention-getting techniques which do not add too much clutter. Be sure that what you highlight is important—do not make the page look overly busy. Chapter 26 of *Information Design* (Easterby and Zwaga, 1984) presents a fuller discussion of structure and presentation of procedural information.

Finally, do not get so caught up in trying to follow all these guidelines that you lose sight of what you really are trying to do—communicate an important message to the user. Keep checking with users to see whether you are on the right track.

Warnings issues related to owners' manuals. Often it is appropriate to use warnings in the text of owners' manuals. Warnings which identify hazards associated with any procedure should be placed at the beginning of the instructions describing that procedure. In some cases, it may be important to repeat specific parts of the warning at the point in the sequence where a failure modes analysis has shown that human failure is likely. A good example of this relates to jumpstarting. It is important to warn about the danger of flames and sparks at the very beginning of the jumpstarting instructions. But it is also desirable to *remind* users not to create a spark by restating this warning at critical points in the procedure—especially when the user is about to attach the final clamp.

All facilitators are interdependent and a facilitator system must be evaluated as a system. Instructions will be more effective when labels and warnings are effective. For each warning in the owners' manual there should be a properly situated warning on-site when possible. Labels should also be placed to help users identify objects and orient themselves when performing functions. All terms must be consistent throughout the system. This seems obvious, but different industries are responsible for the warnings on items like batteries and tires. You must determine what terminology is used on those items so that you can be consistent.

Warnings which will appear in owners' manuals or other brochures or pamphlets must be prominently placed. Normally they should be located at the beginning of any passage which will describe a function or situation in which the hazard may be encountered. The warning should be set off by surrounding it with a border and/or other attention getting device. Typographic features (letter size, font, case, contrast and spacing) to enhance readability should be employed. Warnings in text should be contained in one location—they should not be continued to the next page, especially to an overleaf.

Evaluating instructions

We have already emphasized the need for getting early user feedback about instructions. This is critical in generating effective instructions because the effectiveness of the instructions as a whole depends on the effectiveness of each component statement or diagram.

As soon as you feel that you might be approaching a final version of your instructions, you should conduct a user evaluation. The users that you choose to evaluate the instructions must represent the typical users who will use your instructions. It is important to choose people who do not have the expertise you have and who do not know the technical language of the designers and technicians. You may want to include some experienced users and some who have fairly recently become users, some women and some men, some foreign born and some native English speakers.

Initially, however, try three to five native English speakers of 'average' characteristics who are good readers. Ask them to use the instructions to perform the task. Ask them to talk their way through the procedure, to 'think aloud' and to give reasons why they do what they do. This will give you insight into the mental models they have of the system and their interaction with it. It is very useful to videotape users actually using your instructions to perform a function so you can go back and review it as a team. Finally, ask these users to give good constructive criticism or feedback about the instructions and how usable they were.

Once you have developed instructions which work for this group of users, find a larger group of more diverse users (teenagers, senior citizens, females and males, non-teenage new users, foreign-born users, users who do not read, etc.) and test your instructions on this group of users. Observe them performing the function and record how long it takes them to perform the function, how many times they fail to perform an action step, how many times they perform action steps out of order, how many times they perform actions which are not in the instructions, how many times they perform an action step incorrectly, and how many times they do something which might put them at risk.

Use this data to identify areas where your instructions are not facilitating most users. If you find a problem that seems to be limited to one group of users, try to determine what their special needs are and whether changes could be made in the instructions to support them. Try the new version out on a similar group. Your goal is to facilitate a broad spectrum of users to perform the tasks safely and efficiently. Be especially alert to identify any failures in instructions which may lead users to behave in an unsafe way, either because they do something or fail to do it.

Monitoring and updating the system

In the course of evaluating facilitator systems of any kind, it is not likely that you will find that any facilitator is 100 percent effective. How effective is effective enough? When will you know that you have finished the job of generating effective facilitators?

The answer is that the job does not end. It is an ongoing process. You will use the information you have generated in the evaluation to come up with hypotheses about why the facilitator was not completely effective and how it can be improved. You will revise and evaluate again. The vehicle will change, and you will need to do new failure modes analyses and generate new facilitators. User knowledge will change, and you will need to monitor user preferred terms. You will need to evaluate how those changes affect other facilitators.

Remember that a facilitator system is a *communication system*. Communication systems never go in one direction only. By testing and evaluating facilitators, you are getting feedback from the receiver about the success of the communication. This allows you to evaluate how closely the message received is the same message that was sent. It also allows you to revise and try again.

Manufacturers should monitor all information sources to determine where changes are occurring in the system that could affect the effectiveness of facilitators. This includes in-house resources like the legal and regulatory departments and the engineering and design departments, as well as external sources such as dealers, other manufacturers, accident statistics, and newspapers and reports.

Any time there is a change in the vehicle itself, of course, the facilitator must be updated and that involves function/task analyses and failure modes analysis. New and revised facilitators must be tested just like original facilitators. You cannot wait for accident reports to provide feedback from users. The impact of any change in the system must be evaluated from a safety point of view. Developing facilitators is part of a feedback loop. The loop includes the manufacturer, the facilitator, the user, and the vehicle. Changes in the characteristics of any of those components will affect the entire system and influence the effectiveness of the facilitator.

Changes in users

Users change over the course of their lifetimes. Some new devices may be especially difficult for older drivers to understand and operate. They have a long history of interacting with their vehicle which shapes their expectations and beliefs about any vehicle interaction. Their sensory motor capacities are diminished. They may not be able to read numbers from LEDs as readily as younger drivers, and their responses are slower. They may have habits and preferences which are incompatible with new devices.

User populations as a whole change, as well, although this tends to happen very slowly. Over the last century levels of education have changed dramatically. At least half of all users are now women. More recently, the user population has grown older and there are many more foreign-born users in certain areas of the United States. In addition, we are now exporting vehicles to other countries.

Other suggestions

A vehicle facilitator is anything which allows a user to interact safely and efficiently with the vehicle. This chapter has focused on traditional vehicle facilitators: written warnings and instructions (owner's manuals). There are new devices for communicating in the vehicle: computers with displays and sound and voice generating capacity. There are other routes to user facilitation which might be considered. It should be noted that the impact of any change in a facilitator must be carefully evaluated, both in terms of the facilitator itself and in terms of system functioning. There are terrific new opportunities but it is not always easy to predict the effects of any change.

Auditory facilitators

Auditory facilitators have been in place in vehicles for some years. There is a buzzer which reminds you to fasten your seatbelt; a chime tells you that your keys are in the ignition; the turn signal makes a clicking noise when activated. In more expensive cars, voices tell you to have your engine checked or that you need gasoline. The success of these auditory facilitators has been mixed.

There are some very real reasons to look at auditory facilitators with renewed interest. They are excellent at attracting attention, they are comprehensible to users who do not read, and they do not require visual processing. Because visual attention may be overloaded by the complex displays on the instrument panel and the demands of the driving task, visual information on gauges and dials and warning lights may not be seen. This can be dangerous when the information signals a hazardous situation.

We recommend using auditory facilitators for emergency situations in particular. To produce an effective auditory facilitator, it is again important to consider human characteristics. On the positive side, our sensory system is set up to detect change automatically, so the onset of sound can be very attention-getting when visual attention is overloaded. People automatically attend to stimuli that are unusual or intense, so the use of unexpected or intense sound signals would attract attention. Some sounds have particular learned associations with danger or hazard and will therefore be attended to because they have high informational value.

On the other hand, users soon become habituated to any stimulus they hear frequently or which has no informational value to them and the stimulus no longer attracts their attention. It is not unusual to see people driving long distances with their turn signals flashing because they have become habituated to the auditory signal.

Another type of auditory facilitator which is gaining popularity provides user instructions on cassette tapes. These instructions attempt to guide users in identifying and using various devices and instruct them in carrying out various procedures. While this appears to be a good idea, user response has not been overwhelmingly positive. It has been hard to find the appropriate level of information for a tape because users cannot find information they do need and ignore information they do not need. As a reference tool it is not all that useful.

Auditory facilitators have great potential for alerting users to hazards but their design must be carefully considered. Too many auditory messages or unimportant messages will result in user habituation—then the users simply will not respond to the signal at all. The onset of sound will no longer be unusual and attention-attracting. To be effective as warnings, auditory messages should be researched to determine which ones are most likely to be heard and attended to by most users. They should be used sparingly and only to signal hazards. We definitely do not recommend the 'talking car'.

Training

How efficient a system can be is limited by the efficiency of its weakest element. The weakest part of current facilitator systems is the level of knowledge on the users' part. Most users are uninformed about their vehicles and the way they should interact with them and that makes the job of facilitating by traditional methods extremely hard. Interacting with vehicles is dangerous; many people are killed or injured in interactions with vehicles. In any job setting, we would prohibit untrained persons from performing hazardous maintenance and emergency repairs on complex machinery. A large part of the danger to users lies in the fact that users' perceptions of the hazards and their role in preventing the consequences are inaccurate.

Users must be able to learn to interact safely with their vehicles when performing nonoperating functions. Very few procedures can be learned from reading about them. Hazards are not well understood if people do not have mental models of the mechanisms which underlie those hazards. It is just not reasonable to expect any facilitator to be able to instruct and inform users who are totally without a mental model of the system with which they are interacting.

It is much easier to facilitate users in understanding nonoperating functions when they have some kind of mental model of the hazard situation. When users can understand the hazard mechanism because it is similar to other hazard mechanisms they have actually interacted with, getting them to understand what to do and why is fairly straightforward. One method of increasing the effectiveness of facilitators would be to provide user training. Having trained or knowledgeable users would surely improve the effectiveness of all facilitators.

Final comments

Society is willing to accept the hazards associated with vehicle use, but has given manufacturers and agencies certain safety responsibilities. There is a duty to warn, both legal and moral. There is a duty to instruct.

In this chapter we have presented a method for developing facilitators as part of the dynamic user-vehicle system. Users want well-designed manuals and their safety requires the development of effective warnings. Our central point has been that to create effective instructions and warnings, developers must adopt a user-oriented approach. They must seek to understand the beliefs, attitudes, expectations and knowledge of the people who will have to understand and use the facilitators that they make. A keen understanding of user characteristics is required not only to develop facilitators, but also to maintain the facilitators as the dynamic system evolves.

Needless to say, to take all of these issues into account, facilitator development must be a team effort. Technical experts (engineers, designers, safety experts) can specify what the content of the message should be and writers and graphic artists and others should determine the best way to organize and present that information. Warnings and instructions hardly constitute what is possible with regard to facilitators. Many innovative technologies can and are being used to assist users in their daily interactions with vehicles. The user-centered method of developing facilitators is an approach which can be used to generate effective facilitators no matter what the nature of the facilitator itself. It can even be applied to other aspects of the vehicle system.

The bottom line on system effectiveness will be determined from statistics revealing the rate of death, injury and vehicle damage which occurs due to user action or inaction in performing nonoperating driver functions. Every effort to communicate effectively through existing vehicle facilitators and to develop new methods of facilitating users must focus on the user. If the approach we have advocated is conscientiously and creatively applied, facilitator developers will be able to assist manufacturers to satisfy both their safety responsibilities and the safety regulations which govern them.

References

Easterby, R. and Zwaga, H. (Eds), 1984, *Information Design: The Design and Evaluation of Signs and Printed Material*, New York: John Wiley and Sons.
FMC Corporation, 1990, *Product Safety Sign and Label System*, Santa Clara, CA: Author.
Hammer, W., 1972, *Handbook of System and Product Safety*, Englewood Cliffs, NJ: Prentice Hall.
Klare, G. R., 1974, Assessing readability, *Reading Research Quarterly*, **X**, 62–102.
Kolers, P. A., Wrolstab, M. E. and Bouma, H. (Eds), 1979, *Processing of Visible Language*, New York: Plenum.
Kolers, P. A., Wrolstab, M. E. and Bouma, H. (Eds), 1980, *Processing of Visible Language 2*, New York: Plenum.
Laux, L. F. and Brelsford, J. W., 1990, *Age Related Changes in Sensory, Cognitive, Psychomotor and Physical Functioning and Driving Performance in Drivers Aged 40 to 92*, Washington, DC: AAA Foundation for Traffic Safety.
Leonard, D. C. and McGuire, P. J., 1983, *Readings in Technical Writing*, New York: Macmillan.
Mayer, D. L., 1990, *Recognizability and Effectiveness of Pictorial Symbols Used in Warning Messages*, Unpublished master's thesis, Rice University, Houston, TX.
Mayer, D. L. and Laux, L. F., 1989, Recognizability and effectiveness of warning symbols and pictorials, in *Proceedings of the Human Factors Society: 33rd Annual Meeting*, pp. 984–8, Santa Monica, CA: The Human Factors Society.
Mayer, D. L. and Laux, L. F., 1990, Automative maintenance and safety preparedness among drivers: Aspects of age and gender, in *Proceedings of the Human Factors Society: 34th Annual Meeting*, 584–8, Santa Monica, CA: The Human Factors Society.
Miller, G. A., 1988, *The Challenge of universal literacy*, Science, **41**, 1293–99.
Miller, J. M., Lehto, M. R. and Frantz, J. P., 1990, *Instructions and Warnings: The Annotated Bibliography*, Ann Arbor, MI: Fuller Technical Publications.
National Transportation Safety Board, 1972, *Nonoperating Motor Vehicle Safety Study*, Report No. NTSB-HSS-72-3, Washington, DC: National Research Council.
Rowland, H. E. and Moriarty, B., 1983, *System Safety Engineering and Management*, New York: John Wiley and Sons.
Sanders, M. S. and McCormick, E. J., 1987, *Human Factors in Engineering and Design*, New York: McGraw-Hill.
Schoff, G. H. and Robinson, P. A., 1984, *Writing and Designing Operator Manuals*, Belmot, CA: Lifetime Learning Publications.
Strunk, W. and White, E. B., 1979, *The Elements of Style*, New York: Macmillan.
Szlichcinski, K. P., 1979, Diagrams and illustrations as aids to problem solving, *Instructional Science*, **8**, 253–74.
Transportation Research Board, 1988, *Transportation in an Aging Society*, Special Report 218, Washington, DC: National Research Council.

20

Unintended acceleration: Human performance considerations

Richard A. Schmidt

According to the driver's complaint, he entered his vehicle after a shopping trip, started the engine normally, put his right foot on the brake and shifted from Park to Drive in preparation to depart. He says that the vehicle then suddenly accelerated wildly, with loud engine noises and tire squeal, and that all attempts to halt its motion with brake pedal action were ineffective, the brake pedal going completely to the floor. He states that he steered to avoid several other vehicles, but that the acceleration continued for 25 meters until his car collided with a wall to end the episode. After the accident, mechanical experts are unable to find any problems with the braking system, throttle linkage, or with any other of the vehicle's systems that would account for the unwanted acceleration. The driver is convinced that his foot was on the brake pedal throughout, and bristles at the suggestion that he may have pressed the accelerator pedal by mistake.

This description is a typical account of several thousand accidents that have occurred in the past few decades, and have occurred sufficiently frequently that this type of accident has received a formal name, 'unintended acceleration'. Other labels have been used as well, such as 'unwanted acceleration', 'sudden acceleration', or 'sudden acceleration incidents', but unintended acceleration seems to be the most common term and will be used here. The National Highway Traffic Safety Administration (NHTSA) has defined the phenomenon as 'unintended, unexpected, high-power accelerations from a stationary position or very low initial speed accompanied by an apparent loss of braking effectiveness' (Pollard and Sussman, 1989, p.v.). These episodes occur most frequently at the start of a driving cycle, after the driver has (re)entered the vehicle to begin a trip. The phenomenon has appeared in nearly every make and model vehicle with automatic transmission (not in standard-shift vehicles, except as noted later here), both foreign and domestic, and has been appearing over the last several decades. The earliest report I am aware of concerned accidents of this type in the 1940s, where it was apparently termed the 'runaway motor problem' (Gallaway, 1987). Something very much like this phenomenon also seems to have occurred in the 1930s (Henderson, 1935).

Unintended acceleration does not occur very frequently. One estimate is that approximately 400 vehicle years occur between incidents (ODI, 1989; Wierwille, in press). Yet when they do occur, they often lead to serious accidents, injuries and even deaths, and the incidents tend to be publicized as a result. The existence of such accidents has led various governmental agencies in several countries to investigate them (Marriner and Granery, 1988, in Canada; Ministry of Transport, Japan, 1988; Pollard and Sussman, 1989; Office of Defects Investigation, 1989, both in the US). In addition, several private research firms (e.g., Battelle Corporation, 1987), as well as many of the automobile manufacturers, have examined the problem to determine the causes and to define possible remedies. It has also attracted considerable attention from the print (e.g., Higgins, Job and Pepper, 1986; Tomerlin, 1988) and television media (e.g., the CBS '60 Minutes' broadcast, 25 November 1986).

Largely because the driver typically insists that his or her foot was on the brake pedal throughout (even though eyewitnesses often saw no brake lights illuminated), the focus of the various investigations has been primarily on various mechanical systems whose failure could account for the full-throttle acceleration and combined brake failure. It is beyond the scope of this chapter to document the findings of these mechanical investigations; but it is safe to say that, except for a few isolated cases, no mechanical defect has ever been identified that could account for the facts (see Pollard and Sussman, 1989, for a summary). This is perhaps not surprising given the wide variety of kinds of vehicles involved—with and without cruise control, electronic engine management systems and fuel injection—and having vastly different engines, layouts, linkages and brakes. It seemed unlikely that a single defect, or combination of defects, could have been found in all these vehicles that could have explained all of the unintended acceleration episodes.

Several key facts are mentioned in all of these mechanical investigations. One is that the only way to develop sufficient power for the vehicle to accelerate at the reported rates is to allow large volumes of air into the engine's intake system, which can only be accomplished by opening the throttle plates in the fuel intake system. While such openings can be accomplished by a cruise control system as well as by the accelerator pedal, the fact that many of these accidents occur in vehicles without cruise controls—or in which the cruise control system was deactivated—points to the likelihood that the driver had inadvertently pressed the accelerator. Also, many analyses of the accident vehicle reveal damage to the accelerator pedal, or to switches located under it, which are consistent with the driver having the accelerator depressed with great force at the time of the collision. Second, the braking system in all modern automobiles is sufficiently effective to overcome the engine's full power and bring the vehicle to a stop, often with only minimal increases in stopping distance (e.g., Pollard and Sussman, 1989). This finding led to the suggestion that the driver's foot could not have been on the brake as he or she believed, and that the right foot was on the accelerator instead.

Human factors considerations

Several other lines of evidence supported this general hypothesis that the phenomenon could not be purely mechanical in nature, further implicating errors in human behavior. The analysis of reports of unintended acceleration has led to a number of 'risk factors', or characteristics of *drivers* who have reported having this type of accident. First, the reports are much more frequently from older as compared to younger drivers; one estimate is that the incidence of these episodes for drivers in the 60–70 year old range (corrected for vehicle miles driven per year) is some six times the rate of drivers in the 20–30 year old range (NHTSA, 1987). Another factor is the driver's familiarity with the accident vehicle (independent of the driver's experience generally), with approximately 25 percent of the accidents occurring in vehicles having been driven less than 4000 miles (see also Tomerlin, 1988). This factor perhaps accounts for why there is a high rate of reported unintended acceleration in rental cars, borrowed cars, car washes, parking lots, and so on, where the drivers are not highly familiar with the internal configuration of the controls. Also, there is some tendency for women (in Japan, at least) to have slightly more of these accidents than men (Higgins *et al.*, 1987), and for shorter people to have more of them than taller people, although these trends seem very weak in contrast to the first two above.

These are critically important observations because, if the problem were purely mechanical, it is difficult to understand how such episodes occur differentially in different categories of driver. However, one confounding problem is that these data are on *reported* unintended accelerations, and there is some tendency for different classifications of drivers to have differential rates of complaint (Tomerlin, 1988). However, to the extent that these reports do reflect differences in rates of unintended acceleration accidents, these data suggest that the causes should be understandable at least in part from principles generated in the fields of human performance and human factors.

Evidence of foot placement errors

Most of the evidence currently available suggests that the driver in some way placed the right foot on the accelerator rather than the brake that was intended. If the right foot applied substantial force to the pedal, this would account for the full-throttle acceleration, the perception that the brakes had failed and that the brake pedal could be depressed to the floor, and for the fact that no mechanical defects can be found in the vehicles after the accident. Several lines of evidence point to these kinds of foot-placement errors (called 'pedal misapplications' by Pollard and Sussman, 1989).

Perel's (1976) analysis of accident reports taken from North Carolina in 1974 and 1975 suggest that foot placement errors may be a factor in such accidents. Analyzing key words, Perel found 62 cases in which foot placement problems were involved in the accident. Of these, 12 involved the foot contacting the accelerator rather than the brake, and another 21 involved the right foot slipping from the brake to the accelerator. Also, accident reports taken from 1982–1983 in North Carolina produced 80 statements from the drivers that the accelerator was pressed rather than the brake. It seems clear that such errors do occur, but at a relatively low rate (0.03 percent of the cases in Perel, 1976). Such examples almost certainly underestimate the actual rate of such errors, however. First, these are *admitted* errors, and do not include the unknown number of errors for which the driver has not been so forthcoming.

Second, they do not include any errors the drivers may have made, but which did not result in a reportable accident. The fact that such errors are reported to occur relatively infrequently does not detract from the hypothesis that they are involved in unintended acceleration, as these accidents are also relatively infrequent. On balance, these data suggest that such foot placement errors might be involved in the initiation of unintended acceleration episodes, in spite of the drivers' strong opinions to the contrary.

A second line of evidence comes from more direct measures of driver behavior in simulators or in actual vehicles. In a systematic study of foot-placement errors in a simulator with several different configurations of foot pedals, Rogers and Wierwille (1988; see also Wierwille, in press) found that various kinds of pedal errors were relatively frequent, occurring in one out of approximately 24 foot movements, on average. The kind of error of interest here, where the accelerator is pressed instead of the brake, was relatively rare, however, with only two instances in the entire experiment, and these were immediately corrected when they happened. Drivers also sometimes placed their right foot on both pedals, with 10 such actions reported. Somewhat higher rates of errors were reported by Tomerlin and Vernoy (1988) in a study conducted in various makes of stationary vehicles; 14 instances of subjects pressing the accelerator rather than the brake (out of 258 foot movements), were found. These actions were extremely 'rushed', however, and drivers were generally unfamiliar with the vehicles, both of which could lead to the higher rates of errors found.

Perhaps the best evidence for pedal placement errors comes from Tomerlin (1988), who had drivers perform a series of vehicle maneuvers through cones in an open lot. In one situation, when the driver was preparing to make a reversing maneuver, the idle speed was unexpectedly increased by the experimenter-passenger. For two of the drivers, this resulted in control errors, during which the driver's right foot was pressing on the accelerator. In both cases, the driver was 'trying to stop' by pressing what he or she believed to be the brake pedal. These appear to be the first *observed* instances of pedal misapplication, where the driver believed that the right foot was on the brake when it was actually on the accelerator.

Understanding the processes in unintended acceleration

In the process of attempting to understand what happens in unintended acceleration episodes, it has been useful to consider three separate aspects of the problem. First, how does the driver's foot reach the accelerator pedal, when the driver fully intended to press the brake pedal? Second, given that such an error occurred, why does the driver not detect the error and correct it immediately? Third, why does the driver persist in pressing the accelerator rather than the brake until the accident occurs, and why does the driver not take some other form of evasive action (e.g., switching off the ignition)? In the next several sections, I deal with some of the principles of human performance and movement control that provide answers to these questions, and thus help to remove some difficult barriers to believing the pedal-error hypothesis. The human performance research begins to tell a coherent story about how unintended acceleration episodes occur, and the psychological and physiological processes that contribute to them.

How are pedal errors produced?

An initial problem is to understand how such movement errors could occur, even in drivers who have otherwise been successful, accident-free operators for many years. As background, it will be helpful to review several important concepts from research in movement control that will form the basis for understanding movement errors.

Motor programs

A critical concept in motor control is that of the motor program, a representation of the action stored in the central nervous system that, when activated, causes movements to occur. The program is thought to be responsible for the coordination and timing among the limbs in relatively brief movements (see Schmidt, 1988, Chs 7–9 for a more complete discussion). Motor programs are thought to provide the basic form and trajectory of the movement, but can interact with lower-level reflexes if the limbs are blocked or perturbed in some way, which allows some flexibility. An important point is that, once the action is planned and initiated, the actual production of programmed movements is thought to occur without the involvement of attention and consciousness. Much attention is required in the selection and initiation of a movement program, but once it is initiated it can cause the movement to run its course until the next movement is programmed, controlling the limbs without much attention from cognitive processes.

The functionally 'highest' levels in the CNS are presumably responsible for the selection of which actions to produce, and thus ultimately determine which one of the many different motor programs to run (e.g., a reach versus a throw). Once the program is selected and retrieved from long term memory, it is parameterized to define its superficial features (e.g., to move quickly or slowly), and then initiated, with functionally 'lower' levels in the system being responsible for carrying out the already selected action. One of the advantages of this style of control is that it allows routine, habitual and highly learned actions to be carried out with minimal attention, so that the attentional resources can be applied to other events in the environment, such as steering in a vehicle.

Two kinds of errors

This kind of model provides a basis for several different kinds of errors that humans can make in producing actions (see, e.g., Norman, 1981; Reason, 1979). For the present purposes, though, we can divide these errors into two movement classes (after Schmidt, 1976)—a) errors in selection and b) errors in execution.

Errors in selection. First, if the 'highest' level in the system is incorrect in evaluating the environment, or is otherwise incorrect in selection of the proper action, it can either choose the wrong program or parameterize it incorrectly. If so, the movement that results will be inappropriate for the intended goal, and will be classed as an error. This kind of mistake has been termed an 'error in selection', because the performer has selected the wrong movement to produce. An example is turning right when I should have turned left at the corner near my home.

Errors in selection are probably not very important in understandng unintended acceleration. In these accidents, the driver probably intends to put the right foot on the brake pedal, and this is a proper, correct action under the task of starting the vehicle and shifting into Drive. It is difficult to believe that the experienced driver has made a choice error in such a fundamental, simple, and well practiced aspect of driving.

Errors in execution. However, errors can also occur after the movement has been selected and initiated, through biases and inconsistencies in the 'lower' motor-program levels in the motor system that produce the movement that was chosen by the 'higher' levels. This has been called an 'error in execution' because the errors occur during the production, or execution, of the action. An example is the failure to make a 'bull's eye' on every attempt in dart-throwing, even though the choice of action (hit the 'bull's eye') is correct on every try. In terms of driving, even though the driver has chosen the proper action to make (move the right foot to the brake pedal), it is possible to have errors in the movement's endpoint because the lower-level processes in the motor system can make the foot deviate from its intended path.

For several reasons, it is more reasonable to think that the pedal misapplications occur because of an error in execution, where the proper action was chosen, but lower-level processes interfered with its execution in some way. Next I turn to some ways that these lower-level processes operate to make the movement's trajectory deviate from that which was intended.

Biases in initial positioning

Before an action, the performer processes proprioceptive and visual information about the body's initial position, and programs a movement to a goal based on these analyses. Therefore, one form of execution error can occur as a result of small shifts in the initial position or orientation of the performer. If such shifts are not detected, and the performer programs an action as if he or she were in a different initial position, then the movement that results will have a trajectory biased according to the actual initial position. Such biases can occur for several reasons, such as a disruption in the performer's seating position (Schmidt and Stelmach, 1968), shifts in gaze and/or movements of the head (e.g., Marteniuk and Roy, 1983), or a host of other factors that momentarily disrupt the performer's relationship to and perception of his or her environment. In typing, for example, if one (or both) of the hands are shifted by one key on the keyboard, an experienced typist can produce a whole line of nonsense before a glance at the typed material finally indicates that the hands were in the wrong position.

In terms of unintended acceleration, shifts in the driver's seating position away from the habitual, straight-ahead position—if undetected—would be expected to lead to foot movements that are biased in the same general direction. For example, if the driver is rotated slightly clockwise (as viewed from above) in the vehicle—due perhaps to looking over the right shoulder in a backing maneuver, or simply due to random positioning variability upon entering the vehicle at the start of the driving cycle—a programmed movement to the brake pedal would result in the foot traveling too far to the right. If this bias is sufficiently large, the foot could miss the brake pedal altogether and strike the accelerator instead. Of course, biases could occur in the other (counterclockwise) direction as well, but these would only result in striking the usually large brake pedal too far to its left, or perhaps even missing it altogether. In any case, the accelerator pedal would not be contacted instead.

This account provides a basis for understanding why unintended acceleration episodes may be so strongly related to the start of a driving cycle. Upon the first entry into the vehicle, the positioning deviations from the straight-ahead position are likely to be maximized, and would provide the largest opportunity for the foot movements to be biased directionally. If the error does not occur at this point, then with further driving the operator would tend to shift posture toward the habitual position in the vehicle, thus leading to systematically less movement bias as driving continued.

However, incidents of unintended acceleration have been documented in which the accident has happened well after the start of a driving cycle. In many of these, however, there is often some factor that shifts the driver's position with respect to the vehicle's interior, such as dealing with a teller at a drive-up bank window, or operating a key-card in an automatic gate. Thus, even though the driving cycle may be well under way, there is still a shift in position which can lead to the pedal error through the orientation bias processes mentioned here.

The concept of seating misalignment can also help to understand why unintended acceleration episodes occur with drivers who have had relatively little experience in the accident vehicle. Upon entering an unfamiliar vehicle, the driver would not so easily detect a deviation from the straight-ahead position, and thus would be less likely to correct any misalignment before the foot movement was made. It also helps to understand why older drivers have higher rates of unintended acceleration, as the kinesthetic sensitivity to shifts in body position is well known to decrease with age (Stelmach and Worringham, 1985). Thus, both inexperienced and elderly drivers could fail to detect initial misalignments, but for different reasons, each leading to biases in movement trajectory toward the accelerator rather than to the brake pedal that was intended.

Variability in movement trajectory

A second way that errors in execution can occur is through increases in variability in the processes of movement control. The lower levels in motor system that are responsible for selecting muscles (and the particular motor units within them)—determining how much force will be generated, grading the durations of the contractions and the intervals between contractions—are all governed by relatively 'noisy' processes. Therefore, the values of these functions that are intended by the higher levels in the CNS have a source of variability added to (or subtracted from) them as they are processed in several stages on the way to movement production, making the eventual action somewhat different than intended. If so, then the trajectory of the limb that is driven by the affected muscles will be variable also.

In terms of unintended acceleration, the highest levels in the CNS can have produced the proper choice of action, and yet the noisy processes in the lower levels can cause the foot's trajectory to be different from that intended. If the trajectory takes the foot too far to the driver's right, the accelerator pedal can be contacted instead of the brake that was intended. Next, I turn to some of the evidence for these lower-level sources of variability.

Force variability. As a muscle contracts, it produces force which acts through a tendon to move a bone. The trajectory of a bone is usually controlled by several muscles, and the resultant direction will be changed if any of the muscles' outputs is changed by sources of force variability. The size of these force-variability effects can be seen in experiments in which the performer was asked to produce ballistic 'shots' of force against an immovable strain gauge. The strain gauge activated a trace on an oscilloscope, and the subject attempted to reach a target value on each of a long string of trials. On different sets of trials, the target force value could be changed.

The findings from one such experiment are shown in Figure 20.1 (from Schmidt *et al.*, 1979), where the target force values ranged from 20 to 115 N. Plotted are the within-subject (computed over trials) standard deviations (SDs) of the produced forces as a function of the size of the target force. The inconsistency in the force produced was about 7 percent of the amount of force, and the relationship between force variability and force was approximately linear over the range of forces studied here. This feature holds for both static and dynamic contractions, and is thought to represent relatively low-level processes in the spinal cord or muscle contractile processes. Other experiments show that this function is probably not perfectly linear, but concave downward to approximately 70 percent of the subject's maximum force capability, with a slight downturn as maximum force capabilities are reached (Newell and Carlton, 1985; Sherwood, Schmidt and Walter, 1988). This source of variability is thought to be inherent to the motor system, largely unaffected by prac-

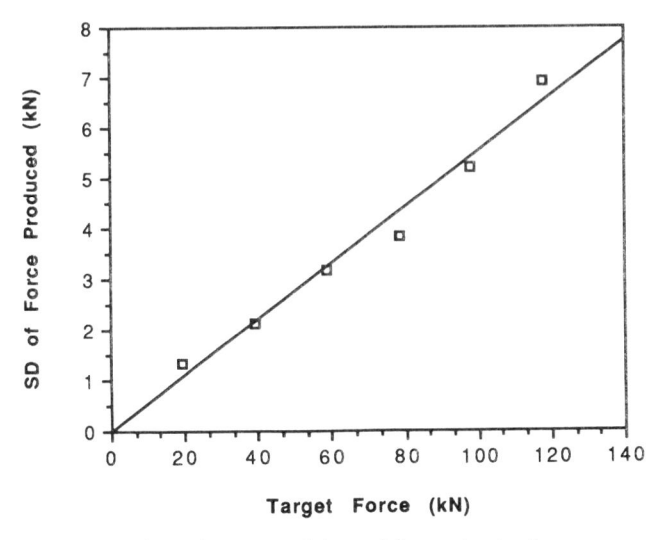

Figure 20.1. Average within-subject variability of forces in simple, uncorrected contractions as a function of the level of force required
(Source: adapted from Schmidt *et al.*, 1979.)

tice; it is one of the limitations that the motor system must 'live with' in producing precise actions.

Temporal variability. Another source of variability is in terms of the durations of the contractions, and in the intervals between contractions, of the participating musculature. This also affects trajectory, because if a muscle contracts too long in a movement, it will bias the trajectory in the direction of its action, contributing another source of variability. We examined temporal variability by asking subjects to move a lever back and forth in time to a metronome, and measured the durations of the force-impulses recorded via a strain gauge attached to the lever. Subjects would produce long strings of actions, attempting to be as consistent as possible. We also varied amplitudes of the movements and the instructed movement time.

The results are shown in Figure 20.2, where the within-subject SD of the force-impulse duration is plotted as a function of the target impulse duration (the instructed movement time). The SD of the impulse durations was approximately 5 percent of the instructed movement time, and the relationship was strongly linear (and nearly proportional). The interpretation is that as the movement is made slower—but not so slow that the actions are produced under feedback control—the programmed intervals among the events in the action are increasingly inconsistent, leading to an increased source of variability in the movement's trajectory. Interestingly, the increases in movement amplitude do not affect this temporal variability. For this reason, these sources of temporal variability are thought to reside somewhat 'higher' in the motor system than the force variability mentioned above, perhaps at the level of the program that defines the temporal structure of the limbs' movements. In any case, these sources of variability are thought to be another fundamental limitation in the system's capability to produce precise timing control.

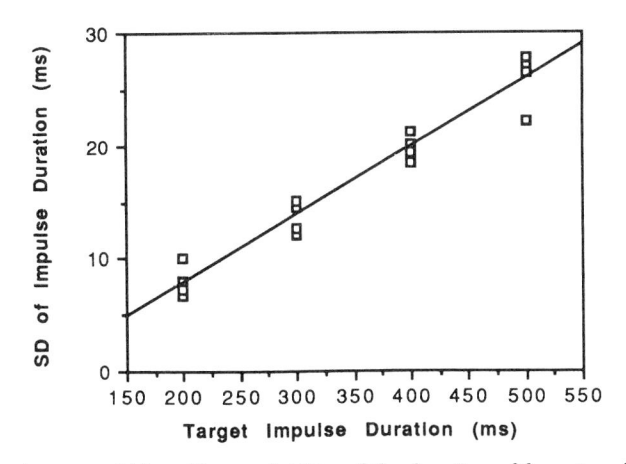

Figure 20.2. Average within-subject variability of the duration of force-impulses in repetitive arm movements as a function of the target impulse duration. Note: The target duration was manipulated by changes in instructed cycle time
(Source: adapted from Schmidt et al., 1979.)

Force and time variability. These two sources of variability are of course combined in the motor system to influence real-world movement behavior. In one example, we asked subjects to make rapid aiming movements of a stylus to a target, where the movement distance and the movement time were controlled by the experimenter. The spread of movement endpoints about the target (termed effective target width, W_e), is plotted for different combinations of movement amplitude and movement time. The spread of movements increases as the movement time decreases, as the movement amplitude increases, or both. The interpretation is that the deviations in the trajectory, caused by the summation of factors affecting force- and time-variability, determine the amount of variability in where the movement lands near the target.

These data also show that the amount of endpoint variability plots nearly linearly with the average velocity (A/MT, in cm/s), providing what has been termed the linear speed-accuracy trade-off for quick actions (Meyer et al., 1988). Meyer et al. have suggested that these processes for quick actions can be combined with feedback-based processes in slower actions, to explain speed-accuracy effects in many classes of human limb movements. Schmidt et al. (1979) and Meyer et al. (1988) argue that these principles of impulse variability are involved in Fitts' Law (e.g., Fitts, 1954), where the speed-accuracy trade-off is manifested as the increase in the time required to make a movement to a target as the target width increases and the movement amplitude decreases. The early, programmed portions of a movement are degraded in accuracy because of impulse variability, and feedback processes are used later in the action to correct for them. See also Welford (1976) and Meyer et al. (1990) for reviews of the research and thinking about speed-accuracy trade-offs, and Drury (1975) for applications to automobile driving.

In terms of unintended acceleration, the notion is that the movement to the brake pedal is a rapid, aimed movement, just as the movements in Figure 20.3 are. The endpoint variability at or near the brake pedal will be influenced by variability in time and force that influence the ways in which the muscles move the bones. Making the braking action more quickly, from a more distant initial position, from an inconsistent initial position, and with a higher velocity, will have the effect of making the endpoint of the movement more variable. If the variability shifts the foot too far to the right on a particular attempt, the result could be contact with the accelerator rather than the brake that was intended.

Why are pedal errors not detected?

Given that the motor system can make trajectory errors even though the correct action was intended, the second major question concerns why these errors are not immediately detected. Recall that the drivers in unintended acceleration episodes typically believe that the foot was on the brake, which indicates clearly that they had not detected the pedal error. There are two related processes that underlie these phenomena.

Automatic movements

The first issue is that, when the motor system produces rapid (e.g., 200–500 ms in duration), stereotyped actions that are well learned and habitual—especially

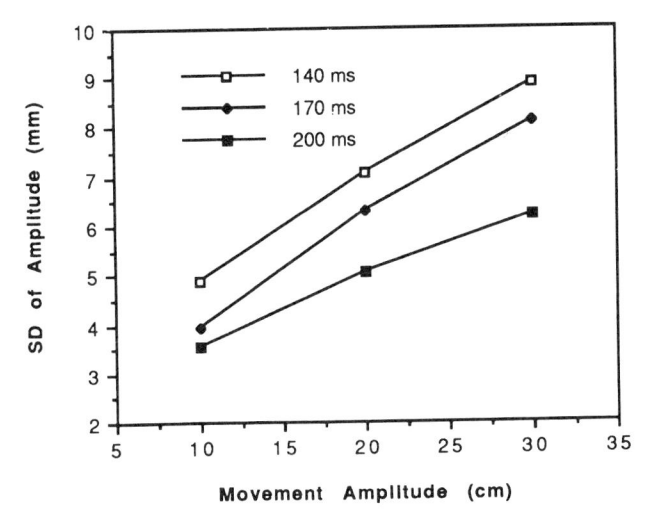

Figure 20.3. Average within-subject variability in movement amplitude in a stylus-aiming task as a function of the required movement amplitude and movement time
(Source: adapted from Schmidt et al., 1979.)

in closed (i.e., stable, unvarying) environments—the system prefers to generate movements that are 'automatic'. That is, the system organizes the action in advance via a motor program, as discussed above, and then the control is passed to lower levels that carry out the action, all without direct control from the higher levels. Because the program level is capable of controlling such action without much sensory input, and because the foot's environment is so stable, the highest levels in the system allow the movement to run off automatically, without having to devote attention to the sensory consequences of the action. Instead, resources can be devoted to other, less stable, aspects of the environment that do require information processing and attention.

In terms of unintended acceleration, the movement to the brake pedal is an action that is highly practiced, very stereotyped, and is performed in a closed, unvarying environment within the vehicle—characteristics that would suggest an automatic control mode effected by a motor program. Under these conditions, if the foot reaches the wrong pedal, the highest levels in the system responsible for error detection and (then) error correction are not aware of the error because the response-produced feedback from the relevant receptors in the leg and foot are simply not processed by the highest conscious levels in the system. This lack of attention to the feedback from the foot is exaggerated by the intense pattern of other stimuli produced by the vehicle's sudden and unexpected acceleration, as I discuss in the next section. Therefore, even the relatively different patterns of tactile and proprioceptive stimulation generated by contact with the accelerator pedal rather than the brake do not alert the driver that an error has been made. It is as if, during these movements, the 'threshold' for detecting kinesthetic information has been elevated to prevent the sensory information from being involved in the action; in the field of motor control, this has often been referred to as the 'gating' of sensory information.

These ideas are consistent with findings that drivers often do not have very clear knowledge about what their limbs are doing in driving maneuvers. For example, drivers usually report that they habitually place their right foot on the brake, and only then shift from Park to Drive (or to Reverse). However, analysis of drivers' actual actions recorded via videotape reveals that the foot movements to the brake and the shift-lever movements to Drive frequently occur more or less simultaneously and with considerable between-subject variability. Roush and Rasmussen (1974) reported that, in 31 percent of the starting sequences, the driver had the right foot on the accelerator when the shift lever was moved from Park; also, 87 percent of the actions had the right foot on the brake by the time the lever had reached Drive. This implies a coordinated movement of both limbs—quite in contrast to the drivers' beliefs about what their hands and feet were doing—and automatic, coordinated movement patterns, about which the drivers have little awareness. If a particular movement to the brake results in a pedal error, reaching the accelerator instead of the brake as discussed above, then the vehicle would accelerate unexpectedly as soon as the shift lever reached Drive (or Reverse).

This lack of attention concerning the movements of the feet in habitual maneuvers provides a basis for understanding the driver's lack of memory about the pedal errors after an accident, coupled with the strong belief that the right foot was on the brake pedal. Typically, the driver's account of what the feet did just prior to the accident seems more related to the driver's impression of his or her *usual* practice than to what actually happened in a particular episode, and so the driver's account of foot behavior must be taken cautiously. These ideas help us to understand how drivers in unintended acceleration accidents can be so certain that their right foot was on the brake, even though it was probably on the accelerator instead.

Efference copy mechanisms

A second explanation of the fact that the driver does not detect the pedal misapplication is based on processes thought to occur in movement perception. A class of theories usualy termed 'efference copy' or 'corollary discharge' (Sperry, 1950; von Holst, 1954) hold that, when a movement is made, a copy of the commands to the musculature are also sent to sensory areas in the CNS. This information allows the highest levels in the CNS to know that an active (as opposed to passive) movement was made, and to know features of the action. Originally, these theories were proposed to explain how, when an eye movement was made, the viewer perceived that the eye moved in a stable world, rather than the eye being stationary in a moving world. These views also provide a basis for a rapid movement perception and evaluation, which can occur prior to feedback being received from the musculature. If this analysis reveals that the movement will be in error, corrections can be made far more quickly, even before the movement has started (Evarts, 1973).

For unintended acceleration, though, the issue is slightly different. Here, the efference copy indicates that a movement to the brake pedal has been initiated. Then, even though the driver does not monitor feedback from the leg (for the reasons mentioned in the previous section), the driver has a basis for knowing the position of the limb. The driver 'knows' that the foot is on the brake because he or she has initiated an action that will take it there. In fact, the program usually does take it there, as the movement is produced in a stable environment, and the movement patterning is relatively simple. If lower-level processes associated with shifts in initial position or force- or time-variability cause the foot to go slightly to the right and contact the accelerator pedal, however, then the driver will be unaware of this error and remain confident that the foot is on the brake pedal. Efference copy views therefore provide one way to understand how the driver can be so certain that the foot was on the brake, even though all evidence points to it being on the accelerator instead.

Why are pedal errors not corrected?

In many laboratory situations, errors are detected and corrected in several hundred milliseconds, and 750 ms is a typical estimate for information processing sufficient to initiate a braking response in a vehicle. Why, then, does the driver in unintended acceleration continue to press on the accelerator for so many seconds (eight to 12 seconds is common, and some accidents have lasted 40 seconds), without taking effective action to stop it? This long-duration lack of control has been particularly difficult to understand, as there are after all several options open to the driver (turn off the key, shift to Neutral or Park, apply the parking brake), and yet these remedies are seldom if ever taken. Why not, and what are the processes that allow the foot-placement error to be converted to a serious accident?

Positive feedback systems

When the driver presses the accelerator pedal rather than the brake, even if lightly, with the right foot, the vehicle responds very differently than the driver intended, moving forward (or backward, if in Reverse gear) slightly. Because the driver believes that the right foot is on the brake, the natural response to the vehicle moving is to apply the brake harder. Pressing the right foot harder continues to actuate the accelerator of course, leading to more acceleration, which leads to pressing harder with the right foot, etc., until full-throttle acceleration is achieved in a few seconds. This situation is in a class of closed-loop control with positive feedback (as opposed to the more usual negative feedback), in which the compensation for the error makes the error even larger. These systems are known to be highly unstable, leading to loss of control quickly if the feedback loop's gain is high. This kind of control is consistent with statements frequently heard in unintended acceleration cases: 'The harder I pressed the brake, the faster the car went'. Also, the solution of removing the foot from the accelerator is not attempted, because the driver already 'knows' that the foot is on the brake, and continuing to press the brake in such situations is perfectly reasonable.

Hypervigilant, or panic, states

As a result of this severe loss of control, the driver is presented with a set of stimuli which results in an extreme emotional reaction known as *hypervigilance*, or what many would term 'panic' (Janis and Mann, 1977). Reviews of evidence in a variety of frightening situations by Janis, Defares and Grossman (1983) reveal that essentially three features are present which lead to this emotional response. First, there is a strong, unexpected stimulus which is startling (see, e.g., Thackray and Touchstone, 1970); in the vehicle it is represented by the sudden, powerful unintended acceleration of the car. Second, this stimulus is perceived as life-threatening, or at least dangerous, to the person or to people in the area; the driver certainly regards unwanted acceleration, often in close quarters, as dangerous. Finally, the situation has a time-stress component in which time to determine an effective solution is quickly running out. In the vehicle, the speed increases continuously, and with each passing moment the situation becomes more and more serious, with increasing demands that some action be taken.

The result of this set of characteristics is a state of 'panic or near panic ... characterized by indiscriminate attention to all sorts of minor and major threat cues as the person frantically searches for a means of escaping from the anticipated danger' (Janis et al., 1983, p. 2). This state continues for many seconds, and is not relieved by any events during the episode; in fact, the situation usually becomes worse with passing time until a collision ends the episode. Hypervigilance is accompanied by several decrements in cognitive functioning and information processing, discussed in the next few sections.

Perceptual narrowing. One of these is the well-known perceptual narrowing, in which the perceptual fields are contracted, allowing effective processing for those stimuli that are central and most relevant, but ineffective response to stimuli that are presented in unexpected places or that call for novel actions (e.g., Baddeley, 1972; Weltman and Egstrom, 1966). As a result, novel actions tend not to be attempted, such as switching off the key or applying the parking brake, and the individual is dominated by habitual, well learned compensations for the perceived threat.

Perseverance. Another feature of panic is that the performer becomes relatively stereotyped in generating solutions. Often, a given solution is tried over and over without success, and then it is tried again. Action is often taken without regard to its consequences, reflecting a failure to think through the problem and the various options that are available.

Distractability. In a panic state, the individual is very distractible, responding to all sorts of cues that direct attention in one way, then to another cue, then to another, with the result that very little processing is accomplished on any one of the cues (Easterbrook, 1959). Even though a novel solution may be tried, attention is distracted by the next cue, and the solution is not completed sufficiently that it can have an effect on the vehicle's travel. There is a related loss in immediate memory capacity (or span), and often there are serious disruptions in movement control even if an effective solution is generated.

Visual capture and visual dominance. One feature of visual information is that it tends to attract (capture) attention, and to dominate other sensory channels. This seems heightened in unintended acceleration, where visual information about up-coming objects that must be avoided tend to dominate attention. This creates a strong bias toward steering movements (often done with remarkably good precision during the episode), but at the expense of processing that would lead to making an action that would end the episode (e.g., turning off the key), or which would lead the driver to realize the pedal misapplication. Perceptual narrowing directs attention toward these visual channels that are perceived as most relevant at the moment.

Freezing. The individual often appears to 'freeze', without taking any action until a collision occurs, a condition found in many panic states. It is unclear, however, as to whether the information processing systems freeze, and the case can be made that there is a kind of hyperactivity that leads to many different actions, none of which is effective. Janis *et al.*'s (1983) view is of the person showing very active (but ineffective) information processing activities on the inside, but with minimal overt action as viewed from the outside.

Static tests in simulators. These hypervigilant states are difficult to produce in laboratory situations, which may explain why so little evidence for pedal errors in unintended acceleration comes from experimental or simulator data. In laboratory tests, either in actual (static) vehicles (Tomerlin and Vernoy, 1988) or in simulators with only minimal movement possible (Rogers and Wierwille, 1988), errors in foot placement are frequently observed, but these are immediately corrected and do not result in the kinds of long-duration, full-throttle conditions that occur in actual unintended acceleration episodes. One reason may be that the features of hypervigilance are missing from these settings, as there is no sustained acceleration, which may reduce the life-threatening qualities of the error. Interestingly, in experiments where realistic movement of the vehicle was present (Tomerlin, 1988), evidence of persistent foot-placement errors as in unintended acceleration has been found. Also, after several actual accidents, drivers have discovered that they had been pressing the accelerator, but did not realize it during the episode because of the stress involved (see Schmidt, 1989).

Generality to other kinds of accidents

Several other kinds of accidents, which do not fall under the definition of unintended acceleration, probably occur by the same general processes that are discussed above. One category involves vehicles with cruise control, and another involves standard-shift vehicles.

Vehicles with cruise control

In several cases involving vehicles with cruise control (both with automatic and standard transmissions), after a trip on the highway during which the cruise control is activated, the driver intends to leave the highway via an exit ramp, disengaging the cruise control in the process by pressing the brake. Several incidents have occurred where the vehicle then accelerated wildly, leading to an accident on the exit ramp, with the driver claiming the brakes failed and the vehicle accelerated uncontrollably. Discounting the possibility that the driver reactivated the cruise control by inadvertently pressing the 'resume' switch on the turn signal, the focus of investigation has often been in terms of malfunctions in the cruise control system itself (e.g., Vehicle Research and Test Center, 1989).

These accidents have not been classified as unintended acceleration, as they do not start from a very slow speed (and at the start of a driving cycle), but even so they may have many of the same processes underlying them. One possibility is that the driver becomes misaligned in the vehicle because the cruise control frees the driver to shift seating positions in the cockpit area. When a brake pedal movement is intended, if the foot moves to the accelerator instead, a positive feedback system is created, in which the vehicle's failure to slow down leads to more right-foot pressure, which only causes the vehicle to accelerate more, and so on; this situation is consistent with the driver's subjective experience that the harder the brake was pressed the faster the vehicle went. In addition, long before the advent of cruise control and automatic transmissions, Henderson (1935) hypothesized that something like hitting the accelerator pedal instead of the brake in various dangerous maneuvers (avoiding a collision, for example) can lead to full acceleration and an episode very much like unintended acceleration (Ruggerio, 1990). Again, these episodes do not begin from a stop, but may have many of the same processes underlying them as unintended acceleration as defined here.

Standard-shift vehicles

An adequate account of unintended acceleration should be capable of explaining all of the many facts of these cases. One puzzling concern has been the finding that unintended acceleration almost never occurs in standard-shift vehicles (except as noted above concerning the cruise control situations). According to a pedal-error view, in the vehicle with automatic transmission, the driver intends to hold the vehicle at a stop by pressing the brake, and the errant foot movement to the accelerator produces the opposite effect to accelerate the vehicle. The error not only produces the beginnings of the episode, but also generates a strong set of unexpected stimuli that are startling and disorienting to the driver, allowing the episode to continue.

Note that this is not the case in the standard-shift vehicle. If the driver intends to hold the vehicle at a stop, and the brake pedal is missed and the accelerator contacted with the right foot instead, the vehicle will not move because the clutch pedal remains depressed by the left foot. The clutch would have been engaged only if the driver intended to move the vehicle. The result is that the right foot misses the brake, causing an elevation in engine speed, but unintended acceleration is not initiated, nor is there the startling and disorienting stimulation from the unexpected movement of the car to contribute to the error. Pedal errors can occur in standard-shift vehicles, but they do not result in unintended acceleration from a stop because of the simultaneous action of the clutch necessary to move the vehicle.

In addition, if a pedal error occurred when the clutch was already engaged (in low shift maneuvers, for example), the driver in a standard shift vehicle has the additional option of depressing the clutch to prevent the unwanted acceleration (Wierwille, in press). In addition, Wierwille has suggested that the existence of the clutch pedal provides additional orientation cues and constraints on driver position that decrease the probability of right-foot errors.

Left-foot brakers

An intriguing observation is that drivers who use their left foot for the brake in vehicles with automatic transmission appear to be under-represented in the unintended acceleration statistics.[1] Under a pedal-error viewpoint, if the driver intended to press the brake with the left foot, it would be unlikely that the foot's trajectory would deviate sufficiently far rightward—across the body's midline—that it would contact the accelerator. Even if it did, the left foot would likely contact the right foot which, constrained by the transmission tunnel to the right, would already be on or near the accelerator. This should provide strong cues for error detection and subsequent correction.

Solutions to prevent unintended acceleration

The weight of evidence, both from the engineering analyses of the accident vehicles and the human-factors principles, suggests that unintended acceleration is the result of pedal misapplications. Several design modifications can be suggested to reduce the likelihood of these pedal errors in future vehicles.

Shift-lock mechanisms

One solution to unintended acceleration, based on a pedal-error view, has been the installation of a 'shift-lock' mechanism. It was initiated in Audi vehicles in 1987, and has been adopted by several other auto makers at this writing, with most others to adopt it in the next few years. Here, a mechanical lock prevents the shift lever from being moved out of the Park position unless there is (light) pressure on the brake pedal. This would appear to eliminate the major cause of unintended acceleration—the application of the accelerator pedal rather than the brake as the shift lever is moved to Drive or Reverse.[2] Particularly with right-foot brakers, if the right foot must be on the brake when the lever reaches either Drive or Reverse, then it cannot be on the accelerator, and the vehicle will not move until the driver releases pressure on the brake and places the foot on the accelerator. This solution would seem to eliminate the problem of unintended acceleration if it is caused by pedal errors, but would not be expected to have an effect on accident rates if the cause were other than pedal misapplication (e.g., some mechanical defect).

It is perhaps too early to be certain, but the data on the shift-lock vehicles suggests that this is a promising solution. One data set concerns Audi automobiles, where a comparison can be made between vehicles in which the shift-lock mechanism had and had not been installed.[3] One analysis of these effects was done by the Office of Defects Investigation (1989), in which cars without shift-locks had 2.5 times the rate of reported unintended-acceleration episodes as did cars with the shift-locks in place. It was estimated that there was an 83 percent reduction in reported unintended acceleration as a result of the shift-lock. One view of these data is that it provides strong support for the pedal-misapplication theory, which insists that shift-lock mechanisms should be effective in eliminating many unintended acceleration incidents. However, there were still a disturbingly large number of unintended acceleration episodes reported in vehicles with shift-locks, which should not be the case if pedal misapplications were the only causes of the accidents. What are the factors here?

One confounding problem is that the shift-lock can have no effect after the transmission has been shifted from Park, and thus unintended acceleration could occur if the driver made a pedal misapplication later in the driving cycle. For example, if the driver stopped the vehicle, placed it in Neutral, and then became misaligned while inserting a key-card in a gate mechanism, nothing would prevent the shift to Drive if, upon resuming the driving cycle, a pedal misapplication had subsequently occurred.[4] Similarly, examples of missing the brake and contacting the accelerator instead could occur during the driving cycle, including examples mentioned earlier with deactivation of the cruise control mechanism. Other factors are present as well, making this analysis somewhat difficult to interpret, such as improperly installed shift-lock mechanisms, biases in the rate of complaints from publicity, and several others, as documented in the ODI report. Overall, though, the shift-lock mechanism must be viewed as a powerful solution for preventing most of the unintended acceleration accidents.

Pedal configurations

Several analyses of the pedal configurations of existing vehicles have been made, in an attempt to correlate particular configurations with the rates of reported unintended acceleration. One argument is that, compared to the 'typical pedal configuration', vehicles in which the brake and accelerator are a) located more to the left, b) have small vertical separation between the accelerator and the brake, and c) have the brake and accelerator relatively close together laterally might induce more pedal misapplications. The problem is that many exceptions can be found, in which vehicles extreme on these pedal configuration variables have both very high and very low rates of unintended acceleration.

Probably the most serious flaw in these analyses is that different classifications of people, as discussed earlier, have differential rates of unintended acceleration, and particular classifications of people are over-represented in certain vehicle makes and models. Older people, for whom reported unintended acceleration rates are higher than younger people, tend to be biased toward larger, more expensive models, making it appear that these makes and models have an inherently larger rate of unintended acceleration than less expensive vehicles that are very similar mechanically. These analyses are further complicated by publicity effects in complaint rates (see Tomerlin, 1988, for a discussion of these effects) and several other factors. As a result, it is difficult to argue that certain kinds of pedal configuration are more or less prone to generate pedal misapplications and unintended acceleration.

Others have suggested possibilities for a redesign of the pedal arrangements to minimize the likelihood of pedal errors, but each kind of modification would have other drawbacks and side-effects (see Office of Defects Investigation, 1989; Wierwille, in press). For example, separating the accelerator and brake pedal horizontally should reduce the likelihood of pressing the accelerator rather than the brake, but would produce increased movement time from the accelerator to the brake in 'panic-stop' situations, which is generally considered an unacceptable trade-off. Also, separating the pedals too far might allow some drivers to get their feet caught between and under the pedals under certain circumstances. In addition, at least one empirical investigation of various pedal configurations has failed to reveal any effect on pedal errors (Rogers and Wierwille, 1988).

Changes in the pedal dynamics, or 'feel', to make the brake pedal feel very different from the accelerator when depressed have also been considered. The evidence is not clear that these kinds of variables are related to unintended acceleration, however. In addition, it is doubtful that pedal dynamics could have major effect on unintended acceleration, particularly if the driver does not process information from the foot pedals at all, as suggested by the pedal-error viewpoint presented here. More work is needed with these variables to be certain about their role in unintended acceleration.

Summary

The available analyses of unintended acceleration have consistently failed to find any mechanical defect that would account for the full-power acceleration and the perceived reduction in braking effectiveness, and which would disappear soon after the accident. As a result, the focus has shifted to human performance principles to understand how unintended acceleration might be the result of human errors of various kinds. The evidence points to pedal misapplications as the major cause, where the right foot contacts the accelerator rather than the brake, commonly (but not always) as the shift lever is moved from Park to either Drive or Reverse. This leads to unexpected acceleration, and further attempts to press the brake pedal only lead to increased accelerator pedal application and further acceleration. This event generates hypervigilance (or panic), which further reduces information processing capabilities that would otherwise allow the driver to remedy the problem. As a result, the episode continues until terminated by a collision. Effective design remedies for unintended acceleration concern the shift-lock mechanism, which prevents the driver from shifting into gear unless pressure is applied to the brake pedal, but redesign of the pedal configurations does not seem to be a promising solution.

Notes

1 This is a difficult trend to document carefully, as the relatively small number of left-foot brakers (about 18 percent of drivers, according to Peacock, 1991) makes the data somewhat unreliable.
2 The adoption of shift-lock mechanisms has angered various 'victims' organizations (e.g., the New York Public Interest Research Group, NYPIRG; see Tomerlin, 1988), because this solution assumes that the fundamental cause of unintended acceleration is pedal misapplication, and denies that the cause is a mechanical defect.
3 Audi began installing shift-locks on its 1987 vehicles and installed them on earlier vehicles in a recall campaign. Thus both types of vehicle are in the population.
4 Later Audi vehicles also lock the lever in neutral unless the brake pedal is activated. This solution should greatly reduce the problem.

References

Baddeley, A. D., 1972, Selective attention and performance in dangerous environments, *British Journal of Psychology*, **63**, 537–46.

Battelle Corporation, 1987, *Unintended Acceleration of 1978–1986 Audi 5000 Cars with Automatic Transmission*, Columbus, OH: Author.
Columbia Broadcasting System, 1986, '60 Minutes', 25 November.
Drury, C. G., Application of Fitts' law to foot pedal designs, *Human Factors*, **17**, 368–73.
Easterbrook, E. A., 1959, The effect of emotion on cue utilization and the organization of behavior, *Psychological Review*, **66**, 183–201.
Evarts, E. V., 1973, Motor cortex reflexes associated with learned movement, *Science*, **179**, 501–3.
Fitts, P. M., 1954, The information capacity of the human motor system in controlling the amplitude of movement, *Journal of Experimental Psychology*, **47**, 381–91.
Gallaway, G. R., 1987, Personal communication, 5 January.
Henderson, Y., 1935, How cars go out of control: Analysis of the driver's reflexes, *Science*, **82**, 603–6.
Higgins, J. V., Job, A. M. and Pepper, J., 1987, Runaway Cars: Man or Machine, *Detroit News*, 13–17 January.
Janis, I. L., Defares, P. and Grossman, P., 1983, Hypervigilant reactions to threat, in Selye, H. (Ed.) *Selye's Guide to Stress Research*, pp. 1–43, New York: Van Nostrand Reinhold.
Janis, I. L. and Mann, L., 1977, *Decision-Making: A Psychological Analysis of Conflict, Choice, and Commitment*, New York: Free Press.
Marriner, P. and Granery, J., 1988, *Investigation of 'Sudden Acceleration' Incidents* (Report No. TP9607E), Ottawa, Canada: Transport Canada.
Marteniuk, R. G. and Roy, E. A., 1983, Head position, eye position, and reaching movements, Unpublished experiments, University of Waterloo, Canada, Department of Kinesiology.
Meyer, D. E., Abrams, R. A., Kornblum, S., Wright, C. E. and Smith, J. E. K., 1988, Optimality in human performance: Ideal control of rapid aimed movements, *Psychological Review*, **95**, 340–70.
Meyer, D. E., Smith, J. E. K., Kornblum, S., Abrams, R. A. and Wright, C. E., 1990, Speed-accuracy trade-offs in aimed movements: Toward a theory of rapid voluntary action, in Jeannerod, M. (Ed.) *Attention and Performance XIII*, pp. 173–226, Hillsdale, NJ: Erlbaum.
Ministry of Transport, Japan, 1988, *Sudden Starting and Acceleration in Automatic Transmission Vehicles*, Japan: Author.
National Highway Traffic Safety Administration (NHTSA), 1987, *1973–1986 General Motors Passenger Cars Sudden Acceleration* (Document EA78-110), Washington, DC: Author.
Newell, K. M. and Carlton, L. G., 1985, On the relationship between peak force and peak force variability in isometric tasks, *Journal of Motor Behavior*, **17**, 230–41.
Norman, D. A., 1981, Categorization of action slips, *Psychological Review*, **88**, 1–15.
Office of Defects Investigation (ODI), 1989, *Alleged Sudden Unwanted Vehicle Acceleration, 1978 through 1986 Audi 5000 Passenger Cars Imported by Volkswagen of America*. Washington, DC: National Highway Traffic Safety Administration.
Peacock, B., 1991, Personal communication.
Perel, M., 1976, *Analyzing the Role of Driver/Vehicle Incompatibilities in Accident Causation Using Police Reports*, (Technical Report DOT HS-801-858), Washington, DC: US Department of Transportation.
Pollard, J. and Sussman, E. D., 1989, *An Examination of Sudden Acceleration*, (Technical Report DOT HS-807-367), Washington, DC: US Department of Transportation.
Reason, J. T., 1979, Actions not as planned, in Underwood, G. and Stevens, R. (Eds) *Aspects of Consciousness*, pp. 67–89, London: Academic Press.
Rogers, S. B. and Wierwille, W. W., 1988, The occurrence of accelerator and brake pedal actuation errors during simulated driving, *Human Factors*, **30**, 71–81.

Roush, G. S. and Rasmussen, R. E., 1974, *Brake Pedal Location Study* (Technical Report PG-34703), Milford, MI: General Motors Proving Grounds.
Ruggerio, A., 1990, Personal communication.
Schmidt, R. A., 1976, Control processes in motor skills, *Exercise and Sport Sciences Reviews*, **4**, 229–61.
Schmidt, R. A., 1988, *Motor Control and Learning: A Behavioral Emphasis*, 2nd edn, Champaign, IL: Human Kinetics Publishers.
Schmidt, R. A., 1989, Unintended acceleration: A review of human factors contributions, *Human Factors*, **31**, 345–64.
Schmidt, R. A. and Stelmach, G. E., 1968, Postural set as a factor in short-term motor memory, *Psychonomic Science*, **13**, 223–4.
Schmidt, R. A., Zelaznik, H. N., Hawkins, B., Frank, J. S. and Quinn, J. T., 1979, Motor-output variability: A theory for the accuracy of rapid motor acts, *Psychological Review*, **86**, 415–60.
Sherwood, D. E., Schmidt, R. A. and Walter, C. B., 1988, The force/force variability relationship under controlled temporal conditions, *Journal of Motor Behavior*, **20**, 106–16.
Sperry, R. W., 1950, Neural basis of the spontaneous optokinetic response produced by visual inversion, *Psychological Monographs*, **74** (Whole No. 498).
Stelmach, G. E. and Worringham, C. J., 1985, Sensorimotor deficits related to postural stability: Implications for falling in the elderly, *Clinics in Geriatric Medicine*, **1**, 679–94.
Thackray, R. I. and Touchstone, R. M., 1970, Recovery of motor performance following startle, *Perceptual and Motor Skills*, **30**, 279–92.
Tomerlin, J., 1988, Solved: The riddle of unintended acceleration. *Road & Track*, **39**, June, 52–59.
Tomerlin, J. and Vernoy, M. W., 1988, *Pedal Errors in Several Make and Model Cars*, (Unpublished Technical Report), Laguna Beach, CA: Public Policy, Inc.
Vehicle Research and Test Center, 1989, *Driver Task and Vehicle Response to a Simulated General Motors Cruise Control, Cruise 3 System, Fast Idle Malfunction*, East Liberty, OH: Author.
von Holst, E., 1954, Relations between the central nervous system and the peripheral organs, *British Journal of Animal Behavior*, **2**, 89–94.
Welford, A. T., 1976, *Fundamentals of Skill*, London: Methuen.
Weltman, G. and Egstrom, G. H., 1966, Perceptual narrowing in novice drivers, *Human Factors*, **8**, 499–505.
Wierwille, W. W., in press, Driver error in using automotive pedals—A review of the major issues, in Peters, G. A. and Peters, B. J. (Eds) *Automotive Engineering and Litigation*, Vol. 4, Eau Claire, WI: Professional Education Systems, Inc.

21

The older driver and passenger

David B. D. Smith, Najmedin Meshkati and Michelle M. Robertson

For older people transportation is a key to a life of quality and personal independence. In technological societies it has come to mean access to the private automobile. In the United States 80 percent or more of all trips that people over age 65 take are in individual automobiles, usually their own vehicle (National Research Council, 1988). Not only is the automobile the transport of choice for this population but many more older people will be drivers in the future (National Research Council, 1988). The twentieth century has seen a dramatic increase in the number and proportion of older persons. By the first half of the twenty-first century the proportion of those over 55 will near one-third of the population for many industrialized nations. While we hear most about this graying of the population and its dependency on the automobile in the case of Western Europe and North America, the trend is world-wide (*Traffic Safety of Elderly Road Users*, 1985). As the older population grows in number, expectation and economic influence, the motivation to design for them will almost certainly increase.

In general, the need to design automobiles and highways ergonomically has been widely acknowledged and increasingly practiced (Wierwille and Peacock, 1989). However, the design approaches and data used to this point are mostly for a younger population. The purpose of this chapter is to define and describe the older user population and to assemble and review selectively, as space permits, the current database applicable to automobile design.

Ergonomic design: Descriptive data on the older population

Ergonomic design begins by identifying and describing the user population and the unique problems they face. The purpose of this part of the chapter is to describe the older driver and passenger.

Population variability

What we perceive as aging is a progressive and variable combination of ontogenetic (e.g. physical strength), historical (e.g. war service), and life events (e.g.

accidental injury; Baltes, Reese and Lipsitt, 1980). The result is a non-uniform set of behavioral and physiological changes with age. This variability is a principal difficulty and a necessary consideration in applying ergonomics at mature ages.

The first source of diversity occurs in the definition of who constitutes this population. Society usually identifies persons aged 65 and above as 'older', 'aged' or 'elderly'. However, younger age markers can be appealed to as having applied relevance; such is the case for commercial airline pilots (Fay, 1978), for the training of air traffic controllers (Mathews and Cobb, 1974), and in definition of the older worker (Smith, 1990). Even for driving there is evidence that one should look at 'the middle years' (45–65) for age implications in driving skills and safety (Planek and Overend, 1973), which may also be true for ergonomics design.

However, specifying 'any age and older' as the age marker can be misleading because persons in their 60s may differ as much from those in their 80s as they differ from persons in their 30s. This between-ages variability has led to classifying the older population into age groups of five, 10, or some number of years and variously designated, for example, as young-old, old and old-old (Neugarten, 1974). Such finer grain analysis is useful in identifying where significant age changes occur in overall norms but may still mask a deceptive kind of age variability, that between persons. Older people in fact show the greatest individual variability of any age cohort. This fact shows the danger in relying on chronological age and the need to have measures of variability that are applied in automotive design for the mature user. Such a mental set is also important because the stereotypes and myths about older people in general and in particular to drivers still are very much with us (Butler, 1975, p. 12).

Within-person variability also is said to increase with age; that is, task performance varies markedly from time to time for the same individual. This variability over time in behavior and skill is especially present in persons with the medical disabilities prevalent at old age, such as stroke, heart conditions, dementia, Parkinson's disease and diabetes. A final source of variability, cohort differences, refers to generational changes occurring for this population. Today's mature drivers and passengers differ from those of 25 years ago, and those 25 years hence. These differences are present in diverse ways, such as economic status, education, home location (increased residency of the elderly in suburban and independent settings) and with today's elderly having higher expectations for leisure, services and technology. Especially relevant to driving in the future will be the increased years of driving experience, the increased number of very old drivers (80+) and women drivers, and the greater health of future elderly (National Research Council, 1988).

Age patterns and problems in driving

An initial step in design is to ask where persons of concern have problems and are likely to benefit from ergonomic intervention. Table 21.1 presents information from numerous sources on established driving patterns, problems and changes with age.

One of the better documented changes that occurs with older age is a decrease in number of miles driven (US Census Bureau, 1984). In the US from a peak of over 12000 annual miles in mid-life (29–54) this drops to 5000 or

Table 21.1. Characteristics and problems of the older driver and passenger

General characteristics or changes with aging:

- The large majority of all trips are in the private auto (National Research Council, 1988);
- Miles driven declines after age 50 and dramatically so after age 60 (US Census Bureau, 1984);
- Those over age 75 are more likely to be passengers than drivers (National Research Council, 1988);
- There is a reduction in driving at night, in bad weather, in rush hour traffic and long trips (Winter, 1984).

Vehicle trip characteristics of older relative to younger ages (Kobayashi, 1984; Rosenbloom, 1988; Traffic Safety of Elderly Road Users, 1985):

- They make *more* non-work trips;
- They make *more* shopping and medical trips;
- They make *fewer* trips for social and visiting reasons.

Reported and observed driving problems with aging:

- *Reported Problems:* Seeing while driving at night; merging and exiting in high speed traffic; reading traffic signs; turning head while backing; reaching seat belt and reading instrument panel (Yee, 1985; Tsander and Herner, 1976);
- *Observed Problems:* Changing lanes, merging, passing, backing, controlling speed, leaving from a parked position, paying attention, turning. (Planek and Overend, 1973).

Traffic citation of older drivers (Huston and Janke, 1986; McKnight, 1988):

- Frequency of citations decreases with age;
- Citations *per mile* increases at about age 69 for violations of right of way, signs, signals and turns.

less by age 70–79. In part this decrease reflects the absence of work mileage, a growing proportion of women drivers, generational differences in automobile use and dependency, and perhaps a recognition of decreased driving skills (Rosenbloom, 1988). However, quite possibly the future old will drive more, reflecting their generation's propensity for auto use and surpass the low mileage of today's elderly. This possible increase enhances concern about their safety, access and convenience in using the automobile.

There is also data (Kobayashi, 1984; Rosenbloom, 1988; *Traffic Safety of Elderly Road Users*, 1985) that suggests that in spite of the low mileage driven by older people, individual trips, though shorter in length, are not appreciably different in number from other ages. These trips do differ in other ways; they are less likely to occur at night, in rush hours and in bad weather (Winter, 1984). US data indicates for persons over 65 some 13 percent of trips are made at night; the national average for all ages is nearly double that (National Research Council, 1988).

Table 21.1 also provides some insight into the specific uses of the private auto by older persons. Data reported by Rosenbloom (1988) from the US Nationwide Transportation Study (Federal Highway Administration, 1986) shows that older persons make more trips for shopping and medical reasons than the younger traveler but somewhat less for social reasons.

One place to identify problem areas for the older driver, with possible design importance, would be to ask them where they have driving difficulties. The major evidence in this area is a joint study conducted by the AAA Foundation and the Safety Research and Evaluation Project at Teacher's College, Columbia University (Yee, 1985). Data was collected by a survey of older drivers aged 55 and over living in urban, suburban and rural areas of 11 states. A total of 446 older drivers were surveyed and responses compared with a sample of 104 younger drivers (30–45). While the results were extensive with a 128-item questionnaire, driving difficulties which older persons reported significantly more often than younger ages can be summarized as in Table 21.1. Two of the factors, *that of seeing while driving at night and reading traffic signs*, seem to involve age factors in visual functions. Another two, *turning head while backing, and merging and exiting in high-speed traffic* are reasonably associated with neck and torso mobility, and in the latter case the slowing of response-speed with age. The last two, *reaching seat belt and reading instrument panel* directly reflect age factors involved in ergonomic design of the automobile.

Another data source for identifying problem areas for the older driver would be to look at traffic citation data. This information is summarized at the bottom of Table 21.1. Consistent with the data on reported and observed problems, older people are reported as most likely to violate traffic laws associated with signs, signals and turns (Huston and Janke, 1986; McKnight, 1988).

A final source of descriptive information about the driving of the elderly and where they have problems comes from accident statistics. Table 21.2 summarizes the statistics concerning accidents and injury for the older person. Older people have fewer accidents and serious injury accidents per capita than do young people. When the statistic accounts for exposure, since older people drive much less, their accident rate per mile is similar to that for young people. Also, they are more likely than younger ages to be killed in the accident or by the injury sustained, to have slower recovery from injury, to spend more time in the hospital and to lose more time from work. There is reason to believe that older persons are aware of and try to compensate for their greater susceptibility to injury (Smith, 1987). This peculiar susceptibility to injury and its consequences should be an important consideration in automotive design focused on the older population. Table 21.2 also gives the major driving situations in which statistics indicate the elderly are predisposed to accidents. They are *backing, changing lanes, merging, disregarding signs and signals, improper turning and in yielding of right of way* (Maleck and Hummer, 1986; McKnight, 1988). These incidents correlate well with the reported driving difficulties and location of traffic citations in Table 21.1.

Disabilities and age

All chronic diseases show an increase in both frequency and severity with age (Blake, 1984). About 85 percent of persons over 65 have a chronic medical

Table 21.2. *Accident and injury characteristics for the older vehicle driver and passenger*

Accident characteristics for the older driver (Brainn et al., 1977; Finesilver, 1969; Maleck and Hummer, 1986; Monforton et al., 1988):

- Frequency of accidents decreases with age until the 70s and then increases;
- Accident *rate per mile* increases in the 70s, approaching that for teenagers;
- Speed is usually not a factor;
- Multi-vehicle involvement increases relative to single vehicle;
- Accident occurs most often in daylight, in good weather and in an urban setting;
- Accident frequently involves right-of-way violations, improper turns and lane changes, backing and parking, ignoring signs and traffic lights.

Injury characteristics for older persons in vehicle accidents (National Research Council, 1988):

- Frequency of serious injury decreases;
- Fatalities increase;
- When injured older persons have slower recovery, spend more time in the hospital and lose more time from work.

condition such as diabetes, arthritis, emphysema or cardiovascular disease (Adams and Collins, 1987). While chronic disease is very prevalent in old age, it is not necessarily associated with functional disability. The most recent statistics for the US (Dawson, Hendershot and Fulton, 1987) indicate that one in four persons over 65, and living in the community, report a disability that interferes with at least one home management activity, such a shopping or doing heavy house work. This disability increases from about one in 10 for the ages 65 to 74 to one in two at and beyond age 85. Also, women are more likely to report difficulty in home management activities than men. The disabled old includes both people disabled late in life and those disabled early in life who are now older. The latter group is the first generation of any size of such people and may number some 20 percent of the disabled old (Codd and Kurland, 1985).

These medical problems are involved in some of the marked decrease in travel by the elderly. Table 21.3 shows the percentage of older persons not driving who gave health and disability as a cause; note that the percentage increases with age. For severe disabilities more than 50 percent of respondents gave health or disability as a cause for not driving (Rosenbloom, 1988). Even where disability is not a barrier to driving, there is reason to believe it can influence free access to mobility, a loss in driving skill and possibly safety. The major medical disabilities studied in regard to age and driving have been cardiovascular and cerebro-vascular disease (Sivak *et al.*, 1981), visual impairment (Johnson and Keltner, 1983), diabetes (Frier *et al.*, 1980) and, recently, Alzheimer's disease (Federal Highway Administration, 1990). For reasons of greatest benefit the disabled elderly may pose one of the primary areas for ergonomics application.

Table 21.3. Effect of health and disability on driving

Age range	% of nondrivers giving health or disability as a cause
60–69	18.7
70–79	22.8
80+	34.2
All ages	19.2

(Source: US 1977 Health Interview Survey Data. Adapted from Rosenbloom, 1988, Table 7, p. 39).

Ergonomics design: Functional data for the older population

Ergonomics (human factors, human engineering) is the discipline that applies knowledge about the capabilities of people to the design of technology and environments. A fundamental limitation on considering older people in such design activity is the scarcity of specific data about this population in the informational and educational database for this field (Smith, 1990). The data is likely to be only descriptive, lacking a quantitative form and without any reference to automotive design. Some appropriate data exists in the gerontological literature, but its collection and translation to application remains largely a task to be done. Some general references on ergonomics and age are of value (Charness and Bosman, 1990; Smith, 1990). Others (Transport Canada, 1986; Yanik, 1988) have dealt more specifically with automotive design and older people. The purpose of this section is to bring this and other information together in one place. Given space limitations, the treatment is not complete but greater coverage is available in the cited references.

Physical and motor changes

The older population as a group differs from younger counterparts in specific musculoskeletal characteristics and related physical abilities. For discussion purposes these differences can be grouped into three categories; 1) body size and shape, 2) range of motion and joint flexibility, and 3) strength and force.

Body size and shape

With age there is a progressive decrease in body height measurements. Other dimensions of stature, such as sitting height, eye height and length of the arms and legs show a similar decline (Stoudt, 1987). These age related anthropometric differences are due to two factors. First is a biological process such as atrophy, compression and thinning of the intervertebral discs of the spinal cord. The second is a secular trend due to cohort differences in nutrition and health in early life allowing the genetic potential to be approached. The relative contribution of these factors has been estimated to be 59 percent biological and 41 percent secular (Borkan, Hults and Glynn, 1983; see also the chapters by Roe and by Kroemer, this volume).

In measures of body size and circumferences the young ages tend to be smallest, increasing in middle age and then decreasing in later years. Middle-aged men (35–54) are about 15 pounds heavier than their younger (18–24) and older cohorts (65–74), while middle-aged women are approximately 16 pounds heavier than young persons but only about three pounds heavier than those aged 65–74 (Stoudt, 1987).

Table 21.4 provides the major sources of anthropometric data available in the English literature on body dimensions for the older population. Two important limitations for use of the information is its paucity at older ages (80+), and the relatively old date of collection for some of the studies. A summary of much of this data and evaluation of its limitations has been provided by Kelly and Kroemer (1990) and by Charness and Bosman (1990).

Range of motion and joint flexibility

Accompanying age is a reduction in the range of motion at various joints (Adrian, 1981). This general age trend is especially impacted by an older person's health and maintenance of physical activity (Harris, 1983). Moreover, the presence of arthritis is a major contributor to joint problems. At age 60 about 20 percent report having this disorder and that amount increases to 50 percent by age 75 (Adams and Collins, 1987). Fully 40 percent of those over 65 report some limitation of their normal activity because of such conditions (National Health Survey, 1986).

The reports of Table 21.4 largely concern static measurements, those taken while an individual is sitting or standing still. What is needed for range of motion and flexibility problems are functional and dynamic measurements, those made while the individual performs a function or operational activity. Charness and Bosman (1990, Table 6) have provided a collection of functional reach measurements from two of the sources in Table 21.4 (Damon and Stoudt, 1963; Roberts, 1960) and one other (Grandjean, 1973). While general anthropometric data of this sort are potentially useful for design, what is most needed are functional measurements obtained in the automotive context. Vehicle entry, exit and seating are likely to be major points of ergonomic application for the elderly. Brooks, Ruffle-Smith and Ward (1974) studied 206 men and women, both elderly and disabled elderly, 80 percent of whom were over age 60 and 54 percent over 70. Tests were carried out in a mock-up of a bus body to identify ergonomic problems of entry/exit, seating and standing. In addition to measures of reach, pulling strength and twisting strength, preference ratings were obtained for seat and step dimensions. Results for seats and steps are summarized in the top part of Table 21.5. Page et al., (1984) investigated the ideal access clearance and seat height for automobiles. Based on a preliminary analysis of six production cars, they developed a functional mock-up to test the critical dimensions in passenger car entry/exit and seating.

Table 21.4. *Major sources of anthropometric information for the older population*

Source	Sample statistics	Kind of data
Damon and Stoudt (1963)	133 ambulatory American men; age 72–91	47 sitting and standing measurements (means, S.D., percentiles) of height, body lengths, breadths, depth and circumferences; group strength and reach.
Roberts (1960) (see also Ward and Kirk, 1967)	78 British community-living women; Mean age 71.65, S.D. 7.51 years	33 sitting and standing measurements of static body dimensions and reach: means and S.D.s.
Damon, Seltzer, Stoudt and Bell (1972)	72 ambulatory American men; age 60–69	51 sitting and standing measurements; means and S.D.s.
Brooks, Ruffle-Smith and Ward (1974)	206 elderly and disabled British men (64) and women (142); 80% over 60 and 16% over 80 years old.	9 static anthropometric measures of weight, heights, and widths. (5th, 50th, 95th percentile) Dynamic measures of reach, pull and twist capacity, step heights negotiated. Preference measures.
Friedlaender *et al.* (1977)	188 healthy American men; age 55–70+ from the Boston Normative Aging Study.	29 body and head measurements; means and S.D.s.
Stoudt (1981) (Summary of U.S. Health Examination Survey, 1960–62)	Probability sample of US civilian, noninstitutionalized population; 265 men and 299 women 65–74; 72 men and 70 women; 75–79.	16 sitting and standing measurements (mean, S.D., percentiles) of height, weight, lengths, breadths and circumferences.
Borkan, Hults and Glynn (1983)	1212 men from the Boston Normative Aging Study (US); Age 20–82; 55 men 60–70+.	44 anthropometric dimensions measured twice 10 years apart. Includes height and arm span measurements (means and S.D.) for 10-year age intervals.
Viitasalo, Era, Leskinen and Heikkinen (1985)	183 men; aged 31–50 188 men; aged 51–55 176 men; aged 71–75	Dynamometer measurements of grip, trunk and knee extension strength and height, weight and skeletal widths and skin fold thickness.

This simulation was then used to test 60 elderly subjects, both able and disabled, to determine the access and seating measurements of greatest acceptance. The results are presented in the lower part of Table 21.5.

Outside visibility is another critical feature for the older driver. It has been argued that decreases in upper body mobility adversely impact older people's

*Table 21.5. Vehicle access and seating dimensions showing high acceptance by older people**

Access and seat dimensions for buses (Brooks, Ruffle-Smith and Ward, 1974):

- a seat height between 43 and 46 cm preferred,
- a seat spacing of 68 cm best for both elderly and disabled elderly,
- a step height of 27 cm was possible with hand holds for both elderly and disabled elderly. A height of 18 cm was negotiated with ease.

Access dimensions for cars (Page et al., 1984):

- a door opening angle of 75°,
- a lateral distance of 140 mm between seat and outer sill edge,
- a seat height above ground of 510 mm,
- a footwell depth not to exceed 50 mm,
- an 'A' pillar to front seat edge of 450 mm,
- a cant rail height above ground of 370 mm and an 'A' pillar to 'B' pillar distance of 980 mm.

Note: * High acceptance = 90 percent acceptance, preference or reported ease of use.

ability to scan to the rear when backing, turning and merging (Malfetti, 1985). It is precisely during these driving activities that older people report and are observed to have difficulties (Table 21.1), a fact that is supported by the accident data (Table 21.2). McPherson, Ostrow, Michael and Shaffron (1988) measured range of motion at seven joint sites (shoulder, elbow, hip, knee, ankle, torso and neck) on 80 young (n = 33; age 20–35) and old (n = 47; age 60–75), both male and female, drivers in an automobile simulator. Older drivers were significantly less flexible at neck, shoulder and torso sites. The study report gives means and standard deviations of joint range of motion for all joint sites studied (see McPherson *et al.*, 1988, Table 14). They also reported that less flexibility for older people was predictive of actual on-road driver performance.

There are numerous potential applications for representative static and functional measurements of the older population, including vehicle entry and egress, control and display arrangement, seat belt design, seat adjustment, mirrors, head rest, roof and roof pillar design to name the more obvious.

Strength and force

Although there are great differences between muscle groups and between individuals, on the whole older people are weaker and can exert less force (Fisher and Birren, 1947; Larsson, Grimby and Karlsson, 1979). As a general statement, muscle strength relative to the late 20s is for age 40, 95 percent, age 50, 85 percent, and age 65, 75 percent (Stoudt, 1987). There are various studies that present in detail the age capability of the individual muscle groups as measured by dynamometer (Clement, 1974; Damon and Stoudt, 1963; Viitasalo, Era, Leskinen and Heikkinen, 1985). Many of these studies provide detailed strength data across the full age span, e.g., see Clement (1974).

Some studies of performance by old people in daily activities, for example, operating faucet handles (Bordett, Koppa and Congelton, 1988) and opening jars (Rohles, Moldrup and Laviana, 1983), have appeared and give useful sample strength data. However, for application it would be most helpful to have measures of performance on various uses of the automobile, such as lifting a vehicle's hood or trunk and taking items from the trunk and back seat, opening doors, operating the stick shift, emergency brake, seat controls, opening the fuel and water cap, rolling down windows, turning knobs and switches and performing other simple maintenance and emergency operations. Unfortunately, for auto design use this information simply does not exist, at least in the open literature, in any more systematic way than outlined here.

Sensory changes with age

There can be little doubt that driving is primarily a visual psychomotor task. It is also the case that almost every aspect of visual performance declines with age (Bailey and Sheedy, 1988). It is not surprising therefore that visual related features should be a primary target of ergonomics and the older driver. For a review of general visual age changes see Kline and Schieber (1985). Staplin *et al.* (1989) have extensively reviewed the visual changes with age that might be related to driver performance. Yanik (1988) provides a highly readable account of visual considerations in the design of autos for older persons.

Sensitivity to glare

The quality of illumination reaching the eye is of special importance for older persons (Kline, 1987). With age, and accelerating after 40, there is an increase in the susceptibility to scotomatic and veiling glare such as is present at night and in fog from oncoming headlights and other bright sources, reflected from mirrors and other surfaces of vehicles and, during the daytime, associated with the sun. In addition to the direct effects of glare on acuity, the older eye takes longer to recover from glare. These impairments with aging are due primarily to changes in function of the lens of the eye causing scattering of light over the retina. A study (Pastalan, 1982) to simulate aging effects on the eye reported glare to be the single most prevalent visual difficulty experienced. In Yee's (1985) survey of drivers aged 55 and over, headlight glare (25 percent), night driving (25 percent) and rain or fog (19 percent) were three of the four most frequent responses when respondents were asked to describe situations in which their driving ability was 'worse than it was five years ago'.

A few studies have looked at glare and age in a driving context (Pulling *et al.*, 1980; Olson, 1988; Olson and Sivak, 1984; Staplin *et al.*, 1989), though none strictly focused on vehicle ergonomics. An obvious problem is glare from oncoming headlights. Pulling *et al.* (1980), using a driving simulator, found what they defined as 'resistance to glare' (a measure based on the headlight and ambient illumination that would cause slowing or other changes in driving behavior) decreased with age. Their data indicated that such resistance to glare decreased by 50 percent every 12 years. The design of headlights and roadways would seem the answer to problems of direct glare but the issue is complicated by the trade-off between needs for headlight illumination and reduction of glare. Several alternatives have been discussed (Olson, 1988; Pulling et al., 1980; Yanik, 1988) including polarizing headlights, headlights that better limit the illumination above the horizontal, glare screens, better road delineation, highway lighting and driving restrictions for persons with very low glare thresholds (see also Sivak and Flannagan, this volume).

Rear view mirrors likewise produce glare from following vehicles. Olson and Sivak (1984) studied, in laboratory conditions simulating a dark rural environment, the forward visual effects of glare associated with rear-view mirrors. Glare values up to one lux had minimal effects for young subjects and somewhat more for the old. Above that level older persons (65+) needed a target luminance about twice that of young subjects for equal visibility. For transient conditions, e.g., where a dark-adapted person is suddenly exposed to glare from a mirror, the initial visibility thresholds for both ages were about double their eventual levels, and reached these levels in about 45 seconds for young subjects and 70 seconds for the elderly. The design of mirrors or headlight design of the following car are obvious answers to mirror glare, but have the same trade-off problems as forward glare. Ways discussed to alleviate mirror glare have been day/night outside mirrors, lowering headlight heights or beam patterns on vehicles with high mounted lights, and designing the night setting to facilitate use and better accommodate rear viewing information needs (Olson and Sivak, 1984; Yanik, 1988).

While it is clear that vision is impaired by glare, its impact on real world driving at any age is not well understood. A correlation with accidents has been reported (Burg, 1967) for glare sensitivity, but the association is weak. Glare sensitivity has been suggested a one of the reasons that older people decrease their nighttime driving (Pulling et al., 1980), and it can be assumed that the associated fatigue and tenseness decrease the pleasure of driving and degrade skills. Given its relevance to the older driver and potential for solutions there would appear to be good reason to explore both ergonomic as well as safety engineering solutions to this problem.

Acuity, contrast and color

As an optical instrument, the human eye loses precision with age; the retina of a 60-year-old receives about one-third and an 80-year-old about one-twelfth the light of a 20-year-old's retina. Light to the retina is decreased by a smaller pupil size (senile miosis), and in the cornea, lens and vitreous body, is scattered by deposits and spectrally is differentially absorbed. In addition, the light sensing cones in the fovea of the aging retina have been reported to decrease in number and otherwise change noticeably in the 40s and 50s. The clarity of vision is further affected by loss in the refractive ability of the lens. This substantial loss in transmissivity and retinal sensitivity theoretically should put the older person at a visual disadvantage for driving tasks and a safety risk on the highways. Added to this problem is an increasing incidence with age of pathological factors, primarily cataracts, glaucoma, retinal pathology and macular degeneration (Pitts, 1982).

Some 40 percent of the older driving population have corrected acuity of 20/40 or worse, the acuity score widely used in the US as a licensing standard. Several large sample (10 000 +) studies (Burg, 1971; Hofstetter, 1976) have been conducted in an attempt to link visual performance to accidents and citation statistics. In Burg's study, correlations were found with static and dynamic acuity; Hofstetter found a propensity for some persons with poor visual acuity to report having multiple accidents compared to persons of better acuity. While these results were statistically significant, the relationship was weak and probably of minor practical importance. In general it has been difficult to link visual tests to safety statistics. The trend today seems more toward relating visual impairments to specific driving tasks than of vision to general accident statistics (Federal Highway Administration, 1990).

For most persons, young or old, changes in visual optics are readily compensated for by correction. Nevertheless, the driving difficulties that older people report (Yee, 1985) more often than younger ages are primarily visual, such as (see Table 21.1) *seeing while driving at night, reading traffic signs*, and *reading the instrument panel*. Some empirical study exists on all of these problem categories.

Yanik (1988) has noted that the bifocals worn by many older drivers have a focal length for reading (35 to 40 cm) that is not well-matched with the distance to the instrument panel (50 to 70 cm). While this inconvenience might be avoided with trifocals or moving the head closer to the display, he suggests that increasing letter size, brightness contrast and luminance level for displays would be helpful for many older drivers.

Poynter (1988) collected data to answer the question of the brightness contrast between letters and background needed at different ages for daytime recognition of instrument panel lettering. The experimental set-up used simulated visual driving conditions and letter sizes typical of current automobile displays. Older persons (X age = 65) required about 2.13 times the contrast needed by young subjects. Equations are given by Poynter for determining the contrast needed for a given letter size, letter/background color combination and age.

While older people would benefit both day and night from a visually well-designed instrument panel, it is under low illumination conditions that they are most disadvantaged. Mourant and Langolf (1976) studied the minimum letter luminance level needed by older drivers (45–67 years), both those who wore and those who did not wear glasses for driving. Letter luminance levels (0.34 to 68.52 cd/m) were tested for 12 letter size/contrast conditions under simulated low-illumination driving. The authors concluded that for the 1:6 stroke-width-to-height letters used in their study, letters needed to be 0.64 cm high to be visible to the majority of drivers over 45 using the then current letter luminance levels of about 1.71 cd/cm^2. The 1.71 cd/cm^2 is the maximum instrument panel luminance value that avoids annoyance or discomfort from glare. More recently Yanik (1988), reviewing studies relevant to instrument panel letter heights and aging issues, concludes that heights between 0.55 and 0.64 cm would be desirable for important markings.

The spectral characteristics of light, that is, its color, plays a meaningful role in the detection, discrimination and visual comfort of information presented to the driver. For the older driver the role of color is even more important because of the changes in color perception that occurs with age. Color recognition is impaired with age, but the relationship is not uniform either for age or color (Kline and Schieber, 1985). Changes principally in the optic media (Staplin et al., 1989) decrease the ability to distinguish the short wavelength blues and greens, and cause objects to appear yellowish and less intense. These issues in color perception become more acute as illumination levels fall near the threshold for functioning of the color-sensitive cones.

In spite of experimental differences, the several studies reported in the literature are reasonably consistent in indicating that blue-green, green and yellow colors are preferred for instrument panel symbols (Galer and Simmonds, 1985; Poynter, 1988). Red and blue colors are generally seen as less attractive, more distracting and harder to read. While instrument panel color is among the better researched areas of automotive ergonomics, still sample sizes are small and do not include the oldest (75+) drivers. In light of the pervasive visual changes that occur with aging it may be useful to extend ergonomic considerations beyond its traditional link with the instrument panel and driving task to other aspects of vehicle illumination; this could include accessory illumination, interior and trunk lighting and illumination that would make it easier for the older person to access and feel safe in using the automobile at night.

Visual field

Traditional perimetry study has consistently verified age decrements in the total visual area as indicated by target detection (Burg, 1968; Wolf, 1967). Burg's study provides mean visual field scores and standard deviations by age and sex from the teens through the 80s for 17 277 people. The mean visual field for this population declined from about 175 degrees at age 35 to less than 140 degrees at ages over 80. Age changes in the ability to move the eyes may also impair the effective visual field. For example, tracking objects across the visual field slows with advancing age (Sharpe and Sylvester, 1978), and the ability to rotate the eyes upward above the horizontal decreases (Chamberlain, 1971). Recent studies (Cerella, 1985a; Scialfa, Kline and Lyman, 1987) of performance on complex visual displays suggest that as people age there is a constriction of the useful field of view.

It is difficult to imagine almost any phase of driving that would not be impacted by such changes in the visual field; examples are detecting signs, roadway markers and pedestrians, awareness of other vehicles around the car, passing, turning and other critical vehicle maneuvers. Many of the problems reported and observed for the older driver in Table 21.1, such as changing lanes, backing, passing and leaving from a parked position, would seem related to the useful field of view. Johnson and Keltner (1983) looked at the visual field and driving records of some 10 000 license applicants. They found drivers with binocular field loss had twice the accident and conviction rates of matched controls. Also the field loss in older drivers (65+) was four times that for drivers aged 60 years or less; many of those with field loss were unaware of it.

Combining visual changes, age-related decreases in neck and torso flexibility and the blind areas induced by auto design may make this aspect of visual functioning one of the more significant for an older person's feeling of confidence and her/his safety as a driver. Study needs to be done on how outside vehicle visibility might be enhanced for older people. Vehicle features that might be considered are mirrors, head rests, roof pillars, and vehicle technology (e.g., low visibility driving assistance, heads-up displays, and guidance and avoidance detection) that would aid the older driver in detection of critical information.

Cognitive changes

Age differences in information processing have repeatedly been observed in the laboratory setting, but the general practicality of these changes for everyday tasks that older people need to perform, including driving, remains in some question and largely unexplored (Salthouse, 1990; Smith, 1990). Easily the most ubiquitous of these changes is the slowing and decreased capacity with age to process information (Birren, Woods and Williams, 1980; Cerella, 1985b; Cerella, Poon and Fozard, 1982). There is a great deal of quantitative data on speed of response for different tasks, different task conditions and different ages (Welford, 1977). While such information could aid design decisions in ergonomics, it is fragmentary and difficult to relate to real world situations. Recently however, Charness and Bosman (1990) using the information processing model of Card, Moran and Newell (1983), have extracted from the gerontology literature constants and regression equations to estimate age effects in information processing, including both speed of response and capacity measures. Though limited in its representation of the aging population and certain aspects of information processing, such a data model may provide a first approximation, quantitative aid in automotive and highway design decisions.

Several studies related to response time and age have been done in a driving context (Korteling, 1990; Olson and Sivak, 1986). Olson and Sivak (1986) studied the perception-response time of unalerted drivers for sighting of an obstacle in the roadway. The results showed a 95th percentile perception-response time, the time from detecting an obstacle until application of the brake, of about 1.6 seconds for both young (N = 49; age 18–40) and old (N = 15; age 50–84) drivers. Using a car-following task on a closed driving course, Korteling (1990) likewise found healthy older subjects (N = 10; age 61–73) were not statistically slower in brake response times than younger controls (N = 10; age 21–43). However, there is reason to believe that cognitive functions more complex than brake reaction time may show greater and practically important age effects, but the evidence is largely from anecdotal and simulated observations (Fell, 1976; Kline, Ghali, Kline and Brown, 1990; Kosnik, Sekuler and Kline, 1990; McPherson, Ostrow, Michael and Shaffron, 1988; Scialfa, Kline, Lyman and Kosnick, 1987).

The issue of cognitive impairment and driving has assumed a special importance because it is thought there are, or will be, a growing number of older persons driving with medical conditions associated with cognitive impairment, such as hemiplegia (stroke), Alzheimer's and Parkinson's diseases and the disabled young who are now old (Federal Highway Administration, 1990). Persons with these disorders often have a desire to continue to drive and driving may be critical in their family life. Even if not drivers, after age 75 when such disabilities are more prevalent, older people increasingly use the private auto as a passenger. The older disabled passenger also has problems in vehicle use, such as entry/exit, fatigue, seat belt use, and maintaining posture. Areas of cognitive impairment and medical disabilities in general are ones in which new and better designed vehicle technology and technological aids, (e.g. automatic adjustment of vehicle packaging, workload reducing aids, low visibility driving assistance, automatic guidance and avoidance) are likely to play an important role. At this time, ergonomic studies and data relevant to the medically disabled elderly and of their use of the automobile is an important challenge that is, for the most part, waiting to be done.

References

Adams, P. F. and Collins, G., 1987, Measures of health among older persons living in the community, in Havlik, R. J., Lui, M. G., Kovar, M. G. *et al.* (Eds) Health statistics on older persons, United States, 1986, *Vital and Health Statistics* (3, No. 25) DHSS Publication, No. PHS 87-1409, Washington, DC: US Government Printing Office.

Adrian, M. J., 1981, Flexibility in the aging adult, in Smith, E. L. and Serfass, R. C. (Eds) *Exercise and Aging*, Hillside, NJ: Enslow.

Bailey, I. L. and Sheedy, J. E., 1988, Vision screening for driver licensure, in *Transportation in an Aging Society: Improving Mobility and Safety for Older Persons*, Special Report 218, Vol. 2, p. 294–324, Washington, DC: National Research Council.

Baltes, P. B., Reese, H. W. and Lipsitt, L. P., 1980, Life-span developmental psychology, *Annual Review of Psychology*, **31**, 65–110.

Birren, J. E., Woods, A. M. and Williams, M. V., 1980, Behavioral slowing with age: Causes, organization and consequences, in Poon, L. W. (Ed.) *Aging in the 1980's*, pp. 293–308, Washington, DC: American Psychological Association.

Blake, R., 1984, What disables the American elderly? *Generations*, **8**(4), 6–9.

Bordett, H. M., Koppa, R. J. and Congleton, J. J., 1988, Torque required from elderly females to operate faucet handles of various shapes, *Human Factors*, **30**(3), 339–46.

Borkan, G. A., Hults, D. E. and Glynn, R. J., 1983, Role of longitudinal change and secular trend in age differences in male body dimensions, *Human Biology*, **55**(3), 629–41.

Brainn, P., Bloom, R., Breedlove, R. and Edwards, J., 1977, *Older Driver Licensing and Improvement System*, Final Report, July, Washington, DC: US Department of Transportation.

Brooks, B. M., Ruffle-Smith, H. P. and Ward, J. S., 1974, *An Investigation of Factors Affecting the Use of Buses by Both Elderly and Ambulant Disabled Persons*, (Contract Report 3140/32), London: Transportation Road Research Laboratory, British Leyland.

Burg, A., 1967, Light sensitivity as related to age and sex, *Perceptual and Motor Skills*, **24**, 1279–88.

Burg, A., 1968, Lateral visual field as related to age and sex, *Journal of Applied Psychology*, **52**(1), 10–15.

Burg, A., 1971, Vision and driving: A report on research. *Human Factors*, **12**(1), 79–87.

Butler, R. N., 1975, *Why Survive? Being Old in America*, New York: Harper and Row.

Card, S. K., Moran, T. P. and Newell, A., 1983, *The Psychology of Human-computer Interaction*, Hillsdale, NJ: Erlbaum.

Cerella, J., 1985a, Age-related decline in extrafoveal letter perception, *Journal of Gerontology*, **40**, 727–36.

Cerella, J., 1985b, Information processing rates in the elderly, *Psychological Bulletin*, **98**, 67–83.

Cerella, J., Poon, L. W. and Fozard, J. L., 1982, Age and iconic read-out, *Journal of Gerontology*, **37**, 197–202.

Cerella, J., Poon, L. W. and Williams, D. M., 1980, Age and the complexity hypothesis, in Poon L. W. (Ed.) *Aging in the 1980's*, pp. 322–340, Washington, DC: American Psychological Association.

Chamberlain, W., 1971, Restriction in upward gaze with advancing age, *American Journal of Ophthalmology*, **71**(1), 341–6.

Charness, N., and Bosman, E. A., 1990, Human factors and design for older adults, in Birren, J. E. and Schaie, K. W. (Eds) *Handbook of the Psychology of Aging* 3rd Edn New York: Academic Press.

Clement, F. J., 1974, Longitudinal and cross-sectional assessments of age changes in physical strength as related to sex, social class, and mental ability, *Journal of Gerontology*, **29**(4), 423–9.

Codd, M. D. and Kurland, L. T., 1985, Polio's late effects, *1986 Medical and Health Annual, Infectious Diseases*, Encyclopedia Britannica.

Damon, A. and Stoudt, H. W., 1963, The functional anthropometry of old men, *Human Factors*, **5**, 485–91.

Damon, A., Seltzer, C. C., Stoudt, H. W., and Bell, B., 1972, Age and Physique in Healthy White Veterans at Boston, *Journal of Gerontology*, **27**, 202–8.

Dawson, D., Hendershot, G. and Fulton, J., 1987, Aging in the eighties: Functional limitations of individuals age 65 years and over, *Advance Data: Vita and Health Statistics*, No. 133, DHHS Pub. No. (PHS) 87-1250, Hyattsville, MD: Public Health Service.

Fay, R. C., 1978, Air safety and the older airline pilot, *Aging and Work*, Summer, 153–61.

Federal Highway Administration, 1986, *Nationwide Personal Transportation Study 1983–1984*, Washington, DC: US Department of Transportation.

Federal Highway Administration, 1990, *Conference on Research and Development Needed to Improve Safety and Mobility of Older Drivers*, August, Washington, DC: US Department of Transportation.

Fell, J. C., 1976, A motor vehicle accident causal system: The human element, *Human Factors*, **18**, 85–94.

Finesilver, S. G., 1969, *The Older Driver—A Statistical Evaluation of Licensing and Accident-Involvement in 30 States and the District of Columbia*, January, Denver, CO: University of Denver College of Law.

Fisher, M. B. and Birren, J. E., 1947, Age and strength, *Journal of Applied Psychology*, **31**, 490–7.

Friedlaender, J. S, Costa, P. T. Jr., Bosse, R., Ellis, E., Rhoads, J. G. and Stoudt, H. W., 1977, Longitudinal physique changes among healthy white veterans at Boston, *Human Biology*, **49**, 541–58.

Frier, B. M., Matthews, D. M., Steel, J. M. and Duncan, L. J. P., 1980, Driving and insulin-dependent diabetics, *Lancet*, **i**(8180), 1232–4.

Galer, M. and Simmonds, G. R., 1985, The lighting of car instrument panels—Drivers' response to five colours, *SAE 850328*, Warrendale, PA: Society of Automotive Engineers.

Grandjean, E., 1973, *Ergonomics of the Home*, New York: John Wiley and Sons.

Harris, R., 1983, Exercise and the aging process, *Annals of the Academy of Medicine: Singapore*, **12**, 454–6.

Hoffstetter, H. W., 1976, Visual acuity and highway accidents, *Journal of the American Optometric Association*, **47**, 887–93.

Huston, R. and Janke, M., 1986, *Senior Driver Facts*, (Report No. 82, 2nd Edn), Sacramento, CA: Department of Motor Vehicles.

Johnson, C. A. and Keltner, J. L., 1983, Incidence of visual field loss in 20 000 eyes and its relationship to driving performance, *Archives of Ophthalmology*, **101**, 371–5.

Kelly, P. L. and Kroemer, K. H. E., 1990, Anthropometry of the elderly, *Human Factors*, **32**(5), 571–95.

Kline, D. W., 1987, Aging and human factors in transportation, paper prepared for the *National Research Council Workshop on Human Factors Research Issues and an Aging Population*, August, Washington, DC: National Research Council.

Kline, D. W. and Schieber, F. J., 1985, Vision and aging, in Birren, J. E. and Schaie, K. W. (Eds) *Handbook of the Psychology of Aging*, 2nd Edn, New York: Van Nostrand Reinhold.

Kline, T. J. B., Ghali, L. M., Kline, D. W. and Brown, S., 1990, Visibility distance of highway signs among young, middle-aged, and older observers: Icons are better than text, *Human Factors*, **32**(5), 609–19.

Kobayashi, M., 1984, Driver behavior and accident characteristics of the elderly drivers in Japan, *IATSS Research*, **8**, 83–91.

Korteling, J. E., 1990, Perception-response speed and driving capabilities of brain-damaged and older drivers, *Human Factors*, **32**(1), 95–108.

Kosnik, W. D., Sekuler, R. and Kline, D. W., 1990, Self-reported visual problems of older drivers, *Human Factors*, **32**(5), 597–608.

Larsson, L., Grimby, G. and Karlsson, J., 1979, Muscle strength and speed of movement in relation to age and muscle morphology, *Journal of Applied Physiology*, **46**, 451–6.

Maleck, T. and Hummer, J., 1986, Driver age and highway safety, in *Transportation Research Record 1059, TRB*, pp. 6–12, Washington, DC: National Research Council.

Malfetti, J. W. (Ed.), 1985, *Drivers 55 + : Needs and Problems of Older Drivers: Survey Results and Recommendations*, Falls Church, VA: AAA Foundation for Traffic Safety.

Mathews, J. J. and Cobb, B. B., 1974, Relationships between age, ATC experience and job ratings of terminal area traffic controllers, *Aerospace Medicine*, **451**, 56–60.

McKnight, A. J., 1988, Driver and pedestrian training, in *Transportation in an Aging Society: Improving Mobility and Safety for Older Persons*, Special Report 218, Vol. 2, pp. 101–33, Washington DC: National Research Council.

McPherson, K., Ostrow, A., Michael, J. and Shaffron, P., 1988, *Physical Fitness and the Aging Driver*, Phase 1, August, Washington, DC: AAA Foundation of Traffic Safety.

Monforton, R., Dumala, T., Yanik, A. and Richter, F., 1988, Accident experience of older AAA drivers in Michigan, in *Effects of Aging on Driver Performance*, SP-762, pp. 1–6, Warrendale, PA: Society of Automotive Engineers.

Mourant, R. R. and Langolf, G. D., 1976, Luminance specifications for automobile instrument panels, *Human Factors*, **18**(1), 71–84.

National Health Survey, 1986, Current estimates from the National Health Interview Survey, United States, 1983, *Vital and Health Statistics*, Series 10, No. 154, Hyattsville, MD: Public Health Service.

National Research Council, 1988, *Transportation in an Aging Society: Improving Mobility and Safety for Older Persons*, Special Report 218, Vol. 2. Washington, DC: Author.

Neugarten, B., 1974, Age groups in American society and the rise of the young-old, *Annals of the American Academy of Political and Social Science*, **415**, 187–98.

Olson, P. L., 1988, Problems of nighttime visibility and glare for older drivers, in *Effects of Aging on Driver Performance*, SP762, pp. 53–60, Warrendale, PA: Society of Automotive Engineers.

Olson, P. L. and Sivak, M., 1984, Glare from automobile rear-vision mirrors, *Human Factors*, **26**(3), 269–82.

Olson, P. L. and Sivak, M., 1986, Perception-response time to unexpected roadway hazards, *Human Factors*, **28**, 91–96.

Page, M., Spicer, J., McClelland, I., Mitchell, K., Feeney, R. and James, J., 1984, Access to standard production cars by disabled and elderly people, in *Third International Conference on Mobility and Transport of Elderly and Handicapped Persons*, Orlando, FL, 29–31 October.

Pastalan, L. A., 1982, Environmental design and adaptation to the visual environment, in Sikuler, R., Kline, D. and Dismukes, K. (Eds) *Aging and Human Visual Function*, New York: Alan R. Liss.

Pitts, D. G., 1982, The effects of aging on selected visual functions: Dark adaptation, visual acuity, stereopsis and brightness contrast, in Sekules, R., Kline, D. and

Dismukes, R. (Eds) *Aging and Human Visual Function*, pp. 131–59, New York: Alan R. Liss.

Planek, T. W. and Overend, R. B., 1973, Profile of the older driver, *Traffic Safety*, **73**(9–10), 37–39.

Poynter, D., 1988, The effects of aging on perception of visual displays, in *Effects of Aging on Driving Performance*, SP-762, pp. 43–51, Warrendale, PA: Society of Automotive Engineers.

Pulling, J. H., Wolf, E., Sturgis, S.P., Vaillancourt, D. R. and Dolliver, J. J., 1980, Headlight glare resistance and driver age, *Human Factors*, **22**(1), 103–12.

Roberts, D. F., 1960, Functional anthropometry of elderly women, *Ergonomics*, **3**, 321–7.

Rohles, F. H. Jr., Moldrup, K. L. and Laviana, J. E., 1983, Opening jars: An anthropometric study of the wrist-twisting strength of the elderly, in *Proceedings of the 27th Human Factors Society Annual Meeting*, pp. 112–6, Santa Monica, CA: Human Factors Society.

Rosenbloom, S., 1988, The mobility needs of the elderly, in *Transportation in an Aging Society: Improving Mobility and Safety for Older People*, Special Report 218, Vol. 2, pp. 21–71, Washington DC: National Research Council.

Salthouse, T. A, 1990, Influence of experience on age differences in cognitive functioning, *Human Factors*, **32**(5) 551–69.

Scialfa, C. T., Kline, D. W. and Lyman, B. J., 1987, Age differences in target identification as a function of retinal location and noise level: Examination of the useful field of view, *Psychology and Aging*, **2**, 14–19.

Scialfa, C. T., Kline, D. W., Lyman, B. J. and Kosnick, W., 1987, Age differences in judgments of vehicle velocity and distance, in *Proceedings of the 31st Human Factors Society Annual Meeting*, pp. 558–61, Santa Monica, CA: Human Factors Society.

Sharpe, J. A, and Sylvester, T. O., 1978, Effects of aging on horizontal smooth pursuit, *Investigative Ophthalmology and Visual Science*, **17**, 465–8.

Sivak, M., Olson, P. L., Kewman, D. G., Hosik, W. and Henson, D. L., 1981, Driving and perceptual/cognitive skills: Behavioral consequences of brain damage, *Archives of Physical Medicine and Rehabilitation*, **62**, 476–83.

Smith, D. B. D., 1987, Human factors issues in safety and security for the older population, paper prepared for the *National Research Council Workshop on Human Factors Research Issues for an Aging Population*, August, Washington, DC: National Research Council.

Smith, D. B. D., 1990, Human factors and aging: An overview of research needs and application opportunities, *Human Factors*, **32**(5), 509–26.

Staplin, L. K., Breton, M. E., Haimo, S. F., Farber, E. I. and Byrnes, A. M., 1989, *Age-Related Diminished Capabilities and Driver Performance*, Contract No. DTFH61-86-C-00044, Washington, DC: Federal Highway Administration.

Stoudt, H. W., 1981, The anthropometry of the elderly, *Human Factors*, **23**(1), 29–37.

Stoudt, H. W., 1987, Changes in the physical characteristics and capabilities of the aged—An overview, paper prepared for the *National Research Council Workshop on Human Factors Research Issues for an Aging Population*, Washington, DC: National Research Council.

Traffic Safety of Elderly Road Users, 1985, Paris, France: Organization for Economic Cooperation and Development and World Health Organization.

Transport Canada, 1986, *Proceedings of the Fourth International Conference on Mobility and Transport for Elderly and Disabled Persons*, Ottawa, Canada: Author.

US Census Bureau, 1984, *Projections of the Population by Age, Sex and Race for the United States, 1983–2080*, No. 952, Series P-25, Washington, DC: US Government Printing Office.

Viitasalo, J. T., Era, P., Leskinen, A. L. and Heikkinen, E., 1985, Muscular strength profiles and anthropometry in random samples of men aged 31–35, 51–55 and 71–75 years, *Ergonomics*, **28**(11), 1563–74.

Ward, J. S. and Kirk, N. S., 1967, Anthropometry of elderly women, *Ergonomics*, **10**(1), 17–24.
Welford, A. T., 1977, Motor performance, in Birren, J. E. and Schaie, K. W. (Eds) *Handbook of the Psychology of Aging*, pp. 450–96, New York: Van Nostrand Reinhold.
Wierwille, W. and Peacock, B., 1989, Human factors and the automobile of the future, *Human Factors Bulletin*, **32**(11), 1–5.
Winter, D. J., 1984, Needs and problems of older drivers and pedestrians: An exploratory study with teaching/learning implications, *Educational Gerontology*, **10**, 135–46.
Wolf, E., 1967, Studies on the shrinkage of the visual field with age, *Highway Research Record*, **167**, 1–7.
Yanik, A. J., 1988, Vehicle design considerations for older drivers, in *Effects of Aging on Driver Performance, Sp762*, pp. 31–41, Warrendale, PA: Society of Automotive Engineers.
Yee, D., 1985, A survey of the traffic safety needs and problems of drivers age 55 and over, in *Drivers 55+: Needs and Problems of Older Drivers: Survey Results and Recommendations*, pp. 96–128, Malfetti, J. L. (Ed.) Falls Church, VA: AAA Foundation for Traffic Safety.
Ysander, L. and Herner, B., 1976, The traffic behavior of elderly male automobile drivers in Gothenberg, Sweden, *Accident Analysis and Prevention*, **8**(2), 81–86.

Postscript: Future challenges for automotive ergonomics

Brian Peacock

The future challenges for the automotive ergonomist stem from changes in the characteristics of vehicle users, changes in vehicle technology, changes in the roadway environment and changes in the broader social environment. In addition, there are the ever present challenges of traffic, weather, darkness and temporal and spatial coincidence.

The automobile user population is probably one of the broadest of all consumer product user populations, at least in the developed countries where almost everyone rides in cars. The automobile designer and his or her lawmaking colleauges currently can do very little to influence who uses what kind of car and under what conditions. The age of entry to the driving community varies from the early to the late teens in different countries. Departure from the driving population may be delayed until the individual's faculties have deteriorated to very low levels. In the middle are drivers with enormous variation in knowledge, experience and driving abilities.

Other dimensions of human variation include size, strength, senses, skills and sicknesses. Economic, occupational and travel pattern diversity further complicate the picture. The vehicle user population also includes passengers, who range from the new-born, through active children to those with devices such as walking aids or wheelchairs. Thus the challenge to the automotive ergonomist is to please most of a very wide variety of vehicle users, most of the time. The level of accommodation may vary widely, according to the importance of the dimension and implications of failure to accommodate. For example, safety related characteristics of vehicles will address the vast majority of users, whereas aesthetic matters will be deliberately targeted at narrower sub-populations.

Historically, this challenge of human variability has been very well met by vehicle designers. There is probably no other consumer product of this order of complexity that can be used with minimal familiarization by any driver in the world. Compare the barriers presented by many other consumer devices and services—such as the VCR and tax laws!

The well recognized trend in developed countries is an increase in the average age of vehicle users. Age, and its concomitant disease processes, brings with it a general deterioration of the functional ability to adapt to situational variations. However, these age effects are contaminated by cohort effects in that groups and generations bring their own typical experiences. The new older generation has grown through the age of rapid developments in electronics, computers and telecommunications and thus adapts more easily to the introduction of these technologies into vehicles than their older colleagues. The ergonomics challenge is to build on the common experience of older people through the design of familiar interfaces that mask the complexities of underlying technology.

Older users place different emphases on vehicle features than the younger generation. These differences are explored and exploited by the marketing communities through differential emphasis on economics, ecology, excitement, safety, service and sophistication. Target market sets are fuzzy, however. The ergonomics challenge is to balance preference dimensions for sub-populations with performance matters for the larger populations that may find age-related effects as the principal limitation. The truism that what is good for the older driver is also good for the younger one is appropriate for many design dimensions.

Thus, the principal human trends that face the designer are changes in the proportions of the driving population who fit into the distinguishable capability and experience categories. The demographic distribution in the United States and other developed countries is moving its center of gravity to the older end. Medical science and public health can be thanked for this change. However this group does bring greater associations with degenerative disease and the associated medications have adverse effects on the driving skills of an increasing number of drivers. This topic of the older driver is addressed extensively in recent issues of the *Human Factors Journal* (1991, Vol. 33, No. 5 and 1992, Vol. 34, No. 1).

The changes in vehicle technology have been driven largely by developments of electronics, which may now represent on the order of 10 percent of the cost of the car. Computers help control the engine, brakes, suspension, throttle, seats, thermal environment, lighting and many forms of vehicle system diagnosis. Other computer-based journey aids include displays, navigation and route planning. Even non-driving related computer applications intrude, including communications devices such as telephones and fax machines. There is practically no limit to the blending of everyday business and domestic devices into the vehicle environment. The obvious human factors challenge is to articulate human resource capacity for attention and other sensory, cognitive and motor faculties under the time-limited process of driving. Given this articulation, features and interfaces must be defined to accommodate the variability in human and situational conditions. Wierwille and Peacock (1988) reviewed many of the developments in this vehicle feature area.

The human operator has learned to accept physical aiding—as in power assisted steering and braking. Also, antilock brakes have been found to provide control levels beyond the capability of drivers in many situations. However, the driver will be faced in the near future with features that replace or aid his or her cognitive capabilities. Already memory systems are to be found in the control of the position of seats, steering wheels, thermal environment and entertainment. Sensory enhancement is found in night vision systems and near obstacle detection systems. Decision advice is available for both service and navigational assistance. The challenge to the designer of these cognitive aids is to anticipate and plan for exceptional circumstances such as system failure or driver override.

Thus the principal vehicle changes that will present human factors challenges are increases in feature content and complexity. The advent of the car radio had its doubters in terms of division of attention, but drivers have learned to time share or to restrict their usage of radio features. The more recent introduction of the car phone also has its critics and there is no shortage of anecdotal evidence that identifies inappropriate time sharing. However, many drivers make considerable successful use of this device, which attests to the general level of spare mental capacity. The move towards hand free devices will certainly address the availability of an important human resource. These two widely available devices—the radio and the car phone—are examples of the continuous incursion of vehicle features into the attentional capacities of drivers. The demonstrable human ability to learn to cope strategically with these devices in the framework of driving provides strong evidence that they are compatible with safe driving behavior and performance. However, the measurement and prediction of interference with primary task performance presents many methodological challenges.

The challenge for the human factors specialists is to articulate the interactions between human performance variability and feature complexity in the context of the time constraints imposed by driving. The solution domain includes the inclusion of even greater intelligence in the vehicle that restricts access to features in inappropriate situations and for untrained drivers. This challenge, like that of contemporary areas of human–computer interaction, presupposes an ability of the 'system' to judge reliably the state of training of the user and the temporal constraints of the driving task. Devices of this nature, such as alcohol detection and performance tests, have been demonstrated for many years, but have so far not achieved widespread application.

A second technological development is that of alternative fuels, such as electricity, for powering cars. Most major automobile companies are working aggressively in this area in anticipation of environmental legislation and potential applications. Electric power, like all new technologies, will bring novel conditions to the driver. Vehicle control, refueling, maintenance as well as ownership issues of purchase, insurance and disposal will all need careful planning and will inevitably produce user generated surprises in the early stages of implementation. Once again, the human factors engineer is faced with the challenge of prediction based on knowledge of driver performance and behavior with current technology, and the limited opportunity for substantial field trials. The analogy with the testing of a new drug by the FDA is worth consideration. Another analogy is the rapid change from typewriters to wordprocessors, that was accomplished with substantial market input leading to differing levels of success of different interface strategies.

Perhaps the greatest changes that present human factors challenges are changes to the roadway and operational environment of the vehicle. Intelligent vehicle highway systems are a reality. Sheridan (1991) presents an outline of many of the issues and the automobile community can add even greater insight to the possibilities and problems. With the advent of IVHS, the driver will not be just an individual controlling a car, but will be an element of a traffic system that aspires to move the focus of control away from the individual and towards the optimization of limited roadway capacity. Of course the driver will aspire to retain a certain level of autonomy and may, on occasion, exhibit either voluntary or involuntary influence counter to that desired by the system. The fictional analogy is to be found in Huxley's Brave New World (Harper and Row, 1932). Analogies that are closer to home may be found with those drivers under the influence of psychotropic drugs or alcohol, or simply those who are in a hurry.

Today, there is relatively little limitation on where any driver or vehicle can go within the roadway network. There are some limitations on large vehicles, single occupant vehicles and inexperienced drivers. Some segregation occurs through natural selection. For example, drivers with diminished eyesight may avoid nighttime driving. However, in the future it is very likely that increasing traffic control and legislative effort will be paid to the interaction of particular combinations of roads, vehicles, times and drivers with the aim of improving the efficiency and safety of the total system. The control of access to and activity on different parts of the roadway network is currently generally accomplished by static, visual road signs, although auditory and tactile devices (road texture and speed bumps) are sometimes used to control speed. In the future the control of access will inevitably involve greater use of electronic devices. This trend will present a whole new opportunity for human factors in the traditional area of displays and controls.

Changes in the social environment may occur with an increasing level of automation and telecommunication that changes work and travel patterns. The 'wired city' predicted by futurologists a few decades ago is now a reality. Practically all forms of informational transaction can take place in the home. Of course, the movement of goods will always need the roads and the social desire for face-to-face information transfer will continue to put pressure on the transportation infrastructure. However, the optional nature of the latter may be influenced by economic factors if the costs of energy rise significantly and traffic congestion increases journey times. The changes in the age structure of communities leads naturally to changes in the social environment. Journey types and distances change, as do vehicle types.

These social changes present many opportunities for the human factors specialist to become involved in the total system planning process. Such a process should address journey needs and population characteristics and design roadways, vehicles and operational controls accordingly. For example, the adoption of the electric golf cart in the retirement communities of the southern United States attests to the interest of at least one population in alternative transportation facilities.

Adverse weather conditions and darkness are ever-present complications for the driver. Furthermore, different drivers may vary in their relative abilities to cope with this variation. For example older people commonly avoid driving at night, but sometimes find this task unavoidable. Similarly, experience with snow, ice, wind, rain or fog may vary considerably. The consequence is that, whereas statistically drivers may avoid adverse situations, the absolute numbers of coincidences of inexperienced drivers and major environmental challenges will continue to be a concern for vehicle and roadway designers and traffic controllers. The development of technologies such as vision aiding, traction control and route planning should reduce the vulnerability of the driver to these environmental demands.

Driving cars is, for the most part, a very useful and safe thing to do. However, the attractiveness of this mode of transport in a temporal and spatially constrained environment will inevitably lead to challenges. Drivers will continue to find themselves, on occasion, in the wrong place at the wrong time. The challenge to the automobile design ergonomist is to design vehicle systems so that they do not add to the variation caused by driver differences. Vehicle systems must be robust and forgiving. The future of road transportation will experience considerable change, but the nature of human behavior and traffic will be more resistant to change. The future employment of the automotive ergonomist is well assured.

References

Sheridan, T. B., 1991, 'Human Factors of Driver–Vehicle Interaction in the IVHS Environment', DOT Report HS 807 837, June.

Wierwille, W. and Peacock, J. B., 1988, 'Human Factors Concerns in the Driver/Vehicle/Environment as the Year 2000 Approaches', FISITA International Conference, Detroit.

Subject Index

Page numbers in bold denote chapters primarily concerned with that subject.

acceleration 143, 299, 431
accidents
   databases 406
   injury characteristics 366, 457
   statistics 456
active restrain systems 153, 154
adjustable driving package 63
advanced indication of breaking 196
age effects 310, 370
age patterns 454
aged drivers 151, 453, 456, 474
ageing **219, 339, 453**
   acuity 463
   attentional demands 311
   braking ability 227
   braking force 228
   cognitive changes 466
   cognitive-motor impairments 219, 466
   colour perception 463
   contrast perception 463
   coordination of movements 228
   crash force tolerance 151
   glare 370
   impaired movement execution 229
   joint flexibility 230, 231
   motor performance 231, 310
   movement execution 226
   movement initiation 219, 224
   movement time 226
   movement trajectory 227
   muscle mass 230
   neuropsychological hypothesis 225
   probability of injury 151
   problems in driving 455
   proprioception 229
   reaction time 220, 225, 340
   response preparation 220, 221
   response programming 223
   response selection 222
   restrain systems 152
   seating dimensions 461
   sensitivity to glare 462
   sensory changes 462
   slowing-with-age phenomenon 340, 403
   speed/accuracy trade-off 224
   visual field 465
   visual performance 170, 311
air bags 147, 149, 151, 152, 156
alcohol 370
anthropometric models 4, 5
anthropometry **1, 11, 99, 117, 431**
   acceptable overlift height 117
   anthropometers 20
   anthropometric measurements 17, 51, 61, 107, 108, 460
   evaluation models 23, 150
   functional measures 14, 22
   hand controls reach envelope 38
   head clearance 135
   human variability 473
   lumbar curve 111
   reach envelope 13, 24
   rear body panel thickness 117
   sources of anthropometric data 18
   spatial trunk design recommendations 132, 138
   static measures 15
anti-lock brake system 232
auditory facilitators 426
automated lateral guidance 376
automated longitudinal guidance 376
automatic information processing 194
automobile
   accident rates 141
   crash sequence 143
   operation 367
   secondary controls 367
   signaling 197
   trunk dimensions **117**
automotive design 474

biomechanical models 4, 5
biomechanics **1, 99, 117, 141, 431**
body base grids 37
body posture 24, 33, 101, 105, 461
body size, **1, 99, 117, 431, 460**
brake lamp location 195
braking reaction times 197

climate 112
cognitive loading 302
cognitive-motor capabilities **219**
combined reconfigurable
    displays/controls 368
communication 366, 425
communication within vehicle 366
computer-aided modelling and design
    1, 43, 79
  accuracy 59
  CAD/CAM models **43**
  cost 58
  field of vision 93
  headlamp performance 189
  man-modelling 1, 47, 63, 66
  man-modelling CAD systems 58
  methodological problems 3
  model animation 2
  models of human-vehicle interface
    1
  SAMMIE modelling **43**
controlled information processing 194
control stereotype evaluation 277
  expectancy 269
  generic controls 291, 293
  manual windows 289
  power door locks 291, 294
  power mirrors 286, 287
  power windows 287
  stalk controls 283, 289, 292
  supplementary cues 269
controls **237, 299**
  hand controls 36
convex mirrors 210
crash injury 144, 151
crash pulses 143
cruise control 445
cushion 107, 112
cushion stiffness 110

daytime running lights 196

design
  automobile design 70
  car signals design 198
  design limits 21
  design of indirect vision 208
  design of visual field 87
  design process 44
  for older drivers 340
  headlamp performance 189
  informational aspects 401
  prototype testing 74
  system design 45, 72
discomfort glare 190
display design
  age effects 340
  body movements 346
  colour discrimination 343
  design for older drivers 340
  display design parameters 340, 368
  glance time at the display 342, 345
  lane position deviation 342, 347,
    352
  time to vocalize response 342, 352
  word complexity 346
displays **237, 299, 321, 339**
  auditory displays 317
  display clutter 312
  display design 339
  head-up displays 199
  placement displays 211
  specialized displays 315
distractibility 444
driver
  eye range 23
  hand control reach 23
  head position 23
driver experience 369, 370
driver mental workload **359**
  conceptual framework 360
  driver performance 383
  embedded secondary task measure
    362
  environmental factors 372
  individual differences 371
  mental workload assessment 361
  physiological measures 375
  primary task measure 361, 373
  secondary task measure 361, 373
  subjective ratings 375
  verbal protocols 375
  workplace design 23

Subject Index

driver traits and states 368
driver vision 85, 88
driving
   environment 372
   performance 341, 431
   population variability 453
   skills **219**
   task elements 364
   vs piloting 300

effect of secondary task 354
elderly drivers **219, 339, 453,** 456, 474
   traffic citations of older drivers 455
electronic control of reflectivity 213

facilitator design and development 405, 414
facilitator failure 410
facilitators **401**
   instructions 401
   user centred design 404
   warnings 401, 414
Federal Motor Vehicle Safety Standard 142, 239
foot placement 433
freezing 445
future challenges in automotive ergonomics 473

generic crash pulse countermeasures 145
geometry of seat design 107
glare 173, 190, 208, 212, 370

H-point machine 25
HANDY-1 384
HANDY-2 391
hand controls **383**
hand controls reach envelope 38
hand movement time 367
head position contours 35
head-up displays 199, 335
headlamps **185,** 188
   ambient mode in vision 187
   beam patterns 189, 191, 195
   computer models 189
   focal mode in vision 187
   functions 187
   HID colour retention 192
   HID colours 192
   high intensity discharge (HID) 191

   illumination 192
   polarized headlamps 190
   visual information processing 187
headlight symbols **237,** 251
human back 106
human geometry 101
human-equipment interface **1**
human linkage system 101
human variation 473
hypervigilant states 443

improving vision 95, 207
in-car glance times 307
in-car navigation system 309
in-car tasks 303
indirect vision **205**
   convex mirrors 210, 214
   design 208
   indirect vision systems 207
   placement of displays 211
   rearview mirrors 205
   visual capacities 205
individual differences 371
information aspects of design 401
   communication system 402
   failure mode analysis 406
   function/task analysis 407
   hazard monitoring 409
   hazard pattern analysis 406, 416
   owner's manual 420
   systems approach 402
   user-centre design of automobiles 404
   warning effectiveness 418, 424
   warnings 418
injury 144, 146, 151, 406, 457
intelligent map-matching augmentation 334
intelligent vehicle/highway system 198, 319, 475
interface models 6
lamp rise time 197
lateral guidance 376
lateral space 108
left-foot breakers 446
limits of crash protection 147
linkage anthropometry 103
locator lines 36
longitudinal guidance 376
loss of braking effectiveness 431
lumbar curve 111

mandatory standards 142
manikins
  H-point 25, 64
  percentile models 21
  three-dimensional models 31, 63
manual control 299
mental workload 360
mental workload indices 363
mirror vision 94
modelling **1, 43, 237, 383**
Motor Vehicles Manufacturing
  Association 30

National Highway Traffic Safety
  Administration 141
navigation **321,** 365
  cognitive driving requirements 322
  cognitive operations in navigation
    322
  driver characteristics 322
  driver cognitive mapping 322
  driver spatial ability 322
  ETAK Navigator 327
  information requirements 324
  moving-map display 326
  navigation reference frames 322
  paper maps 324
  recoverable navigation waste 331
  route guidance 328
  safety in navigation 329
  spatial navigation aids 328
  traffic management systems 321,
    332
  traffic management technology 331
  traffic navigation systems 322
  vehicle location 334
  vehicle navigation aids 326
  verbal navigation aids 328
night blindness 173
night driving 167
night myopia 174

occupant collision 146
occupant packaging **11, 99**
  interior space 11, 29
  occupant envelope 11, 24
  tools for occupant packaging 25,
    100
  vehicle spatial dimensions 27

occupant packaging practices 31, 99
  accommodation tool reference
    point 37
  body base grids 37
  H-point machine 31
  hand controls reach envelopes 36
  locator lines 36
occupant protection **141**
occupant velocity 144, 150
older driver disabilities 456
older driver medical problems 457
older drivers 453, 456, 474
older drivers and passengers 453
older drivers characteristics 458
  anthropometric dimensions 460
  body size and shape 458
  joint flexibility 459
  range of motion 459
  seating dimensions 461
  strength and force 461
  vehicle access 461
overlapping fields of view 214

panel clutter 312
passive belts 156
passive restrain systems 150, 153, 154
pedal configurations 448
pedal errors 440, 443
  corollary discharge mechanism 442
  efference copy mechanism 442
  habitual manoeuvres 442
  kinesthetic information 441
pelvic geometry 105
perceptual narrowing 444
perseverance 444
placement of objects 118
population stereotypes 269
positive feedback 443
postural comfort 72, **99**
postural stress 100
pressure distribution 110
problems in driving 455, 456
processing resources 363

reaction time 197, 220, 225, 340
reflectivity and glare 212
reflectivity/electronic control 213
restrain forces 146
restrain systems 100, 149, 150, 152
ride comfort 107

risk taking 152
run-away motor problem 431

seat
   back adjustments 108
   comfort 31, 99, 110
   contours 111
   design **99**
   design criteria 99
   memory 112
   position/selection 33
seating **43, 99, 141**
   chair height 39
   cushion size 107
   driver selected seat position 33
   ellipse 33, 84
   head position contours 35
   locator lines 36
   lumbar curve 111
   pelvic geometry 105
   seat back 40, 108
   seating arrangement drawing 40
   seating dimensions 461
   seating package 26
   seating reference point 32
shift lock mechanisms 447
signal lamps **185**
   ambient illumination 195
   brake lamps 193, 197
   daytime running lights 196
   detection and identification 195
   front lamps 193
   indicator location 195
   lamp rise time 196
   reaction time 197
   side marker lamps 193
   spatial separation 195
   systems approach 198
   turn and back-up lamps 193
   unalerted observer 186
   vehicle signalling 193
signal words 415
simulated driver performance 383
   knowledge-based system 391
   learning of panel usage 386
   multicomponent computer model 384
   simulated hand control 383
   simulation of driver cognitive activities 391

sitting comfort **99**, 110
sitting position 104
social environment 366
Society of Automotive Engineers 142
spatial cognitive aspects of navigation 321
spinal curvature 106
standard-shift vehicles 446
static comfort 110
static test in simulator 445
statistical confidence limits 279
stereotypes in vehicle control/design 269
   angle of presentation 270
   control coding 271
   control colour coding 271
   control labelling 271
   control operation 269
   control shape coding 271
   direction-of-motion stereotypes 269, 273
   driver expectancy 270, 274
   driver workload 270
   expected/dominant stereotypes 270
   general stereotypes 275
   handedness 276
   natural stereotypes 270
   plasticity 272
sudden acceleration 431
sudden acceleration incidents 431
symbol discriminability 250, 253, 256
symbol evaluation 263
symbol meaning 247
symbol presentation 248, 257
symbol production methods 258
symbol rating 260
symbol testing 261
symbols discriminability 251
symbols for controls **237,** 240
   automotive symbols 240, 242, 246
   ease of symbol learning 248
   heater symbols 259
   ISO standard (2575) 241
   image overlap representation 255
   labelling for controls 237
   labelling for displays 237
   national differences on symbols 239
   response times 250, 254
symbols for displays **237**
systems design 45

task measures
  primary 373
  secondary 373
testing of occupant protection 148
time sharing of tasks 363
tolerance to crash forces 151
traffic accidents 366
traffic citations of older drivers 455
training 427
trunk design 117
trunk floor dimensions 132
trunk lip thickness dimensions 132
trunk measures 119
types of information 413

ultraviolet headlamps 193
unintended acceleration **431**
  causation 432
  errors in execution 436
  errors in selection 435
  foot placement errors 433
  force and time variability 440
  force production variability 438
  human factors considerations 432
  initial positioning biases 436
  movement control 435
  occurrence 432
  temporal variability 439
  variability in movement trajectory 437
unwanted acceleration 431

vehicle design 79
vehicle
  accelerator foot plane 40
  accelerator heel point 39
  back angle 40
  ball of foot 40
  body base grid 39
  chair height 39
  dimensioning procedures 24
  dimensions 37
  entrance 29
  foot angle 40
  location technology 334
  operation and monitoring 367
vehicular guidance requirements 365
vibration 107
vision **79, 161, 185, 205, 299**
  ambient vision 188

colour sensitivity 79
cones 167
eye fixations 306
eye glance 307, 308
eye movements 79, 207, 307, 315, 375
eye position/location 83, 88, 207
eye structure and function 164
eyellipse 33
focal vision 188
indirect vision 205
nodes 167
visual acuity 79, 168, 206
visual adaptation 169
visual alertness 186
visual capture 444
visual control tasks 302
visual dominance 444
visual field 79, 93, 205, 465
  buses and trucks 94
  designs consideration 87, 93
  direct field of view 82, 89
  direct measurement 90
  evaluation 93
  field of view 79, 81, 208, 214, 465
  limits of visual field 86
  laboratory testing/measurement 91
  mirror field of view 82, 89, 94
  off-road vehicles 95
  overlapping field of view 214
  visual demands 299, 301
  visual loading 299
visual monitoring 79, 210
visual needs of drivers 80, 206
visual perception 175, 206
  expectancy 179
  eye glance duration 307
  glance times 309, 310
  information gathering 299
  mirror vision 94
  perceptual limitations 180
  perceptual problems in driving 179
  perceptual-response stages 178
  visual illusions 176
  visual information 175
visual performance
  adaptation 169
  ageing 170
  aniseikonia 174
  anisocoria 175

Subject Index

cataract 174
colour blindness 172
detection 185
disability glare 186
discomfort glare 186, 190
identification 185
illumination levels 169
measurement 163
perceptual trap 177
vehicle design 79
viewing distance 181
visibility vs conspicuity 186
visual capacities 205
visual defects 171

visual sampling 305
visual target position 168
visual sampling strategy 318
visual stimulus 162, 206
visual task performance
   drive models 303
   sampling models 304, 306
   visual sampling 305
visual training 317

warning effectiveness 418
workload reduction 376
workplace anthropometry 13
workplace modelling 65